Evolution and Water Resources Utilization of the Yangtze River

Jin Chen

Evolution and Water Resources Utilization of the Yangtze River

Jin Chen
Changjiang Water Resources Commission
Changjiang River Science Research Institute
Wuhan, Hubei, China

The edition is not for sale in the Mainland of China. Customers from the Mainland of China, please order the print book from: Changjiang Press (Wuhan) Co.,Ltd.

ISBN 978-981-13-7874-4 ISBN 978-981-13-7872-0 (eBook)
https://doi.org/10.1007/978-981-13-7872-0

This Springer imprint is published by the registered company Springer Nature Singapore Pte Ltd.
The registered company address is: 152 Beach Road, #21-01/04 Gateway East, Singapore 189721, Singapore

Preface I

The Yangtze (Changjiang) River, which originates from the snow-capped mountains and runs thousands of kilometers into the East China Sea, is the longest river in Asia and the third longest in the world. With its mainstream flowing through China's 11 provinces, autonomous regions, or municipalities, the river is the cradle of the Chinese civilization and has supported its socioeconomic development.

Since it came from the ancient times, it has frequently undergone, sometimes great, changes. In the past trillions of years, it has experienced headward erosion, uplifting, and cutting, cut through the Wu Mountains, traversed the Yunmeng Mountain, and connected the Han River and thousands of large and small lakes into the river. Influenced by the monsoon climate, she suffers flooding in summer and low water in winter, demonstrating the complexity and variability of the environment. Low in the east and high in the west, the watershed of the river has various vegetative landscapes and provides habitats to hundreds of fish species, including rare species such as the Yangtze River dolphin (or baiji) and the Chinese alligator. As a result, the river has provided mankind with an abundance of natural resources.

Water is essential to life. In the past thousands of years, people have lived by water bodies in order to perform various activities, namely, slash-and-burn farming, construction of hydraulic projects for flood control, and hydroelectric power generation. One of the projects was the Three Gorges Project, which led to the "rising of a vast lake within high gorges." Upon seeing the lake, we felt that was as "amazing as a World Wonder."

However, with the rapid socioeconomic development in China, the following ecological and environmental problems emerged in the Yangtze River basin: deforestation, soil erosion, landslides, sedimentation, river-lake separation, wetland loss, endangered species, water pollution, cyanobacteria outbreaks, etc.

For survival and development, human transformation of nature appears to be indisputable. Without the Dujiangyan Water Conservancy Project, there would have never been the prosperity of the Chengdu Plain. However, it is difficult to achieve harmonious existence between human and nature. In recent years, research and discussion of human activities on their living environment have become a hotspot in science and society. Undoubtedly, human activities have been affecting or altering certain natural processes, and the consequences are self-evident. However, the 5000-year history of human civilization is only a split second compared to the history of the earth's evolution, which has spanned billions of years. Thus, human activities are only a part of the earth's evolutionary process. Therefore, without the knowledge of the evolutionary history of the earth, one cannot appropriately understand the impact of human activities, making it impossible to mitigate the impact or develop an appropriate plan for the future.

With a broad vision and new ideas, the author of *The Yangtze River: Evolution and Water Resources Utilization* has discussed the relationship between the natural revolution of the Yangtze River and the human utilization of water resources from a multidisciplinary perspective, including geology, physical geography, ecology, hydrology, and hydraulic engineering. Additionally, the author has presented the definition, connotation, and research scope of the watershed ecosystem. This book has important theoretical and practical values for evaluating the evolution of the Yangtze River ecosystem and assessing the environmental impact of hydraulic projects. However, to know and understand the Yangtze River, a lot of work will still need to be done. I hope that the author will continue his efforts in this area not only to fill the gap in China's watershed ecosystem research but also to provide scientific and technical support for the protection and integrated management of the ecological environment of the Yangtze River watershed.

President, National Natural Science Foundation of China Chen Yiyu
Member of Chinese Academy of Sciences
Signed in Beijing in October 2012

Preface II

The Yangtze River has the largest watershed area in China. It encompasses south-western, central, and eastern China with a total drainage area of approximately 1.8 million square kilometers. The river traverses the entire three-step topographic staircase that stretches across China from the west to the east. As the world's third longest river, its mainstream travels a total distance of more than 6300 kilometers through the following provinces, autonomous regions, or municipalities, Qinghai, Tibet, Yunnan, Sichuan, Chongqing, Hubei, Hunan, Jiangxi, Anhui, Jiangsu, and Shanghai, and its tributaries extend to the following eight provinces or autonomous regions: Gansu, Shaanxi, Henan, Guizhou, Guangxi, Guangdong, Fujian, and Zhejiang. As the Yangtze River basin possesses rich and varied river landforms, complex ecosystem, and enormous renewable water resources, it facilitates water supply, irrigation, hydropower generation, navigation, etc. and sustains the socio-economic development within the watershed. Due to continuous population growth and uneven spatial and temporal distribution of the precipitation and runoff within the watershed, the river basin has experienced frequent flood and drought hazards throughout history. People in the river basin have begun water resources development activities such as flood control, drought relief, river regulation, and water diversion for irrigation since the Spring and Autumn Period. Especially, since the 1950s, the construction of large river and lake improvements projects and water resources development projects nationwide, as well as within the Yangtze River basin, has ensured that China's rapid socioeconomic development maintains to advance and people's standard of living within the watershed continues to be improved.

With the completion of the Three Gorges Project and other control reservoirs in the twenty-first century, the flood control situation in the middle and lower reaches of the Yangtze River has significantly improved, and a flood control engineering system has been established. This system is based on dikes coupled with control reservoirs as a major control means and retention/diversion zones as the final safeguard measure. Additionally, further improved reservoir, irrigation, water supply, and hydropower facilities within the river basin continue to meet the needs of the rapid socioeconomic growth within the river basin. Therefore, great achievements have been accomplished in the development and utilization of water resources and the regulation of rivers and lakes within the Yangtze River basin.

With China's rapid socioeconomic development, increased improvements of people's standard of living, and public concern about ecological and environmental issues and the desire to improve the quality of life, the public has become more aware about the impact of water resources development and utilization projects on the ecological environment of the river basin while reaping the benefits of the hydraulic projects. The concern is natural and reasonable, and constructive comments will be beneficial for scientific development, conservation, and management of the water resources within the watershed. Presently, a large number of hydroelectric projects are being constructed in the mainstream and major tributaries in the upper reaches of the Yangtze River, and the demand for water resources is increasing in response to the economic development within the river basin. It is necessary to comprehensively evaluate the impacts of the construction of cascade reservoirs on the ecological environment of the basin and take comprehensive measures to mitigate adverse effects. Therefore, it is necessary to objectively and fairly evaluate the benefits of water resources development and the impacts of hydraulic engineering projects on the ecological environment, as it is very important in the future development, management, and protection of the river basin. To ensure an objective, scientific, and comprehensive analysis and evaluation of the environmental impacts, it is necessary to objectively analyze the previous and existing conditions of the basin-wide water resources and the aquatic environment of the Yangtze River using a multidisciplinary perspective. For hot issues of the Yangtze River, we should use dialectical materialism coupled with historical materialism to present research results and conclusions, which will stand the test of history.

Mr. Chen Jin, the author of this book, has engaged in research and scientific experiments in water conservancy for a long time, focusing on key technical topics associated with the development, utilization, and protection of water resources of the Yangtze River. To address hot issues associated with regulation, development, utilization, and protection of the Yangtze River, the author has thoroughly overviewed the history of river regulation and water resources development throughout the evolution of the river using a multidisciplinary perspective based on years of observation and research. The author has further objectively evaluated the history and role of present-day hydraulic projects in the socioeconomic development and the effects on the ecological environment in the Yangtze River basin. Moreover, the author has presented many new viewpoints and recommendations on regulation of rivers and lakes, development of water resources, and ecological and environmental protection of the river basin based on the existing major ecological and environmental problems within the Yangtze River basin. All of this will provide scientific and technical support for the comprehensive regulation of the Yangtze River and comprehensive management of water resources, as well as important informative value for development, regulation, protection, and management of the water resources in the Yangtze River basin.

郑守仁

Chief Engineer, Yangtze River Water Conservancy Commission Zheng Shouren
Member of Chinese Academy of Engineering
Signed in October 2012

Preface III

The Yangtze River, the longest river in China, possesses one-third of the country's water resources, and the watershed generates one-third of the country's gross domestic product and is home to one-third of the country's population. Moreover, it supplies water to north China through the South-to-North Water Diversion Project. As a result, the river is the primary water source in China, supporting the ever-growing socioeconomic development in the Yangtze River basin and north China. For flood control, food security, and navigation safety, people in the river's watershed have continuously constructed dikes and dams, reclaimed farmland from lakes, diverted water for irrigation, and excavated navigable canals for transportation over the past 2000+ years. Since the 1950s, an additional 40,000+ dams have been constructed on the mainstream and tributaries of the river for flood control, irrigation, water supply, power generation, and navigation. While these water conservancy and hydropower projects have provided benefits to civilization, they have affected the continuity and the natural hydrological process of the river, resulting in some negative effects on the ecological environment. Thus, it is undoubtedly meaningful to study the relationship between the evolution of the river and the utilization of its water resources. It will help one not only to analyze how and to what extent human activities will affect the ecological environment of the river but also to improve our ability to effectively protect and scientifically manage the water resources of the river.

The evolution of the Yangtze River is the result of tectonic movement, climate change, biological evolution, and human activities. Tectonic movement has created the undulating continental topography and helped to form mountains, hills, plains, lakes, rivers, and other landforms. This resulted in the elevational difference between the continent and the sea, which provided the geologic and geomorphic conditions for the formation of rivers. Precipitation, runoff, and soil erosion, under certain conditions, have also helped form rivers, lakes, and water systems. All kinds of erosion and denudation have resulted in mountains to be carved into canyons or leveled into plateaus. The Qinghai-Tibet Plateau, Jinsha River Valley, and the Three Gorges are the result of interactions among tectonic movement, soil erosion, and stream hydrodynamics. Soil erosion caused by hydraulic force, gravity, and wind power has resulted in soil loss and sediments, while floodwaters have transported some of the sediments and nutrients into streams, estuaries, and the sea. The atmospheric circulation and seasonal climate change have not only generated large storms and floods but have also provided hydrodynamic conditions for the river's evolution. Periodic climate changes have resulted in alternating ice ages and interglacial periods, causing the sea level to drastically rise or fall, the slope gradients of rivers to considerably change, new streams to emerge, and some old streams to disappear. The vegetative and forest coverage conditions and sediment characteristics not only affect the physical and chemical properties of stream water but also determine the characteristics of flow, sediment generation, and the evolutionary patterns of stream channels. Even before the advent of mankind, the Yangtze River had undergone great changes, but the timeframe that the changes occurred in was very expansive.

A comparison of the natural evolution of the Yangtze River and the impact of the human social activities indicates that the timeframe for natural evolution is larger than the human activity timeframe. However, the impact of human activities is larger and is amplified when the activities are more recent due to higher productivity in current times. The timeframe and space required for the natural evolution of streams vary greatly. For example, it takes several years for floodland to evolve, which is a short timescale; it takes decades or centuries for a river-lake relationship to change, which is a medium timescale; and it takes tens of thousands of years to hundreds of thousands of years for a river channel to change its course or a stream capture to occur, which is a large timescale.

The impact of human regulation, development, and utilization of water resources on the evolution of the Yangtze River has occurred in a very small timeframe. As human productivity has been exponentially increasing, the impact of human activities has obscured the natural evolution of the river. Since thousands of years ago, human beings have started slash-and-burn farming, reclaimed farmland from deforestation, colonized land from alluvial plains and lakes, constructed dikes, etc. All these human activities have gradually resulted in the separation of lakes from rivers and narrower stream channels. Moreover, the present construction of cascade hydroelectric projects has significantly altered the stream's hydrological process and channel characteristics. Soil erosion, massive discharge of wastewater and sewage, removal of riparian vegetation, and river bank hardening and utilization have changed the physical and chemical properties of water bodies and damaged riparian ecological buffering, which have diminished the ecological function of rivers. Important future subjects for regulation, development, and protection of the Yangtze River include how to scientifically develop and regulate the river and lessen human impact on the river ecosystem and how to protect the water resources and aquatic environment of the river in accordance with strict water resources management standards.

Starting from the evolution of the Yangtze River and the regulation of streams and lakes and then using an integrated perspective involving geological evolution, physical geography, aquatic ecology, water resources, and hydraulic engineering, the author has discussed the relationship between the natural evolution of the Yangtze River and the human utilization of its water resources; presented the basic concept of large river watershed ecosystems; preliminarily evaluated the relationship between watershed evolution and utilization of water resources at various time and spatial scales; and introduced new methodologies and ideas for the utilization of water resources, protection, and comprehensive management of the aquatic and ecological environment in the Yangtze River watershed. The author has a broad vision and is future-oriented and innovative. The book has significant theoretical value for research on the evolution of large rivers and sustainable utilization of water resources. Additionally, it has important practical value for the comprehensive management of water resources in the Yangtze River watershed. I hope that the author will continue his studies of the water resources and aquatic environment of the Yangtze River to make a greater contribution toward the protection and management of the watershed's ecological environment.

Professor, China Institute of Water Resources and Hydropower Research
Member of Chinese Academy of Engineering
Signed in Beijing in October 2012

Wang Hao

Abstract

This book delves into the impact of the historic evolution of the Yangtze River system, the human regulation of rivers and lakes, and the development of water resources on the watershed ecosystem. It also presents a strategy for the sustainable utilization and protection of water resources. This book accomplishes these tasks by combining the natural evolution of the Yangtze River and the development of water resources and other human activities. Additionally, the book examines these topics with an integrated perspective involving geology, physical geography, biology, environmental science, and hydraulic engineering. The main contents of the book include the formation and evolutionary process of the Yangtze River system and its aquatic environment in geoscientific history; the characteristics of the Yangtze River system and water resources; the existing conditions of the Yangtze River watershed ecosystem and rare aquatic organisms; the history, existing conditions, and future trend of the development/utilization of the Yangtze River water resources; the impact of water resources development on the Yangtze River ecosystem; and the relationship between development/utilization and protection of water resources. This book can be used as a reference for researchers and managers regarding numerous topics, including the evolution of the Yangtze River ecosystem and its aquatic environment, the evaluation of water-resources development and utilization, the ecological and environmental protection of the Yangtze River, and the comprehensive management. It can also be used as a textbook or reference book for undergraduate students, graduate students, scientific researchers, and even the public to understand the evolution and existing conditions of the Yangtze River.

Introduction

The Yangtze River, the longest river in China, plays an important role in and has a profound impact on the socioeconomic development and ecological environmental protection of China. A comprehensive and objective evaluation of the ecological and environmental conditions of and the impact of human activities on the Yangtze River should start from the evolutionary history of the river, followed by a multidisciplinary perspective that integrates geology, physical geography, ecology, environment, hydraulic engineering, and ethics of nature. The author's purposes of preparing this book are to understand the past, comprehend the present, and foresee the future.

As I have lived by the Yangtze River since my childhood and engaged in research of the river for 30 years, I have a deep connection with the river and its mysteries. I have inspected or walked through most of the river's representative sections and regions, had the honor to visit the source of the river twice, and personally witnessed the location, from which the first drop of water for the river comes–Jianggendiru Glacier and the broad bend of the Tuotuo River. Then, after driving along the bank of the Tongtian River for more than 100 km, I visualized the raging water of the Tongtian River under the blue sky and white clouds. I also traversed the magnificent Jinsha River Valley and the Three Gorges; visited the four rivers south of the Jingjiang River, which have formed a complex water system, and the Dongting Lake area; and beheld the boundless Poyang Lake in the summer and the linear-featured wetland landscape in the winter.

When one travels by ship from Wuhan, which is in the middle reaches of the Yangtze River, to Shanghai, he will see a broad river in one segment and then a narrow water surface in another section within the middle and lower reaches of the river. In fact, when the river is narrow, this is where tributaries converge into the channel of the river, and when it is broad, one may be in a large curvy segment where the water surface appears as if it were unrivaled in its width. Due to the fact that the dominant water current in the main channel swings between the left bank and the right bank, the ship has to proceed by the left bank for a while and then by the right bank for another moment so as to follow the major water current since there is only one deep navigable channel beneath the water flow that only the captain of the ship knows. When one reaches the Yangtze Estuary, it does not appear to be a river, but instead a sea. The water level and flowrate of the river at the estuary are impacted by tides, and the river appears to be boundless at the estuary. Therefore, the estuary is both a river and a sea. Of course, this is only a perception. After reviewing previous documents and research about the Yangtze River, I have found that many more profound mysteries have yet to be unveiled and even multiple scholars' qualitative explanations are still very inconsistent. Therefore, to quantitatively describe the evolutionary process of the Yangtze River, many scientific and technological issues will need to be studied.

For survival and development, human beings started to regulate the Yangtze River and instinctively explored and utilized its water resources. Naturally, human socioeconomic development has always paralleled the history of modification, development, and utilization of the river, because the human utilization of the river and its water resources is always accompanied by the effects on the natural evolution of the river and the ecological environment. The impact of human activities on the evolution of the Yangtze River began about 10,000 years ago in the agricultural period of the Neolithic Age (Holocene), while Dayu (Yu the Great) began flood control about 5000 years ago. The construction of hydraulic structures and dikes started in the Spring and Autumn and Warring States periods more than 2000 years ago, but the installation of hydropower stations did not commence until the early twentieth century. The construction of large water resource utilization projects such as large hydropower stations and reservoirs, water diversion projects, and large irrigation facilities started in the 1950s.

As the Yangtze River basin is located in a humid area with an abundant precipitation, a well-developed water system, a high river runoff, and an enormous discharge into the sea, biological species in the Yangtze River have adapted to the natural hydrological characteristics. However, large-scale human water uses within and outside the river channel have significantly changed the natural characteristics of the river flow. Therefore, we will need to rethink our perception of the evolutionary pattern of the Yangtze River and its ecosystem in the context of human activities and influence. We also need to analyze and predict if this trend will be able to sustain the human use of the water resources and a robust ecosystem.

As the evolutionary investigation of the Yangtze River and the impact of water resources utilization involve multidisciplinary cooperation, a comprehensive analysis of these issues requires a broad and systematic integration of knowledge. However, modern scientific research has become divided into narrower and more specific fields with quite different research ideas and methods. As a result, the use of a simplified comprehensive method of analysis may not be professional, or lack depth and quantification. Many of the issues discussed may be only at the conceptual and qualitative level. Nonetheless, the author believes that comprehensive analysis and discussion will be beneficial for river developers, researchers, and managers to explore ideas and especially useful for eliminating public misunderstanding of the development and utilization of the Yangtze River. Therefore, this book is intended to overview the evolution of the Yangtze River and the impact of water resources utilization from an integrated multidisciplinary perspective. The academic value is not high, but I hope, it will be rather educational and exploratory.

This book has cited or referenced to research results from a large number of predecessors and researchers in various fields. Although references are listed, maybe not everyone is mentioned in the text. I hereby express my great condolences. If there are descriptions of some results, the results shall belong to the quoted data. The book is intended to present the research results; therefore, the quoted data may not be complete or well-organized, but this does not affect the author's thought and beliefs presented herein. Due to limitations of the author's extent of knowledge and availability of data collection and analysis, there may be many deficiencies, or even errors or omissions, in this book. As a result, readers are welcome to provide valuable comments.

Special thanks are extended to the two members of the Chinese Academy of Engineering, Zheng Shouren, Chief Engineer of the Yangtze River Water Conservancy Commission, and Wang Hao, Professor of the China Institute of Water Resources and Hydropower Research, and one member of the Chinese Academy of Science, Chen Yiyu, President of the National Natural Science Foundation of China, for their advice and support during the preparation of the book. While they have been extremely busy, they still dedicated time to prepare prefaces in which they have highly valued the author's research results. I also gratefully acknowledge the funding agencies for this project: the Hubei Provincial Important Book Publishing Foundation and the Yangtze River Scientific Research Institute. Since this book is on the national priority list of important publications for China's twelfth 5-year planning period, the Hubei Provincial Important Book Publishing Foundation provided funding. The Yangtze River Scientific Research Institute also provided funding through one of the Ministry of Water Resources' Public Sector Research and Special Projects – "Control and Management, and Pilot Study for Total Water Use Quantity of the Yangtze River Watershed." Special thanks are also given to the Changjiang Publishing House for their deliberate planning, design, and editing of this book. And lastly, special thanks go to my wife, daughter, and parents for their mindfulness and support of my work and allowing me to spend substantial personal time on research and writing that I otherwise would have spent with them.

I thank Prof. Yang daoqiong for his time and effort in translating this book into English. He was my college classmate and a hydrogeologist. He has been in the United States for more than 30 years and is now a Certified Environmental Manager in USA. He not only translated my books accurately, but also corrected many mistakes in the Chinese version. His professionalism is greatly appreciated.

Wuhan in October 2012 Jin Chen

Contents

List of Figures

List of Tables

List of Acronyms and Abbreviations

CISPDR	Changjiang Institute of Survey, Planning, Design, and Research
CITES	Convention on International Trade in Endangered Species of Wild Fauna and Flora
cm	Centimeter(s)
CPPCC	Chinese People's Political Consultative Conference
CWIIWH	Convention on Wetlands of International Importance especially as Waterfowl Habitat
EIA	Environmental impact assessment
FMCC	Four major Chinese carps
Ga	Billion years ago
GDP	Gross domestic product
GIS	Geographic information system
GPS	Global positioning system
GW	Million kilowatts or gigawatts
GWCP	Gezhouba Water Control Project
Ha	Hectare(s)
IPCC	Intergovernmental Panel on Climate Change
IUCN	International Union for Conservation of Nature and Natural Resources
JPWCPDI	Jiangxi Provincial Water Conservancy Planning and Design Institute
ka	One thousand years ago
kg	Kilogram(s)
kg/m^3	Kilograms per cubic meter
km	Kilometer(s)
kWh	Kilowatt-hours
m	Meter(s)
m^2	Square meter(s)
m^3	Cubic meter(s)
mg/L	Milligrams per liter
mm	Millimeter(s)
Ma	One million years ago
mm	Millimeter(s)

m/s	Meter(s) per second
m^3/s	Cubic meter(s) per second
MW	Megawatts
MWR	Ministry of Water Resources
MWRWP	Ministry of Water Resources and Electrical Power
PMP	Probable maximum precipitation
PRC	People's Republic of China
RS	Remote sensing technology
TDS	Total dissolved solids
TEU	Twenty-foot equivalent units
TGD	Three Gorges Dam
TGP	Three Gorges Project
TGR	Three Gorges Reservoir
UNFCCC	United Nations Framework Convention on Climate Change
UNESCO	United Nations Educational, Scientific, and Cultural Organization
WARI	Wuhan Aquatic Research Institute of the Chinese Academy of Sciences
WWF	World Wide Fund for Nature
YRBPO	Yangtze River Basin Planning Office
YRFCSRH	Yangtze River Flood Control and Drought Relief Headquarters
YRHEA	Yangtze River Hydraulic Engineering Administration
YRWCC	Yangtze River Water Conservancy Commission
YRWRC	Yangtze River Water Resources Commission

Chapter 1
Hot Issues of the Yangtze River

Abstract The Yangtze River provides very important ecological functions and services for China. Being the longest river in China, the Yangtze River is so prominent that any development and utilization activities on the river attract global attention. This chapter introduces the natural evolution of the Yangtze River and the utilization of water resources with four biggest hot issues of the Yangtze River. The first issue concerns itself with the following questions: has the development and utilization of the Yangtze been excessive, how should the relationship between utilization and protection be evaluated, and where is the equilibrium point of the utilization and protection? The answers to these questions depend on China's level of socioeconomic development and the cognitive level of the majority. The second issue is associated with where the true source of the Yangtze River is, how people have gradually recognized the source of the river over the past thousands of years, why the Tuotuo River was finally determined to be the source, and what the controversies have been. Next, the third issue discusses how we should objectively respond to existing controversies regarding the planning, construction, and management of the world's largest hydraulic project (the Three Gorges Project) while also describing the project's main functions, role, and current social and ecological environment problems. Finally, for the fourth issue, the chapter provides an overview of the impact of human activities on the natural environment of the Yangtze River and discusses the current evolutionary phenomena as well as future trends of the Yangtze River in the context of climate change.

The Yangtze River, the longest river of China and one of the mother rivers of the Chinese nation, has not only been the cradle of the Chinese civilization for thousands of years, but it also currently shoulders the heavy burden of supplying water for production and life for more than 400 million people. Moreover, it has maintained the rich and colorful land, as well as the ecological cycle of the large river basin. As for the understanding of the Yangtze River, some aspects are commonly understood, such as the distribution of the aquatic system, hydrological characteristics, and important protected species, but other aspects are uncertain or disputable, such as where the balance point between watershed development and conservation is and what kind of river ecosystem should be preserved. There are still some unanswered questions, such as the long-term impact of the Three Gorges Project (TGP)

J. Chen, *Evolution and Water Resources Utilization of the Yangtze River*,
https://doi.org/10.1007/978-981-13-7872-0_1

on the ecology and environment of the Yangtze River, the response of the river's ecosystem to the impact, and the future evolutionary trend of the Yangtze River. To begin with this book, this chapter discusses several hot issues of the Yangtze River.

Keywords The Yangtze River · Changjiang River · Evolution of river system · Basin ecosystem · Water resources utilization · Floods and drought · Ecological and environmental protection · Basin management

1.1 Balance Point Between Utilization and Protection

The debate over regulation, development, utilization, and ecological environment protection of the Yangtze River has always been a hot issue of great concern to all circles of the society, and there are differing views between the public and experts, and among experts in different fields, which reflect the diversity in perception regarding the development and protection of the Yangtze River. This is healthy for a society. Variation of opinions bears a positive meaning because it indicates that people are paying more attention to the protection of the river's ecological environment, and the constant questioning from the general public makes water resource developers more cautious and careful, so that river managers pay more attention to the comprehensive and coordinated scientific management.

In ancient times, people who lived in the Yangtze River basin were mainly faced with three basic survival issues that they had to resolve: flood hazards, food shortages, and transportation difficulties. As such, almost all of the hydraulic control works constructed by our ancestors were aimed to address these three issues. For example, constructing dikes was to prevent flooding; reclaiming farmland from lakes and constructing canals and irrigation systems were for food production; and excavating navigable canals were for transportation. However, human productivity was low during these times, and disturbance to the natural environment was also small. As a result, the overall human impact on the ecological environment was relatively small compared to now, and most rivers had a healthy ecological environment. Thus, people did not have any consciousness for the protection of the ecological environment until their survival, safety, food, and clothing needs had been met. It was not until World War II did the concept of protecting the ecological environment begin to emerge and gain global attention.

However, ancestors had still different opinions among themselves regarding the strategy and methodology of regulating rivers. In prehistoric times, Gun and Yu used different approaches to flood control. The former used the "blocking" approach that proved to be a failure. By contrast, Dayu (or Yu the Great) changed to the "directing" approach, meaning "to give floodwater a way out" into low-lying areas and eventually into the sea, which was successful. Yu the Great has been widely praised for several thousands of years for his flood control achievements and work ethic. For example, classical stories have it that during the flood control period, he "passed by his own home thrice without entering." Later he became a king and his achievements were commended.

From the level of human productivity and people's living environment at the time, Gun's idea for floodwater control was not completely unreasonable. In ancient times, there were not many human settlements, and building walls or dykes required the least amount of work and was the easiest way to protect settlements from flooding. Although this method might not work for large flood events, it could be effective for small- and medium-sized flood occurrences. Therefore, it is incorrect to conclude that Gun's method was totally unreasonable, for at least it was somewhat economically reasonable. Regulating streams is a systematic project, and it is difficult to have an absolutely optimal solution. There may be only a relatively good or reasonable solution; therefore, we should not completely exclude other solutions or reasonable elements thereof.

Li Bing and his son have been praised for their design and construction of the Dujiangyan Project. Their thoughtful and environment-friendly design still serves as an important reference today. Their use of a low weir system was based on the scientific and technical knowledge and construction capability available at the time. Because it was not possible to build a high dam, the low weir alternative was implemented to incorporate local conditions. The excavation of the Beijing-Hangzhou Grand Canal has also been admired for solving the problem of transportation between North and South China, but to think of the fact that the canal was the result of hard work worth tens of millions of people's hundreds of years, one realizes that the project was constructed with the gigantic sacrifice of innumerable working people. Because dike construction along the Yangtze River and around large lakes within the river watershed has continued for more than 2000 years, most of their foundations were constructed in ancient times. Dikes not only provide safety to human lives and food production and force floodwater to return to the river channel but also increase the channel's storage and passing capacity so that the floodwater flows through the waterway as human desired. In the ancient times, when there were no issues about the ecological environment, dike construction was considered effective, although nowadays they appear to restrict the connection between the river and its floodplains, affect the lateral evolution of the river, and have far-reaching impacts on the evolution of river-lake relationships. From the viewpoint of changing the river flow, the Dujiangyan Project and the Grand Canal Project are water transfer projects across multiple river basins. The former transferred water from the Min River into the Tuo River, and the latter connected the Yangtze, Huai, Yellow, and Haihe Rivers. From the present-day point of view, both projects have had adverse effects on the ecological environment. However, what the ancient people did for river-regulating projects was in accordance with their socioeconomic conditions and their basic needs in their time, so we are not entitled to "make irresponsible remarks." Regarding historical events and historical figures, we should use the viewpoints of dialectical materialism and historical materialism to perceive them. We should not simply blame the ancient people and claim that they did something wrong. When problems associated with people's life safety, and food and clothing supplies are not resolved, talking about the protection of the ecological environment is like building a castle in the air. This is because people only seek a higher quality of living, such as physical health, environmental aesthetics, spiritual pleasure, and awareness of environmental protection, when problems relating to their basic safety, food, and clothing supplies are resolved. In fact, before the Industrial Revolution, there was little global variation in productivity and standards of living, and human activities had little impact on the natural characteristics and ecological environment of rivers.

In the eighteenth century, mankind began to enter the era of industrialization. With rapid advancement in science and technology and greatly improved productivity, the increased ability of human beings to transform nature and excessive exploitation of natural and water resources had an increasingly apparent impact on the earth's ecological environment. In the early twentieth century, people began to recognize the impacts of water pollution and soil erosion on human health and the environment and began to pay attention to the ecological environment. Moreover, people started to contemplate if human activities should be regulated and whether or not a limit should be established for the development and use of natural resources. After World War II, many developed countries first recognized that the ecosystem and the natural environment should no longer be harmed. Soon after the 1960s–1970s, the world began to pay attention to the ecological environment. Obviously, if the ecological environment was destroyed, not only human living environment and human production but also human health (e.g., introduction of diseases) would be affected. This would cause the quality of living to decline, which was not consistent with the goals of human efforts and sustainable development. In the early 1990s, people started to become aware of the means of achieving sustainable development by doing things cautiously, leaving our future generations with the space and environment for survival and development, and not completely depleting the resources of future generations.

Looking back at history, we need to think about whether human activities were correct in the past. From the viewpoint of historical materialism, we may say they were correct. When taking into account the limited knowledge of science and technology and poor human understanding at the time, all human activities associated with river regulating and flood control are understandable. However, from the present point of view, some activities are considered unscientific, even overreaching, such as deforestation, reclamation of farmland from lakes, etc., which have caused large numbers of forests and lakes to disappear. This not only damaged the environment but also made floods more intense and rapid. With a continuous decrease in the storage capacities of low-lying lakes and wetlands, people who lived in the low-lying areas felt unsafe. Limited by their knowledge of scientific and technological advancement at the time, some of the ancient people did not fully recognize the issues. Looking back at these experiences and lessons, one must hope that we will not repeat similar mistakes in the future. Therefore, we will need to balance the relationship between development and protection for sustainable development.

From the present conditions of the Yangtze River, the source area of the Yangtze River, and the headwaters of its main tributaries, such as the upper reaches of the Jinsha River and the mainstream of the Chishui River, the Yangtze River is still in a relatively natural state. In other sections of the river where hydropower stations and reservoirs have been built or will soon be built, the processes of water flow and sediment transport have undergone substantial changes even though the total water flow has not changed. However, the overall conditions of the ecological environment in the Yangtze River watershed are still good. This is because the utilization rate of the water resources has barely reached 20% with a small consumptive quantity, the total water discharge into the sea has minutely changed, and the water quantity regulated by the reservoirs is less than 25% of the annual amount of runoff. Presently, the prominent areas of conflict are the hydroelectric development in the upper reaches versus the protection of rare fish species, the development of small hydroelectric projects in tributaries versus the preservation of stream continuity, the regulation of the middle and lower reaches of the river versus the preservation of the wetland ecosystem, the existence of lakes and reservoirs versus the water pollution problems occurring in riparian zones of streams, etc. These problems will require special attention, in-depth research, and effective management.

Determining the balance point between watershed development and protection depends on the conditions of socioeconomic development, people's standard of living, and the values of the government and the majority of the society. It also involves people's views on nature and the social psychology. In any society, some people choose to live a simple, easy, and even poor life, while some other people choose to live a human-centered life that involves becoming rich first and then considering the ecological environment, or some choose to live in a seminatural, half-developed environment. Most people want to live in a relatively natural and robust ecological environment, and only a few people want to live in a completely man-made environment, such as a space station. We cannot deprive people of the right to choose a way of living, and the government and society generally choose the balance point between development and protection according to the will of the majority. But of course, we should leave living space for those who want to live a simple life and preserve the natural environment. Human efforts to pursue economic development to become rich as soon as possible are somewhat contradictory to preservation of the ecological environment or the desire for healthy and happy living. However, we should understand that if there were no place for large wildlife or important wild plants to survive, the quality and environment of human life would not be any good, or there would be little potential for sustainable development. Therefore, human beings should choose their own way of life prudently, scientifically develop ways to utilize natural resources, and protect the environment and natural ecosystem to an extent that is practically possible.

1.2 Source of the Yangtze River

As to where the source of the Yangtze River is, the answer has changed several times over the past thousands of years. The earliest belief was that the Yangtze River originated from the Min River, while there was another belief that the source was the Jialing River, but it was later believed to be the Jinsha River. Currently, there is an official conclusion about the source of the Yangtze River (Yangtze River Water

Conservancy Commission [YRWCC] 2003). It originates from the Tongtian River with three sources: the Tuotuo River as the true source, the Dangqu River as the southern source, and the Chumar River as the northern source. In ancient times, it was normal to make mistakes because there were neither accurate maps nor scientific means for measuring the length, water volume, or direction. Moreover, there was no unified and publicly accepted standard for determining the source of a river. Therefore, even though there are still some debates about source-determining standards (Li and Song 2010), the understanding of the river source area is now consistent. The unanswered question is whether the Tuotuo or Dangqu River (Dangqu means marsh river in Tibetan) is the true source.

1.2.1 Ancient People's Understanding

The written records for the Chinese civilization began during the Xia and Shang Dynasties as early as 3000–4000 years ago. During this period, the Yangtze River basin was sparsely populated with poor traffic conditions, and there were no scientific means for measurement. Therefore, the ancient people had a very superficial understanding of the Yangtze River's source. In the period of 8000–3000 years ago, China was in a warming period with abundant precipitation since the early Holocene. Consequently, the Yangtze River channeled through the Three Gorges; the Han River ran out of Danjiangkou; and the water in the middle and lower reaches of the Yangtze River flowed into alluvial plains, such as floodplains, lakes, marshes, aquatic systems, etc. Also at that time, the Jianghan Plain and the Dongting Lake area were all lakes, marshes, and floodplains, such as the historical Yunmeng Marsh and Pengli Marsh, and people could not even tell whether the Yangtze River or the Han River was the mainstream of the Yangtze River (Zheng 1998). This might be attributed to the fact that the Yangtze River was very narrow at the mouth of the Three Gorges, or the Nanjin Pass, with a water surface of only 200–300 meters (m) wide. However, in the middle reaches of the Han River, the water surface was 1000–2000 m wide. Thus, people could not "differentiate Yangtze from Han" at the time. In fact, people did not pay attention to where the Yangtze River's source was either. Although the Nanjin Pass was very narrow, the water was deep, and the flow had a high velocity, while the middle reaches of the Han River was wide but shallow. In actuality, the former had a much higher flowrate.

By 3000–2200 years ago in the Spring and Autumn and Warring States periods, China's famous classical history book *Book of Documents ● Tribute of Great Yu* was the first to record the source of the Yangtze River and stated that "the Min Mountains leads the Yangtze River and the Tuotuo River is located to the east," meaning that the Yangtze River originates from the Min River. However, the book was not referring to the Min Mountains at the eastern margin of the Qinghai-Tibet Plateau in western Sichuan, but instead to the Bozhong Mountains which is southwest of the present-day Tianshui, Gansu (Shi 2001). Thus, the book actually meant the source of the Yangtze River was at the western source of the Han River in the upstream of the Jialing River. Nonetheless, the Min River had widely been considered as the source. This was because ancient people revered the classical book, and no one performed scientific research or measurement. Therefore, over the past 2000+ years, the argument that the Min River was the source of the Yangtze River had remained almost unchallenged. In the Northern Wei Dynasty, the famous geographer Li Daoyuan in his *Commentary on the Water Classic* argued that although the Shengshui (now Jinsha River) was already known to be a part of the Yangtze River, it was a tributary of the Yangtze River and that "the Min Mountains lead the Yangtze River," meaning the Min River was the mainstream. The fact that the ancient people considered the Min River as the source of the Yangtze River was not completely unreasonable, and it has been inferred that the belief stemmed from the following three reasons:

1. The discharge in the Min River is very large with an annual average runoff of 106.5 billion cubic meters (m^3), which is the largest of the Yangtze tributaries, and it is only 39.5 billion m^3 less than that measured at the Yibin Station on the Jinsha River (average amount of annual runoff at 146 billion m^3) that is part of the Yangtze River's mainstream. Additionally, there are no areas with large rainstorms in the Jinsha River basin, while the Min River basin has a well-known rainstorm area in western Sichuan. And in terms of maximum peak flow, the Min River is even larger than the Jinsha River. Historic records indicate that the maximum peak flood flow once reached 51,000 cubic meters per second (m^3/s) at the mouth of the Min River – Gaochang Station – while it was only 36,900 m^3/s at the mouth of the Jinsha River, Pingshan Station.
2. The lower reaches of the Jinsha River above Yibin are mostly narrow valleys in mountainous area with water surfaces ranging in width from 150 to 200 m, while the width of the water surface in the lower reaches of the Min River is 400–1000 m, giving the impression that the Min River is wider and has larger flows.
3. Because the Jinsha River flows rapidly through many dalles with many obstacles, the river is generally not navigable. In contrast, the lower reaches of the Min River have had boats constantly passing the prosperous inhabitants since ancient times. If we do not use scientific measurements, it is easy to mistake the Min River as the mainstream and the Jinsha River as a tributary. Therefore, it is understandable that the ancient people made this mistake.

In fact, the ancient people before Xu Xiake, a travel writer and geographer in the Ming Dynasty, had been more naïve as well as more imaginative than the modern people in their understanding of the source because of their lack of hydrological and geographical knowledge. For example, the *Classic of Mountains and Seas*, written during the Spring and Autumn and Warring States Periods, claims that "there is a stone gate beneath the Jishi Mountain, and the river water originates from it." This reflected the common misperception that the source of the Yellow River was the spring water originating from subsurface flow. Although this perception was correct to an extent, it still led to an incorrect conclusion. During the Han Dynasty, after Zhang Qian's official mission to the western regions, the source of the Yellow River was extended to the east of Yutian County, Xinjiang; then it would flow through Lop

Nor; and then it would underflow to Qinghai. Greatly influenced by this perception, Li Daoyuan did not determine that the Jinsha River was a segment of the mainstream of the Yangtze River.

According to historical documentary records, the ancient people in the early Eastern Han Dynasty knew that the Jinsha River was long and had a distant source, but it was regarded as a tributary of the Yangtze River. It was Xu Xiake who had a work ethic comparable to that of respected field scientists today that challenged the classic authoritative statement for the first time. In his article "Travel along the River Headward to the Source," he argued that "for the source of the Yangtze River, the Jinsha River must be at the top of the list." Although much of "what Xu Xiake knew had been known to his predecessors," Xu Xiake was still beyond his predecessors' knowledge and was clearly aware of a simple principle to determine the source of a stream: the farthest tributary should be the true source. Moreover, he used this principle to refute the old statement that the Min River was the true source of the Yangtze River and established the new perception that the Jinsha River was the true source. He wrote: "I followed the Min River from Chengdu to Yibin for less than 500 kilometers (km), while the Jinsha River flows through Lijiang, Yunnan, and the Wumeng Mountains to Yibin for more than 1,000 km. Should a close source be chosen over a distant one as the true source? No!" "Therefore, as for the true source of the Yangtze River, the Jinsha River should be at the top of the select list." "In fact, the Min River joining with the Yangtze River is analogous to the Wei River converging with the Yellow River. Both are tributaries. While the Min River is navigable, the Jinsha River winds down through remote mountains and narrow valley with dalles. In addition, the Jinsha River is unnavigable and there are no travelable roads along the river banks." Xu Xiake analyzed why the ancient people considered the Min River as the source of the Yangtze River. The most moving part of Xu Xiake's scientific work was that he performed all the fieldwork while eating in the wilderness and sleeping in a camp. Not only this, but he also made one of his most important scientific contributions to the discovery of the Yangtze River's true source through his arduous fieldwork. His discovery was not because he visited the Jinsha River, but instead, he inferred through the compilation and analysis of the data collected from his fieldwork (Zhu 1991).

Because of his unofficial status, Xu Xiake's findings were not accepted by the central government/society until 200–300 years later when it was recognized by the Qing Emperor Kangxi and the scholars. With the national unification in the early Qing Dynasty, some Western scientific technologies, such as cartography, were introduced into China. Emperor Kangxi, in his book and article, respectively, titled *Collection of Articles by Kangxi in His Spare Time* and "Yangtze River Source," expressed his appreciation of "Travel along the River Headward to the Source" by Xu Xiake and officially recognized Xu Xiake's argument regarding the source of the Yangtze River. During Emperor Kangxi's reign, in order to effectively rule the Qinghai-Tibet area, he once sent a special envoy to the source area of the Yangtze River in an attempt to further investigate the source and the river system distribution. However, due to inclement weather in the river's source area, high-altitude-related oxygen deficiency, and poor transportation conditions, the envoy could not reach the hinterland after they came to the area of the river source. They could only leave disappointed, "sigh while looking to the source," and confirm an old conclusion that "the river source is like a broom and spreads out widely." Even so, the basic characteristics of the complex interactions between the river network and palustrine wetlands were briefly described, but no determination was made as to which was the mainstream. However, they concluded that the water system in the source area was complex and intricate. Between Kangxi's 47th and 57th years of rule (1708–1718), the Imperial Court organized scholars together to complete the *Kangxi Provincial Atlas of China*, by using the modern surveying technology for the first time. The map showed the approximate location of the water system in the upper reaches of the Tongtian River, which provided the preliminary determination of the geographic locations for the sources of the Yangtze and Yellow Rivers. In Qianlong's 26th year of rule (1761), Chi Zhaonan, in his *Framework for Waterways*, provided a more detailed description of the water system in the source area of the Yangtze River, including the rivers presently known as the Buqu, Gaerqu, and Tuotuo Rivers. He considered the Buqu to be the true source, while the Gaerqu and the Tuotuo Rivers were tributaries. From the Qing Dynasty to the Republic of China period, some Chinese people and foreigners dived deeply into the source area for the purpose of exploration or scientific investigation. However, because of the complex geographical conditions, inclement weather, natural environment, and lack of effective measurement techniques, they only roughly understood that the source of the Yangtze River included the Dangqu, Buqu, Chumar, and other rivers.

1.2.2 Modern People's Understanding

During the late Qing Dynasty and the Republic of China period, works covering the water system in the source area of the Yangtze River began to increase, though none of them provided more detailed information than Chi Zhaonan's *Framework for Waterways* published during the Qianlong years. *Introduction to China's Geography*, which was published in 1946, is a representative of the works and states that "the Changjiang River is also known as the Yangtze River, originating from the southern foot of the Bayan Har Mountain in Qinghai … with a length of 5,800 km, and being the longest river in China. There are two sources, south and north sources, in the upper reaches within Qinghai. The south source is called Muruusu, and the north source Chumar." Because the Yellow River originates from the northern foot of the Bayan Har Mountains, and the Yangtze River starts from the south of the mountain, the document indicates that "the two rivers originate from the same mountain" and "the Yangtze and Yellow Rivers are sister rivers." Geography textbooks for primary and secondary schools were written on the basis of the publication that erroneously indicates that the length of the Yangtze River was 5800 km and that it is the world's fourth longest river. As a result, this information had circulated a great deal, leading to a broad acceptance of this incorrect statement, which would remain unchallenged until sometime after the founding the People's Republic of China (PRC). The

Muruusu River, which is now called the Tongtian River, is separated from the Yellow River by only the Bayan Har Mountains. However, there are three source rivers in its upper reaches. The Buqu and Gaerqu Rivers had been considered as the true sources of the Yangtze River because these two rivers originate from the northern foot of the Tanggula Mountains and run parallel to the Tongtian River. However, the lengths and flowrates of these two rivers are less than those of the Tuotuo and Dangqu Rivers. Therefore, it was incorrect to consider these two rivers to be the source of the Yangtze River, and consequently, the Yangtze River Basin Planning Office (YRBPO) designated the two rivers as tributaries of the Dangqu River in 1978.

Before the 1970s, there was no accurate topographic map generated from survey data for large areas west of the Qinghai-Tibetan Highway. As a result, it was believed that the true source of the Yangtze River was the Gariqu (Gaerqu) River that originates from the main peak of the Tanggula Mountains – the snowcapped Geladandong Mountains. In 1970, the Lanzhou Military Region used the aerial photogrammetry to prepare a 1:100,000-scale topographic map for the area extending from the Jinsha River's segment in the Dege Prefecture of Sichuan Province up to the source area of the Yangtze River. The results provided the YRBPO, which was renamed the Yangtze River Water Resources Commission (YRWRC) in 1982, with the data that allowed them to perform an investigation into the source area. In the summer of 1976 and 1978, the YRBPO sent two separate teams to the source area to conduct detailed investigations. The results of these investigations indicated that the source of the Yangtze River extends into the Tanggula and Kunlun Mountains in the Qinghai-Tibet Plateau where there are at least ten streams of varying sizes, of which there are three larger ones: Chumar, Tuotuo, and Dangqu Rivers. Of the three rivers, the Chumar River is the smallest in flowrate and often dries up in the winter; it cannot be considered as the true source of the Yangtze River. In terms of the drainage area and water volume, the Dangqu River is the largest. However, in accordance with the principle of "the source is the farthest and flows in the same direction of the Yangtze River," the Tuotuo River, which is 5–6 times smaller in flow volume and 18 km longer than the Dangqu River, is considered the true source of the Yangtze River. The Tuotuo River has two upper sources: the east and west branches. The east branch originates from the southwest side of the Geladandong Mountains (at an elevation of 6621 m); the west branch originates from the western side of Gaqiadirugang Snow Mountains (at an elevation of 6513 m). The east branch is slightly longer than the west branch; therefore, the true source of the Yangtze River should be the east branch. The upper segment of the east branch is a very large glacier (Jianggendiru Glacier). Melted water from the glacier forms a stream that feeds into the beginning of the Yangtze River. On January 13, 1978, based on the results of the two site visits, the Xinhua News Agency announced the following conclusion of the river source investigations: "How long is the Yangtze River? Where is its source? The results of the investigations by the YRBPO have showed that: the source of the Yangtze River is not in the south foot of the Bayan Har Mountains, but is the Tuotuo River originating from the west side of the main peak of the Tanggula Mountains – the snow-capped Geladandong Mountain. The Yangtze River is not

5,800 km long, but it is 6,300 km long, longer than the Mississippi River of the United States, only shorter than the Amazon River in South America and the Nile River in Africa." The next day, the Associated Press reported: "The Yangtze River has replaced the Mississippi River to become the world's third longest river." Since then, all countries in the world have accepted this conclusion (Shi 1983). Figure 1.1 is the map of the water system in the source area of the Yangtze River that was determined by the YRWRC. The figure indicates that the Tuotuo River flows in the same direction as the Tongtian River. Although the Dangqu River has a larger drainage area, the river flows in an opposite direction from the Tongtian River. Figure 1.2 shows the source of the Tuotuo River – the Jianggendiru Glacier – where the first drop of water for the Yangtze River comes from. Figure 1.3 shows the Chumar River – the north source of the Yangtze River.

Although the YRWRC officially confirmed the true source of the Yangtze River, unofficial folks still cast some doubts. With the scientific and technological advancement, especially in the past 10 years, scientists and explorers have had greatly increased interest and enthusiasm in the investigation and exploration of the river's source. If it is assumed that several tens of people from outside the area had already reached the source area located in a remote, lifeless glacier 30 years before, then several hundreds of people today have been to the source area, some of which even have been there several times. The following are summaries and remarks for several representative expeditions and investigations:

1. In 1985, an American expedition team led by Ken and Jan Warren was permitted to explore the source area of the Yangtze River by rafting. In order to complete

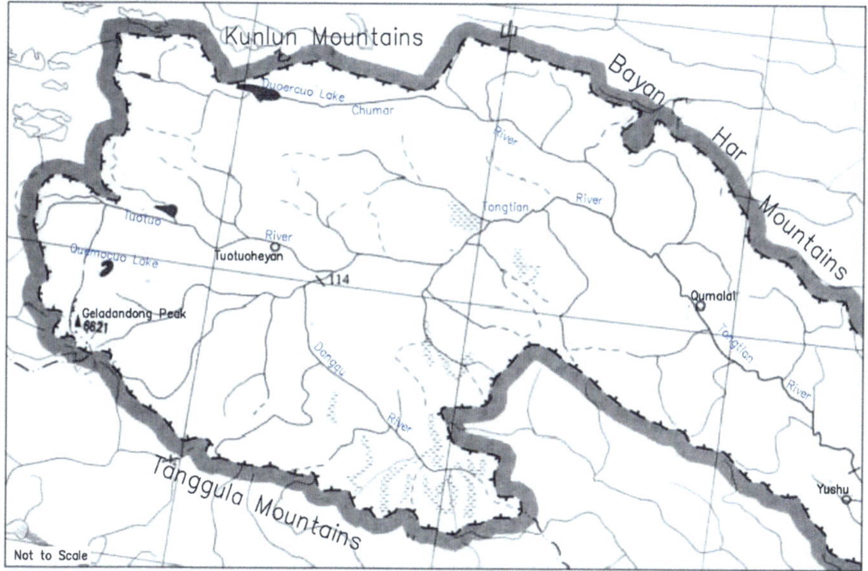

Fig. 1.1 Map of water systems in source area of Yangtze River

Fig. 1.2 Jianggendiru Glacier where first drop of water of Yangtze River comes

Fig. 1.3 North source of Yangtze River: Red Chumar River

the first rafting of the Yangtze River before the American team, Yao Mao, a Chinese youngster, went rafting alone in the Yangtze River in June 1985. Unfortunately, he died in the expedition. In October 1985, the Sichuan Provincial Geographic Society set up the headquarters for a scientific expedition and formed a 54-member Yangtze River Scientific Drafting Expedition Team consisting of rafting members, scientists, journalists, security personnel, and logistics staff. Eleven of the 54 members were research personnel from the Chinese Academy of Sciences – Institute of Geography at Chengdu, Sichuan – and other research organizations, and 15 were rafting specialists. Led by Tang Bangxing and Zhu Jianzhang, the team members had an average age of 25. On June 3, 1986, the Yangtze River Scientific Drafting Expedition Team left Chengdu, Sichuan, for Lhasa, Tibet, and then proceeded northward toward the Tuotuo River, the source of the Yangtze River. On June 16, 1986, the rafting expedition and scientific investigation of the river source officially began. In the source area at an altitude of 4500 m, they collected more than 1000 of various hydrological and geological samples and captured more than 3000 photographs. The team investigated the source area of the Yangtze River and questioned why the Tuotuo River had been considered as the true source of the Yangtze River when they found that the Dangqu River exceeded the Tuotuo River in length, water quantity, drainage area, and development of the river system.

2. In October 2008, organized by the Qinghai Province and led by Liu Shaochuan, a researcher at the Chinese Academy of Sciences, the "Three River Source Scientific Investigation Team" considered the south source as the Yangtze River's true source, which was based on their findings obtained from their investigation. They used modern advanced technology tools such as the global positioning system (GPS), geographic information system (GIS), and remote sensing (RS) technology to measure the length of the Tuotuo River (348.63 km) and that of the Dangqu River (360.34 km), 11.71 km longer than the Tuotuo River. The team concluded that the Dangqu River should be the true source of the Yangtze River "solely based on the length." Their conclusion obtained support from Professor Li Zhiliang, Chief Engineer and Deputy Captain of the Survey Team for the Topography of the Source Area from the Lanzhou Military Region in the 1970s (Li and Song 2010).
3. In October 2009, a "Water for China" Investigation Team consisting of 14 geologists and explorers with Yang Yong as the Team Leader, the Chief Scientist of the Hengduanshan Research Institute and the Deputy Director of the Professional Commission of the China Desertification Research Foundation, conducted an investigation of the Yangtze River's source. They confirmed that the Gaqiadiru Glacier was farther away than the Jianggendiru Glacier as previously determined; therefore, the Yangtze River would have been about 10 km longer if the Dangqu River was considered as the source. However, they argued that "the Dangqu River cannot be considered as the true source since it is a marsh itself with a limited water flow, primarily originating from groundwater." Yang Yong argued that "60% of the source water is from the Tuotuo River on the southwestern side of the Geladandong Mountains. The topographic map shows that the Tuotuo River flows from the west relatively straight toward the east with its origin at a relatively high-altitude glacier. This flow direction is consistent with the mainstream direction of the Yangtze River. The origin of the Dangqu River is located in a relatively lower-altitude swamp area receiving groundwater from the southeast and has a large bend where the precipitation is relatively high, and the catchment area is large. Therefore, the flow of the river system is naturally large. However, its alignment is not consistent with the mainstream of the Yangtze River." In addition, the length of the Tuotuo River is almost equal to that of the

Dangqu River. It is also controversial that the length of the source glacier is not counted toward to the length of the river. They believed that it was appropriate to consider the Tuotuo River as the true source of the Yangtze River, which was in agreement with the view of the YRWRC.

4. In October 2010, with the help of the Bureau of Water Conservancy of the Qinghai Province, the YRWRC organized a 100-member team to inspect the source area for the third time. The Exploration of the Jianggendiru Glacier (5400 m altitude) included 23 members from the YRWRC including Commissioner Cai Qihua, 15 members from the Qinghai Bureau of Water Conservancy, and the driver from Xining Jiaming Outdoor Sports Service Company. The author also had the honor to participate in the scientific expedition and successfully reached the source. The following points are a summary of the expedition: ① The monument for "the Yangtze River Water Resource Comprehensive Investigation" was erected at the top of the Jianggendiru Glacier, which is above the source. It was the first eternal imprint of the people from the YRWRC on the Jianggendiru Glacier. ② Nearly 100 water conservancy experts, including China's survey and design masters, chief scientists, academic leaders, and young talents in various disciplines, were organized into an investigation team to perform a comprehensive study of the hydrology, water resources, aquatic ecology, aquatic environment, geography, glaciology, meteorology, geology, and geospatial information transformation of the river's source area. ③ The investigation reconfirmed the determination that the Yangtze River has three sources with the Tuotuo River as the true source.

1.2.3 Discussion

There is no unified standard for determining the river sources of the world. Usually, researchers consider factors such as direction, length, water quantity, and basin area. However, the main criterion for determining the river source is the river length. But in terms of length, the Tuotuo River is actually not much different from the Dangqu River. Moreover, the straight-line distance from the Jianggendiru Glacier, the source of the Tuotuo River, when compared to that from Xiasheriaba Mountain, the source of the Dangqu River, is farther from the Yangtze Estuary at the sea. Furthermore, the Jianggendiru Glacier is at a higher elevation (more than 6000 m) and is representative of the characteristics of the Qinghai-Tibet Plateau where the source of the Yangtze River is located. The Jianggendiru Glacier is also the world's highest source of large rivers. Geladandong, meaning "high-pointed peaks" in Tibetan, at an elevation of 6620 m, and Jianggendiru Glaciers at an altitude of 6548 m possess two half circular arc-shaped large glaciers from the south to the north. The south glacier is 12.5 km long and 1.6 km wide, with 2-km-long ice towers in its tail. The towering icy snow mountains and crystal clear white glaciers are the source of the Yangtze River or the place where the first drop of water of the Yangtze River comes from. However, the Dangqu area is located in the southeast of the river source. The source of the Dangqu River is the Xiasheriaba Mountain, meaning "yellow mountain" in Tibetan, at an elevation of only 5395 m. The straight-line distance from the Xiasheriaba Mountain to the Yangtze Estuary at the sea is nearly 400 km shorter than that from the Jianggendiru Glacier. Even though the volume of water at the mouth of the Dangqu River or the entrance into the mainstream of the Yangtze River is relatively large, the Dangqu River receives water from many small streams originating from alpine marshes and has no clear mainstream. Moreover, the Dangqu River does not flow in the same direction as the Tongtian River or the mainstream of the Yangtze River. Therefore, it is not appropriate to consider the Dangqu River as the true source of the Yangtze River.

As a result of climate change and other factors, the glacier at the source of the river is constantly changing. For the 6300-km-long Yangtze River, the length of the source river being plus/minus several km to more than 10 km is basically within the error range. Every year, and even every quarter of the year, the length of the river changes a few km. Therefore, the author doesn't believe it is necessary to discuss the source of the Yangtze River because of the length being several km different. With the very rapid advancement of modern science and technology, it is easy to measure the source river through aerial remote sensing and the modern surveying and mapping technology. A few extra km of length will have no effect on the status of the Yangtze River. Earlier survey results found that the peak of Mount Everest was 4 m (at an elevation of 8848 m) higher, and later survey results indicated it was not that high (only at 8844 m). Although the peak is 4 m lower, which might be a result of the ice cap melting, it is still the world's highest peak. No matter which river is the source of the Yangtze River, the status of the Yangtze River as the third longest in the world and the longest river in China will not change. The view of the Yangtze River having three sources can be considered as a combination of perspectives from all groups.

1.3 Three Gorges Project

The Three Gorges Project (TGP) is not only the largest hydraulic project on the Yangtze River but also the largest hydraulic project in China and possibly in the world. Because of its gigantic scope and the mass relocation of people as a result of the project, its impacts were great. The project has always faced various controversies or doubts since the beginning of its engineering planning, feasibility study, and design, as well as up to its construction and operation. Debates not only demonstrated technical democracy and public participation but also provided important information for the Central Government to make scientific decisions. Early debates were at the professional and governmental levels, mainly involving technical and ecological issues, as well as socioeconomic issues. Since the completion of the TGP, the debates have mainly been at the nonprofessional public level and primarily related to the natural integrity and social psychology. Influenced by the Western anti-dam movement, some people, including "angry youngsters," have developed a rebellious tendency and "opposed every dam." However, from democratic decision-making and scientific/technological advancement standpoints, public discourse was needed and in fact had positive effects. Ultimately, it minimized adverse effects and promoted continuous improvements in planning, design, construction, operations, and management.

1.3.1 Construction History of the TGP

The TGP was originally envisioned by Mr. Sun Yat-sen, the founder of the Republic of China. In his book *International Development of China* (published in 1919), he stated that "the rapids should be dammed up to form locks that would enable crafts to ascend the river, as well as to generate water power" in the paragraph that began with the Yangtze River "above Ichang enters the Gorges." Based on this idea, in the mid-1940s, the Nationalist Government of China signed a contract with the United States Bureau of Reclamation to use United States funds to build the Three Gorges hydroelectric station and invited John Lucian Savage, the Bureau's chief engineer and a world-renowned dam engineer, to visit China. After he surveyed the Three Gorges area, he prepared the "Proposal for the TGP on the Yangtze River," which indicated that the TGP was feasible, and made the arrangement for the preliminary work. Thereupon, the Chinese government sent a group of technical personnel to the United States to receive training and participate in the design process of the TGP. However, the contract was fruitless due to the following Chinese civil war that ensued.

When the PRC was founded in 1949, a large flood event occurred in the Yangtze River. In 1950, to aid the Central Government in controlling flood hazards and developing the water resources of the Yangtze River, the Yangtze River Water Conservancy Commission (YRWCC), later renamed as the YRBPO, was established to be specifically involved in the planning and regulation of the Yangtze River basin. In 1954, a 100-year extremely large flood event occurred in the Yangtze River basin and caused enormous economic loss. This event accelerated the planning and investigative work for regulating and developing projects on the Yangtze River, including the TGP. Through the cooperation among the governmental departments related to geology, electricity, and transportation, the Ministry of Water Resources (MWR) and the YRWCC made the arrangement to conduct the surveying, investigation, design, and scientific research work for the TGP. At the time, the main purpose of the TGP was flood control. It was proposed that a 200+-m-high dam would be constructed to completely solve the flooding problems in the middle and lower reaches of the Yangtze River. It was also proposed that a 30-million-kilowatt (GW) hydroelectric station would be installed to generate hydroelectric power. However, the disadvantage of the proposal was that most of Chongqing would be submerged and large numbers of people would have to be relocated.

In 1956, Chairman Mao Zedong authored the famous poem "Swimming" after he swam across the Yangtze River in Wuhan. In the poem, he stated that "walls of stone will stand upstream to the west, to hold back Wushan's clouds and rain, till a vast lake rises in the high gorges" and visualized a blueprint for the TGP. After the high dam plan for the TGP was proposed, there were many different opinions from various water conservancy departments. As a result, in January 1958, the Central Government held a conference in Nanning, Guangxi, where Chairman Mao and other leaders listened to different opinions on the TGP.

In August 1958, a follow-up conference was held in Chengdu, Sichuan, where the Central Government decided that it would be necessary to take a proactive and prudent approach to the TGP because of the varying opinions. They clearly pointed out that Chongqing shall not be inundated and the dam shall not exceed 200 m in height and urged the study of dam alternatives with lower heights. Thus, all alternatives evaluated were in favor of a 200-m-high dam with a 25 GW hydroelectric station. But as a result of the natural and man-made disasters in the early 1960s and the destruction of the 10-year Cultural Revolution, the planning and design work of the TGP could not move forward until the mid-1970s when the construction of the Gezhouba Water Control Project (GWCP) began. The GWCP was considered part of the TGP because it would provide a reverse regulating reservoir. Thereafter, the planning and research work for the TGP was resumed.

In the 1980s, the Ministry of Water Resources and Electrical Power (MWREP) realized the difficulties of relocating large numbers of people for the 200-m-high dam plan and instructed the YRBPO to evaluate various lower dam alternatives, in which the Central Government would choose the best one. In 1984, the YRWRC presented a feasibility study report for an alternative with a normal operational water level of 150 m. The alternative would have a hydroelectric station with an installed capacity of 18 GW, provide certain flood control capability, and improve the navigation conditions for hundreds of km of the Chuan River's mainstream channel. The proposal was reviewed by more than 360 experts and governmental officials under the leadership of the State Planning Commission. In April 1984, the State Council approved the feasibility study report but instructed the dam crest elevation be raised by 10 m in order to provide additional storage capacity for extremely large flood events which would lower the flood hazards in the middle and lower reaches. In the same month, the preparatory work for the construction of the TGP began. In September 1984, the government of Chongqing Municipality communicated with the State Council and requested a raise in the normal operational water level to 180 m for the TGP, so that 10,000 tonne ships could reach the Chongqing Port. The Ministry of Transportation backed Chongqing's claim. As a result, per request of the State Council, the State Planning Commission and the State Science and Technology Commission organized experts to perform an additional evaluation of the design water level. During this period, objections to the construction of the TGP emerged, and even the proponents for the construction of the TGP presented different opinions about the water level of the reservoir and approaches to the project's development. Accordingly, the State Council decided to comprehensively re-evaluate the TGP in April 1986. By November 1988, 14 special topic reports had been prepared. In 1989, the YRWRC prepared a second "Feasibility Study Report for the TGP" and concluded that it was better to go through with the construction of the TGP sooner, rather than later. The report recommended a construction plan that was a "one-stage development, one-time construction, multi-staged water storage that would require the continuous relocation of people." The report recommended that the final dam height be 185 m and the normal operational water level be 175 m. Additionally, it suggested initiating preparatory work for the project construction.

On April 3, 1992, the fifth session of the National People's Congress reviewed and passed the "Resolution on the TGP," and the project officially started in 1993. In 1997, the river closure was completed for the TGP. On June 1, 2003, the reservoir began to store water. On June 10, 2003, the reservoir water level was up to 135 m. In July 2003, the first generator unit started to send power into the grid, and the project began to generate revenue. In 2006, the phase three cofferdam was demolished, and the entire dam started to block water. On October 28, 2006, the reservoir water level rose to 155.68 m with a corresponding storage capacity of 23.3 billion m^3, and the tail of the backwater reached the Tongluo Gorge in Chongqing. On September 28, 2008, the major part of the dam was completed, and the reservoir entered the normal storage stage to naturally operate at water levels of 175 m–145 m–155 m on an experimental basis. The water level rose to 172.8 m on November 5, 2008; dropped to 145.87 m on September 15, 2009; rose to 170.87 m on October 31, 2009; and reached 171.4 m on November 24, 2009. On September 10, 2010, the reservoir started to store water with a starting water level of 160.2 m, and the water level rose to the normal storage level, which is 175 m, for the first time. By 2010, the hydroelectric station had generated an accumulated electricity of 500 billion kilowatt-hours (kWh) since the first group of generator units were put into production in 2003. Presently, the entire project has been completed, except for the vertical ship lift on the left bank, and the reservoir has entered the full beneficial stage. Meanwhile, the operational plan of the reservoir has also been in the process of continuous optimization. Figure 1.4 is a panorama of the TGP.

Fig. 1.4 Panorama of TGP

1.3.2 Constant Improvements with Questioning and Debates

Because the TGP is a world-class mega project, it has impacts on a wide range of aspects. It is natural that various questions and doubts arose regarding the project planning, design, construction, and the regulating strategy of the project operation and engineering, economic, and social issues. The vast majority of opinions were brought up for the better of the country and the project and had positive effects. Most of the suggestions belonged to different academic viewpoints or opinions, which helped promote the scientific planning and design of the project. As Pan Jiazheng, a member of both the Chinese Academy of Sciences and the Chinese Academy of Engineering, said, to some extent, "opposing opinions are the greatest contributors to the TGP." Various doubts and debates had made the planning, design, construction, and operation of the project more cautious and more respectful of science, and in fact, the beneficial opinions and suggestions from opponents had been fully taken into account in the design and research of the project.

The timeframe for debates on the TGP can be divided into three stages. The first stage was the early planning stage of the project, that is, before the 1980s; the second stage was from the comprehensive evaluation to the construction of the project, namely, the 1980s to 2008; and the third stage was the later stage of the test operation of the project, that is, after 2008.

1.3.2.1 First Stage of Debates

The focus of the early debates was mainly on issues associated with the flood control strategy of the Yangtze River, national economic distribution, engineering techniques, adequacy of economic strength, and macro strategic issues about the timing and sequence of construction. Two major opposing sides were Lin Yishan, Commissioner of the YRBPO, and Li Rui, Minister Assistant of the Ministry of Fuel Industry and Director of the Bureau of Hydropower. In fact, they represented two different sectoral expert groups with different overall opinions. After the 1954 Yangtze River flood, the Central Government instructed the YRBPO to present a framework to ultimately solve the flood problems in the middle and lower reaches of the Yangtze River with a flood control system that would utilize the TGP as the core of the system. In June 1956, Chairman Mao envisioned "a vast lake to rise in high gorges" in his poem "Prelude to Water Melody ● Swimming." Chairman Mao and other key leaders of China basically agreed to move forward the planning and research work for the TGP. As to how and when to construct the project, experts from the MWREP and the Ministry of Fuel Industry had different opinions at the time. The disagreements were reflected in two 20,000-character articles prepared by Lin Yishan and Li Rui, which were published in *China Water* (fifth and sixth of 1956) and *Water Power* (ninth of 1956), respectively. Both discussed the various aspects of regulation strategies for the Yangtze River, the construction of a flood control system, and the TGP. At the Nanning Conference held by the Central

Government in January 1958, Chairman Mao let Lin Yishan and Li Rui present their arguments. From the present-day's point of view, both sides had reasonable and unreasonable aspects in their presentations. The plan presented by Commissioner Lin Yishan determined that it was necessary to construct regulating reservoirs in the mountainous areas, such as the TGP, to solve the flood problem of the Yangtze River. It was concluded that the first purpose of the TGP would be flood control and that the construction of the project could fundamentally solve the flood control problem in the middle and lower reaches of the Yangtze River; secondly, the project would significantly promote China's socioeconomic development by providing great improvements in power generation, navigation, and water supply. The argument about the purposes and socioeconomic role of the TGP was basically correct. However, due to the influence of experts from the former Soviet Union and the ideology of the Great Leap Forward, the shortcomings of Lin's proposal were the following: ① The project was too large in scale, such as the high water level in the range being between 190 m and 235 m, which would submerge most of Chongqing. ② The capital cost was beyond the country's financial capacity based on the national economic conditions. ③ The electricity generation would account for too high a proportion of China's electricity grid capacity and would result in an overload to the grid. In fact, China's total electricity consumption was only 18 billion kWh in 1957, while the TGP would provide more than 100 billion kWh. Therefore, it was not technically feasible for such a single, centralized power station for a stable grid. These shortcomings were one of the major reasons for Li Rui's opposition to hurriedly start the construction of the TGP. At that time, due to the fact that China's engineering expertise was inadequate, and the Central Government had limited financial resources, there was a real problem associated with the overall capabilities for the construction of such a complex, huge project. As for the flood control strategy, Li Rui advocated to first use dikes and flood diversion/retention zones in the middle and lower reaches of the Yangtze River, then to consider building reservoirs on tributaries (such as the construction of the Wuqiangxi Dam on the Yuan River), and finally to consider constructing a reservoir on the mainstream of the Yangtze River. Li Rui's point of view mainly focused on the national economic distribution, technical capability, and relocation of local people, while Lin Yishan's point of view did not consider these practical issues and mainly extended to technical aspects. Due to the officials' positions at that time, Li Rui was at the Central Government and had a more macroscopic and comprehensive view, while Lin Yishan was responsible for the work at the YRBPO, which meant he mainly had a technical point of view and relied on the Central Government to make socioeconomic decisions. From the analysis of the situation at that time, Li Rui's opinion was correct and played an important role in the suspension of the TGP by the Central Government. In the Chengdu Conference held in August 1958, the Central Government decided to take a positive and prudent approach to the TGP, clearly declaring that Chongqing should not be inundated, the reservoir water level should not exceed 200 m, and lower dam alternatives for the project should be evaluated to determine the upper limit of the project. This decision played a beneficial role in the ensuing planning and design of the TGP. Later, due to the "Great Leap Forward," natural disasters, and other factors that occurred in the 1960s, the national technical and economic capabilities could not support the TGP. Consequently, the focus of regulating the Yangtze River was shifted to the Han River, which was relatively easier to deal with. As a result, the construction of the Danjiangkou Project began. During the Cultural Revolution, there were no planning or designing activities for the TGP.

1.3.2.2 Second Stage of Debates

After the reform started and China's door was reopened to the world, all aspects of the country got back on track, and large numbers of intellectuals returned to their technical posts to fully utilize their professional skills yet again. All experts were no longer afraid of being disorderly put on "a notorious designation" and had the conditions to fully express their views. The state government not only advocated scientific democracy but also promoted democracy in the decision-making process for major projects by listening to the views of all sides and taking different opinions more seriously. As the planning and design work for the TGP was moving forward in an orderly manner, debates among experts with different opinions from various fields followed. The majority of the people with different opinions were those who had professional expertise, and their life experience and courage to express their different views were worthy of admiration. The construction of the TGP involved not only the idea, strategy, and technical issues regarding regulating the river, and its environmental effects, but also involved the consideration of social issues such as the relocation of residents and the economic issues such as financing. The main topics of the debates were the strategy of the Yangtze River's flood control, sedimentation, relocation of local residents, ecological environmental impact, engineering design, and construction-related technical issues. The following are the main issues in this phase of the debates:

1.3.2.2.1 Sedimentation and Issues Regarding Reservoir's Long-Term Use

A relatively conspicuous opponent of the project was Professor Huang Wanli. Born in a noble family, he was sent to the United States at a young age for high-quality education. After returning to China, he dedicated himself to regulating rivers and schooling water conservancy personnel. Especially in the case of the Sanmenxia Project, he did not surrender to pressure and boldly voiced his views against the project, which demonstrated his scholarly independent thinking and disposition of adhering to the truth. He was wrongly classified as a right-winger in 1958 and had since had the right-winger designation for 22 years and experienced a long-term unfair treatment. Because of his life experience, he is very worthy of admiration and sympathy. While his views and recommendations on the Sanmenxia Project were fairly accurate, his view on the TGP was somewhat biased and incomplete and presently seems to be incorrect.

Professor Huang Wanli's knowledge about the basic conditions of the Yangtze River and the TGP was far less than his understanding of the Yellow River and the Sanmenxia Project. He had worked in the upper reaches of the Yellow River in Gansu and for the YRWCC. Therefore, he had a better understanding of the Yellow River. The Sanmenxia Project began construction in 1955 when he was 44 years old and a level II professor at Tsinghua University. It was his golden age of teaching and academic research. As a result, he was one of the authoritative figures in China's water conservancy community and had a profound understanding of the Yellow River and its sedimentation issues. His prediction about the operational consequence of the Sanmenxia Project was very far-sighted.

However, he had not been engaged in teaching or research for 22 years since he was classified as a "right-winger." He was 69 years old when his "right-winger" designation was removed in 1980. Due to his long leave from teaching and scientific research, he did not have much knowledge about the basic information of the Yangtze River and the status of the planning, research, and design for the TGP. Therefore, he objected to the construction of the TGP by primarily following his view about the Sanmenxia Project on the Yellow River. He was an expert in sedimentation and was primarily concerned about the bedload of large particles from main tributaries in the upper reaches of the Yangtze River and especially the sedimentation problem for the Chongqing waterway and the Three Gorges Reservoir (TGR). As a result, he was concerned that the sedimentation issues would worsen the flood problems in Chongqing, cause the navigable channel and the reservoir to be filled with sediment deposits, and consequently cause the Three Gorges Dam (TGD) to be abandoned, just like the Sanmenxia Dam, which caused the Wei River to be blocked with sedimentation. His concern was not totally unjustified. During the 1950s' Great Leap Forward campaign, the Yuzui Hydroelectric Station was constructed upstream of the Dujiangyan facility on the Min River. Due to not fully considering the impact of sedimentation on the reservoir and the river regime of the downstream Dujiangyan facility, the hydroelectric project failed. In 1963, Premier Zhou Enlai ordered to blast the dam. Thereafter, the riverbed gravel load in the upper Yangtze had become a significant concern to departments of water resources and hydropower and the YRWRC.

As a matter of fact, because of the lessons learned for the failures of the two hydroelectric projects at Sanmenxia on the Yellow River and Yuzui on the Min River, the sedimentation problems were thereafter considered as a key technical issue for planning, design, construction, and operation of reservoir projects, and a lot of field observations, laboratory experiments, and research were performed. The TGR was no exception, as the experience and lessons learned from the Sanmenxia Project and concerns of experts such as Huang Wanli and Lu Qinkan about sedimentation problems were taken seriously, which was demonstrated by various research organizations performing scientific experiments for different alternatives to evaluate the long-term impacts of the sedimentation issues and the reservoir's operation. All these efforts were the results of opposition from Huang Wanli and other experts. However, the basic information of the Yangtze River and the monitoring data of the past 9-year operation of the TGR indicate that the TGP is quite different from the Sanmenxia Project in the following aspects: ① Although the total amounts of sediment generated and transported by the Yangtze River are very large, the sediment concentration in the water is not as high as the Yellow River. A comparison of historical data indicates that the average sediment content at the Yichang Station on the Yangtze River (before 1985) was 1.2 kilograms per cubic meter (kg/m^3) based on data collected over many years, which is 30 times lower than that of the Sanmenxia Station on the Yellow River, which was 35 kg/m^3. ② In the past 20+ years, due to combined reasons associated with sediment interception by upstream tributary reservoirs, soil conservation, natural forest restoration, reduced precipitation in traditional sand-producing areas, and human sand quarrying from river channels, the amounts of sediments entering the TGR were reduced to less than half, from 493 million tonnes/year (1950–1985) to 377 million tonnes/year (1991–2000) and then to 205 million tonnes/year (2003–2008 after the TGR started to store water). As reservoirs have been continuously constructed in the upper Yangtze, the trend of sediment reduction at the TGR will continue. ③ Sediments are divided into suspended and bedload types. The former consists of fine particles flowing with water that can be discharged into the downstream channel. The latter consist of large particles, most of which are forced to move along the bottom of the river and are deposited in river channels upstream of the reservoir and the reservoir area. However, data collected at the Yichang Station indicates that the average total annual amount of sediments were 535 million tonnes/year (1950–1986), of which only 8.78 million tonnes/year were bedload, only accounting for 1.64% of the total sediments. In the recent 20 years, the amounts of both suspended and bedload sediments have been reduced, and the total amounts of the bedload can be considered negligible when compared to the huge dead storage capacities of the reservoirs in the upper Yangtze. Therefore, it is anticipated that the bedload sediments will have no impact on the long-term operation of the TGP. ④ The TGR has an operation pattern that "stores clear water and discharges turbid water." During the flood season, the water level at the reservoir is low and the flow is rapid. This means that water entering the reservoir is discharged into the downstream channel, and the river runs similarly to a natural stream. After the balance between scouring and sedimentation is reached, basically all amounts of sediments entering the reservoir are released from the reservoir with the discharging water. ⑤ The TGD has 23 low-level outlets, 7 sediment-scouring outlets, and 32 water entry inlets to the hydroelectric station. All of these outlets and inlets are at elevations far below the reservoir's lowest operational level of 145 m and can be used to timely discharge sediments from the reservoir according to the amounts of incoming sediments and conditions of sedimentation in the reservoir. It has been proved that the sediment problem at the TGR is not as serious as predicted by Huang Wanli, and the condition is much better than what was anticipated during the planning and design phase.

The field observation data (Xu 2012) indicates that the feasibility study and design of the TGP used the data collected during 1956–1990. Even during 1991–2002, before the reservoir started to store water, the amounts of sediments in the upper reaches of the Yangtze River and all its major tributaries, except the Jinsha River, had decreased, with the Jialing River having the largest decrease. Since 2002, the entire upper Yangtze and its major tributaries have had reduced total amounts of sediments. During 2003–2011, there was a much better sedimentation condition than expected, and the average annual amount of sediments entering the reservoir was 201 million tonnes, which was 58% lower than that in 1990 and 43% less than that during 1991–2002. Also, during 2003–2011, the average annual amount of sediments deposited in the reservoir was 140 million tonnes, which was only 40% of the theoretical value used in the feasibility study. The reservoir's ratio of sediment discharge over deposition was 24.9%. As for the most concerning bedload sediment issues raised by Huang Wanli, according to the observation data collected at the Cuntan Station during 1991–2002, the average annual total amount of bedload sediments was 258,000 tonnes, which was about 0.08% of the total suspended sediments during the same period. During 2003–2011, the average annual amount of bedload sediments was only 17,000 tonnes, which was a 95% reduction when compared to the amount during 1991–2002, while the average annual amounts of bedload pebble were 43,000 tonnes, which was 70% less than that during 1991–2002. There are many reasons for the large reduction in amounts of sediments entering the TGR in the past 30 years. These major reasons include soil conservation, natural forestation projects, interception by reservoirs in the upper Yangtze, sand quarrying on river channels, and changes in spatial distribution of rainstorms.

1.3.2.2.2 Relocation Issue

The issue of relocating people was one of the biggest restricting factors for the construction of the reservoir. With the dam constructed, the resulting reservoir would submerge a large area, causing huge land losses and forcing many people to relocate, which was the most critical issue that could impact the decision as to whether or not the TGP should be constructed. The reservoir would be 600 km long when the water level is at 175 m, causing 28,000 hectares (ha) of farmlands and orchards along both sides of the reservoir to be submerged. During the feasibility study stage of the project (1980s), the population living below the inundation level was 725,000. When taking the natural population growth and accompanying relocation factors into consideration, it was projected it would relocate 1.18 million people. In 2007, when the water level reached 175 m, the actual resettlement was 1.24 million people from the reservoir area, which was a gigantic social project. Only in China could such a huge problem of relocation and resettlement be solved. It was not only a very important social issue but also involved ecological/environmental issues such as the environmental carrying capacity of the reservoir area. China took the advantage of its social system and the roles of governments at all levels to the maximum extent possible in order to successfully solve the problem of relocation and resettlement. Beginning in 1985, the pilot project for relocation and resettlement went on for 22 years until 2007, when all the work associated with relocation and resettlement was completed.

Before the 1980s, it was not difficult to relocate people from the reservoir area because the socioeconomic development in mountainous areas was slow and people's standard of living was low. Since the people living in the reservoir area had a few fixed assets, the relocation cost was low. A considerable number of relocated people improved their living standards by resettling in new houses and obtaining relocation compensations. From a resettlement point of view, the construction start time of the TGP was very opportune. If the project had started 10 years later, the decision to start the project might have been impacted by a larger number of people to be relocated and an associated higher cost of resettlement. In means of the resettlement, the one-time compensation approach had also been changed to a developmental resettlement that required that the living standards of relocated people be higher than before. Presently, the post-construction plan of the TGP has begun to resolve issues associated with post-resettlement assistance and the development of relocated people.

1.3.2.2.3 Ecological and Environmental Issues

The TGP is one of the earliest large projects in China for which a comprehensive environmental impact assessment (EIA) was performed, and the EIA work was conducted for nearly 10 years between 1979 and 1988. The lead agencies were the Bureau of the Yangtze River Water Resources Protection and the Chinese Academy of Sciences, and more than 40 universities and research institutes participated in the EIA. Later, the Chinese Academy of Sciences and Canadian ecologists were involved in research, evaluation, and consultation associated with the EIA work. It should be noted that the experts involved in the EIA at that time were multidisciplinary, representative, and authoritative. The EIA made a systematic and comprehensive analysis of the ecological and environmental impacts that might potentially result from the TGP. At that time, the conclusion of the EIA was: the TGP would have some beneficial and some adverse effects on the ecology and the environment. The major adverse effects could be mostly reduced through implementing mitigation measures, and ecological and environmental issues would not affect the feasibility of the project's construction.

The beneficial effects of the TGP on the ecological environment included the following: ① Floods in the upper Yangtze would be effectively controlled, and the flood control capacity of the middle and lower reaches, especially the Jingjiang River section, would be greatly improved. As a result, potential ecological and environmental consequences caused by flood or waterlogging hazards would be effectively lessened or eliminated. As the sediment entering the lake would be reduced, sedimentation and shrinking of Dongting Lake would be slowed down. ② The project would increase the flowrate in the middle and lower reaches of the Yangtze River during the dry season, improve the water quality of the Yangtze River during the dry season, and create conditions for a water supply source of the South-to-North Water Diversion Project. ③ Utilization of hydraulic potential to generate electricity, when compared to using coal-fired power generation, would reduce emissions of large amounts of carbon dioxide and other pollutants.

The adverse effects of the TGP on the ecological environment were as follows: ① The reservoir would inundate farmland, resulting in the relocation of cities, towns, and large numbers of people, which would aggravate the prominent conflicts between human and land because the environmental carrying capacity of the reservoir area was limited. ② Before the construction of the TGP, the annual discharge of industrial and domestic wastewater into the reservoir area had been over 1.0 billion tonnes, and serious pollution plumes had developed in the local river sections near urban areas along the river. If the pollution sources were not controlled, they would aggravate the pollution in the local waters of the reservoir area. ③ The TGP would change the structure and function of the aquatic ecosystem in the reservoir area and the middle and lower reaches of the Yangtze River; the survival conditions for some rare and endangered species would be potentially affected; and the natural reproduction of the "four major Chinese carps (FMCC)" (black carp, grass carp, silver carp, and bighead carp) might be adversely affected. ④ After the TGP is completed, its operation would cause scouring and sedimentation changes in the middle and lower Yangtze, might potentially cause the low-lying farmland in the middle Yangtze to become gleyic and swampy, and would also result in sedimentation changes in the Chongqing section, affecting the existing water supply and drainage facilities. ⑤ After the TGD is built, some cultural sites along the river would be submerged, and some natural landscapes would be affected. ⑥ The TGP would have some impacts on the local geological hazards and human health.

The TGP started the impoundment for power generation in 2003; construction was basically completed in 2008 and is currently in the experimental operation of the normal storage water level. Based on the past 9-year operation, the real beneficial effects are in agreement with those as assessed in the EIA. The adverse effects ①, ②, ③, and ⑤ predicted in the EIA are the same as observed, but the predicted adverse effect ④ has never occurred, and effect ⑥ has been minimal. Overall, the predicted TGP impacts on the ecological environment are largely accurate, and the present conditions are better than predicted. According to the real observation results, the TGP has the following three most noticeable impacts on the ecological environment:

(1) The number of drifting eggs produced by migratory fish species in the middle and lower reaches has decreased significantly. Despite the fact that fishermen have overharvested, according to investigations conducted during May–July in 2003–2006 on present conditions of spawning sites for the fish species that lay eggs after migration into the middle reaches of the Yangtze River, there are 13 migratory species that lay eggs in the middle reaches, 8 of which are economic species: black carp, grass carp, silver carp, bighead carp, mandarin, barbel chub, bream, and bronze gudgeon. When compared to the 1970s, the number of species decreased by more than 10. In the more than 300 km section between Yichang and Chenglingji in the middle reaches, there are more than 10 spawning sites for the FMCC, indicating little change in the geographic distribution of the sites when compared to the conditions before the TGR started its impoundment. However, during 2003–2006, 1.44 billion eggs were laid by major migratory

fish species in the middle reaches of the Yangtze River, 1.08 billion of which were laid by the FMCC, which is a significant decrease when compared to that before the reservoir started its impoundment. The number of total eggs laid by the FMCC during 2003–2006 was only 42.82% of the average annual number of eggs laid by the FMCC during 1997–2002 and 56.88% of 2002, just before the reservoir began its impoundment. The situation has remained unchanged in the last 2 years. The main reason for this is that all the FMCC have spawning requirements for certain ecological and hydrological conditions, such as water temperature, duration of water level increase, and the rising rate of water level. The FMCC prefer the water temperature range of 18.6–25.5 °C when laying eggs, the continuous rise of the water level for 4–7 days, and a daily rate about 0.3 m/day. Since the operation of the TGP, such hydrological conditions have been affected or weakened.

(2) While the overall water quality of the TGR is good, algal blooms have occurred frequently because the concentrations of nitrogen and phosphorus in the water bodies of the Yangtze River have been relatively high; the enlarged surface area of the reservoir has reduced the flow velocity of the entering water; and the water quality at some reservoir bays where tributaries flow into the reservoir has been poor. For example, algal blooms occurred in the reservoir area 13 times in 2008, involving 11 tributaries such as the Xiangxi, Shennongxi, and Daning Rivers. The dominant algae species that have caused algal blooms to occur were *Cyclotella* of the diatoms, *Peridinium* of the dinoflagellates, *Chlamydomonas* and *Pandorina morum* of the green algae, and *Microcystis* of the cyanobacteria. Moreover, the construction of chemical industrial parks in Chongqing and other upstream areas and large numbers of cargo vessels passing through the reservoir area have increased the risk of water pollution accidents and put great pressure on the environmental protection of the reservoir area.

(3) If the TGR starts its impoundment on October 1, a drought event in the middle and lower reaches of Yangtze River basin would result in an early occurrence of the dry season in Dongting Lake, Poyang Lake, and the mainstream in the middle and lower Yangtze. This would cause adverse impacts on water use, the ecology, and environment in the areas. If an optimal impoundment schedule is not adopted, the impoundment at the TGR may lower the water level by a maximum of about 2 m at Chenglingji (outlet point from Dongting Lake to the Yangtze) and by a maximum of about 1 m at Hukou (outlet point from Poyang Lake to the Yangtze), which would result in an early start of the dry season in the two lakes and other impacts on the lacustrine wetland ecosystem in the middle and lower Yangtze. Table 1.1 shows recent occurrences of the lowest water levels and their dates in the Xiang River in Hunan. Except for 2007, all occurred after the TGR started its impoundment. Occurrences of extremely low water levels at the Xiang River have certain apparent correlation with the impoundment of the TGR. The water level at Chenglingji has an obvious impact on maintaining the water level of the mainstream in the lower reaches of the Xiang River. A lower water level at Chenglingji can cause an increase of the hydraulic gradient between the lower reaches of the Xiang River and the water

Table 1.1 Lowest water levels in Changsha section of the Xiang River during 2003–2009

Year	2003	2004	2005	2006	2007	2008	2009
Water level (m)	25.24	25.68	25.60	25.49	25.15	25.17	24.80
Date	November 1	November 4	November 9	November 9	December 14	October 25	October 26

surface at Dongting Lake, increase the water flow velocity, and make the water level lower in the Xiang River even when the discharge is still the same. Therefore, the Yangtze River Flood Control and Drought Relief Headquarters (YRFCDRH) has optimized the operation schedule for the TGR to incorporate drought relief in the middle and lower reaches of the Yangtze River into the operation schedule of the reservoir. When it is extremely dry, the reservoir can be operated to provide emergency water supply to the downstream. In the meantime, based on the incoming water and sediment from the upstream of the reservoir and the drought conditions in the middle and lower reaches, the starting water level of impoundment can be appropriately raised, and the impoundment start time can be moved to mid-September, while the minimum discharge is maintained in September–October to meet the basic downstream requirements for navigation and water use. All of this can mitigate the impact of the impoundment at the reservoir on the middle and lower reaches of the Yangtze River and the two lakes.

1.3.2.2.4 Air Defense Problem of the Dam

If there is a war, the public would be concerned about the safety of the TGD and the impacts of a dam break on the safety of the downstream. In the design of the TGP, war defense and protection from terrorist attacks were fully considered. The following addressed the concerns:

1. A gravity dam was chosen, which is the best dam type for anti-attack capabilities.
2. Preemptive reservoir regulation: the losses resulting from a dam break can be reduced through increasing discharge and lowering the reservoir's water level. The TGP has large sluices, where the reservoir's water level can be lowered in a short time. In terms of the TGR itself, its storage capacity is small compared to the total storage capacity of the Yangtze River during a large flood event. In extreme adverse circumstances, an instantaneous dam break would only cause a regional disaster. The TGR is 600 km long with an average width of 1.2 km, making it resemble "a ditch of water," rather than "a basin of water." Even if the dam breaks, the water will flow through a time process and will not cause the Jingjiang Dike to break or "half of China, including the people in Jiangxi, Jiangsu, Zhejiang, Hubei, and Hunan, to be inundated like fish and turtles." In other words, the devastation of a dam break is not comparable to the impact of a

nuclear attack on important cities with tens of millions of people, important military facilities, or industrial bases. During the design of the TGP's structures, a combination of wartime and peacetime operations were considered. The dam has deep and shallow outlets with large discharge capacities that can be used to lower the reservoir's water level within a short time. For example, it will take no more than 7 days to lower the reservoir's water level from the normal storage level of 175 m to 135 m using 40% of the maximum discharge capacity. Also, the Nanjin Pass, which is about 38 km downstream of the dam, is only 200–300 m wide, and the 20-km-long narrow canyon of the upper section will play a restraining role on the water from a dam break. Dam break model experiments indicate that a dam break flood resulting from a sudden attack will not have a disastrous impact on the Jingjiang River section in the lower Yangtze due to the restraining effects provided by the narrow canyon. The water level of the reservoir in wartime can be controlled at 145 m and, when necessary, can be further lowered to 135 m or even lower, which can significantly reduce the losses resulting from a dam break.

1.3.2.2.5 Major Technical Problems of Dam Design

Many technical problems emerged during the planning and design of the TGD, and many different opinions were received in the design phase. All of these issues were resolved through lots of scientific research, consultation with experts, and continuous efforts on design optimization. Moreover, this problem-solving process has advanced China's technology for the design and construction of hydraulic projects.

(1) Dam and Hydroelectric Station Structures

Although the TGD is not very high, the design of the dam's structures was complex because the amount of floodwater in the Yangtze River is large, especially when a large flood process is coming. Therefore, extraordinary discharging capacities needed to be designed to include water diversion and passing facilities for construction purposes and floodwater discharging facilities for operation purposes, which required multiple layers of large discharge outlets in the dam. The structural safety and floodwater discharging capacity of the dam were designed based on a 1000-year flood event and were checked using a 10,000-year flood event (98,800 m^3/s) plus 10% of the flood flowrate of 124,300 m^3/s. In the end, 23 deep outlets and 22 sluices were installed in the gravity dam in an alternating pattern along the dam alignment. Bottom diversion conduits for construction use were placed directly beneath the sluices. During the construction phase, the combined discharge capacity of deep outlets and bottom diversion conduits were required to reach 70,000 m^3/s, and during the project's normal operation, the total discharging capacity is 100,000 m^3/s with the joint operation of the deep outlets and sluices plus the discharge capacities of trash rack outlets, sediment-scouring outlets, and the hydroelectric plant's inlets. The large sizes and high quantities of discharging outlets and sluices in the dam resulted in a high ratio between openings and solid concrete sections and a complex structural stress distribution.

The placement of the hydroelectric plant on a downstream slope resulted in the dam foundation in that section to be exposed to the air and form a 67.8-m-high steep side slope. Consequently, an approximately 100-m-high concrete gravity dam had to be placed on the top of the sloping foundation. The low-dipping-angle structure of the bedrock beneath the dam foundation constituted a potential unfavorable sliding surface, which would be the critical sliding surface for the stability against deep sliding. There were geological defects in the dam foundation that would need to be reinforced, but the reinforced dam foundation would become a part of the dam's water blocking structure, which would exceed the general dam's design standard for structural stability. The horizontal and vertical positions, extent, and shape of the long low-dipping-angle structural plane of the deep bedrock mass in the dam foundation were determined through field explorations. In situ prototype shear tests coupled with many laboratory tests determined the shear strength of the low-dipping-angle structure surface. The stability factor of the deep dam foundation against sliding under the worst-case scenario in terms of operation and rock parameter values was improved through comparing a variety of calculation methods and geo-mechanical structural model tests to ensure the safety of the dam operation.

The inlet structure of the hydroelectric plant is a single small trumpet-like shape. In order to reduce the inlet's tensile stress, structural measures, such as grouting transverse seams and locally placing transverse seam seals further downstream, were taken. The steel penstock in the dam and the concrete dam body were constructed as a joint bearing structure, and the steel-lined concrete diversion penstock behind the dam was also constructed as a joint load-bearing pipeline, which successfully solved structural design problems of the dam and the plant.

(2) Navigation Structures

Navigation is an important service function of the Yangtze River. It was not only required that one of the main functions of the TGP was to improve the navigable conditions of the river channel but also to maintain navigation during most of the construction period. Therefore, it was necessary to design and construct a temporary navigable channel, excavate a navigable channel on the left bank, and design and construct a rapid ship lift facility at a later stage. The main navigation facilities of the TGP are two five-level ship locks, both of which are located in deep excavation pits of the left bank with bedrock between the two locks as a partition wall. Separation structures were used as entries and chambers of the locks whose walls were lined with reinforced concrete. The design deviated away from the tradition that large ship locks usually use gravity structures, and a new type of lock wall structures was presented through research. The wall structure maintained the partitioning rock mass and used the lining reinforced concrete and the rock mass to jointly bear loads. The side slopes and the partitioning barriers not only include natural rock mass, but also parts of the navigation facilities, which required high standards of stability and anti-deformation.

(3) Electromechanical Design

Before the construction of the TGP, China had not had the capability to design or build large turbine-generator units of 700 megawatts (MW). Through the TGP, foreign technologies began to be introduced. After understanding and innovating the introduced technologies, China developed its own capabilities to design and build large hydroelectric turbine-generator units. Because of the large capacity for an individual unit, highly variable hydraulic head, high sediment contents in the passing water, and frequent unit start-stop switches, the turbine-generator units of the TGP would require excellent performance and many other characteristics for safe and stable operation. As a result, major scientific and technological studies were successfully performed on the turbines to determine major parameters, primary dimensions and structures, auxiliary equipment, standards for stiffness and strength, etc. Domestic manufacturers optimized the hydraulic design. Through model experiments, they successfully lowered the hydraulic pressure fluctuation amplitude of the turbine and solved the technical problems regarding "special pressure pulsation peak zone" associated with high hydraulic head and high position load zones, which improved the operational stability of the unit. Domestically for the first time, they proposed the quantitative test parameters based on the fluctuational amplitude of the pressure caused by the hydraulic head and load zoning in the measured parts of the turbine's tail draft pipe and leafless zone. In order to widen the stable operational zone of the unit with a high hydraulic head, the manufacturers set the maximum capacity for the generators as 108% of the rated capacity and reserved a path for a forced air supplement. These research results have laid the foundation for domestic independent research and development of the 700 MW gigantic hydraulic turbine-generator units and promoted the technical advancement of the industry. The results were also adopted by other large hydroelectric stations in China. Presently, China has the capability to build hydraulic turbine-generator units of one GW, which is the world's largest capacity for one unit. This achievement should be greatly attributed to the TGP's platform for development and practice.

1.3.2.2.6 Project Capital Issues

The economic issue in the construction of the TGP has always been the focus of the project debate and justification. Before the reform, China's economic strength was limited. Even if all the national capital had been collected, it would still have been difficult to accommodate the huge financial needs for the TGP. Based on China's economic development and financial strength, it was correct to not start construction until after the 1990s. Although China's economy had experienced rapid growth before 1992, the national construction fund was still limited, and a market-driven financial system had yet to be established. One of the reasons for many early opponents in the hydroelectric units and experts to act against the TGP was the project's extraordinarily large capital investment. They were worried that once the project began, the start of other hydroelectric projects would be impacted to a very large extent. Some experts advocated that hydroelectric projects should first be constructed on the Jinsha River or tributaries of the Yangtze River before the TGP. Huang Wanli and Li Rui, among the expert opponents, advocated the construction of reservoirs in the tributaries of the upper Yangtze to replace the TGP. After Deng Xiaoping's speech in South China in early 1992, China began to build a socialist market economy. The Central Government decided to form a company with an independent legal entity – Three Gorges Development Corporation. Through a multichannel financing approach, including state capital investment, TGP Construction Fund, corporate bonds, market financing, and bank financing, the capital issue of the TGP was resolved effectively, which did not only affect the construction of other projects but also promoted the all-around development of China's economy. It then appeared that the formal starting time of 1994 for the TGP was an appropriate choice.

According to the cost estimate made during the preliminary design at the end of May 1993, the total investment was 90.9 billion yuan at the price of the time, of which 50.9 billion yuan was for the construction of the designed project (including the Maoping River Protection Project) and 40 billion yuan for relocation. After considering the rising price and the financial interest during the construction period, the total investment in the TGP was 203.9 billion yuan. According to the 2007 audit results from the National Audit Office, as of the end of 2005, China Three Gorges Project Corporation had raised 122.031 billion yuan of the construction capital (including resettlement projects), including 62.322 billion yuan from the state funds (accounting for 51%), 62.138 billion yuan from the TGP Construction Fund (thereafter referred to as the Three Gorges Fund), 41.397 billion yuan from debt funds (mainly loans from the National Development Bank and Three Gorges Corporation bonds), and 18.312 billion yuan from power generation revenues and other services. The accumulated dynamic investment for the TGP was 64.2 billion yuan (excluding resettlement projects), while the static investment was 42.585 billion yuan, which accounted for 85% of the total static investment of 50.09 billion yuan. By the end of 2010, the TGP had had an accumulated static investment of 128.06 billion yuan, accounting for 95% of the total estimated investment, of which 48.226 billion yuan had been used for pivot project components, 29.6 billion yuan for transmission and transformation components, and 50.234 billion yuan for resettlement projects.

In fact, the TGP has been generating electricity and making economic benefits since 2003. For example, in 2010, the TGP hydroelectric station generated electricity of 84.37 billion kWh, corresponding to a revenue of 21.1 billion yuan from the power generation alone when using a price of 0.25 yuan/kWh for delivery to the power grid. Presently, the revenue from the hydroelectric generation by the TGP is adequate to provide funding for the continuous construction of late-stage components of the TGP, as well as funding for the development and construction of several cascade hydroelectric stations in the upper reaches of the Jinsha River sponsored by the Three Gorges Project Corporation without the need for the state to invest any capital. It has been proved that the rapid development of the national economy and the establishment of a socialist market economy have solved the problems of the construction fund for the TGP while leaving the construction of other projects undisturbed.

1.3.2.3 Present and Future Doubts

As of now, all components, but the ship lift facility, of the TGP have been completed. The reservoir has reached the design water level for three consecutive years. Major public concerns are the long-term impacts of the TGP on the ecological environment and whether some of the major natural disasters in the basin are related to the project. The following are several relatively prominent issues:

1.3.2.3.1 Reservoir-Induced Seismicity

The issue as to whether the TGP would induce earthquakes has been one of the key issues for the project's design, research, and evaluation and has also been one of the issues to which the public has been paying more attention. Based on analysis of the seismic monitoring data collected since the TGR started its impoundment, the following conclusions can be made: ① Although there were frequent small earthquakes in the early stage of the impoundment, seismic events were dominantly small to micro earthquakes with magnitudes below 3, accounting for 99.4% of the total number of earthquakes. The strongest earthquake recorded had a magnitude of 4.1 that is estimated to be equivalent to an intensity below III at the Three Gorges based on the location of the earthquake, which is much lower than the design standard of intensity VII for the TGD and other pivot structures. ② The majority of the seismic events occurred within the predicted potential hazard zone or peripheral edge for reservoir-induced earthquakes and are mainly concentrated within 10 km of both banks of the reservoir. ③ Based on field investigations, most earthquakes were mine-type and karst-type earthquakes caused by the reservoir water flooding abandoned mines and developed karst zones.

The TGD and TGR are located in an area that has relatively low-intensity earthquakes. Historical records indicate that the largest earthquake had a magnitude below 6.0, which is estimated to be equivalent to an intensity below V at the Three Gorges based on the location of the earthquake. The seismic design standard for the dam, power plant, ship lock, and other structures was an intensity of VII, which provides a large safety factor. According to the long-term observations of existing reservoirs in China and abroad, although reservoir-induced earthquakes are common, most of them are low-magnitude earthquakes, and there are no records of large earthquake occurrences. The largest reservoir-induced earthquake in China had a magnitude of 6.1 that occurred at the Xinfeng River Reservoir in Guangdong Province. All other earthquakes had a magnitude below 5.0. The probability that the TGR would induce large-magnitude earthquakes is very low, because the water surface of the reservoir is long; additional pressure from the water is dispersed; and the reservoir is in a geologically stable area. The Wenchuan Earthquake that occurred in the upper reaches of the Yangtze River had nothing to do with the TGP, because the Qinghai-Tibet Plateau and the transition zone between the first step and second step of the Yangtze River basin are in one of China's earthquake danger zones that is several hundred to one thousand km away from the TGR and at least the Sichuan Basin is located between the TGR and the earthquake danger zone. Moreover, they are located in different tectonic belts. Therefore, the small earthquakes induced by the TGR could not affect the upstream region, so there is no scientific basis to argue that the TGP caused large earthquakes in other areas.

1.3.2.3.2 Amount of Greenhouse Gas Emission from TGR

Global warming has become an issue of common concern. According to the fourth report from the Intergovernmental Panel on Climate Change (IPCC report), as a result of human activities, greenhouse gas emissions such as carbon dioxide (CO_2), methane (CH_4), and nitrous oxide (N_2O) have increased markedly since the Industrial Revolution and have caused the global average temperature to rise by about 0.74 °C. There is no scientific basis to infer that there are large amounts of greenhouse gas emissions from Chinese reservoirs, including the TGR, from the fact that large amounts of greenhouse gas emissions have been released from the Brazilian Balbina Reservoir in South America. According to an analysis, the geographical locations and reservoir characteristics between Balbina and the TGR are very different. The Balbina Reservoir is very special in that it has a small installed capacity and low power generation, but it boasts a large reservoir area (equivalent to the area of Luxembourg). Other additional special characteristics include the following: ① The reservoir is in the tropical rainforest area with high temperatures and fast water vapor exchange rate. ② One-third of the reservoir area has a depth less than 4 m, and a large area of the reservoir's bottom is exposed to the atmosphere, indicating it is a plain-type reservoir. ③ Since the cost of harvesting the forest in the reservoir area was higher than the revenue, the reservoir was not cleared before the reservoir started its impoundment, resulting in more than 100 million tonnes of forest foliage being submerged in the reservoir. As a result, the reservoir area is replete with drowned dead trees and other dead plants. ④ The reservoir's energy density (installed capacity/reservoir area) is very low, and the unit submerged area over power generation is high.

When the water level of the TGR reaches the normal level (175 m), the reservoir water surface is 1084 km^2 in area, 600 km in length, 1–2 km in average width, and more than 20 m in average depth. This is a typical mountain-type reservoir. The net inundated land area by the reservoir was only 632 km^2 (original water surface area was 452 km^2). The original forest cover in the reservoir area was about 20%, and the vegetation was primarily sparsely distributed shrub colonies. Before the reservoir started to store water, large-scale surface clearance was performed, and not much forest vegetation was really submerged. From the operational schedule of the reservoir, the TGR is a seasonally regulated type. The reservoir water surface area of

1084 km^2 mainly occurs during November–January. From January to June, the water level gradually declines from 175 m to 145 m. During the flood control season between early June and late September, the water level is generally maintained near the flood control level of 145 m with a reservoir water surface area of about only 735 km^2. The main function of the reservoir is to discharge floodwater and sediments, operating in a manner that the amount of water entering the reservoir is the same as the amount discharged. During the flood season, because the reservoir water flows at a very high velocity and the sediment content is high, it is difficult for the aquatic plants to survive. Therefore, when compared with the major reservoirs abroad and based on recent field observation results from the TGR and its characteristics, its contribution to the increase of greenhouse gas emissions is very small and the impact on climate change is negligible. In fact, the reservoir's generation of hydroelectric energy reduces emissions because it replaces fossil-fueled electricity. The TGP's hydroelectric power generation of more than 84.7 billion kWh per year is equivalent to more than 50 million tonnes of raw coal, resulting in an annual emission reduction of 100 million tonnes of CO_2, 1 million tonnes of sulfur dioxide, 10,000 tonnes of carbon monoxide, 370,000 tonnes of N_2O, and large amounts of wastewater and waste residue. These numbers have fully proved that the Three Gorges hydroelectric station generates green energy in nature. Therefore, the propaganda pushed by many Western anti-dam extremist organizations that claim the TGR emits a large amount of greenhouse gases has no scientific basis.

1.3.2.3.3 Effects of Three Gorges Dam on Atmospheric Circulation and Climate

In recent years, global warming is an indisputable fact, frequently leading to extreme weather conditions and hydrological events, such as the unusual drought in Sichuan-Chongqing in 2006, the extremely large rainstorm in Chongqing in 2007, the drought in Southwest China from 2009–2001, and the drought in the middle and lower reaches of the Yangtze River in 2011.

The Yangtze River basin boasts an area of 1.8 million km^2 and an average annual precipitation of more than 1000 millimeters (mm), where 70% of the precipitation comes from the circulation outside the basin, mainly from the ocean. The precipitation in the middle and lower reaches of the basin is from the southeast where the Pacific Ocean and South China Sea are located, while the precipitation of the upper reaches of the basin is mainly from the South China Sea and Indian Ocean. Less than 30% of the precipitation in the basin results from the water circulation of the continent itself. The vast majority of the 30% continental precipitation results from the evaporations of the alpine snowmelt, underground and wetland water, land surface, and water surface in the upper reaches of the Yangtze River. About 60% of the continental precipitation results from the evaporation of land surface and water surface, or about 18% of the total precipitation in the basin is from continental water cycle. In the continental water cycle, evaporation over continent accounts for 90% and that from water surface (water surface area accounts for about 3% of the continental area) accounts for about 10%. Therefore, the water surface area within the entire Yangtze River basin contributes about 1.8% to the total precipitation of the Yangtze River basin. On the 50,000+ km^2 water surface area in the Yangtze River basin, natural lakes and wetlands account for 90%, while the reservoir water surface area accounts for 10%. Thus, all 40,000+ reservoirs within the Yangtze River basin contribute about 0.18% of the total precipitation. Therefore, the maximum water surface area of the TGR is only 1087 km^2 (when the water level is at 175 m), which is still less than one-fourth of the total water surface area within the basin. Therefore, the maximum contribution of the TGR to the total precipitation of the Yangtze River basin is less than 0.045%. Although the above calculation is not necessarily the most accurate, the order of magnitude of the calculation should not be a problem. As a result, the TGR has a negligible impact on the precipitation within and the water moving to the Yangtze River basin, which are at most limited near the vicinity of the reservoir area and will never affect a larger region or the climatic characteristics of the whole basin.

Some argued that the Three Gorges is the primary channel for the southeast warm and humid air to flow to the frequently drought-occurring Sichuan Basin and that the constructed dam has blocked the transport pass for the air flow, causing frequent droughts in Sichuan and Chongqing. However, from the atmospheric circulation, the main cause of extreme dry weather events are changes in weather conditions between the ground surface and the sky at about 5500 m above the ground. The TGD is only 185 m high and the dam crest elevation is 185 m. The dam is miniscule when compared to 5500 m height and does not affect the transport pass for the air flow between the sea and inland or has no significant impact on the atmospheric circulation in the Yangtze River basin. The precipitation in the Sichuan Basin is mainly from the water vapor produced by the Indian Ocean and the South China Sea, but not the air from southeast. Moreover, the Sichuan Basin is a traditionally arid area in the Yangtze River basin. In the long history before the existence of the TGD, severe droughts still occurred frequently. The famous Dujiangyan Project was constructed to solve the problems associated with agricultural production and droughts in parts of the Tuo River watershed and the Fu River basin in the Chengdu Plain and the Sichuan Basin. Therefore, there is no reason to link the TGP with the extreme weathers or hydrological events in the Yangtze River basin, especially in the upper reaches of the river and southwest region.

1.3.2.3.4 Possible Debatable Issues in the Future

Although the TGP has been completed, debates over the TGP have not ended. The main controversy is centered on the long-term impacts of the project on the ecological environment. Because the current environmental evolution in the middle and lower reaches of the Yangtze River has not stabilized, the biological evolution will need more time to complete. Possible future debates will be focused in the following areas:

(1) *Ecological Environment of Reservoir Area.* Due to the construction of the reservoir, the aquatic environment of the Three Gorges section has changed a great deal. As the water surface area has expanded and the flow has slowed down, can the algal bloom problem at the reservoir bays and the tributary entries be effectively controlled? The ecological and land use issues within the water fluctuation zone are also a major topic for the aquatic environmental protection in the reservoir area. It is worth paying attention to the question as to whether the good water quality in the reservoir area can be maintained for a long time. Since filling the TGR, there have not been many very large flood events. It is difficult to determine the long-term sedimentation issues in the reservoir. It is worth monitoring if the effective storage capacity of the reservoir can be maintained for a long time and the long-term navigation and safe operation of the Chongqing Port can be maintained while the sediments continue to accumulate in the tail area of the reservoir. Because there are many historically unstable slopes in the TGR area, it is a subject of long-term monitoring to see if the slopes will be stable under the conditions of the new water level fluctuations. It remains to be seen how the succession of the aquatic organisms will proceed as a result of the environment change from a rapid flow situation to a slower flow environment in the reservoir area.
(2) *Evolution of River Channel and River-Lake Relationship in Middle and Lower Reaches.* It will take more than a century to have a relatively clear answer as to how the evolution of the river channel/river-lake relationship in the middle and lower reaches of the Yangtze River, and the estuarine environment under the reduced sediment conditions will proceed. It will need many observations over a long period of time to monitor the evolution of the river channel/the river-lake relationship, dike stability, riparian zone utilization, habitat change, and evolution of aquatic organisms in the middle and lower reaches of the Yangtze River.
(3) *Service Life of TGP.* In a few decades, the dam structure and electromechanical equipment will need to be updated. In 100–200 years, the sedimentation in the TGR should be close to a balance. In 200 years, how much of the effective flood control capacity and beneficial storage capacity will remain? In 500 years, it remains to be seen whether the dam concrete will experience aging problems such as alkali aggregate reactions, carbonation, dissolution, etc.
(4) *Relationship Between Construction and Operation of Cascade Reservoirs in Upper Yangtze and Operation of TGR and Relationship Between Storage and Release of Water from a Group of Reservoirs.* It is a long-term process to adjust the function and status of the main control reservoirs of the Yangtze River basin in a joint operation and to optimize the joint operation.

Currently, the doubts about the TGP are not of simple engineering technology. However, they involve public participation in the understanding of the ecological environment and even involve the sociology and the social psychology. There is still a lot of work to do, including the circulation of scientific knowledge, publicity, and education in order to help us better understand how to let the shareholders and the society share the enormous socioeconomic benefits of the TGP. Additionally, it would increase societal and public participation in the management of the TGD, help the public understand the TGP, and obtain their support of the project.

1.4 Evolution Trend of the Yangtze River

1.4.1 Impact of Human Activities on Yangtze River's Evolution

Presently, China is still rapidly growing economically, and many water resources development projects are still underway in the Yangtze River basin, such as the following projects that are under construction: cascade hydroelectric stations in the upper Yangtze's mainstream and tributaries, South-to-North Water Diversion Project, new water source development projects, utilization projects of riparian zones, bridges across rivers, and development and utilization of lacustrine wetland areas in the middle and lower Yangtze. These activities will have long-term and cumulative effects on the evolution of the Yangtze River system and its ecosystem.

Currently, the Yangtze River system has a total of approximately 30,000 km of constructed dikes in length, of which about 3900 km are along the banks of the mainstream of the river. These dikes are mainly located in the low-lying flood protection areas where the river channel and lakes have historically undergone intense changes. Obviously, for the safety of flood control and the need of human land use, the unchecked evolution of the river channel or adjustment of the river system is not allowed. The changes of the river course are limited mainly in the river channel between the dikes along both sides of the river and major changes occurred on floodlands and floodplains, braided channels, and mid-channel sandbars/islands. The flood process has been the main cause of the river's evolution. With the completion of regulating reservoirs (e.g., TGR) and their role of regulation, the large flood process has been generally evened. As a result, sediment transport and river channel changes in terms of severity have also been lessened. The discharge of clear water to the downstream of the reservoir for a considerable period of time will cause a long-term and gradual progression of the scouring process toward the downstream. In case of a 100-year large flood event, large changes will occur to the adjustment of the river channel and the river-lake relationship.

There are 47,000 reservoirs in the Yangtze River basin with a total storage capacity of more than 250 billion m^3 and the total beneficial storage capacity of more than 120 billion m^3. Of the 47,000 reservoirs, 166 are larger ones with a total storage capacity of 190.8 billion m^3 and a total beneficial storage capacity of 98.3 billion m^3, accounting for 76% of the total storage capacity and 81% of the total beneficial storage capacity of all the reservoirs in the basin, respectively. These reservoirs have not only changed the longitudinal continuity and natural hydrological characteristics of the river but also changed the pattern of scouring and sedimentation in the channel. Large amounts of sediments have been deposited in the reservoirs, resulting in an even more uneven spatial distribution of sediments and great changes in the river landform and aquatic environment of the river channels in the reservoir areas when compared to the conditions before the reservoirs had been constructed. Moreover, the natural landform of mountainous valleys and the rapid-flowing aquatic environment will gradually become seminatural semi-man-made mountainous reservoir landforms and relatively static aquatic environments; large water fluctuation zones will result from seasonal changes; the number of the aquatic organisms that prefer rapid flows will be greatly reduced; and the aquatic organisms that love static aquatic environment will become the dominant biological community in each reservoir area. The long-term and long-distance scouring process downstream of each reservoir will cause the river channel and riverbed to change. The channel in the middle and lower reaches of the Yangtze River will experience a sedimentation-scouring rebalance process that may take more than 100 years. During this process, transport of sediment, inorganic ions, and nutrients will decrease, resulting in a long-term and slow effect on the evolution of the river channel and the river-Lake relationship in the middle and lower reaches and the estuarine ecosystem.

The impact of human activities on the evolution of the Yangtze River will need further observations, and no conclusion can be easily reached in a considerably long time. The TGR was completed a short time ago, and its water level reached the normal water level of 175 m for the first time in 2010. As many large hydroelectric plants are being or will be constructed in the upper reaches of the Yangtze River, the impact of the reservoirs on the ecological environment of the Yangtze River basin will need to be further observed and manifested. The TGR was in a pilot operation stage during 2008–2012. During this time period and a period of time thereafter, there was a need not only to test the safety of the dam and reservoir but also to monitor and analyze the impact of the reservoir operation on the ecology and environment of the Yangtze River. The plan for the reservoir operation was also in the process of optimization with the goal to maximize the comprehensive benefits of the TGP and minimize the impact on the ecology and environment.

1.4.2 Evolution Phenomenon of the Yangtze River

Although the impact of human activities has been enormous and the main manifestation of the current evolution of the Yangtze River has been caused by human activities, the Yangtze River basin has still demonstrated some slow natural evolution no matter how small the magnitude of this evolution is. It may take a long time before human can notice the evolution, but it is still worth our attention. There are some trends of changes in the following areas:

1.4.2.1 Changes in Climate and Hydrological Conditions

Under the influence of global climate change, the temperature in the Yangtze River basin has had a warming trend in the last 50 years (Xu and Ma 2009), especially since the 1990s. According to the data from 147 meteorological stations in the

Yangtze River basin, the temperature increased by an average of 0.33 °C in the 1990s, while the temperature soared by an average of 0.71 °C during 2001–2005. A much higher temperature increase occurred in the source area of the Yangtze River where the recent 10-year average temperature was 1.42 °C higher than that in the 1960s and 1.12 °C higher than that during 1959–1999. In the source area of the Yangtze River, glaciers appear to have shrunk; the active layer of the permafrost has increased by 0.5–1.2 m; the palustrine wetland and vegetation in the plateau have experienced some degree of degradation; and the regional ecosystem has become more fragile.

The rise in temperature has led to frequent occurrences of extreme meteorological and hydrological events in the watershed in recent years, such as the large drought in Sichuan and Chongqing and rare low water level during the flood season in the summer of 2006, frequent rainstorms in Sichuan Basin in 2007, snow and ice disaster in Hunan and Hubei in 2008, 3-consecutive-year droughts in Yunnan during 2009–2011, and drought and ensuing flood in the middle reaches of the Yangtze River in 2011. The Yangtze River basin is trending warmer slowly.

Although the overall amounts of precipitation and runoff in the Yangtze River basin have not changed significantly in the past 50 years, the change in runoff in the source area and some tributaries of the Yangtze River is very large, exceeding the normal range of change. For example, ① the summer rainfall has increased, and the winter snow has decreased in the source area of the Yangtze River in the last 10 years, resulting in a decrease in area of glaciers and snow cover. However, the runoff amount of the Tongtian River has not decreased, but the distribution of the runoff has changed significantly during the year, which has aggravated the soil erosion in the source area. ② The Min, Jialing, and Han Rivers, which are major north bank tributaries of the Yangtze River, have been trending down apparently in the amount of runoff over the past 20 years. In the recent decades, the runoff from the upper reaches of the Min River has been less than normal and 26% less than the 1930s. According to the statistical data collected from the Jialing River during 1954–2003, the average annual precipitation in the watershed of the Min River was 828 mm during 1991–2003, 116 mm less than that before 1990, or a 12% decrease. During the same period of time, the average annual amount of runoff at the Beibei Station decreased from 70 billion m^3 to 54.1 billion m^3, a decrease of 15.9 billion m^3, or a 22.7% decrease. ③ In the Dongting Lake and Poyang Lake areas, although the total amount of precipitation and runoff have changed little, the seasonal distribution has been more uneven, and the incoming water in the last 10 years has been seriously less during fall and winter. For instance, the aquatic systems of Dongting and Poyang Lakes have continuously experienced less incoming water during dry seasons for many years, and the Xiang and Gan Rivers have suffered extremely low water levels for many years.

1.4.2.2 Geology, Geomorphology, and River Regime of the Yangtze River Basin

Seismicity and tectonic movements in the upper Yangtze are still in a relatively active period of time. Natural erosion and denudation events, such as flash floods, landslides, mudslides, and soil erosion, have occurred frequently in the mountainous areas where there are few human activities. These disastrous events can be attributed to tectonic movements, as well as human activities (hydroelectric stations, highway construction, quarrying and mining, etc.). It's hard to avoid them. There are more than 10,000 debris flow gullies in the mountainous area in the upper reaches of the Yangtze River basin and the river basins in southwest China and more than 220,000 collapsed hillocks in the red soil area in South China (Sun 2011). Based on observations and inspections, although the amount of soil erosion has been reduced significantly since the 1990s, the amount of average annual soil loss was 2.313 billion tonnes in the Yangtze River basin during 1950–1995 but 1.373 billion tonnes during 1996–2005 (Li, 2009), which can be attributed mainly to the combined effects of reduced human activities, natural forest protection, soil conservation, and sediment interception by reservoirs. The amount of average annual sediment transport was 224 million tonnes through the hydrologic stations in the lower reaches of the Yangtze River during 2001–2005, and the difference between the sediment yield and the channel sediment transport in the basin was 1.149 billion tonnes. Based on the data collected in 2005, it is estimated that 30 million to 40 million tonnes of sand were quarried from river channels in the year. Based on a rough estimate, the amount of sediments deposited in creeks and ditches at foothills, river channels, lakes, and reservoirs each year is about 1.1 billion tonnes. This redistribution of the sediments has slowly changed the landform and river regime of the Yangtze River basin.

In the middle reaches of the Yangtze River, Dongting Lake and the Jianghan Plain have been in the stage of tectonic sedimentation since the Quaternary and the current tectonic sedimentation rate can reach 5–10 mm/year (Zhang et al. 2003). Five hundred years ago, the Jingjiang Dike along the north bank of the Yangtze River completely prevented sediments of the river from entering the Jianghan Plain, and sediments seldom reached the plain from the Han River. This has made the difference in elevation between the Jianghan Plain and the flood level of the Yangtze and Han Rivers become increasingly larger. Consequently, the flood hazards and drainage conditions in the Jianghan Plain have become worse and caused the floodwater in the upper Yangtze to flow into the Dongting Lake area where the elevation is relatively higher, but not to the lower-lying Jianghan Plain, which made it controversial as to how to allocate and schedule the use of the flood diversion/retention zones along the north and south banks of the Yangtze River. In the past, because the deposition rate of sediments entering Dongting Lake exceeded the tectonic sedimentation rate in the Dongting Lake area, people could not feel the influence of the tectonic sedimentation. With the meander cutoff of the Jingjiang River and the construction of the TGP, the amount of sediments entering Dongting Lake through the three outlets has been reduced greatly, and the Dongting Lake area and the Jianghan

Plain have experienced and will continue the sinking process, which will have long-term impacts on the future flood control situation in the middle Yangtze.

In the Yangtze Delta area and Yangtze Estuary, the average amount of annual sediments passing the Datong Station on the Yangtze River was 486 million tonnes before the 1980s. The amount of sediments entering the sea has decreased by one-third in the last 15 years. The average amount of annual sediment transport was only 130 million tonnes during 2005–2010, which will significantly reduce the estuarine sediment replenishment for land development. If the current trend continues for a long time, not only the expansion of land resources in the estuary area of Shanghai and Jiangsu will be affected, but coastline erosion may also occur.

1.4.2.3 Lacustrine Wetlands and Aquatic Organisms

Presently, due to poor natural conditions in the source area of the Yangtze River and the upper reaches of the Jinsha River, there are not many living organisms, and the ecosystem is fragile. The river system and the lacustrine ecosystem are mainly affected by global climate change and rare human activities. Therefore, the ecosystem can remain the original state for a long period of time. However, the shrinking glacier area and the thickened active layer of the permafrost will have a far-reaching effect on the ecological environment of the source area of the Yangtze River.

In the middle and lower reaches of the Jinsha River, and tributaries of the Yangtze, such as Yalong and Dadu Rivers, large cascade hydroelectric stations are now being and will be constructed and operated in the future 20 years. The number of aquatic organisms that prefer the rapid-flowing conditions will gradually decrease, and aquatic organisms that love the static aquatic environment will significantly increase with the completion of the reservoirs in the upper reaches. Although there are no plans to construct dams between Yibin and Yichang on the Yangtze River or a few tributaries such as the Chishui River to preserve some of the rapid-flowing aquatic environment and habitats, the continuous rapid-flowing section is too short. Moreover, because people have the habit to eat wild fish, excessive fishing occurs sometimes. If no stringent measures are taken to ban fishing and prohibit eating wild fish, the surviving conditions for rare and endemic fish species in the upper Yangtze will be seriously affected, and many rare and endemic fish species will probably face the fate of extinction.

As the middle and lower reaches of the Yangtze River and the freshwater lakes thereof are located in the typical East Asian monsoon region with abundant precipitation and runoff, plenty sediment, and nutrients transported from the upper reaches, they have good natural conditions and are historically the richest areas of China's ecological diversity. Because there have been most human activities and interference in the region, the ecosystem has been affected by human activities in the past 2000 years, and not much natural ecological environment has remained. However, due to the superior natural conditions, there are a lot of seminatural and semi-man-made environments. It is worthy of concern and action as to how to limit or reduce human activities and preserve some habitats to provide wildlife with a basic living environment.

The fishermen in the Poyang Lake area have a unique habit of production activities, which is the so-called chopping autumn lake (Liu et al. 2011): to facilitate fishing, at the end of autumn and early winter, fishermen dig a ditch to gravity drain the water from a low-lying area, place a fish trap cage at the exit of the drain ditch, and gradually lower the water level of the low-lying land; when the water level gets lower, they deepen the ditch until the water on the low-lying land is largely drained dry so as to "catch fish as the pool is dry." In this way for a period of time, the dish-shaped depression is inadvertently lowered, and a pond with a certain area of water is preserved. Many migratory birds like to return to the dish-shaped depression habitat for food until the water is drained basically dry, and cranes, storks, and other migratory bird species are forced to fly away and find new habitats. If there were no impacts from human activities, the water level in the depression would gradually be lowered, and new transition zones would develop constantly between water and land with sufficient food. However, the wintering habitat that would be best for migratory bird species (especially cranes, storks, etc.) could not be maintained for a long time. This shows that as long as human beings do not actively interfere with or destroy the biological environment, some seminatural and semi-man-made environment may also provide habitats for wildlife. In fact, with improved human awareness of the ecological protection, as long as they do not intentionally kill or destroy, some plants and animals, such as birds, will spread to urban areas. Although the biological structure may not be reasonable, a certain number of species may be maintained. Therefore, for the terrestrial ecosystems, as long as space and time are given, and interference and destruction are minimized, it is possible the ecosystem will be restored. Aquatic ecosystems require hydrological connections and lake-river connections. However, constructed sluices and dikes have disconnected lakes from rivers, resulting in decrease in numbers of migratory and semi-migratory aquatic organisms.

In the estuarine, deltaic, and coastal area of the Yangtze River, where protective conditions for the wetland ecological system are better than the middle and lower reaches of the river due to the complex water system and broad water surface, the main problem is the increasingly serious water pollution. The extensive use of the shoreline, shipping traffic congestion, and overharvesting of fishery resources have caused severe depletion of the fishery resources in the Yangtze Estuary and offshore. Thus, the balance of the estuarine ecosystem has been affected, and the function of the ecosystem has been degraded, which has ultimately affected the sustainable utilization of the estuarine and offshore resources in China and forced the Chinese fishermen to go fishing beyond the coast and even on international waters.

1.4.3 Future Evolutionary Trend of the Yangtze River

The Yangtze River basin is located in an area with a large total amount of the water resources. The present and future utilization rate of water resources can be controlled at a reasonable level. Moreover, the extent of the Yangtze River system is large. Due to the restriction of dikes and the regulation of reservoirs, the structure of the water system will not change drastically. The impact of human activities in general can be basically determined. The general manifestation of the evolution, or the natural properties, of the Yangtze River will not change significantly. If effective managerial and protective measures are implemented, the overall pattern of the ecologic environment of the Yangtze River will not change considerably. However, if the development and utilization of the Yangtze River are not scientific or controlled, irreversible changes will occur to the ecological environment.

The future evolution of the Yangtze River depends on the transformation of China's economic development model. If China continues the extensive development model and does not establish protective reaches or sections, improve water quality, or conserve soils, the ecosystem of the Yangtze River basin will bound to continue to be degraded. However, the ecological environment of the Yangtze River basin can become better and can even be restored to the condition of the 1980s if the resource-conserving and environment-friendly economic development model is adopted, in which the sectors of high-technology and modern services are the mainstream of the future economic development; urbanization rate is more than 80%; the level of agricultural intensification is high; the intensity of human activities in the forest and rural areas is lowered; the awareness of governments at all levels and citizens for the ecological environmental protection is further improved; and more efforts are devoted to the ecological restoration. If the self-healing capability of the ecosystem is improved, the ecological environment will gradually be restored and will be on track to the sustainable watershed-wide socioeconomic development, even if there are disasters such as floods, droughts, earthquakes, landslides, mudslides, and human interference.

References

Li Z (2009) Current situation and dynamic change of soil erosion in China. China Water (7):8–11

Li Z, Song Y (2010) Let the statement of the Yangtze River source return to reality. Area Res Dev 29(2):139–144

Liu C, Tan Y, Lin L etc (2011) Effects of water level variation on migratory birds' habitats in Poyang Lake. J Lakes Sci 23(1):129–135

Shi M (1983) On the determination of the true source of the Yangtze River. Geogr Res 2(I):32–33

Shi M. 2001. Past Dynasties' knowledge about the source of the Yangtze River, Yangtze River Hist Q (3)

Sun H. 2011. Soil erosion investigation in China and amendment to the water and soil conservation law of the People's Republic of China, China Water (12):45–46

Xu Q (2012) Study on reservoir sediment deposition and downstream scouring since impoundment of three gorges reservoir. Yangtze River 43(7):1–6

Xu M, Ma C (2009) Study on the vulnerability and adaptability of climate change in the Yangtze River basin. China Water & Power Press, Beijing

Yangtze River Water Conservancy Commission (2003) Yangtze River history (volume I), water systems. China Encyclopedia Publishing House, Beijing

Zhang R, Liang X, Duan W et al (2003) Systematic analysis of the evolution of Dongting Lake and formation and development of floods. China University of Geosciences Publishing House, Wuhan

Zheng Z (1998) History of China's water conservancy. In: China Commercial Publishing House. Beijing, China

Zhu Y (1991) Xu Xiake was the discoverer of the true source of the Yangtze River. Study Hist Nat Sci 10(2):182–185

Chapter 2
Evolutionary Process of the Yangtze River

Abstract From the examination of the Yangtze River's evolutionary history, tectonic movement was determined to be the original driving force for the geomorphology of the river. Additionally, climate change, especially the alternate occurrences of the fourth ice age and interglacial periods and the change of land vegetation, was an important part of the main forces that drove the development of the Yangtze River's alluvial plain in the middle and lower reaches. The current river characteristics of the Yangtze River have resulted from the combination of the natural evolution and human activities. The agricultural development since the Neolithic age and the dike construction over the last 2,000 years have had far-reaching impacts on the channel in the diddle and lower reaches and the river-lake relationship. Based on tectonic movement and climate change, this chapter discusses the evolutionary process of Yangtze River's formation and ecosystem since the uplift of the Qinghai-Tibet Plateau. Additionally, this chapter emphatically analyzes the important time nodes formed by the Yangtze River, such as the channeling through the Jinsha River and the Three Gorges, and the eventual formation of the overall west-east landform pattern of the Yangtze River. Next, this chapter summarizes the formation and evolutionary process of the Yangtze River, Dongting Lake, Poyang Lake, Yangtze Delta, and Tai Lake and finally performs a section-by-section analysis of the basic characteristics of the Yangtze River channel.

The timescale, or design base period, considered by hydraulic engineers in the watershed planning and hydraulic engineering design, is generally over a period of several decades to 100 years or so. In preparing a plan for flood control or drought relief, historical flood or drought events for up to 2,000 years need to be investigated. However, geographers and landscape ecologists need to consider the Quaternary period or more than two million years in analyzing the geomorphology of a river and the evolution of a regional landform. Geologists, paleontologists, and paleoclimatic scientists need to examine rocks, biological fossils, and glacier evolution on a timescale that needs to include geologic ages or tens of millions to hundreds of millions of years in studying the evolution of the earth and living things. The time and geometrical scales for different fields of study vary greatly. Therefore, it is necessary to use various timescales and different perspectives to study the river-ecological environment relationship and investigate the long-term effect of hydraulic

J. Chen, *Evolution and Water Resources Utilization of the Yangtze River*,
https://doi.org/10.1007/978-981-13-7872-0_2

projects on rivers so as to objectively and comprehensively understand the essence of the issues and predict the trend of the river evolution. Some of the Yangtze River's rare and endemic fish species of concerns might first appeared on the earth tens of millions of years or even billions of years ago. According to the results of paleontological studies, marine fish species flourished on the earth as early as in the Devonian period 409–362 million years ago (Ma), and freshwater fish species, such as the Chinese sturgeon and other rare fish species of the Yangtze River, thrived in the Cretaceous period 135–65 Ma. As Chinese sturgeons appeared in those early ages, they were called "living fossils." However, more studies will need to be performed on the relationship between the evolution of the Yangtze River and the evolution of aquatic organisms to answer the questions as to when the rare fish species started in the Yangtze River; where their early habitats were; and how their habitat environment is evolving at present and will evolve and be protected in the future. The beginning and evolution of living organisms in the Yangtze River basin are directly related to geographical, climatic, hydrologic, and habitat evolutions. Therefore, to protect the Yangtze River, it is necessary to understand the history of the river and analyze the current condition and predict the future condition of the river based on its history so as to better protect and manage the river.

Keywords The Yangtze River · Changjiang river · Evolution of river system · Basin ecosystem · Water resources utilization · Floods and drought · Ecological and environmental protection · Basin management

2.1 Evolutionary History of Earth and Organisms

The evolution of the earth, from inorganics to organics and from simple organism to complex organism, is closely related to the changes of the earth's environment. Investigations of the relationship between the tectonic movement and the evolution of organisms can lead to the finding of the relationship between environmental change and biological evolution.

The evolution of the earth and organisms is the result of the evolution of the universe and the solar system. According to available results of scientific research, the universe is generally believed to be emerge following the Big Bang about 13 billion years ago (Ga). The solar system, where the earth exists, formed about 4.6 Ga. The formation of the solar system included the birth of the earth and the moon. When the earth emerged, it was a fireball resembling the sun. About 500 million years later, it gradually cooled down, forming a solid crust and the original ocean. About 3.8 Ga, original continents appeared on the earth. Eukaryotic cells (Eukaryotes) emerged 2.4 Ga, and the atmosphere was enriched with oxygen 1.8 Ga. Multicellular life appeared on the earth about 1.4 Ga. Marine life flourished 600 Ma, and terrestrial organisms and animal fossils emerged 550 Ma. Using the formations of many biological fossils as boundaries, geologists have divided the geological period into two major eons: pre-Phanerozoic and Phanerozoic.

2.1.1 *Pre-Phanerozoic*

Even since the emergence of many biological fossils on the earth (Phanerozoic), great changes have taken place to the earth's distribution of continents and geologic strata. Thus, it is very difficult to sketch out the evolutionary history of the earth in the pre-Phanerozoic eon. To understand the relationship of life's emergence with the formation of the universe/the solar system and the evolution of the earth's environment, and to explore the possibility of extraterrestrial life, there is a need to understand the earth's evolutionary history in the pre-Phanerozoic eon. Despite the scarcity of available information on rocks and fossils for that period, astronomers, geologists, geophysicists, and paleontologists from all over the world have sketched out the earth's formation and early evolution.

2.1.1.1 Hadean (4.6–3.8 Ga)

According to available research results, the earth formed about 4.6 Ga. At the beginning, the earth was a "fireball," and its internal and external were very hot. About 500 million years later, the earth gradually cooled down, causing the formation of the original solid crust and the original oceans. During this period the atmosphere contained almost no oxygen but 80% of carbon dioxide. Most of the earth's surface was submerged by seawater. About 4 Ga, the earth began to form the primordial crust (the early basalt, which is now presumed to be present, was the rock of the earth formation period). Due to great changes that occurred on the earth thereafter, no geological or material relics of that period or records about any rock or life of that period have been found so far, which is why we want to retrieve rock specimens from Mars and other alien planets where there are still no signs of life and whose geological environment may still resemble the condition of the earth's pre-Phanerozoic eon. The earth environment of the Hadean Eon was inferred by scientists according to the current condition of the solar system and the formation theory of celestial bodies, and the accuracy of the inference needs further verification.

2.1.1.2 Archean (3.8–2.5 Ga)

The Archean is the oldest eon with geologic records of the earth. Although the available geological records of this period are a few and unorganized, 70% of the substances in the existing crust were separated from the mantle during this period. The crust gradually thickened, and rocks were mainly of magmatic and metamorphic origins, but sedimentary phenomenon began to occur in the crustal rocks during this period. As the atmosphere contained no oxygen or was in anoxic conditions, there were mainly silicate sediments with very little carbonate deposition while the atmosphere was still full of carbon dioxide. Presently available records indicate that the oldest rock is 3.8 billion years old. For example, 3.8 billion-year-old rock was discovered in flint of the Barberton Fig Group in South Africa, and the existence of unicellular organisms was also discovered. As a result, it can be inferred that the earliest life on the earth might be born 3.8 Ga, but no earlier rocks have been found. If older rocks are found, life might be present on the earth much earlier and might even exist at the beginning of the earth. During this period, life and oxygen began to appear. Since life and oxygen are interdependent, both appeared almost simultaneously and interinfluenced. The Archean strata are mainly located in north China, which is the earliest continent present in China.

2.1.1.3 Proterozoic (2.5 Ga–570 Ma)

During this period, as the crust was in an active period, sedimentation occurrence was widespread and ancient land cores appeared in the form of islands. There are more geological records available that indicate there were non-metamorphic rocks and biological fossils. With the rapid increase of photosynthetic organisms (algae) in the ancient ocean, the oxygen content in the atmosphere increased significantly. In the late Proterozoic, the atmosphere was transformed into a mixture of carbon dioxide, nitrogen, and oxygen from a volcanic gas-filled ambiance (mainly carbon dioxide). It is estimated that the atmospheric oxygen content at the time was at least 7% of the present atmospheric oxygen level and the majority of life changed from the original sulfur-absorbing type to the oxygen-absorbing form. The global impact of the hydrosphere made the carbonate deposits reach the peak, causing the formation of dolomite and dolomitic limestone. Meanwhile, the earth's rainfall gradually changed from the initial acidic water (a large amount of carbon dioxide in the atmosphere) into weak alkaline water. The deposition process of carbonate rocks consumed large amounts of carbon dioxide in the atmosphere, which gradually transferred the carbon from the atmosphere to the rocks, enabling the earth to develop a biochemical environment conducive to the rapid development of life.

The Proterozoic in China was mainly located in two major regions: north and south. The north region extends from Tarim to north China and southern northeast, and the south region is mainly in the eastern part of the Hengduan Mountains, including one section in the west edge of the Yangtze Platform and a second section in the Huangling anticline in the Three Gorges area.

Most of the pre-Phanerozoic living things were simple shell-free organisms. After their death, they became organic molecules through biological decomposition and were difficult to form fossils. Therefore, there are very few available records of biological fossils for that period. Minerals that had begun to form in the earth's crust during the pre-Phanerozoic period include iron, copper, uranium, gold, phosphorus, and natural gas.

The characteristics of the earth in the pre-Phanerozoic can be summarized as follows. The atmospheric environment and geological, physical, and chemical properties had changed greatly; the earth's environment had transformed from inorganic to organic, or the carbon in the atmosphere had deposited in rocks, and the content of oxygen and nitrogen in the atmosphere had increased; surface water had changed from acidic to weakly alkaline; and the sedimentary process of silicate rock had transformed to the sedimentary process of carbonate. All these changes were leading to the environmental transformation conducive to the rapid expansion of life on the earth. Therefore, it can be said that if the earth had not experienced the transformation of the physical and chemical environment nearly 4 Ga during the pre-Phanerozoic, there would be no rapid development of living things in the Phanerozoic.

2.1.2 Phanerozoic

Large numbers of organisms appeared on the earth's crust after the Cambrian 570 Ma. The Phanerozoic starts from Cambrian to Quaternary. Since the beginning of the Phanerozoic, the earth's biological evolution has been progressing, accompanied by lots of rock deposits and the formation of large amounts of biological fossils and fossil fuels. Geologists and paleontologists can study the history of the geologic evolution and biological evolution according to the sequence of geological deposition (i.e., sedimentary rock) and biological fossils and provide a reliable basis for the development of geology, paleoenvironment, paleoclimate, and paleontology.

2.1.2.1 Cambrian (570–510 Ma)

The biological evolution underwent an explosion in the Cambrian period (Cambrian explosion or radiation). Advanced creatures appeared on the earth, and hard-body animals prospered, resulting in large amounts of biological fossils preserved. For example, in the early Cambrian, there had already been nine phyla of animals, protozoa, archaeocyatha, porous animals, coelenterate, annelids, arthropods, mollusks, brachiopods animals, and echinoderms, amounting to more than 900 species.

During the Cambrian, a large marine transgression occurred in China. The Yangtze area (or the present middle and lower reaches of the Yangtze River) was a stable shelf sea and was basically in the seawater. Its west side was the Sichuan-Yunnan paleocontinental uplift zone, and its south side was the original Jiangnan paleocontinent. Most of south China was part of the ocean and continents emerged as islands. The Cambrian ended with the first mass extinction of organisms.

2.1.2.2 Ordovician (510–439 Ma)

During the Ordovician, marine invertebrates were the most prosperous, and the sea-continent boundary was similar to that in the Cambrian, while the scale of the marine transgression was greater. The Yangtze River basin consisted only of the Sichuan-Yunnan paleocontinent and the Jiangnan Island arc, and the Yangtze area was an epicontinental sea. The minerals generated in the crust of the earth during this period included iron, manganese, gypsum, limestone, petroleum oil, and natural gas.

The end of the Ordovician was marked by the second mass extinction on the earth that led to the disappearance of about 85% of biological species, presumably due to a global ice age. About 440 Ma, large swathes of glaciers cooled oceanic currents and atmospheric circulations. Consequently, the temperature of the entire earth fell; glaciers froze surface waters; sea level dropped sharply; and the original rich marine ecosystem was severely damaged.

2.1.2.3 Silurian (439–409 Ma)

During the Silurian, vascular plants appeared; terrestrial plants appeared on land for the first time; and organisms began to expand from ocean to continent. During this period, the Yangtze region and the Himalayan area were still part of the ocean, which was the stable sedimentary period of the platform, and the sea started to retreat at the end of the Silurian.

2.1.2.4 Devonian (409–362 Ma)

During this period, vertebrates experienced a rapid growth, and fish species flourished. The evolution from invertebrates to vertebrates was a great leap in the development of animals, resulting in the evolution of the most advanced animals – fish species in waters. Because the oxygen content in the water was much lower than in the air, the most advanced animals living in the water were only fish species. Mammals such as dolphins mainly inhaled oxygen from the air for survival, which again explained the importance of oxygen for living things, especially advanced animals. Terrestrial plants also flourished during this period, and small-scale forests appeared.

The close of the Devonian period is considered the third mass extinction of the earth's history, and the marine life was seriously depleted. In south China, a continent emerged in the early Devonian and experienced many marine transgressions later. Most of the Yangtze River basin was still in the sea.

2.1.2.5 Carboniferous (362–290 Ma)

This period was marked by the unprecedented development of terrestrial animal life with two marine transgressions and one marine regression and was the most famous geological history of the coal-forming period. It has now been proved that one-half of the earth's coal was formed in this period, which is why the period named Carboniferous. Naturally, it was only after land had experienced the unprecedented development of plant organisms could large amounts of fossil fuels such as coal, oil, and natural gas be formed through physical and chemical actions such as geological deposition. However, it should be realized that fossil fuels had been gradually formed through physical and chemical actions for hundreds of millions of years. If mankind relies only on these fossil fuels, the fuels may be depleted in thousands of years. Therefore, human beings must conserve energy, reduce emissions, and vigorously promote the use of renewable energy so that the human society can sustainably develop.

2.1.2.6 Permian (290–250 Ma)

The Permian was also a major coal-forming period and witnessed the further development of terrestrial plants, flourishment of amphibians, primitive reptiles and insects, and development of cartilage fish species.

The fourth extinction event occurred at the end of the Permian. It is the most severe known extinction event in the history of the earth, with more than 95% of all species becoming extinct, of which 90% of marine species and 70% of terrestrial vertebrate species died out; trilobites, sea scorpions, and important coral groups were wiped out; and terrestrial synapsid community and many reptiles became extinct as well. Most scientists believe that the sea-level decline and occurrence of the continental drift in the Permian were the possible causes of the most severe extinction of species. During the period, all the earth's major landmasses joined into a single ancient supercontinent known as Pangaea; the biologically rich shoreline experienced a sharp decline; and continental shelfs underwent a substantial shrinkage. The extinction of many species was due to the loss of habitats. More seriously, after shallow continental shelfs were exposed, the organic matters originally buried in the seabed were oxidized. This process consumed lots of oxygen from the air and released carbon dioxide. The oxygen content in the atmosphere likely decreased significantly in a short period of time, which was very detrimental to animals that lived on land. As temperatures rose, sea level rose, resulting in a catastrophe to many terrestrial creatures, and the ocean became anoxic. The fact that geological investigations identified large amounts of sedimentary organic-rich shale to be in geologic formations representative of the period is the best evidence of this biological catastrophe.

2.1.2.7 Triassic (250–208 Ma)

The Indochina movement in this period caused the Yangtze platform to uplift; the paleogeographical pattern of the Triassic was still with land in the north and the sea to the south; and the cartilage fish species began to decline. The Triassic ended with the fifth mass extinction of the earth. An estimated 76% of species, mainly marine organisms, disappeared in the extermination.

2.1.2.8 Jurassic (208–135 Ma)

During this period, the biological community began to prosper; wetlands and forests and other terrestrial ecosystems flourished; reptiles-dinosaurs and bony fish species boomed; mammals began to appear; and volcanic eruptions occurred frequently.

Paleogeographical landscapes of China in the Jurassic were quite different from those in the Triassic. The sea area had shrunk to southwest China and the Qinghai-Tibet area. At the northern and southern ends of east China, two short-term marine transgressions occurred at different times. During the Jurassic, in the entire Sichuan Basin, even including the Panxi region, there were great lakes, or the former "Bashu Lake" and "Xichang Lake" called by ancestors. Bashu Lake may be the largest freshwater lake in the paleo-Yangtze River system, which encompassed almost all of the present-day Sichuan Basin and extended southwestward to include Qionglai, Mingshan, Tianquan, Baoxing, Rongjian, and Hanyuan Counties and covered an area of about 200,000 km^2. Several large lake basins might be connected and eventually drained westward into the Tethys. Lake shores provided habitats for dinosaurs, and the Sichuan Basin mainly experienced lacustrine and fluvial depositions. At that time, the Chuan River system was still of an outflow type (Zhang 1995). Large amounts of dinosaur fossils were discovered at Dashanpu in Zigong, Sichuan Province, confirming the characteristics of the lacustrine environment at the time. The Zigong dinosaur site includes almost all known dinosaur species during the Jurassic period about 205–135 Ma and is the world's best place for collection and exhibition of the Jurassic dinosaur fossils.

2.1.2.9 Cretaceous (135–65 Ma)

In the early Cretaceous period, a marine transgression occurred, and three major sedimentary areas emerged: Tethys area, western area and eastern area, or Xinhua Xia area in the southern part of south China. At the end of the Cretaceous, a large marine regression occurred and caused a vast area north of the Yalu Zangbu Sea Trough to transform to land. During the time, all the Tethys area in China except the Himalayan area and the western Tarim edge had experienced marine transgressions in the Cenozoic era, and the expansion of marine sedimentation completely ended. In the Jurassic and Cretaceous, the Jianghan Plain and Dongting Lake Plain expanded first through marine sedimentation and then through both lacustrine and fluvial sedimentation, and the total thickness of sedimentation in Yunmeng Lake (a paleo-inland water system) was up to thousands of m. During the Cretaceous, advanced angiosperms emerged; freshwater fish species prospered; the Teleostei replaced the Holostei; and bird species evolved from reptiles.

At the end of the Cretaceous, the sixth mass extinction occurred on the earth, when dinosaurs that had dominated the earth since the Jurassic died out and more than one-half of the biological groups and more than 75% species became extinct. The cause of this catastrophe may have come from outer space and volcanic eruptions. Presently, what most scholars have speculated is at the end of the Cretaceous, an asteroid rain or multiple meteor rains caused the global ecosystem to collapse. The collisions sent massive amounts of gas and dust into the atmosphere. As a result, the sunlight could not penetrate to the earth, and the global temperature dropped dramatically. Such dark clouds obscured the earth for years (several years to millions of years). Plants could not get energy from the sunlight, and algae in the sea and forests gradually died. With the basic link of the food chain destroyed, large numbers of animals died of starvation, including dinosaurs. Naturally, the extinction of biological species, especially the disappearance of dinosaurs, also provided an opportunity for the eventual appearance of mammals and human.

2.1.3 Some Revelations

During the span of a human being's lifetime, the geologic and geomorphic changes of the earth appear to be relatively static, but on a geological timescale, the geomorphic changes of the earth are very active. From the global surface that was once almost completely submerged in an ocean to the emergence of the paleocontinent and then to the formation of the five continents and four oceans, great changes took place in the earth's plates, continental geological landform, physical and chemical compositions of the atmosphere, climatic environment, etc. The theory of continental drift, proposed by German geologist Alfred Wegener, has solved the difficult problem associated with finding species similarities on separate landforms that seemed to fit together and revealed that the landmasses of the earth have a great capacity for activities. Although the majority of the earth's surface in ancient times was an ocean, we should appreciate the constant changes of the sea and continent and marine sedimentation, which not only left us with many records regarding geologic strata and biological fossils but also provided today's human beings with a lot of biofuels: petroleum oil, natural gas, and coal. Without the sedimentation of ancient biological species, there would be no major fuel resources available to mankind today.

Although the exploration of the geologic evolution may not directly help understand the evolution of the river geomorphology, it can help us understand the background for the geological, environmental, and ecological evolution for the formation of rivers and help us study the basic laws of climate and hydrological changes. From the biological evolution and succession process in the Phanerozoic, we can obtain the basic characteristics or law of the earth's biological evolution.

2.1.3.1 Progressiveness

The biological evolution conforms to Darwinism. The environment evolves from inorganic to organic, and biological species evolve from low level to advanced level and from simple to complex. They are continuous evolution processes from imperfect to perfect. The biological evolution always involves the gradual change from a low level to an advanced level. Meanwhile, in order to adapt to environmental changes, physical characteristics, biological features, and food chain structures will change, but there is no big change in the gene.

2.1.3.2 Stages

The biological evolution is not uniform or at an equal rate but consists of alternate occurrences between slow quantitative changes and rapid mutations. The biological catastrophe caused by external factors or biological explosion in geologic history is a biological mutation. The biological evolution occurs in stages. Although a biological extinction caused large numbers of species to disappear, it eliminated dominant species, creating space and conditions for the appearance of new organisms. Each of the earth's six mass extinction events was followed by a rapid development period of new species, in which biological species flourished with a large-scale succession, which showed the mutagenicity of the biological evolution, and the new species were more advanced and more complex.

2.1.3.3 Irreversibility

Under natural conditions, in the process of evolution, biological species did not duplicate the types and traits of disappeared species. The disappeared species have not reappeared, or species did not go backward, revealing the irreversible nature of extinct species. Of course, perhaps in the future with the development of gene preservation and cloning technology, human beings may have the means to reproduce the extinct organisms or species. However, from a natural point of view, this development will not happen. For instance, the extinct dinosaurs will not reappear on the earth. If a certain biological species can survive a large extinction event and traverse a longer geological period, the species is often called a living fossil and has great protective values, such as the Chinese sturgeon.

2.1.3.4 Instantaneity of Biological Dispersion

In the absence of geographical barriers or obstacles, the time for biological migration and dispersion may be very rapid. In terms of geological timescales, it is instantaneous. The time of migration and dispersion can be neglected. For example, since the ape-man became *Homo sapiens*, they have spread around the world in less than 100,000 years.

2.1.3.5 Emergence of Endemic Species Due to Conditions of Geographical and Physical Isolation

If a region is isolated by a natural barrier to the outside world (such as the ocean, mountains, etc.), it is possible to evolve an endemic species or new species. For example, since Australia is isolated from the other continents, its biological species have basically been in a separate evolutionary state and created many unique biological species that have never seen on other continents, such as marsupial animals.

2.2 Formation of the Geomorphic Setting of the Yangtze River

The first section briefly discusses the emergence and evolutionary process of the earth and the earth's biological species. From the view of geological timescales, the earth's atmosphere, tectonic plates, and biosphere have undergone great changes. The early rivers have experienced cycles of appearances and disappearances for many times, leading to tremendous difficulties in exploring the evolution of the earlier rivers. In fact, world's modern river water systems began after the Tertiary. Therefore, the earliest time involving the study of modern rivers' beginning and evolution is generally the Tertiary.

2.2.1 The Yangtze River in the Tertiary (65–2.48 Ma)

Before the Tertiary, there had been many rivers on the earth, but since the Tertiary, great changes have taken place in topography and landform of the earth. As a result, it is difficult to find any trace of the early river settings. It is not clear if the settings of the early rivers, which had been mainly inferred from records of sedimentation, are accurate. The most notable geological feature of the Tertiary is the steady development of continents. For example, areas of continental plates became stable, and large areas of terrestrial sedimentation and terrestrial rock formations emerged, creating a relatively stable geographical environment for rivers to form. If there was a marine transgression for a long time, the original rivers would disappear, or if a large geological uplift or subsidence occurred, the rivers would also change. Only a relatively stable continent existed would rivers have the geographical environment of stable development.

The initial form of modern rivers gradually developed after the Tertiary. Although major freshwater fish species now still living in the rivers may have emerged before the Tertiary, the species were close to the present-day species, such as the migratory Chinese sturgeon that had appeared in lakes and rivers in east China. In the Tertiary, land forests, grasslands, and other vegetation thrived, and mammals prospered. Human also appeared in the late Tertiary.

The Tertiary continental landform of China was close to that of the modern times. During this time, the territory of China already became landmass; the sea area was limited to the west, and marine transgressions intermittently occurred in the east. Land facies were mainly sedimentary basins. As the surrounding mountains rose, basins sank. As mountains were weathered and denuded, sediments were transported to basins or deposited in lakes. Sediment deposits were up to thousands of m thick in some areas. The most obvious characteristic of the Tertiary in China was the change of the western landform. The main features are as follows. ① In the middle Eocene (40–50 Ma), the Himalayan area was still in the sea which disappeared in the late Eocene. ② The Himalayan area's absence of the Oligocene and most of the Miocene indicates that the orogeny began in the middle and late Miocene. It is estimated that the mountain uplifted to about 1,200 m at that time. Since then, China's overall topography has appeared in the initial form of high-west and low-east terrace landform. ③ In the Tertiary, as the eastern Qinghai-Tibet Plateau was uplifting, serious denudation also occurred; clastic materials did not deposit in the Yunnan-Sichuan region, but, instead, they deposited in South Asia and adjacent waters through outflow rivers, indicating that the upper reaches of the Jinsha River flowed southward into the South China Sea at the time.

The Tertiary was the period when the Qinghai-Tibet Plateau developed, and the three steps of China's staircase landforms emerged. The landform of the Yangtze River basin also formed in this period. In the meantime, the Jinsha River and the Chuan River had large areas of lacustrine deposition, but the Yangtze River had not formed. The Yangtze River basin had a drainage divide near the Three Gorges where the Chuan River ran to the southwest while the Jinsha River flowed to the south into the South China Sea. At that time, the Sichuan Basin was shrinking due to the lack of lacustrine sediments and became a denudation area. The denuded material flowed southward into the Tethys or ancient lakes in the southwest area. The middle and lower reaches of the Yangtze River drained the surrounding mountains and flowed to the central lakes, marshes, and alluvial plains. The low-lying areas of the middle and lower reaches of the Yangtze River were alluvial plains where there was no fixed main river channel. The water systems were complex; all the systems were interconnected; and all flowed eastward eventually into the East China Sea. The Tertiary Yangtze River was divided into three major sections: upper and middle Jinsha, lower Jinsha and the Chuan River, and the lower reaches of the Yangtze River and the river systems on the alluvial plains. The water and sediments in the first two sections deposited locally and then flowed southward into the South China Sea; the water and sediments in the third section deposited in the lakes and marshes in the middle and lower reaches of the Yangtze River and then flowed eastward into the East China Sea.

2.2.2 The Yangtze River During the Quaternary

The Quaternary (2.4 Ma–present) is not only the most recent period in geologic history but is also the period of human production and expansion in the world. The earth's environment during this period can be characterized as follows. ① The widespread sedimentation of land facies shaped the modern topography, landform, and soil characteristics. ② There were many glacial and interglacial periods, which had great effects on the modern landform, rivers, soils, and biological successions. ③

During the interglacial periods, the earth was warm; precipitation was high; flood events occurred frequently; and river and lake systems developed, resulting in large alluvial plains, lakes, and marshes. ④ In China, the intense neotectonic movements, especially the uplift of the Qinghai-Tibet Plateau, shaped the three steps of China's landform staircase. ⑤ As the northward movement of the Indian Ocean monsoon was blocked, the northwest inland area became arid, resulting in the formation of large deserts and Gobi. ⑥ The evolution and succession of mammals were rapid; a new species evolved on an average of 5,000–15,000 years; apes and humans appeared on the earth; and human beings first emerged in the tropical African region and migrated through the Middle East to all over the world.

The most important effect of the Quaternary on rivers was climate change or the alternate occurrences of ice ages and interglacial periods. The change in range of cold and warm temperatures was large; there were many sediment deposition events of land facies; floods generated alluvial plains; sea level rose and fell dramatically; and the modern landform and river evolution were impacted greatly. Table 2.1 shows the occurrence times of the Quaternary glacial and interglacial periods.

Table 2.1 indicates that ice ages and interglacial periods occurred alternately and the cycles were uneven. The cyclic changes may be related to the earth's orbit in the Milky Way, cycles of solar activities, and extraterrestrial influences.

Table 2.2 shows important time nodes of the Quaternary evolution of the Yangtze River.

Corresponding to the interglacial period, six marine transgressions occurred in the eastern coastal region of China since the beginning of the Quaternary. During the early and middle Quaternary, there were three large-scale marine transgressions, and the scale of the other marine transgressions is unknown. There have been three more obvious marine transgressions since the late Pleistocene with the transgression that occurred early in the late Pleistocene being large in scale. The marine transgression front reached the Baiyangdian area in north China and Weishan Lake in northern Jiangsu, but the scale of the transgressions was not the same. By the late

Table 2.1 Quaternary ice ages and interglacial periods

Ice age	Time (Ma)	Features	Interglacial period	Time (Ma)	Features
First	3–2.4		First	2.4–1.2	A marine transgression occurred, and fluvial and lacustrine sedimentation developed
Second	1.2–0.85		Second	0.85–0.75	A marine transgression occurred in the east
Third	0.75–0.55		Third	0.55–0.25	Lasted a long time
Fourth	0.25–0.15		Fourth	0.15–0.07	
Fifth	0.07–0.01	In the glacial maximum period, China's sea level declined below continental shelf's water depth of 130–160 m. Land shelfs of the Bohai Sea, Yellow Sea, East China Sea, and South China Sea basically became landmass			

Table 2.2 Quaternary evolution of Yangtze River

Geologic epoch		Time (ka ago)	Features of the Yangtze River basin
Late Holocene	Late	2	
	Middle	3	Hydraulic works were constructed
	Early	4	Yellow River and Yangtze River civilizations appeared, e.g., flood control by Yu the great
Middle Holocene	Late	5	Yunmeng marsh sedimentation occurred; uplift occurred; 5,000–6,000 years (5–6 ka) ago, Tai Lake was still a low-lying plain and appeared alternately as a lake and land; rice planting, livestock raising, and pottery making emerged in the Yangtze River basin
	Middle	6	As climate began to be wet, Dongting Lake surface expanded; until Qin and Han dynasties, cultural sites or tombs were rarely found in the hinterland of the lake area
	Early	7	Human beings began to grow rice; Poyang Lake underwent cutting by river networks; Neolithic age human activities occurred
Early Holocene	Late	8	Dongting Lake was of a plain landform cut by river networks; four rivers (Xiang, Zi, Yuan, and Li) joined at the Dongting Plain and flowed into the Yangtze River; Dongting Lake had not yet appeared like it appeared later; and human activities began in the lake area
	Middle	9	Jingjiang River channel submerged in the lake without apparent riverbed shape, and many water bodies overflowed collectively toward southeast
	Early	10	As temperature increased, sea level rose, impacting middle and lower Yangtze; there were many alternating scouring sedimentation occurrences, forming barrier lakes, such as Chao Lake, Poyang Lake; and Yunmeng marsh formed
Late Pleistocene	Late	20	Last glacial occurrence caused sea level to decline by about 150 m; seawater retreated from North Jiangsu; shoreline moved to 600 km east of the Yangtze Estuary; entire East China Sea and Taiwan Strait became land; ancient Yangtze River east of Wuhu was abandoned; Yangtze River joined East China Sea through Nanjing; and marine transgression occurred at Tai Lake
	Middle	80	Ziyang people appeared; in late Pleistocene, main channel of the then Yangtze River moved southward to the present Yangtze River channel; original abandoned ancient river had gradually expanded and joined the broad Yangtze River water surface at the southern edge of the Jiujiang Basin due to the effect of subsidence since the Holocene; and a large area of lakes or Pengli Marsh emerged as documented in the *Tribute of Yu*
	Early	150	Qinghai-Tibet Plateau had risen to above 3,000 m; affected by last interglacial period and warm climate, sea level rose; seawater backflowed upriver, reaching Zhenjiang area; coastal areas such as Subei Plain and Hangzhou Bay had once been submerged in the sea and Changyang people emerged

(continued)

Table 2.2 (continued)

Geologic epoch		Time (ka ago)	Features of the Yangtze River basin
Middle Pleistocene	Late	200	Affected by the uplift of the Qinghai-Tibet Plateau, the Yangtze River and its tributaries were cut deeply; and areas of ancient Dongting Lake and ancient Poyang Lake were very small and were cut by river networks
	Middle	350	Lacustrine sedimentation accelerated; ancient Dongting Lake, Nanyang Basin, ancient Nanxiang Lake, ancient Poyang Lake, and ancient Subei Lake shrank; unified ancient Yangtze River basically formed; Yangtze River ran from eastern Wuhu, across waist of Maoshan Mountain, through Tai Lake and into the East China Sea
	Early	500	World reached its highest temperature in 900,000 years. Glaciers at the source of the Yangtze River accelerated melting, and precipitation also increased; the Qinghai-Tibet Plateau uplifted intensely; eastern plains declined; Yangtze River's three-step staircase landform basically formed; and Yangtze River channeled through the Three Gorges area
Early Pleistocene	Late	730	Neotectonic movement intensified; the Qinghai-Tibet Plateau rose to over 2,500 m; the Jinsha River ran through; lakes in the Jinsha River watershed shrank to a river environment; and the climate was cold
	Middle	150	Northern loess widely developed; climate was relatively warm; and the Jinsha River captured a stream to flow through
	Early	2,400	The Yangtze River was in section-independent development stage; there were ancient Jinsha River section, section of ancient Yangtze River between Yinbin and Yichange, and section of the ancient Yangtze River downstream of Hukou; Yangtze River basin was in lacustrine development period, including development of Hanjiang basin, ancient Dongting Lake, Nanyang Basin, ancient Nanyang Lake, ancient Poyang Lake, and ancient Subei lakes; and climate was cold

Pleistocene, sea level appeared to have declined significantly with occurrences of obvious low sea level phenomena. The continental shelf of the sea became land, and a land bridge appeared in the Bering Strait, through which ancient human migrated from Asia to the Americas. Sea level in the East China Sea was 100–120 m lower than the present and the continent extended 400 km eastward into the ocean. Japanese, Chinese Taiwan, the Philippines, South China Sea islands, and Malay Peninsula were integrated with China's mainland. The Japan Sea and the South China Sea were inland seas (lakes). The Yangtze River and the Yellow River joined in Jiangsu and flowed to the Ryukyu Islands area before entering the sea. In the marine regression period, the sections' gradients of the Yangtze River, the Yellow River, and other rivers running from west to east into the sea significantly increased, and the dynamic force of the headward erosion was extremely strong, resulting in the great power to cut through mountain canyons and create deep channels due to scouring. The Yangtze River ran through the Three Gorges area and left a 40- to 60-m-deep channel, which was the result of the strong impact of the marine regression. Sea level did not return to the current shoreline position until the early Holocene, eventually forming East China's continental marginal sea. In the early and middle Quaternary, marine transgressions in China did not involve large areas, and depositions of mixed marine and lacustrine sediments occurred in the continental marginal sea. In the late Quaternary, especially in the Holocene, relatively complete marine sediment depositions developed. Thus, on a tens of thousands of year timescale, there was a periodicity of natural evolution for the global climate change that has caused sea level to rise and fall considerably.

2.2.3 Geomorphology of the Yangtze River Basin

A river has a life cycle, and the length of the cycle depends on a combination of tectonic movements, climate change, terrestrial vegetation, and sea-level variation. As long as there is precipitation and elevation difference on land, there will be rivers. The runoff that is generated from a concentrated precipitation event in mountainous areas after soil is saturated will seek the most suitable way to flow to lower places, generally to the low-altitude plains and lakes. Because the storage capacity of lakes is limited, the water will overflow and eventually flow to the ocean. If a stable deep channel is created via scouring, a river will develop and drain to the sea. The earth's plate drift and tectonic movement can lead to the orogenesis. However, under the action of rainfall, large wind, and freezing/thawing, mountains will undergo erosion processes, resulting in denudation such as landslide, debris flow, and soil erosion. The prolonged erosion process can gradually wipe out mountains. The Qinghai-Tibet Plateau was the product of the interaction between uplift and erosion. The vast plain on the plateau was the result of long-term erosion. Large amounts of sediments are transported through river channels to the low-lying plains, forming alluvial plain and estuarine deltas and pushing land gradually into the ocean, which is the basic law of the natural evolution of landform, and floods and rivers play an important role in forming mountainous regions, alluvial plains, and landmasses. Theoretically, when the gradient between the river source and the estuary becomes zero, the river will disappear. Therefore, a river also has a life expectancy, but its life is related to tectonic movement, land erosion, and slope, and its life expectancy is generally tens of thousands of years to tens of millions of years. It generally takes tens of millions of years for large rivers to proceed through formation to disappearance, while the process for tributaries and small rivers is only tens of thousands of years to millions of years. Flying waterfalls in mountains and V-shaped valley sections, which are often observed, are parts of a relatively "young" river. The U-shaped, wide, and shallow channels in the middle and lower reaches of the Yangtze River, which resulted from slow water flow and deposition of large amounts of sediments, are generally parts of a "middle-aged" river. If human beings do not construct dikes or roads along the river banks in the low-lying sections of a river, the river is prone to change its course or overflow, and a new river will develop to flow into the sea. The middle and lower reaches of the Yellow River are the typical “middle-aged” river sections.

After the earth entered the Tertiary period, the tectonic movement of the European Plate entered the new Alps movement, while Asia entered the Himalayan movement period. The global continental distribution and extent were basically stable during this period. All of world’s major rivers began their processes of formation during that time. The formation of the Yangtze River was closely related to the Himalayan movement and the formation of China’s three-step staircase landscape. Although sections of the river or lakes may have existed before that time, it was quite different from today’s Yangtze River, and it was not the true Yangtze River.

Based on available research results (Fan and Li 2007; Mo and Pan 2006; Zheng 2003; Kong 2009), the Qinghai-Tibet Plateau was a result of continental collision along the convergent boundary of the Indian Plate and the Eurasian Plate. The collision started no later than 65 Ma and completed about 40–45 Ma, and during this time period, the Qinghai-Tibet Plateau started its post-collision process, and the ocean in the region retreated drastically. Significant tectonic movement was the post-collision volcanic event with high to extremely high content of potassium in the volcanic flow. The event started in the Qiangtang area of north Tibet and the three source areas of the Yangtze River. Thereafter, the volcanic activity center gradually migrated outward, but an almost intermittent period of magmatic activities occurred in the Gangdise belt during 40–26 Ma. There were only sporadic magmatic activities originating from the granite of the middle and upper crust of the earth at about 30 Ma. During 24–18 Ma, magmatic activities peaked. During 25–10 Ma, two volcanic events with high to extremely high content of potassium in the volcanic flow occurred in succession from west to east and originated from the lower lithosphere. During 18–12 Ma, a volcanic event occurred in the adakites that contained copper-containing porphyry and originated from the thickened lower crust or earlier subduction zone. These three quasi-simultaneous post-collision tectonic-magmatic events occurred mainly during 20–10 Ma, and the north-south-trending graben system also developed during this time period.

The significant uplift of the Qinghai-Tibet Plateau occurred mainly in the post-collision period, and most scholars have accepted the episodic uplift model. According to the model, the altitude of the Qinghai-Tibet Plateau was mainly caused by three uplift events that occurred about 18 Ma, about 8 Ma, and about 3.6 Ma, respectively. The ground surface in the Qinghai-Tibet area was at an average elevation of less than 1,000 m above mean sea level in the middle and late Pliocene (12 Ma or so) and started strong uplifting from the late Pliocene and early Quaternary (3.6 Ma or so).

Since the 1990s, foreign scholars have also presented different views on this model. Some of them believe that the uplift of the Qinghai-Tibet Plateau was a gradual process, but the uplift rate significantly increased in the late Cenozoic. As to when the acceleration began, there are major different views. Some scholars argued that the Qinghai-Tibet Plateau reached its maximum height at 14 Ma and that a collapse resulted from an east-west stretching event, and thereafter the average elevation began to decrease. Many people believe that the Qinghai-Tibet Plateau reached today's height 8 Ma based on an increase of gush flow in the Arabian Sea, which indicates that the Indian Ocean monsoon started or intensified and that the Pothohar Plateau turned dry and vegetation changed from forest to steppe. These phenomena indirectly indicate that the Qinghai-Tibet Plateau blocked the Indian Ocean monsoon from moving northward and intensified the monsoon in the surrounding area.

In the late 1970s, the Chinese Academy of Sciences organized many scholars to perform 30 years of scientific investigations and research on China's Qinghai-Tibet Plateau (Zheng 2003). Many studies were conducted on lithospheric geophysics, tectonics, rock uplift dating, denudation rate of intrusive blocks, and other issues, and comparisons were made between the data obtained from the Cenozoic geologic strata and the information derived from lake-core samples. Recent studies have revealed that since the collision of the Indian Plate with the Eurasian Plate, the uplift of the Qinghai-Tibet area has been in a multistage, heterogeneous, and unequal rate process. The Qinghai-Tibet area experienced three uplift phases in the Cenozoic that occurred during 45–38 Ma, 25–17 Ma, and 3.6 Ma, respectively. Two leveling processes also occurred. The average altitude of the plateau as a result of the first two uplift processes might not exceed 2,000 m. An extensive surface leveling event occurred 3.6 Ma, and it is estimated that the plateau altitude was below 1,000 m. The strongest uplift movement occurred in the late Tertiary and early Quaternary, and since then, the main body of the plateau has been experiencing neotectonic movement to form the present landform. Thus, from the geological angle, the Qinghai-Tibet Plateau is so "young," and it is still the most active area of tectonic movement.

From the paleoclimatic change in China, the Qinghai-Tibet area in the early Tertiary was mainly experiencing the differentiation caused by the planetary wind system. About 3.6 Ma, as the altitude of the plateau uplift reached the critical height adaptive to the baroclinic atmosphere rotation, monsoon began to form in the plateau. Since 3.6 Ma, the Qinghai-Tibet area as a whole has begun the "Qinghai-Tibet movement" manifested by strong uplift, the collapse of the main planation surface, and the formation of large faulted basins. About 2.5 Ma, the uplift reached the "dynamic critical height" (about 2,000 m), and the planetary wind system changed from the one dominated by the wind flow climbing over the plateau to the one dominated by flowing around, and the plateau monsoon changed from a shallow and thin system to a deep and thick system, forming the modern monsoon pattern in China. Meanwhile, loess began to accumulate in north China. The uplift of the Qinghai-Tibet Plateau had a significant impact on the climate of the East Asian monsoon, and the impact on winter monsoon was much greater than that of the summer monsoon. About 0.8–0.5 Ma, the Qinghai-Tibet Plateau rose to the elevation of about 3,000–3,500 m above mean sea level, and the mountain area was up to 4,000 m. During this uplift, the climatic cooling in the area was consistent with the cooling caused by the global orbital transformation known as the upheaval of the middle Pleistocene. During this period, the main body of the plateau entered the ice cycle; large-scale glacial effects appeared; and the glacier area was over 500,000 km^2. The climate of the plateau changed abruptly, which was marked by stronger plateau winter wind and greatly weakened summer wind. At the same time, Western China entered a cold and dry climate, and the precipitation in Central and Eastern China increased significantly to 2–3 times the current level. The snow accumulation in high plateaus formed a strong cold source, and the area of desert, Gobi, and loess in northwestern China expanded. Over the past 150 ka, with the advent of the interglacial period, snow melting, and decreased surface reflectance, the Qinghai-Tibet Plateau has become an atmospheric heat source, possessing the climatic characteristics of the modern plateaus. About 0.15 Ma, a strong but uneven tectonic uplift occurred in the plateau, and mountains at the edge of the plateau became a barrier to intrusive warm and humid airflow. According to records for ice core samples collected from Guliya, during the last glacial maximum, the average temperature in the plateau cooled down by 7 °C; the fluctuation range in high temperatures was the most intense over the past 150 ka; the precipitation was only 30–70% of the present level; and the area of glaciers expanded to 350,000 km^2.

Mountains in the Qinghai-Tibet Plateau are high, and many geological sections are exposed. The Qinghai-Tibet Plateau appears to be a geological museum and is the best area for geological scientists to conduct scientific research. Research results of the Qinghai-Tibet Plateau should be comparatively complete and reliable. The middle and lower reaches of the Yangtze River have also experienced many changes in geological ages. However, due to the low-lying terrains, the geologic information has been buried by the Quaternary colluvium and flood alluvium. The river-lake relationship has also varied greatly. Before the Yangtze River ran throughout its entire course, the middle and lower reaches of the river had mainly been occupied by alluvial plains with many large lakes and marshes. Many water systems had connected with lakes, but there had no single main river channel. The full channel connection and formation of the Yangtze River were being completed under the premise of the formation of the Qinghai-Tibet Plateau. It is the formation of the three-step staircase of China's land surface descending from west to east that provided geological, geomorphic, and external dynamic conditions for the Yangtze River to run through from west to east. Figure 2.1 shows the current geomorphic zoning of the Yangtze River basin that can be divided into three major geomorphic regions: deep cutting plateau area of the upper reaches, moderately cutting mountainous area in the middle reaches, and hilly plains in the middle and lower reaches. The three regions can also be divided into 15 secondary geomorphic subregions.

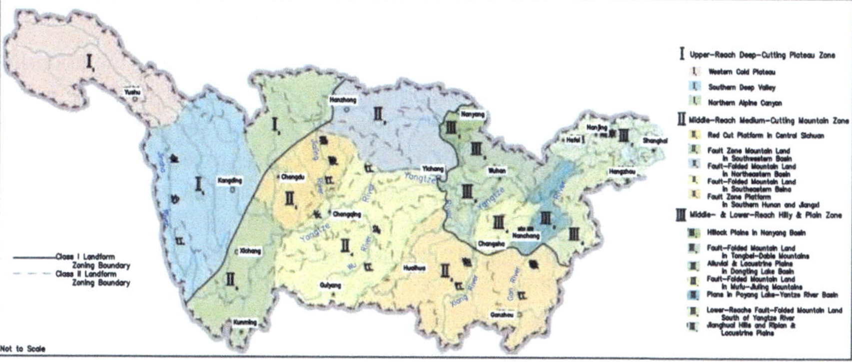

Fig. 2.1 Geomorphic zoning of the Yangtze River basin

2.3 Evolution and Connection of the Upper Yangtze

The river sections in the upper, middle, and lower reaches of the Yangtze River had formed separately in the middle and late Tertiary. However, the connection of the three reaches occurred much later. The approximate true connecting time is about 2–0.1 Ma, but the more accurate time is still in dispute. There are two key troubling questions about the Yangtze River that need to be answered. First, how did the Jinsha River sharply turn at Shigu from south to northeast; then cut through the Tiger Leaping Gorge at Sanjiangkou; and then again sharply turn to flow southward until Jinjiang Street; and then flow eastward with the formation of the First Bend of the Yangtze River? Second, when did the Chuan River channel through the Three Gorges area into the Jingjiang River and the Jianghan Plain?

Since the Cenozoic, major changes have taken place in the upper reaches of the Yangtze River with lots of geomorphic evidence of fluvial and lacustrine sedimentation, such as wind gaps, dry valleys, terraces, and associated fluvial sediments. Located in a tectonic uplifting area, rocks are well exposed and have become an important basis for studying the evolution of the Yangtze River. Meanwhile, the geomorphic evidence of fluvial sedimentation and lacustrine deposits from surrounding lakes has been altered by tectonic activities in varying degrees, and exposed uplifted surface has been susceptible to alteration by the surficial erosion process, which has influenced the interpretation of the ancient geographical environment. This may be the main reason that a consensus cannot be reached as for the timeframe of the evolution and connection of the Yangtze River.

The landform of fluvial and lacustrine sedimentations is the direct evidence of the evolution of the water system, but it is not the only evidence since it is susceptible to modification at a later time. Therefore, the landform may not be the best evidence of the river evolution. Instead, the relatively continuous, better preserved estuarine and marine sediments have relatively completely recorded the evolutionary information of the upper and middle reaches. Through tracer studies of the sediment source, the information about finding the "source" from "convergence" can help investigate the time when the connection of the Yangtze River occurred and may become an important means to study the evolution of the Yangtze River and the timeframe for the river to run through its entire course.

Based on the results from the electron microprobe chemical dating method, the preliminary determination of the age variation of U-Pb contained in monazite in the late Cenozoic strata at the Yangtze Estuary indicates that the Yangtze River basin underwent significant alterations from the late Pliocene to early Pleistocene. According to the initial estuarine strata position of the materials originating from the Qinghai-Tibet Plateau, it has been preliminarily determined that the time when the Yangtze River first ran throughout its entire course should be after 2.58 Ma. Herein at least the upper time limit has been determined, and the connection of the Yangtze River occurred after the Quaternary. This conclusion is basically consistent with the study results of fluvial sedimentary geomorphology.

In recent years, techniques in microprobe dating of single crystalline grains of clastic minerals, such as U-Pb dating and H_f isotope dating on clastic zircon that have been improved and expanded rapidly (Jia et al. 2010), are not only highly accurate but also economy and fast. The techniques have made it possible to perform relatively large-scale analyses and testing and hopefully will revolutionize research on high-resolution tracing in complex watersheds and tectonic activities in source areas of rivers. If the techniques are used to study the evolution of the Yangtze River basin and tectonic activities in the source area, it is likely that great progress will be made.

2.3.1 *The Time When the Jinsha River Channeled Through*

The development and evolution of the Jinsha River system basically reflect the tectonic activity and gradient forming process of the southeastern margin of the Qinghai-Tibet Plateau. The formation of the gradient reflects the uplift process of the Qinghai-Tibet Plateau. For example, most major tributaries of the Jinsha River initiated and developed along fault zones, indicating the formation and evolution of the upper Jinsha River were accompanied by the uplift of the Qinghai-Tibet Plateau.

The first detailed scientific argument for the evolution of the Jinsha River was presented by Barbour. He believed that the First Bend of the Yangtze River was caused by the capture of the Jinsha River; the Jinsha River originally flowed from north to south, from Shigu through Baihanchang-Jiuhe-Jianchuan Rivers to the Red River and eventually into the South China Sea. He provided four pieces of evidence. ① The north-east turning angle of the Jinsha River at Shigu is unusual. ② The intersection angles between the main channel of the Jinsha River and its major tributaries are too large. ③ There are reverse river terraces. ④ The landform of Jianchuan Valley is far from what a small river could create, and there must have been a large river. After this, as to whether the First Bend of the Yangtze River and the Jinsha River resulted from captures, scientists expressed their own views. Li believed that Baihanchang was the result of glaciation encroaching; Yuan assumes that the Shigu landform can be interpreted as entrenching meander; Ren and others conducted a detailed study on Jianchuan Valley and presented the evidence to support the argument of capture by Barbour; He and others and Zhao refuted the capture argument; and Zhang and others, Yang et al. and Ming et al. supported the argument of capture by the Jinsha River. Kong (2009), based on the fact that major tributary valleys maintain great amounts of lacustrine deposits – Xigeda stratum, determined the time of the lacustrine sediments to be between 1.58 and 1.34 Ma using the cosmogenic nuclide burial dating method. Studies of U-Pb age distributions within fluvial sands underlying the Xigeda layer indicate that before the formation of the ancient Lake of Xigeda, the river flowed from east to west between Panzhihua and Taoyuan, which is opposite to the current flow direction. With the formation and cut-through of the paleo-lake of Xigeda, the Jinsha River started to flow from west to east, which occurred after 1.34 Ma. Zhu et al. (2008) conducted a sedimentary sequence study of the Yuanmou Basin and considered that the lacustrine and fluvial deposition occurred 4.9–1.4 Ma, and after 1.4 Ma, the Longchuan River cut through the Yuanmou Lake, and lacustrine deposition came to an end, which indirectly proved that the Jinsha River began to flow eastward at 1.4 Ma, and the Longchuan River became a tributary of the Jinsha River. According to the above comprehensive analysis, it can be preliminarily determined that the time when the original southward flow of the Jinsha River turned to an eastward flow is about 1.4 Ma or later.

The upper reaches of the Jinsha River existed as early as the Tertiary, while the middle and lower reaches of the river formed and connected with the upper reaches through sequential captures after successive uplifts of the Qinghai-Tibet Plateau, Western Sichuan Plateau, and Yunnan Plateau. Since the lower reaches of the Jinsha River channeled through, relatively intense deep cuttings have occurred with an average cutting rate of 80–85 centimeters (cm) per ka. Some of the river sections changed their courses due to channel blockages. Therefore, short-term cutting rate may be as high as hundreds of cm/ka. In the last tens of thousands of years, the sedimentation in the lower Jinsha River has generally been covered. In some deep channel sections or ponds, the sediment thickness can be dozens of m up to hundreds of m. Based on the channel blockage caused by accumulated soil masses from a slope failure at Jinpingzi, the fastest sediment accumulation rate could reach approximately 5 m/ka. The aforementioned successive uplifts of the plateaus not only led to the westward headward erosion of the western tributaries of the Chuan River but also caused the convergent flows in western Sichuan and the Sichuan Basin to have gradually changed to converge and flow eastward and the Chuan River to have a reverse flow. Before the lower Jinsha River channeled through (from Shigu through Hongwen to Yibin), from Yibin to the present-day area where the Chuan River flows through Chongqing, Changshou, and other places, the original convergence may flow in the direction from the Jialing River to Yibin. Only after the sections of the Jinsha River downstream of Yibin channeled through did the Jinsha River (Shigu Hongwen) water flow into the East China Sea.

Based on the above analysis and after considering the comparison of the mineral compositions in the terrace deposit samples of the upper reaches, the following basic conclusions on the changes of the upper reaches of the Yangtze River system can be made. Before the lower reaches of the Jinsha River channeled through (Shigu, Hongwen, Yibin), the area was divided into several river systems, and the systems did not have connections to one another in water flow and material replenishment, and the Jinsha River may had a sharp turn toward south in the vicinity of Shigu and Hongwen into the Baihanchang-Jianchuan Valley and then drained through the Red River into the South China Sea. At 1.4 Ma, the Jinsha River began to flow eastward, and the Chuan River also redirected to the east. Figure 2.2 is the map to show the modern Jinsha River system.

Fig. 2.2 Present Jinsha River system

2.3.2 *The Yangtze River Channeled Through Three Gorges*

Located in the middle reaches of the ancient Yangtze River basin, the nearly north-south-trending Wu Mountains separated the ancient Yangtze River system. The genesis and current flow direction of the lower reaches of the Chuan River and the Three Gorges river section have been extensively studied and debated for more than 100 years. Li, Ye, and others considered the Three Gorges were secondary valleys. After the river systems on the eastern slope of the Huangling anticline captured the river systems on the west slope of the anticline, the Yangtze River channeled through the Three Gorges area. Li, Ren, and others believed that the Three Gorges section was flowing before the initiation of the Yangtze River, and it flowed through as early as the end of the Cretaceous or the east section of the Three Gorges was a flowing river before the initiation of the Yangtze River. Yang and Tang (1999) argued that the mountainous area in the Three Gorges area was a regional drainage divide during the middle Cenozoic; after its western paleo-system drained into the ancient Tethys Sea and later most of Sichuan and Yunnan drained into the South China Sea; and the Yangtze River did not channel through the Three Gorges area until more than 1.0 Ma.

The Shennongjia and Wu Mountain ranges, where the Three Gorges are located, had once been the regional drainage divides among the water systems of the Pacific Ocean, the Tethys Sea, Indian Ocean, and the South China Sea. The Jurassic, Cretaceous, and Tertiary systems are well exposed in the area west of the Three Gorges. The newest formations of deposits formed by regional water flow convergence are the thick fluvial and lacustrine sedimentary formations of Xigeda and Yuanmou groups dated 1.0 Ma. Alluvial deposits associated with the easterly flow of the Jinsha River is the Ya'an gravel layer in the Yangtze River valley near Yibin, and the gravel layer is dated to the middle Quaternary. In the area east of the Three Gorges, the alluvial-fan coarse-grained floodwater deposits associated the Yangtze River channeling through the Three Gorges area are located east of the Huyatan at Yichang and dated about 1.0 Ma. Moreover, the three valley sections in the Three Gorges area are an anticlinal mountain or cuesta formed by a right-angled limestone intersection, indicating that a river capture had once occurred in the area. Therefore, it can be concluded that the Yangtze River channeled through the Three Gorges area was the result of a river capture that occurred more than 1.0 Ma.

Based on W. Penck's mountain development model and the average denudation rate of 0.084 mm/year (Zheng and Li 2009), the initial time for denudation and downward cutting in the Three Gorges area should occur earlier than 32.4 Ma, or the initial formation time of the Yangtze River in the Three Gorges area was about 32.4 Ma. Yang et al. (2006) conducted a study on the geochemical characteristics of sediments and U-Pb dating on monazite from samples collected from a 319-m-deep boring drilled at the Yangtze Delta and found that sediments from Pliocene and Quaternary have different provenances. Therefore, Yang et al. (2006) argued that the sediments from the Pliocene were from nearby acidic substances, while the Quaternary sediments were derived from a faraway source of more alkali-base materials and proposed that the Yangtze River developed from a small river into a large river system with the eastern margin of the Qinghai-Tibet Plateau as its source in the early Pleistocene no later than 1.18 Ma. Xiang et al. (2007) performed an investigation of the sediments near Yichang in the western Jianghan Basin and considered that there existed an ancient lake in Yichang 0.75 Ma, and the ancient lake was predated by an alluvial-fan and alluvial-delta environment. Moreover, they analyzed the provenances of lacustrine deposits and fluvial sediments from the Yangtze River and concluded that the Yangtze River channeled through the Three Gorges area after 0.75 Ma and that the material west of the Three Gorges was brought to the vicinity of Yichang thereafter. Zhang et al. (2008) analyzed the lithologic characteristics, magnetic parameter values, and magnetic mineral characteristics of the sediments collected from two boreholes at Xingou Town and Zhoulao Town in the sedimentary center of the Jianghan Plain, and the results indicate that the content of coarse grains and stable magnetic minerals in the sediment at depths near 110 m of the boreholes obviously increased and the magnetization rate, saturated isothermal residual magnetism, and nonmagnetic hysteresis magnetization rate of the sediment increased sharply. All this indicates that a relatively large adjustment in the water system of the Jianghan Plain occurred near the depth of 110 m and the sedimentary environment and material in the Jianghan Plain changed significantly. The position of the layer may coincide with the layer when the Yangtze River channeled through the Three Gorges area 1.17–1.12 Ma determined by the ancient geomagnetic dating.

Based on the aforementioned studies, it can be preliminarily determined that the time when the Yangtze River channeled through the Three Gorges area is about 1.1–1.2 Ma or slightly later and should not be much different from or slightly later than the time when the Jinsha River changed its course.

2.4 Evolution of the Middle and Lower Yangtze

2.4.1 Mainstream of the Middle and Lower Reaches

Before the Yangtze River channeled through the Three Gorges area, the middle and lower reaches of the Yangtze River had been a mainly lake-marsh system consisting of alluvial plains formed by inflows from the areas bordered by surrounding drainage divides; lakes had been connected by complex and wandering water systems, but no defined channel had formed to drain into the sea. After the Yangtze River channeled through the Three Gorges area, under the influence of the water and sediments from the upper Yangtze River, the upper Han River, and other tributaries, the amounts of water and sediments in the middle reaches of the Yangtze River increased markedly, and many alluvial plains, lakes, and marshes developed. During 0.5–0.1 Ma, due to rising temperatures in the globe and the Yangtze River basin, precipitation and runoff increased greatly, and the middle and lower reaches of the Yangtze

River suffered flooding hazards, resulting in the formation of large lakes and marshes and the prototype of the ancient Yangtze River. About 0.1 Ma, because the climate turned cold, the water levels in the East China Sea and the mainstream of the Yangtze River declined, and the river discharge capacity increased. Consequently, lakes and marshes shrank prominently, and headward erosion in the ancient Yangtze River intensified, which facilitated the formation of the Yangtze River's mainstream channel, the connection of the inner river-lake systems in the middle and lower reaches of the Yangtze River, and the formation of the Yangtze River that drains into the East China Sea. Therefore, the embryonic Yangtze River formed between 1.5 Ma and 0.1 Ma. During this time period, the Chinese sturgeon and other migratory fish species might live in the mainstream and large lakes connected to the river in the middle and lower reaches of the Yangtze River.

Since the Yangtze River channeled throughout its entire course, most of the river channel in the upper reaches has run through canyons and mountains with both banks consisting mostly of bedrock. Therefore, the channel has been relatively stable. However, since the river channel in the middle and lower reaches has been affected by climate change and the rise and fall of sea level, the evolutionary range of the river channel has been relatively large, and the river-lake relationship has undergone many great changes. For example, the formation and disappearance of Yunmeng Marsh, several expansion and disappearance processes of Dongting Lake, the formation and vanishing of the ancient Pengli Marsh, and the enlargement of Poyang Lake, all of which have reflected the large changes of the early channel in the middle and lower reaches of the Yangtze River.

The mainstream in the middle and lower reaches of the Yangtze River has developed following the direction of two groups of fractures and has connected with many tectonic basins and several lake systems centered at tectonic basins. The prominent features of the water systems' structure in the middle and lower reaches can be summarized as follows. Many rivers first flow into lakes and then converge into the mainstream of the Yangtze River. For example, the following major tributary water systems converge from the left bank of the Yangtze River, Jianghan Lake group, the lake system behind the natural dike north of the Yangtze River in Anhui, Hongze Lake-Shaobo Lake, etc., and the following major tributary water systems converge from the right bank of the Yangtze River: Dongting, Poyang, Shijiu, and Tai Lakes.

The three large lakes in the south bank of the Yangtze River and the lake group in the Jianghan Basin were formed by stored water from the Yangtze River when its water level was high during the warm climate on the Mesozoic and Cenozoic inland tectonic basins. After the tectonic subsidence of these basins had occurred, centripetal water systems developed with basin lakes as centers, and the easterly river flow made the basins be connected in series (mainstream of the Yangtze River) or in parallel (such as the Yang, Xia, and Yong Rivers on the former Jianghan Plain). There are many small lakes in the middle and lower reaches of the Yangtze River, and most of them are the lakes behind the natural dikes of the Yangtze River. Specifically, when the water level in the mainstream of the Yangtze River was high, the tributary estuaries were blocked by the natural dikes of the Yangtze River and became lakes by stored water.

During the period of the low sea level in the last glacial maximum, due to deep cutting into the river channel in the middle and lower reaches of the Yangtze River, the water level declined greatly, which had once resulted in the outflow from the lakes along the river and caused low-lying areas of lake basins to become low-lying river systems (e.g., Dongting Lake). In the last glacial maximum, the river channel was cut to depths below −60 m, −55 m, -42 m, and −15 m at Shanghai, Nanjing, Wuhu, and Shashi, respectively, and the average water surface elevation of the Yangtze River was 20–40 m lower than the current level. During the low sea level of the late Pleistocene glacial period, the development of the ancient deep channel started from the Yangtze Estuary upward. The channel was cut into the Pleistocene sedimentary layer between the estuary and the proximity of Jiujiang, and the bottom of the ancient channel at the estuary was below −60 m. From Jiujiang upward, most of the ancient deep channel was cut into the bedrock before the Tertiary, and the Jingjiang section of the ancient channel was cut into the Pleistocene gravel layer. Generally, the bottom of the ancient deep channel was 5–25 m lower than the current riverbed, and the ancient Jingjiang River channel was cut to the elevation of −10–0 m, but the ancient channel in the middle and lower reaches of the Yangtze River was gradually filled with sediments during the warm period of the Holocene. Therefore, the lakes along the river had once been drained and dried up and become low-lying land of river networks. According to a large number of archaeological discoveries, many cultural sites of the late Paleolithic and Neolithic periods were identified in these low-lying areas, indicating that many of the central areas of the lakes have evidence of past human activities. It was 0.07–0.01 Ma, or the last ice age, when sea level fell significantly, and the areas of lakes connecting to the lower reaches of the Yangtze River were the smallest. During the prolonged Paleolithic period, human beings were able to migrate from the mountain front or the high hillock to the low-lying areas of the river to live and produce. These areas are now submerged by lakes or are so often plagued by floods that human beings cannot settle there year-round.

Since the beginning of the Holocene, under the influence of the global sea level rise, the water level of the mainstream of the Yangtze River from the estuary upward has been rising, resulting in the low-lying areas along both banks of the middle and lower reaches of the river gradually unable to drain and consequently becoming lakes. The time for the low-lying areas to have become lakes depends on the elevation difference between the bottoms of the low-lying areas and the local elevated water level of the Yangtze River. Generally speaking, the low-lying areas along the banks in the lower reaches and estuary of the Yangtze River turned into lakes before those of the middle reaches of the Yangtze River, and the deeper low-lying areas became lakes earlier than the shallow ones. Since then, as the continuous rise of the water level of the Yangtze River has caused the water levels of the lakes to gradually elevate, the areas of lakes have increased. By 6 ka to 4 ka ago in the middle of the Holocene (Atlantic high-temperature period), due to the impact of global warming and precipitation increase, the area in the middle and lower reaches of the Yangtze River changed to have a subtropical climate. As a result, the Jianghan lake group was at its heyday for a while, and the lake area between Guangxingzhou and Chenglingji was all connected to the Dongting Lake water system to form the historic "Yunmeng Marsh." At this time, a widespread, but discontinuous, uniform layer of dark gray clay deposited in the Jianghan Plain, indicating that "Yunmeng Marsh" was not a single marsh but an intermixture of lakes and marshes with land and water. Around 4 ka to 1 ka ago, as the climate changed to dry and cold, the lake area of the Jianghan Plain shrank, and "Yunmeng Marsh" began to disintegrate gradually due to cold and dry climate, long-term deposition of sediments, and the impact of human activities. About 2 ka ago, the mainstream of the "embryonic Yangtze River" changed its course to the current Yangtze River channel, and the Dongting Lake basin started to drain into the Yangtze River again through the opening between Guangxingzhou and Chenglingji and following the Diaoxian River.

Since the end of the Pliocene, the Yangtze River basin upstream of Yichang has experienced intense uplift. As a result, the Yangtze River has carried large amounts of sediments originating from the rising mountains to the depression area of the Jianghan Plain. The decreased longitudinal slope of the riverbed has led to deposition of large amounts of sediments, which provided material requirements for the formation and development of the meandering and aboveground Jingjiang River. Especially since 2 ka ago when sediments were retained within the diked channel for deposition through man-made dikes and floodland reclamation, the development of the meandering Jingjiang River has accelerated, and the riverbed has continuously risen as sediment deposited to form the aboveground channel. For example, the Jingjiang riverbed between Shishou and Jianli is 2–3 m aboveground, resulting in the current dangerous flood control conditions of high dikes and high floodwater levels. Flood hazards of the Jingjiang River have become one of China's grave dangers, especially of the lower reaches of the Jingjiang River.

The meandering lower reaches of the Jinjiang River started to form in 1500 or so. In about 1650, the first natural meander cutoff occurred. In the past 400 years, channel meandering and meander cutoff have alternately occurred repeatedly, and the meandering creep of the lower Jingjiang River has gradually accelerated. Before the dike break occurred at Ouchikou in 1860, the lower reaches of the Jingjiang River had generally been relatively straight. However, during the 50 years from the 1860 dike break to 1910, the riverbed became meandered, and natural meander cutoff occurred frequently, such as meander cutoff events in the following river sections: Dagong Lake, West Lake, Yueliang Lake, Guchangdi, and Chibakou. During the 50 years from 1910 to 1960, the diversion ratio of Ouchikou decreased gradually, and natural meander cutoff occurred only once in the Nianziwan section (in 1949). Especially after the implementation of two artificial meander cutoff projects (Zhongzhouzi in 1967 and Shanchewan in 1969) from 1966 to 1972 and one natural meander cutoff event, the curvature was once reduced to 1.93. However, in less than 25 years up to 1995, although many artificial river regime control projects were implemented, the curvature increased to 2.16. In the near future, the meandering creep and bank collapse potential in the lower reaches of the Jingjiang River will

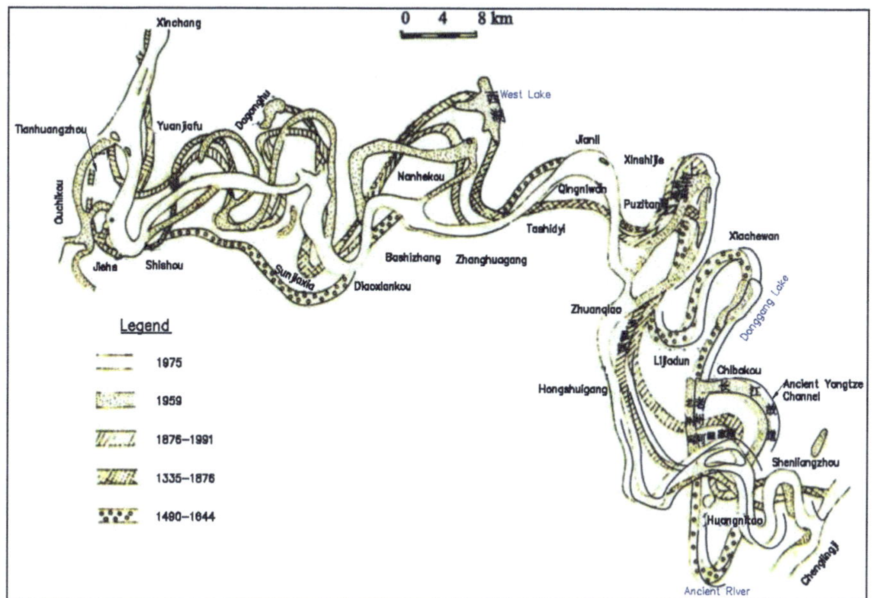

Fig. 2.3 Diagram showing the evolution of the lower Jingjiang River in past 500 years

likely go up. Lin believed that the reduced discharge and discharge variation range after the dike break at Ouchikou would be the main cause for the formation of meandering lower reaches of the Jingjiang River.

The southerly move of the entry point of the lower Jingjiang River into the Yangtze River was related to the blockage of the northern diversion from the Jingjiang River. From the Ming Dynasty to the 1960s, the general trend of the lower Jingjiang River was the continuous increase of the curvature. However, during the time, due to occurrences of a series of natural meander cutoff events, the curvature had twice become smaller. Since the 1960s, artificial meander cutoff projects and natural meander cutoff events have reduced the curvature, increased the floodwater and sediment discharge capacity of the lower Jingjiang River, and reduced the diversion amount of water and sediments from the Yangtze River to Dongting Lake. Figure 2.3 is a diagram to show the evolution of the lower Jingjiang River in the past 500 years. The diagram indicates the lower Jingjiang River has changed its course frequently.

In the past 5 ka, in addition to climate change, sea level rise, and tectonic activities, human activities have caused major impacts on the evolution of the mainstream in the middle Yangtze, and most of the impacts occurred inadvertently, such as:

① *Agricultural Arable Land Expansion*. Historically, frequent wars in north China forced a large number of people to relocate to the middle and lower reaches of the Yangtze River for economic development, which resulted in the destruction of large areas of natural vegetation, increasingly serious soil erosion, and declining water holding capacity. Consequently, the influx of sediments into rivers and lakes has increased; the downslope water concentration time has become shorter; and the water levels of rivers and lakes have risen.

② *Construction of River Dikes*. The purposes of constructing the river dikes are to control the floodwater of the Yangtze River from spreading freely and to protect the crops and residents in the low-lying plain areas along the Yangtze River. However, the dikes indirectly caused the decrease of the sediment quantity discharging to the plains along the banks, which intensified the riverbed deposition, and increased the floodwater level and the annual fluctuation range of water levels. The construction of the Jingjiang Dike is the essential cause of increasingly rising water level and development of southerly diversion of water in the recent millennium.

③ *Land Reclamation from Lakes*. The narrowed riverbed and smaller lake surface areas have lowered the ability of lakes and wetlands to store or regulate floodwater; increased the annual fluctuation range of water levels; caused scouring of lakebeds; and increased sediment quantity from lakes to the river.

④ *Channel Meander Cutoff and River Regime Control Projects in the Lower Jingjiang River*. These projects straightened the lower Jingjiang River straighter and improved its ability of discharging sediments, but at the same time, they resulted in less quantity of water and sediments into Dongting Lake through the diversion outlets and steeper and higher peak flood flow and more reverbed sediment deposition in the lower Jingjiang River. Because the overall gradient of the middle and lower reaches has not changed, a local slope increase in the upper reaches was bound to lower the slope in the lower reaches and force the discharge-storage conflicts to transfer to the lower reaches. Therefore, the regulation of the river channel is a systematic project and needs to comprehensively analyze the whole river channel or relatively long sections, or otherwise it will cause new problems. The meander cutoff of the lower Jingjiang River was one of the key causes for the reduction of water and sediment quantities into Dongting Lake and higher riverbed due to sedimentation in the lower Jingjiang River in the 1960–1970s.

In short, the changes of the Jingjiang River in recent 5 ka are mainly manifested by the blockage of the northern diversion channels (Yang, Xia, and Yong Rivers) from the Jingjiang River, the southerly movement of the lower Jingjing River channel, and the development of free meanders and the formation of the southern diversion channels (Ouchi, Songzi, Hudu, and Huarong Rivers) from the Jingjiang River (Yang 1989). The changes are also the history of Yunmeng Marsh's shrinkage and disappearance, Dongting Lake's expansion and shrinkage, in which the late-stage change occurred mainly under the influence of man-made dikes, land reclamation from lakes, and river regulation.

2.4.2 Evolution of Dongting Lake

The basic layout of Dongting Lake was created by tectonic movements. The lake was an intermixture under the interaction of rivers and impacts from human activities. The ancient Dongting Lake basin was originally a graben formed by the displacement of a land block moving downward between two parallel faults that trend northeast-southwest and the other two parallel faults that trend northwest-southeast due to intense crustal movements in the early Yanshanian period. A series of secondary sags and bulges have formed in the graben of Dongting Lake because of the interaction between the two pairs of faults and the difference in rise and fall of the earth's crust. The lake is surrounded by the Jiuling uplift on the east, the Wuling uplift on the west, the Xuefeng uplift on the south, and the Huarong uplift on the north. The Huarong uplift separates the Dongting depression from the Jianghan depression. The modern Dongting Lake basin has developed on the basis of the ancient lake basins. The process can be divided into two phases: the natural evolution phase and the intermingled evolution phase of human activity-natural effect.

2.4.2.1 Natural Evolution Phase

From the beginning of the Quaternary to the Holocene, the Dongting Lake basin was affected by neotectonic movement. As intermittent rises occurred in the surrounding mountains, the lake basin descended. In the early Pleistocene, the depression range was the largest, and faulting activities surrounding the lake basin were intense. During the middle Pleistocene, the sedimentary extent was the largest, but the sedimentary center was not as apparent as the early Pleistocene, and the sedimentary center migrated to the southwest. During the late Pleistocene, depression activities basically ceased. From the late Pleistocene to the early Holocene, the Dongting Lake area was a plain landscape transected by river networks due to the expansion of Yunmeng Marsh in the northern Jingjiang River basin and the rise of the Dongting Lake basin.

2.4.2.2 Intermingled Evolution Phase of Human Activity-Natural Effect

In the early and middle Holocene, the Jianghan Plain north of the Jingjiang River was part of a large depression that included Yunmeng Marsh, while the Dongting Lake region was still a plain transected by river networks without a large lake, which provided the conditions for the early human activities in the lake area. The human activities in the Dongting Lake area can be traced back to the Neolithic age. The following successive cultures were discovered in the Dongting Lake area: Pengtoushan, Zaoshixiaceng, Daxi, Chujialing, and Longshan in the middle reaches of the Yangtze River. In the surrounding areas of Dongting Lake and the hilly areas, the site of the Daxi culture and 45 sites of earlier cultures were discovered. Sites of the succeeding Chujialing culture were identified to be mainly located in the northwestern lake area and the western five counties or cities, but no Chujialing cultural sites have been discovered in the eastern or southern lake area, indicating that in about 5 ka ago, human activities in the western lake area (the mountain front culture) were much more active than in the eastern lake area (the plain culture) and the hinterland of the lake area was not conducive to human activities because of flooding. Archaeological findings indicate that the Longshan culture in the lake area had been very prosperous and cultural sites were identified throughout the areas around the lake and the hinterland of the lake. Evidently, most of the Dongting Lake area had already been suitable for human activities, and, as a result, the area of Dongting Lake shrank very quickly. Cultural sites for the Shang and Zhou dynasties to the Warring States period were identified to be scattered in the peripheral area of the lake, especially mostly concentrated in the lower reaches of the Li River. No cultural layer of Zhou or Shang has been discovered in the hinterland of the lake. It is speculated that the water surface was vast, and many Neolithic human paddy fields had been in the bottom of the lake.

From the Neolithic age to the period of the Shang and Zhou Dynasties and the Warring States, human beings had a very low activity capacity, basically still in the stage of understanding nature and adapting to nature. When the lake water rose, human beings retreated, and when the lake water declined, people moved in. As human beings did not have enough ability to transform nature, the human impact on the environment was very small. Moreover, the lake area was sparsely populated. Therefore, the evolution of Dongting Lake was mainly driven by natural processes, and human activity was superimposed upon the forces of nature to have an impact.

During the Han and Jin Dynasties, great changes took place. Human activities began to play the utmost role in the evolution of Dongting Lake, mainly manifested as follows: vegetation removal which had resulted in increased soil erosion and sedimentation; land reclamation from the lake which had changed locations for sediment deposition and caused large amounts of sediments to be deposited in the floodways, riverbeds, and the lake, resulting in higher floodwater levels; and the construction of dikes along rivers and surrounding the lake which had changed the natural direction of water flow and exacerbated the river-lake relationship. With the massive influx of population into the central plains due to wars, deforestation for land reclamation and land colonization from the lake had resulted in vegetation removal from mountain slopes, increased soil erosion, and serious sedimentation in the lake area. During the Yonghe years of the Eastern Jin Dynasty, the first dike – Golden Dike – was constructed on the Jingjiang River in southeast Jiangling City. From then on, more dikes had been constructed along the Jingjiang River, resulting in a narrower river channel, relatively higher floodwater levels, and a more complex river-lake relationship. At that time, there were large-scale land colonization activities in the lake area. Up to the Liang Period of the Southern Dynasty, all counties in the current lake area except Nan County had been established, and the order of county establishment was gradually extended from the lakeside to the hinterland of the lake area, indicating that the reclamation activities at that time was a step-by-step-into-the-lake process. During the Tang and Song Dynasties, due to the intensification of human activities, vegetation in the Dongting Lake drainage area was significantly removed; the sedimentation problem was aggravated; and human land reclamation activities had moved deeper into the lakes area, which had reduced the area of Dongting Lake. The total area of the lake at the time was only 3,300 km^2, reduced by almost half when compared to that during the Han, Jin, and Southern and Northern Dynasties. In the Yuan and Ming Dynasties, because of frequent breaks of the Jingjiang Dike, floodwater volume into Dongting Lake increased, and the lake area expanded. The rulers of the Yuan Dynasty changed Song Dynasty's blocking to diversion of floodwater. In Jiangling, Shishou, Jianli, and other counties, six openings were created from the river dikes, three of which in Yanglin, Songxue, and Diaoxian "diverted the river water southward and made all land within 50 km become part of Dongting Lake." In the Ming Dynasty, the people of the Dongting Lake area were overwhelmed with taxes, went bankrupt, and moved away from their land in the lake area. Consequently, nobody maintained the berms surrounding polders; farmland was abandoned; and the lake was expanded to an estimated water surface area of approximately 5,600 km^2. In the early Qing Dynasty, the rulers proactively encouraged and supported the construction of berms to develop polders from lakes. As a result, the population in the Dongting Lake area grew rapidly, and land colonization in the lake area reached another peak. During the reign of Yongzheng and Qianlong, land reclamation in the lake area reached a stage that "no soil would remain unclaimed." In the reign of Daoguang, land colonization in the lake area reached the climax so that "70–80% of the previous water area had gone." The rapid shrinking of the lake area hindered the outflow of the lake water, lowered the storage capacity for floodwater, increased the lake sedimentation rate, and elevated the floodwater level. Moreover, as the Jingjiang Dike was getting higher and thicker by the year, "a soil mountain has been created and the blocked water has appeared like a mountain as well." The river-lake relationship reached a potentially dangerous state so that modification had to be made. Eventually, during the reign of Xianfeng and Tongzhi, dike breaks occurred at Ouchi and Songzi successively, resulting in the formation of four new openings to divert the water flow southward, which caused a new river-lake relationship to form and the amounts of diverted water and sediments into the lake to increase substantially. During the last 30 years of the nineteenth century, on the one hand, the increase in incoming water made the Dongting Lake surface area lager; on the other hand, the increase in sediments rendered the lake basin shallower. In the short time, the water level of the lake was elevated, and at the same time, the lake area was enlarged to about 5,400 km^2.This was the last "rejuvenation" of Dongting Lake that has since started an accelerated shrinking process.

Since the beginning of the twentieth century, as the four rivers connecting the Yangtze River to Dongting Lake had discharged large amounts of floodwater into Dongting Lake and resulted in the expansion of the lake, large quantities of sediments had been brought into the lake. With additional sediments from the system of four rivers (Xiang, Zi, Yuan, and Li Rivers), the total amount of average annual sediments deposited in the lake had been about 100 million tonnes, leading to a shallower bottom of the lake and the expansion of the sandbank on the northern bank. With the expansion of the northern dike of the Jingjiang River, dike breaks or abandonments occurred occasionally on the south bank, forcing Dongting Lake to move southward, and at the same time increasingly more land-reclaiming berms had been constructed rapidly. From 1918 to 1931, about 267,000 ha of polders were developed, which is equivalent to the entire natural lake area of the current Dongting Lake. Polders along the north bank were expanded southward continuously and gradually got increasingly closer to Chishan, dividing Dongting Lake into two parts: east and west. During the southerly expansion of the sandbank on the north bank, due to the impact of the converging flow, the sandbank turned to expand eastward and then northeastward. As a result, Datong Lake was formed by the expanding sandbank and separated from East Dongting Lake. At the same time, the former Wanzi Lake in Yuanjiang and Hengling Lake in Xiangyin County expanded to be connected due to the abandonment of polders, leading to the formation of South Dongting Lake. In 1949, the total area of Dongting Lake was left to be 4,350 km^2.

2.4.3 Evolution of Poyang Lake

The Poyang Lake basin was the Poyang Lake depression in tectonic terms. It is located in the north margin of the Jiangnan anticlinorium and formed in the Mesozoic. The Yanshan movement caused the area surrounding the depression to rise with strong folding and formed the mountains, but the basin itself fractured and subsided into a graben-type fault sag. The east-west-trending structure, the Huaxia structure, and the new Huaxia structure constituted the basic layout of this area. During the early Pleistocene, the lake area may be a basin among rolling hills with a small and fragmentary stratigraphic range, and a complete drainage system had yet to form. During the Pleistocene, the local neotectonic movement was intense, and the basin was expanding, which deposited a two-dimensional structure consisting of red clay and gravel strata in a mesh pattern, and at this time, the ancient Gan River began to develop. In the late Pleistocene, the Poyang Lake basin began to develop from a hilly basin to a larger catchment basin, and the Gan River water system was beginning to take shape. It was not until the Holocene that a larger basin formed, laying the foundation for today's Poyang Lake basin. But not until after the middle Holocene, the Poyang Lake basin was still a valley basin running from south to north downstream of the Gan River, and the wet area was confined to the graben at the mouth of the northern basin, and the vast southern area was still a fluvial sediment accumulation zone (Su 1992).

In the past considerably long period of time, ancient Pengli Lake had been considered the predecessor of Poyang Lake, and, as a result, the time of the formation of Poyang Lake was considered much ahead of the actual time. Investigations have revealed that the ancient Pengli Lake and Poyang Lake were not the same one. The ancient Pengli Lake should be located in the Wangjiang depression in the area of Jiujiang, Susong, Huangmei, and Wangjiang downstream of Wuxi. During the middle Holocene, the warm climate with abundant precipitation in the middle and lower reaches of the Yangtze River and the associated high sea level had tidal currents extend deeply into the upstream area, and low-lying areas were often filled with river water and turned into lakes. The ancient Pengli Lake might form or be expanded at this time. Therefore, Pengli Lake was recorded in the *Tribute of Yu*, and the *Records of the Grand Historian of China* indicated that in 106 BC, during his southern tour, Emperor Wu of the Western Han Dynasty "once went boating in the river from Xunyang to Zongyang through Pengli Lake." Xunyang and Zongyang of the Han Dynasty are now all located on the north bank of the Yangtze River. *The Classic of Water* also recorded that "the Mian River (now Han River) runs southward to northern Shaci County of Jiangxia and then southward into the Yangtze River. The Mian River converges into the Yangtze River, then flows eastward through Pengli Lake, and proceeds northeastward out of Juchao County," and "the Gan River runs out of Nanye County of Yuzhang, northwestward through eastern Gan County ···, then northward through western Nanchang County, eastward through western Pengli County, and then from north into the Yu River." All these records indicate that the ancient Pengli Lake should be part of the riverbed of the current Yangtze River and on the north bank of the river, that there were no large water bodies within the Poyang Lake basin, and that the water area (seasonal) in the graben at the mouth of the lake might only be the tail of Pengli Lake. During the period of more than 2,000 years from the Dayu era to the Eastern Han Dynasty, the Yangtze River channel had been migrating continuously, and the main body of the ancient Pengli Lake had been gradually shrinking into Lei Pond and Lei Creek. Eventually the lake has evolved into current Longgan and Daguan Lakes, and the water area in the graben at the mouth have gradually been expanding southward and taken the name of Pengli.

The formation of the Poyang Lake basin can be traced back to the Mesozoic. The ancient Pengli Lake formed in the middle Holocene, while Poyang Lake formed in about 400 AD or only about 1,600 years ago, and, therefore, it is a relatively "young" lake. During the Six Dynasties (221 AD–589 AD), water systems in the middle and lower reaches of the Yangtze River had changed frequently, because the mainstream of the Yangtze River moved southward to cause Poyang Lake to expand and the expansion reached the peak in the Tang Dynasty. In the early Tang Dynasty, the east-west width of Poyang Lake was 50 km, and north-south length was 150 km with an area of 6,000 km^2, which was the largest area of the lake.

Since the formation of the Poyang Lake basin, there had not been enough accumulated water for a long period of time to form a lake, but it had just been a water-passing basin. Seismic activities indicated that the neotectonic movement was still active and a small locale in the lake area was indeed a depression caused by earthquakes, but not all the vast lake area. Global warming, sea level rise, and the resulting marine transgression could not only make the low-lying water area become a lake but could also lead to the migration of the river. All this was only the indirect cause of the formation of the lake. Poyang Lake is a south-high and north-low water-passing lake with a surface difference up to 11 m. If the mouth had not been blocked, the lake would have continued to discharge water, and a large lake would have never formed.

2.5 Evolution of the Yangtze Estuary

The development of the Yangtze Estuary and Delta area was the combined result of the river water flowing into the sea, sediment transport, sea level changes, and tidal variations. On a long timescale (over 10 ka), the development has been mainly affected by tectonic activities and the alternate occurrences of ice ages and interglacial periods. On an intermediate timescale (one hundred to one thousand years), the development has been affected by the continuous transport and deposition of sediments from the Yangtze and the Yellow Rivers. On a short timescale (1 year to decades), the development has been affected by the upstream runoff, sediment transport, and tides. Since the beginning of the Quaternary, there have been many marine transgressions and regressions in the estuary. The marine transgressions in the Holocene were mainly manifested by the retreat of the Yangtze Estuary and the Qiantang Estuary, as well as the submergence of the ancient delta ground. The northern shoreline of the Yangtze Estuary retreated to today's northern Yangzhou, ran northward around Shugang, and might extend to the eastern area of today's Shaobo Lake-Gaoyou Lake area. Within the shoreline had once been the Xiashu loess hillock where an ancient city called Guangling had been located as well.

Driven by the Holocene marine transgressions, the ancient Yangtze Estuary had been retreating and reached the vicinity of Yizheng west of the modern Zhenjiang and Yangzhou of Jiangsu about 6 ka ago. From any small-scale image, a V-shaped print with the open end toward the east can be clearly seen on the ground in Zhenjiang near Yangzhou, and the two legs of the V-shape are the two wings of the triangle-shaped estuary of the ancient Yangtze River. Between the two lines eastward are the increasingly broader floodland, sandbar, and sandbank, which are the top part of the Yangtze Delta.

For Tai Lake, West Tai Lake and East Tai Lake were two separate waters in the early times (William and Liu 1996). They were not integrated at least 5 ka ago. About 5 ka ago, West Tai Lake changed from a saline water lake to a freshwater one, while East Tai Lake started to develop in 6,500 years ago. About 5–6 ka ago, a group of lakes appeared to be scattered all over the Tai Lake Plain because the body of hills within the lake plain blocked the tidal water. Since then, the water level had been declining, and the lakes had started to shrink, while large areas had become dryland. More than 130 sites of the Liangju culture were identified in East Tai Lake, Wujiang, Qingpu, Kunshan, Suzhou, Wuxi, Changzhou, and the Hangjiahu Plain. Since the Liangju culture mainly flourished in 4–5 ka ago, the modern Tai Lake should form around 3.7 ka ago.

During more than 2 ka since the first century BC, there were three stages of periodic sea-land transition in the Yangtze Delta, each of which had undergone 700 years or so, and its first 300+ years were the time when the land was retreating and the sea advancing landward, and the latter 300+ years were the time when the land was advancing seaward and the sea retreating. In the period of land retreat, the mainstream of the Yangtze River was encroaching northward. During the marine regression, the mainstream of the river was moving southward. The existence of the

700-year cycles has been verified with historical and geological finds and can be analyzed from the special geologic environment of the Yangtze Estuary, the north-south swing of the Yangtze River mainstream, the fluctuation of sea level, and the periodicity of historic climate.

2.5.1 Special Geological Environment of the Yangtze Estuary

The Yangtze Estuary is located between the Yellow Sea and the East China Sea, and its sediment deposits are not only from the middle and upper reaches of the Yangtze River but also from the sediment-laden flow along the coast of the Yellow Sea, including large amounts of sediments moving with the southerly flow along the coast from the Yellow River to the estuary, especially when the Yellow River flowed from the north to the south and captured the Huai River to drain into the sea. This sediment deposition pattern was the main reason for beaches and sandbanks to first form along the northern coast near the estuary during the development of the Yangtze Delta, but this pattern was more prominent before the Tang Dynasty. After the Tang Dynasty, there were two important periods of this sediment deposition pattern in the Song and Qing Dynasties, respectively. The amount of sediments from the upper and middle reaches of the Yangtze River were gradually increasing after the Tang Dynasty, which can be mainly attributed to the population growth, production development, and land reclamation in mountainous areas in the Yangtze River basin after the Tang Dynasty.

2.5.2 North-South Swing of the Mainstream Riverbed in the Yangtze Delta

During the period of the northerly movement of the mainstream riverbed, the land along the north bank retreated extensively and rapidly due to side erosion by the water flow of the river, while the south bank was relatively stable. The main reason for this was that the south bank was restricted by the nearby ancient coastline, while the ancient coastline of the north bank was along the line of Yangzhou, Taizhou, and Hai'an. The area between this line and the river bank was the young delta plain consisting of historic alluvium and was more vulnerable to erosion. It is believed that this condition was also related to the difference in neotectonic activities between the north and south banks of the mainstream Yangtze River, but this was not necessarily the main reason, because historically the Yangtze River channel had recurrent northerly and southerly movements with the general trend to move southward. It may be mainly related to the north-to-south flow along the coast of the Yellow Sea and the Coriolis force. The modern land expansion pattern in the northern coastal area near the Yangtze Estuary clearly reflected the direction of the composite forces of the river flow and seawater flow, but it is hard to state that it reflected the trend of the neotectonic activity lines.

2.5.3 Historical Fluctuation, Rise, and Fall of Sea Level

The advance and retreat of land in the delta area reflected the historic cyclical rise, fall, and fluctuation of sea level. Without this cyclicity, the changes of land and sea could not have recurred periodically, and the rise and fall of sea level have directly affected the water level changes of the Yangtze Estuary. Based on the earth's horizontal component of the centrifugal force resulting from the earth's rotation, rivers in the northern hemisphere have always experienced erosion on right banks. That is why the riverbed has exhibited a strong southerly rolling landscape since the Yangtze River channeled from the Xiling Gorge through the Three Gorges down to the Yangtze Estuary. However, as sea level rose, seawater intrusion into the estuary general intensified, which undoubtedly increased the erosive effect on the north bank.

In each of the early preceding three stages, the delta retreated northward. Therefore, it can be proved that in each of the early cycles, the rise of sea level corresponded to the time when severe coastal disasters occurred frequently. In addition to large-scale land submergence and slope failures along the coast and the river banks, not only large areas of marshes and low-lying land were inundated but also enormous loss of lives and properties occurred along the river, which was the time of disasters. Many historical documents show that the Han Dynasty, the Tang Dynasty, and the Yuan-Ming and Qing Dynasties were periods of intense marine disasters. For example, in the Han Dynasty in 47 BC, 146 AD, 170 AD, and 171 AD, catastrophes occurred on the coast of the Bohai Sea. In the Tang Dynasty in 726, 751, and 752, marine disasters occurred in the Yangtze Estuary area (mainly in the areas of Yangzhou and Zhenjiang) and the southern coastal areas of Shandong Peninsula. During the Yuan and Qing Dynasties in 1373, 1539, 1665, and 1696, major disasters occurred in the delta area on both south and north banks of the Yangtze River and along the coast, resulting in human deaths ranging from thousands up to tens of thousands or hundreds of thousands. Chongming Island was submerged five times during the high sea level period of the Ming Dynasty between 1420 and 1583.

2.5.4 Historical Periodic Climate Changes

In the *Preliminary Study of Climate Changes in China over Past 5,000 Years*, Zhu Kezhen analyzed the climate changes in the past 2 ka from the phenological point of view and opined that "the climate was primarily warm during the Han Dynasty and was cold from the Three Kingdoms to the Northern and Southern Dynasties; and it was warm during the Tang Dynasty and was cold during the Yuan Dynasty. The climate was getting cold from the thirteenth century to fourteenth century, was coldest in the seventeenth century, and second coldest during the nineteenth century…". The conclusion obtained from the evolution of the historical geological environment of the Yangtze Delta is largely similar to that of Zhu Kezhen, but the climate in the early Western Han Dynasty and the seventeenth century was the opposite from the geological point of view. Since sea level was low from the Warring States Period to the following early Han Dynasty, the climate was cold. However, since the global sea level was high during the seventeenth century, it is inexplicable if this period is considered the coldest time. From a geological point of view, warm climate should precede the rise of sea level, and cold climate should precede the decline of sea level. Of course, in the warm climate, we cannot rule out that other natural factors, such as volcanic eruptions with excessive dust in the air, would have impacted solar radiation and resulted in a temporary low-temperature climate.

Since the Warring States Period (475 BC to 221 AD), the Yangtze Estuary has bifurcated into South Branch and North Branch. With the formation of new islands, the South Branch bifurcated repeatedly, and the "maternal body" has continued to evolve to the present state. With islands joining the bank, the North Branch has receded and has been replaced with a new north branch evolved from the South Branch. So far, the North Branch has succeeded for three generations with the first two called as the ancient North Branches.

The formation of the North Branch and the development of Chongming Island were inseparable. Chongming Island began to appear in the early Tang Dynasty and did not develop into a large alluvial island until the late Ming Dynasty. This development process can be divided into three phases: initial development, multiple sandbar variation, and formation of a stable island core. The North Branch formed during the stable period of the island core whose formation was closely related to sediment replenishment from the Yangtze Estuary. The seventeenth to eleventh centuries (from early Tang Dynasty to late Northern Song Dynasty) was the period of the initial development. With the formation of the Hudou-Dongbu Island chain and their connection to the bank, as well as the rise and expansion of the south bank of the river estuary due to sedimentation, the estuary narrowed from 180 km to 120 km, which was close to the 98 km width between today's Nanhui and Qidong. This period also coincided with the peak land reclamation period in the Tang Dynasty. At the same time, soil erosion intensified in the Yangtze River basin, and sediment contents in the river water was high. The river water with high sediment contents made the sediment movement active, and the estuary became even narrower, resulting in more concentrated tidal energy and more forceful tidal current. With the addition of the Coriolis force, the tidal current paths became disintegrated during the tidal rise or fall. The flow velocity in the distributary area was slow and conducive to the formation of alluvial islands in the area of sediment deposition, and the formation of alluvial islands counteracted with the flow of water to force the flow paths to swing and thus caused some existing alluvial islands to fall and other new alluvial islands to rise. It is during the turbulent period of the rise and fall of islands did the initial Chongming Island emerge. According to the *History of Zhengde Years* of the

Ming Dynasty, Chongming Island first developed as two separate sandbars – Dongsha and Xisha – during the Wude Era of the Tang Dynasty (618–626) or approximately 1.3 ka ago. At that time, vegetation was flourishing, and the salinity was relatively low at the Yangtze Estuary, namely, the river flowrate was large with high contents of sediments, favorable for the development of the nascent Chongming Island in the environment of sediment scouring and deposition. During this period, Chongming Island had newly formed, and the North Branch had yet to form. From the eleventh century to the early sixteenth century (late Northern Song Dynasty to middle Ming Dynasty), it was a multi-sandbar transition period. During this period, more sediments from more sources were transported to the Yangtze Estuary, and new sandbars formed frequently. At the Yangtze Estuary, sediments were not only from the Yangtze River basin but also from the Yellow River, which was brought to the Subei coast by tidal currents. Sediment supplies from multiple sources resulted in the frequent formation of new sandbars and the expansion of existing sandbars. According to two historical sets of Chongming maps prepared during the Qing Emperors Kangxi and Yongzheng eras, from the Southern Song Dynasty to the Yuan and Qing Dynasties, more than 40 sand islands formed. The three representative sand islands were Sansha, Pingyang, and Changsha. If the sand islands formed during the Tang to the Southern Song Dynasties are counted, more than 60 sand islands formed. Sand islands were densely distributed throughout the entire Yangtze Estuary and formed an environment consisting of multiple sand islands and multiple estuarine branches with intercrossing water flows, which has become a huge sediment interception gate system at the estuary.

From the beginning of sixteenth century to the early seventeenth century (from the middle to the late Ming Dynasty), the island cores were stable. During the Zhengde years (1506–1521), a small sandbar appeared more than 30 km southwest of Yaoliusha Sand Island. The sandbar was surrounded by many unnamed sedimentary islands where the water flow was slow and the wave was weak. Due to the favorable water and sediment environment, the small sandbar had not changed in position since its emergence from water and had constantly grown with sedimentation. The sandbar later merged with Wujiasha Sand Island and Xiangsha Sand Island that had emerged during the middle Wanli years (1573–1620) of the Ming Dynasty and continued to grow and form the central part of Chongming Island – "island core" – which also provided necessary conditions for the formation of the North Branch. At that time, the egg-shaped island core was not connected with Pingyang Sand Island that was located to the west. Between them were Pingansha Sand Island, Fuansha Sand Island, and a series of north-south-trenching waterways. As a result, the Yangtze Estuary was divided into two separate waterways, independently discharging into the sea, and possessed its first-stage branching characteristics with the formation of the embryonic North Branch and South Branch. Figure 2.4 is a map showing the Yangtze Estuary.

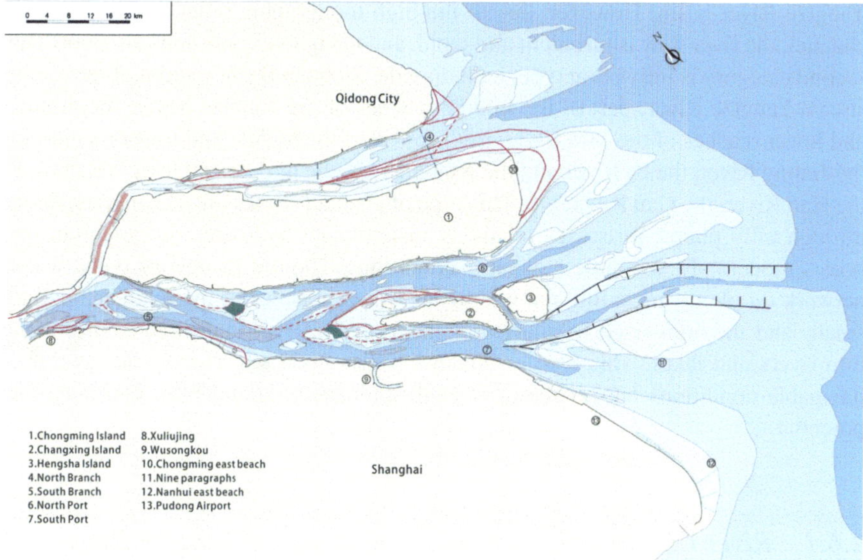

Fig. 2.4 Map of the Yangtze Estuary

2.6 Characteristics of the Modern Channel of the Yangtze River

In the Yangtze River basin, plains (below 200 m in elevation) make up only 19.18% of the total area and mountainous and hilly areas 80.82%, of which alpine areas at elevations of 3,000–5,000 m account for 24%, medium-height mountainous areas at elevations of 1,000–3,000 m 22.64%, low-height mountainous areas at elevation 500–1,000 m 15.48%, and hilly areas at elevations of 200–500 m 18.7%. Based on the morphological features of the river channel, the mainstream of the Yangtze River and its tributaries can be divided into three types of channel. The first category is canyon type, including the following mainstream sections, the Jinsha River and the Three Gorges, and the following tributaries, middle and lower reaches of the Yalong River; the upper reaches of the Min River and its tributary, the Dadu River; the upper reaches of the Jialing River and its tributary, the Bailong River, the Red (Chishui) River, the Wu River, the Qing River, and the Yuan River; and the upper reaches of the Han River. This type of channel flows through the Qinghai-Tibet Plateau, the Yunnan-Guizhou Plateau, and its peripheral mountainous area, Qinling Mountains, Daba Mountain, etc. This type of channel accounts in drainage area for more than one-half of that of the Yangtze River basin and can be characterized by relatively deep cutting, large topographic relief, and rich hydroelectric resources with the hydroelectric potential accounting for more than 70% of that of the entire

Yangtze River basin. However, due to the high topographic relief for this type of channel, the river flow is turbulent and rapid, and navigable conditions are poor. The second category is hilly plain type, including the Sichuan Basin section of the mainstream Yangtze River; part of the middle reaches of the Yangtze River; the middle and lower reaches of the Min River, the Tuo River; the middle and lower reaches of the Jialing River, the Zi River and the Xiang River; the middle and lower reaches of the Han River, the Gan River; etc. This type of channel comes out of canyons, flows through hilly plains, possesses relatively insignificant hydroelectric potential, but boasts moderately favorable navigable conditions. The third category is the water network on plains, including the middle and lower reaches of the mainstream and small- and medium-sized tributaries of the Yangtze River that directly discharge into rivers and lakes. This type of channel boasts good waterways and favorable navigable conditions but possesses a small drainage area and little hydroelectric potential.

2.6.1 River Channels in the Source Area of the Yangtze River

The river channel from Jianggendiru Glacier that is the source of the Tuotuo River to the Zhimenda Hydrologic Station on the Tongtian River is the source area of the Yangtze River. The section is 1,217 km in length, accounting for 19% of the total length of the Yangtze River; 137,700 km^2 in drainage area, accounting for 7.7% of the total area of the Yangtze River basin; and an average elevation of 4,500–5,000 m. The source area is bordered by the Shandong section of the Kunlun Mountains on the north; the middle and east sections of the Tanggula Mountains on the south; the Hoh Xil Mountain, Ulan Ula Mountain, and Zulkenwula Mountain on the west; and the Bayan Har Mountain on the east. The source area is entirely blocked by mountains on the south, north, and west with only the east open where there is a narrow pass from the Tongtian River canyon so that water can flow out toward the east-southeast. The source area is generally a basin-canyon landform surrounded by mountains. Moreover, the topography of the plateau is flat and broad with poor drainage conditions. As a result, large plateau marshes and fanlike water systems formed in the source area, including relatively large streams such as the Chumar River, the Tuotuo River, and the Dangqu River that possesses Gaerqu and Buqu creeks. The marshes are mainly located in the relatively wet Dangqu River basin in the eastern and southern source area.

The geomorphic characteristics of the river channels in the source area of the Yangtze River are closely related to the climate and hydrology of the area. To understand the characteristics of the rivers in the source area, the features of the regional climate, hydrology, and subsurface need to be analyzed.

2.6.1.1 Low Precipitation but High Water-Retention Capacity in the Source Area

In the source area, the average annual temperature is about 0 °C, and the precipitation is generally in the snowfall form. Due to the low temperature, the snowfall can remain on ground for a long time. In the mountains, some snowfall melts into the glacier or becomes a permanent snow cover. In the summer, the snowmelt flows partly into the river channels and into the palustrine wetlands in considerable amounts as well. Moreover, human water utilization and consumption are very little. Therefore, due to the development of the water systems and wetlands and rich water resources, the three-river source area is often known as "China's Water Tower," even though the precipitation is low in the source area.

2.6.1.2 Important Role of Permafrost in Water Circulation and River Evolution in the Source Area

The thickness of the permafrost in the source area of the Yangtze River may reach 30–120 m in the mountains and 10–40 m in plateaus and basins. Underground ice reserves are hundreds of billions of m^3 with 1–4 m of seasonal melting depth. The existence of the permafrost limits the downward cutting of the channels in the plateaus, and the permafrost plays the role of soil stabilization, channel bottom protection, and seepage control, which has rendered the river channel in the source area shallow and wide. The channel width ranges from hundreds of m to thousands of m. Thus, the melting and thawing of the permafrost have become typical geologic disasters in the source area. The abnormal melting of the permafrost can cause the groundwater level to decline, lead to the degradation or desertification of grassland, and, at the same time, can cause the riverbed to undergo accelerated downward cutting. Therefore, the source area is very sensitive and fragile to global warming, and the impact of global warming on the ecological and environmental system is enormous.

2.6.1.3 Characteristics of River Channels in the Source Area

The Tuotuo River, the true source of the Yangtze River, is 346 km in length and 17,600 km^2 in drainage area. The river has a U-shaped wandering shallow channel with crisscrossing tributaries and many sandbars with a braided flow. Across the channel are terraced banks, and smooth and rounded mountains covered with grass in the summer and snow in the winter. The elevation of the Tuotuo River ranges from 5,400 m at Jianggendiru Glacier, the origin of the Yangtze River, to 4,470 m at Rangjibalong (at the mouth into the Dangqu River) with an average gradient of 2.69‰. The Dangqu River, the south source of the Yangtze River, is 352 km in length and 30,786 km^2 in drainage area with an average gradient 1.65‰ that is much smaller than that of the Tuotuo River. Therefore, there are large alpine marshes and meadows in the water system. Due to the existence of permafrost, deep cutting of the river channel is restrained. As a result, the channel is shallow and wide, and the water flow is divergent and braided, demonstrating the typical characteristics of a wandering river and marshes in the plateau. The Chumar River, the north source of the Yangtze River, is 515 km in length and 20,800 km^2 in drainage area. When compared with the other two sources, the water system is undeveloped with many aeolian sand dunes. It mainly receives snowmelt from the southern slope of the Kunlun Mountains. There is not much vegetation on either bank, and soil erosion is serious. As a result, the riverbed and the flowing water appear red. There are not many tributaries, but streams are connected with lakes. The river channel is wide and shallow, and the water flow is divergent. There are many sandbanks throughout the Chumar River. Figure 2.5 shows the Moque Mountain standing on the riverside of the Tuotuo. The figure also shows the typical wide valley and rounded hilltops in the source area of the Yangtze River.

The Tongtian River, from Rangjibalong to the river mouth at Batang, is 828 km in length. The section between Rangjibalong to the mouth of the Chumar River is the source area of the Yangtze River and the upper reaches of the Tongtian River. This section is 278 km long with an average gradient 0.9‰. The river section is similar in channel shape to the Tuotuo River. When the river flows out of the canyon into the broad section, a wandering type river emerges; the water flow swings in the wide and shallow riverbed; and the channel is braided. The lower reaches of the

Fig. 2.5 Moque Mountain standing on the riverside of the Tuotuo

Tongtian River from the mouth of the Chumar River to the mouth of the Batang River is 550 km with gradients of 1.1–1.5‰. The river section is the canyon type in a mountain area with the channel width of 50–200 m. In the source area, the natural conditions are poor, and the ecological environment is fragile. Moreover, in recent years, human activities have increased in the vicinity of Qinghai-Tibet Highway, and disorderly sand quarrying is rampant in the river and on the banks along the Tongtian River. As a result, the problem associated with the ecological environment has become increasingly prominent. The area should be protected as a water source conservation area and a fragile ecological environment right now and in the future.

2.6.2 River Channel in the Upper Yangtze

At the mouth of the Batang River, the Tongtian River converges into the Jinsha River that had been known in ancient times as Shengshui or Lishui River because the sand in the river appeared to be yellow or to be gold until the Yuan Dynasty. The Jinsha River is 2,290 km long, accounting for 36% of the total length of the Yangtze River, and 362,000 km^2 in drainage area, accounting for 20% of the watershed area of the Yangtze River. The elevation difference of the Jinsha River is 3,300 m with an average gradient of 1.45‰. The mainstream of the Jinsha River runs through deep canyons in mountain areas; has a narrow riverbed, a turbulent and rapid flow, and a very varied flow direction; and possesses high hydropower potential. Its upper reaches are from the mouth of the Batang River to Shigu in Lijiang City, Yunnan; the middle reaches start at Shigu and end at Xinshi Township, Sichuan; and the lower reaches extend from Xinshi Township to the mouth of the Min River at Yibin.

The structure of the upper reaches of the Jinsha River can be characterized by parallel flowing main tributaries that converge to the mainstream of the Jinsha River with a similar flow direction. From the mouth of the Batang River to Dilikong, the deeply cut V-shaped river channel is straight with many rapids due to complex terrain conditions on its course, and the water surface is 60–80 m wide. From Dilikong to Dengke, the river channel is broad with floodlands, islands between braided branches, and sedimentary terraces, and the water surface is 150–200 m wide. From Dengke to Benzilan, the river channel becomes deeply cut again, and the water surface is 150–200 m in width. From Benzilan to Shigu, the river channel is broad again, and the water surface is 200 m in width. The deeply cut river channel runs through mountain areas where the valley reaches up to 1,500–2,000 m in depth.

In the middle reaches from Shigu to Lunan, the channel of the Jinsha River is broad. The Jinsha River enters the Tiger Leaping Gorge at Daju. The cliffy gorge consists of the Yulong Mountains with a peak elevation of 5,595 m on the right bank and the snowcapped Haba Mountains with a peak elevation of 5,396 m on the left bank. The difference in elevation between the highest mountain peak and the lowest bottom of the valley is 3,000 m. The gorge is 16 km long with an elevation difference of 220 m and an average gradient 13.8‰. At the narrowest section, the water surface is only 30 m wide, and the flow velocity reaches up to 10 m/s. This section of the Jinsha River has the highest elevation difference in the shortest segment. From where the river runs out of the Tiger Leaping Gorge to Sanjiangkou, the Jinsha River takes a sharp turn, which is known as "the First Bend of Yangtze River." Upstream of the mouth of the Pudu River is the most dangerous rapid, Laojun Rapid, of the Jinsha River. It is 4,600 m long and 41 m in elevation difference with an average gradient of 9.7‰ and the maximum flow velocity of 9.7 m/s. This is the largest obstacle on the Jinsha River course where the river channel changes from a large U-shape at Daju to a V-shape at Xinshi Town.

In the lower reaches of the Jinsha River, the river channel is broad, and the elevations of the banks are mostly below 500 m in a hilly area. Riverbed sedimentation is significant with gravel on the bottom. Broad terraces are present on the river banks, and the water surface is 150–200 m wide with a slow and stable water flow. The common features of the right bank tributaries in the lower reaches of the Jinsha River are the following: ① As for the change of the flow direction, there are many hook-shaped transitions, often called a hook water system. ② Some of the sources for these tributaries are lakes and water-diverting passes, and some of the tributaries have very obvious right-angle turns. ③ Some of the tributaries are suspected to have changed their directions of flow, such as the Longchuan and Xiaojiang Rivers. Therefore, in the tributaries from the right bank of the Jinsha River, captures may have occurred between the upstream and downstream sections at the right-angle turning points. The capturing rivers are the ones that changed their directions to converge into the Jinsha River at a right angle, and the captured rivers were the many original tributaries of the relatively large rivers that had flowed from the north to the south. Due to the impact of the north-south sloping topography of the Yunnan Plateau, the upper parts of these tributaries flow toward the south or toward the east. After the lower reaches of the Jinsha River channeled through, the intense deep cutting on the mainstream of the Jinsha River led to the development of the direction-changing rivers on the right bank.

The Jinsha River flows out of the mountain area into the western edge of the Sichuan Basin at Xinshi Town. The river section between Xinshi Town and Yibin is 109 km long and transitions from a canyon channel in the mountain area to a wide and shallow channel in the hilly area. The water surface of the section widens from 100 m to 400 m with the widest at 500 m, and the gradient of the water surface changes from 1.0‰ upstream of Xinshi Town to 0.44‰. Due to the low velocity of the water flow, many relatively large areas, up to 1–2 km long, of sandbanks consisting of gravels and cobbles have formed, and even three large mid-river sandbars have developed.

The Chuan River is 1,040 km long from Yibin to Yichang with an average river gradient of about 2‰ and a drainage area of 530,000 km^2. The river channel is alternately wide and narrow along its course, which is typical of the hilly areas. The water surface of the canyon section is 200–300 m in width with the narrowest section at only 100 m wide. The water surface in the broad valley section is 600–800 m wide with the widest up to 1,500–2,000 m. There are many sandbars and sandbanks in the section of the river, of which 96 are large- and medium-sized ones. On average, there is one large- or medium-sized sandbar or sandbanks every 4 km with the largest up to 2–3 km in length. There are also 18 mid-channel islets and 10 mid-channel sandbars. The largest mid-river sandbar is about 1,000 m wide and 2–3 km long. Most of these sandbars and islets are the spawning sites and habitats for rare and endemic fish species in the upper reaches, and some have developed into islands and have been reclaimed as farmland.

The most prominent features of the structure of the Chuan River system and its tributaries are that the mainstreams of its main tributaries are mostly capturing streams and the secondary tributaries to the main tributaries may be captured rivers, especially the secondary tributaries to the main tributaries on the right bank of the Chuan River, which developed mostly following structural alignments and formed earlier. The mainstream of the tributaries on the right bank of the Chuan River might form relatively later. However, on the left bank of the Chuan River, secondary tributaries mostly converge into the main tributaries in a similar flow direction, and only the main tributaries in the source area may have captured eastward flowing secondary tributaries. Therefore, the main tributaries on the left bank of the Chuan River should be the ones that formed relatively earlier, and the fact that they converge into the Chuan River in a reversed flow direction indicates the Chuan River had once flowed in an opposite direction. These features also indicate that the Chuan River used to flow westward before the Three Gorges area was channeled through and there is evidence indicating the tributaries on the right bank of the Chuan River had flowed southward into the South China Sea.

2.6.3 River Channel in the Middle Yangtze

Before the construction of dikes in the middle and lower reaches of the Yangtze River, the low-lying areas had been floodplains. Historically, they had been all marshes, such as Zhengze Marsh in southern Jiangsu, Pengli Marsh in north Jianxi, and Yunmeng Marsh in the Jianghan Plain. Some of these large marshes have become part of lakes, and some have been converted by sediment deposition into lakeside plains, most of which have been reclaimed into polders or paddy fields by the construction of berms.

Some of the existing sandbars and sandbanks in the mainstream of the middle and lower reaches of the Yangtze River formed earlier, but most of them formed during the Tang and Song Dynasties. Especially since the Ming and Qing Dynasties, the formation, expansion, integration, and rise of sandbars and sandbanks have had a certain relationship with the construction of dikes along the Yangtze River. The dikes have limited the development of floodplains and have inevitably led to the gradual evolution of the sandbars and sandbanks between the dikes along both banks.

Over the past 2 ka, due to the construction of dikes in the middle and lower reaches of the Yangtze River, sediments from the upper reaches could not be naturally transported into floodplains. As a result, the mainstream of the Yangtze River has been raised by about 20 m, and various changes have occurred such as the development of meandering channels, sandbars, and sandbanks, river channel braiding, and the formation of lakes at the mouths of tributaries behind the dikes. Under normal conditions, the normal thickness of sediment deposition in floodplains should be less than the difference between the floodwater level and the low water level of the river section. However, a comparison of the measured maximum variation of the annual water levels and the thickness data of the known Holocene floodplain deposits in the middle and lower reaches of the Yangtze River in the past 5 ka indicates the latter is much higher than the recently observed maximum annual variation of water levels, and due to the construction and reinforcement of the dikes along the river, land reclamation from lakes along the shores and the removal of vegetation, the maximum annual fluctuation of water levels in the middle and lower reaches of the Yangtze River, has substantially higher than that in the past (Yang 1989). Although the constructed dikes have improved the river channel's storage capacity of floodwater, they have increased the annual fluctuation of the water level and the floodwater level of the river. While the dikes have provided human beings with safety and large area of farmland, they have increased the risk of flood hazards, which constitutes a conflict that is difficult to reconcile.

The middle and lower reaches of the Yangtze River are closely related to lakes. In the past, all lakes had been connected with the Yangtze River, and the water in the lakes were mainly from the flow of the tributaries or floodwater during the wet season. The water of the lakes had eventually drained into the Yangtze River. The cycle of lake water exchange was relatively short and was 59, 20, 264, and 127 days for Poyang, Dongting, Tai, and Chao Lakes, respectively. Due to the connection to the Yangtze River, the first two lakes had much shorter water exchange cycles than the last two lakes. Because of the construction of dikes and large river regulation projects in the recent 2 ka, the river channel of the middle and lower reaches of the Yangtze River has been largely stable, and the lakes, except Dongting and Poyang, in the middle and lower of the Yangtze River are basically separated from the river or connected only seasonally by sluices or outlets.

The middle and lower reaches of the Yangtze River are about 1,900 km long from Yichang to the estuary, of which the section between Yichang to Zhicheng is a straight or slightly curvy transition from the mountainous channel to the alluvial plain waterway. The river channel from Zhicheng to Ouchikou, also known as the Upper Jingjiang River, is curvy. The section from Ouchikou to Chenglingji, also known as the lower Jingjiang River, is of a typical meandering type. The channel from Chenglingji to Xuliujing is of a braided type. The channel downstream of Xuliujing is the estuary section of the river. At present, there are more than 40 braided river sections from Chenglingji to Jiangyin downstream of Zhenjiang, and the combined length accounts to 70% of the total length of the Yangtze River. The braiding is one of the significant features of the river channel in the middle and lower Yangtze.

Although flood often threatens human beings, it plays an important role in maintaining the flow capacity of the river channel, transport of sediments and nutrients to the lower reaches and the estuary, and generation of floodland habitats, and it is necessary to maintain the flood process at a certain scale and frequency for the ecology and environment of the river. The channel-forming flowrate (also known as the flowrate with a corresponding water level of the floodplain elevation) in the middle reaches of the Yangtze River is 29,000 m^3/s from Yichang to Zhicheng, 27,000 m^3/s from Zhicheng to Shashi due to diversion to Songzi and Taipingkou, 22,000 m^3/s at Jianli due to diversion to Ouchikou, 36,000 m^3/s from Luoshan to Hankou due to the inflow from Dongting Lake, 40,000 m^3/s downstream of Hankou due to the inflow from the Han River, and 45,000 m^3/s near Datong due to the inflow from Poyang Lake. In the Yangtze River, the channel-forming discharge is close to the flowrate with a corresponding water level of the floodplain elevation. It is a basic requirement to maintain the channel-forming flowrate in the mainstream of the Yangtze River for the health of the river. If floodwater does not reach the floodplain for many years, not only the river channel will shrink but also the channel's capacity to discharge and store floodwater will be severely affected. Moreover, it is unfavorable to the ecological environment of the riparian wetlands.

The mainstream in the middle reaches of the Yangtze River includes the following 16 sections: Yizhi, upper Jingjiang River, lower Jingjiang River, Yueyang, Jiayu, Paizhou, Wuhan, Yejiazhou, Tuanfeng, Huangzhou, Daijiazhou, Huangshi, Weiyuankou, Tianjia Town, and Longping.

2.6.3.1 Yizhi Section

The river section between Yichang and Zhicheng in the middle Yangtze is about 61 km long. Restricted by the low mountains on both sides, the channel is straight or slightly curvy and consists of the straight segment from Yichang to Yunchi and three curvy segments of Yidou, Baiyangdian, and Zhicheng. The Yichang-Zhicheng segment is close to and downstream of the Gezhouba Dam. Due to the impact of the back water from the Qing River, the mid-river Huya sandbar formed as a node. This node divides the channel into the upper and lower subsegments: Yichang subsegment from the Gezhouba Project to Huya sandbar and Yidou subsegment from Huya sandbar to Zhicheng. The channel from the Gezhouba Project to Huya sandbar is 24 km long and is the only spawning site for the Chinese sturgeon. Along both sides of the channel are low mountains and hilly areas, and the channel is straight to slightly curvy. The upper subsegment from Zhenchuanmen to Wanshou Bridge is slightly curvy, and Yichang sandbank is on the left bank, and the deep channel follows the right bank. The middle subsegment from Wanshou Bridge to the tail of the Yinzhi Reservoir is straight and braided with the mid-river Yinzhi sandbar, and the deep channel is near the left bank. When flood comes, the sandbar is inundated, and when the flow is intermediate, it is exposed above the water to divide the Yangtze River into the left and right branches. The left branch is the primary channel, and the right branch is the secondary one. During the dry season, the right branch dries. The lower subsegment from the tail of the Yinzhi Reservoir to Huya sandbar is straight. Linjiang sandbank is on the left bank, and the deep channel is near the right bank. The Yidou segment from Huya sandbar to Zhicheng Bridge is about 37 km long. The tributary Qing River converges from the curvy Yidu segment on the right bank.

This segment can be divided into Yuchi and Yidu subsegments. The upper Yunchi subsegment is straight, and the channel is 900–1,200 m wide. The mainstream runs adjacent to the left bank near the upper end of the Huya sandbar and gradually transitions to the right bank at Yangjiafan. The mainstream follows the right bank until it is near Yunchi Port and then switches to the left bank at Yanbanchong. The deep channel continues into the lower Chadian subsegment where the Qing River converges from the right bank. The Nanyang moraine is located in the middle of the river. When the water flow is in the low and medium range, the mainstream follows the left branch. During flooding, the mainstream follows the right branch.

2.6.3.2 Upper Jingjiang River

The upper Jingjiang River extends from Zhicheng to Ouchikou with a length of 175 km, and the river channel is slightly curvy and braided. It consists of three north-facing curvy segments of Jiangkou, Shashi, and Haoxue and three south-facing curvy segments of Yangxi, Woshi, and Gongan and straight transitional segments between curvy segments. There are mid-river sandbars within the curvy segments. The sandbars include the following, from upstream to downstream, Guan, Dongshi, Liutiao, Jiangkou, Huojian, Mayang, Sanba, Jincheng, Tuiqi, etc. The average curvature of the channel is 1.72. The smallest radius of curvy river segments is 3,040 m and the largest 10,300 m. The widest channel is located at the upper tip of Tuqi sandbar and is about 3,000 m, and the narrowest channel is located at Tieniuji in Haoxue where it is only 740 m wide. The gradient of the water surface ranges from 0.04‰ to 0.06‰. The gradient is relatively higher during the wet season and relatively lower during the dry season. On both sides of the upper Jingjiang River are dikes. Along the south bank is the Main Yangtze Dike, and along the north bank is the Jingjiang Dike. In the curvy segments in Shashi and Haoxue, there is no or a narrow floodplain between the dike and the channel where the deep mainstream runs near the bank. As a result, the flood control situation is very grim.

2.6.3.3 Lower Jingjiang River

The meandering lower Jingjiang River extends from Ouchikou to Chenglingji and is about 173 km long. The riverbed is composed of medium and fine sands underlain by a gravel layer extending to a great depth below the riverbed surface. The right bank consists of hilly terraces with strong resistance to scouring, and the left bank is alluvial plains and consists of an upper clayey soil layer underlain by a sand layer with weak scouring resistance. The main evolutionary features of the river channel can be described as follows. The curvy concave bank continues to collapse with slope failures, and the convex bank remains to experience sedimentary deposition. As a result, the top of the curve continuously moves toward downstream. When curvy segments have developed to a certain degree, natural meander cutoff occurs under certain flow, sediment, and riverbed boundary conditions. In the past 100 years, natural meander cutoff events have occurred at many locations. From the late 1960s to the early 1970s, artificial meander cutoff projects were implemented at Zhongzhouzi and Shangchewan, and a natural meander cutoff event occurred at Shatanzi, which shortened the channel by 78 km. The meander cutoff projects have somewhat improved the flood control capability and navigable conditions in the Jingjiang area. However, after the meander cutoff events, due to various reasons and the inability to control the favorable river regime in time, bank failures occurred frequently in the lower Jingjiang River, and the river channel was lengthened. Consequently, the amounts of water and sediments from Sankou to Dongting Lake were reduced. In 1983, a systematic river regime control project began in the lower Jingjiang River. After more than 10 years of the implementation of the river regime control project, most of the seriously failing bank segments were preliminarily controlled, and the overall river regime became preliminarily stable. However, bank failures in some river segments were still serious, affecting flood control and navigation safety. For example, in 1994, the river channeled through the Xiangjia sandbar in the Shishou river segment, causing the river regime to adjust. As a result, severe bank failures occurred at the north gate of Shishou City, threatening the safety of the dike. After the 1998 large flood of the Yangtze River, the Central Government stepped up the regulation of large rivers. Since the implementation of the Yangtze River Dike Project during 1999–2002, the overall river regime of the lower Jingjiang River has maintained stable.

2.6.3.4 Yueyang Section

The Yueyang section follows the lower Jingjiang River section and connects to the downstream Luxikou section with a total length of 77 km. The section consists of the upper segment of Chengyang (from Chenglingji to Yanglinshan), middle segment of Jiepai (from Yanglinshan to Shimatou), and the lower segment (from Shimatou to Chibishan). This section is between the Jingjiang River and the outflow exit of Dongting Lake in the middle reaches of the Yangtze River and runs through Yunxi District of Yueyang City and Linxiang City, Hunan and Jianli County, Honghu City, and Chibi City, Hubei. The section is straight and braided. The channel is alternately narrow and wide with a lotus root shape. On the left bank are vast alluvial plains, and the right bank are intermittent low mountains and hilly areas. The bedrock outcrop along the Yangtze River has formed the prominent nodes at Chenglingji, Bailuoji-Daorenji, Yanglinshan-Longtoushan, Luoshan-Yalan, and Chibi and has played a major role in the shape of the river channel and the stability of the river regime. In most wide river segments, there are mid-river sandbars, including the following (from upstream to downstream): Xianfeng, Nanyang, Xinyu, and Nanmen.

2.6.3.5 Luxikou Section

The approximately 23.3-km-long Luxikou section extends from Chibishan to Shijitou. The section has a typical goose head-type braided channel and consists of the upper, middle widening, and lower segments. The upper segment is slightly curvy. From Junminjie, the channel starts to widen where the mid-river Xinzhou and Zhongzhou sandbars formed. Near Caijiadun, the channel starts to braid where Xinzhou and Zhongzhou sandbars divide the river into the left, middle, and right branches. The right branch is relatively straight; the upper segment of the middle branch flows northeastward. To the entrance of the left branch, the flow gradually turns toward east-southeast and merges with the right branch at Liujiapeng. The left branch flows north 30° west and runs out at an almost right angle with the influx water. Then the flow of the left branch gradually turns to the right, and after running around the Zhongzhou sandbar for more than a half circle, it runs at an angle of south 25° west near the exit and converges into the middle branch. The flow direction of the left branch has a great change at the inlet, up to 240°. The Zhongzhou sandbar is located between the left and middle branches, and the right branch is near the right bank and flows north 70° east. After the tributary Lu River (Hongmiao River) converges, the right branch flows through Luxikou to Liujiapeng and mergers to the middle branch. The merged right and middle branches become a single slightly curvy channel that flows eastward toward the downstream.

2.6.3.6 Jiayu Section

The 30.2-km-long Jiayu section extends from Shijitou to Panjiawan and is of a typical straight, slightly curvy, and braided type. Huxian, Baisha, and Fuxing sandbars are located in a straight alignment in the middle of the river. The left branch, the primary branch, is relatively stable and has been widened due to scouring in recent years. The right branch, the secondary branch, is in a declining state due to sedimentation. The right branch of the Fuxing sandbar only flows during the flood season and does not flow during the medium or low water seasons.

2.6.3.7 Paizhouwan Section

The 72-km-long Paizhouwan section extends from Panjiawan of Jiayu County, Hubei, to Shamaoshan in Caidian District of Wuhan City. The plan view of the channel is an "Ω" shape. Its neck of the curvy segment (from Huakou to Shuangyao) is only 4 km wide, and the curvature of the segment is 15.9. The mid-river Tuanzhou sandbar is located in this segment, and the elevation at the upper tip of the sandbar is 26–28 m and lower tip 22–24 m. The ground surface of the sandbar is mainly muddy loam. The left branch is the primary channel, and the right branch (Paizhoujia Waterway) is the secondary one. During the dry season, the flow through the right branch is very small, but it diverts a relatively large flow during the flood season.

2.6.3.8 Wuhan Section

The approximately 70.3-km-long Wuhan section is from Shamaoshan of Caidian District to Yangluozhen of Xinzhou District of Wuhan City. The water in the section flows in the southwest-northeast direction. The roughly 19.9-km-long upper segment from Shamaoshan to Dunkou is straight; the approximately 11.1-km-long middle segment from Dunkou to the Yangtze River Bridge is also straight; and the nearly 35.3-km-long lower segment from the Yangtze River Bridge to Yangluo is slightly curvy and braided with the mid-river Tianxing sandbar. Currently, the right branch is the primary channel, and the left branch is the secondary one.

The mainstream of the section flows from the Tieban sandbar to the vicinity of Longchuanji where it converges. Thereafter, it flows in the main channel to the vicinity of Xiaojunshan where the deep main channel follows the right bank. Near Dunkou the deep main channel transitions from the right to the left and continues along the left branch of the Baisha and Qianzhou sandbars. After the Yangtze River Bridge, the deep main channel follows the Wuchang deep channel along the right bank to the proximity of Xujiapeng where it flows through the right branch of the Tianxing sandbar.

2.6.3.9 Yejiazhou Section

The approximately 28-km-long curvy Yejiazhou section is from Yangluo to Niji. Its evolution is mainly manifested in the changes of scouring and sediment deposition in the Mu'e sandbank, and the change of the main channel is small. In 1953, the Mu'e sandbar connected to the bank of the river to form a sandbank, which transformed from a curvy braided river into a slightly curvy single channel. In the past 50 years, the river regime has been relatively stable with little change in the mainstream, banks, or the deep main channel. The Yejiazhou section has not changed much since 1930. Houziji and Guanyin Hill on the right bank force the river flow to the left bank, resulting in bank failures in the areas of Yejiazhou and Wangjiapu. Since the riprap revetment, the bank conditions have improved. The main change of this section is on the Mu'e sandbank. The 1953 survey map indicates that the Mu'e sandbar was basically connected to the left bank to become a sandbank, and the river was still maintained a single channel. Since 1949, although the sandbank in this section has changed its shape, the main channel has still been slightly curvy with good flow conditions.

2.6.3.10 Tuanfeng Section

The approximately 28.8-km-long Tuanfeng section is from Niji to Sanjiangkou and has a three-branch braided channel. The channel is wide and shallow with multiple branches, and the deep mainstream has changed greatly. The braided Tuanfeng section has a goose-head plan view, and the mainstream swings cyclically and repeatedly between the right and middle branches. After the confluence, the mainstream continues to swing around. Its evolution is mainly manifested by the change of sandbars. During the course of its evolution, sandbars are separated and integrated frequently. During 1976–1981, Lijia and Dongcao sandbars merged together during the dry season, and a small new sandbar, named Xinyu sandbar, emerged between Dongcao and Luohu sandbars due to sediment deposition. As a result, the following five sandbars coexisted in the river section: Renmin, Lijia, Dongcao, Xinyu, and Luohu. The primary channel is located between Renmin and Lijia sandbars. Between 1981 and 1987, the right branch continued its development, and its width expanded from 900 m to 1,600 m, while the original middle branch gradually became narrower. Since 1987, the right branch has been gradually moving toward the left. The Tuanfeng section is one of the braided sections in the middle and lower reaches of the Yangtze River with greatest changes in sandbars. In the past 100 years, the goose-head-shaped river channel in the section has not changed much, but the mainstream has swung repeatedly and cyclically between the right branch and the middle branch. Consequently, the channel and sandbars have changed accordingly.

2.6.3.11 Huangzhou Section

The approximately 29.5-km-long Huangzhou section is from Huangboshan to Yangji and has a slightly curvy braided channel. The left bank of the section is a stable curvy convex and is far away from the mainstream. The right bank has the following natural nodes (from upstream to downstream), Huangboshan, Xishan, Longwangji, Yanji, etc., which play a controlling role on the river bank. The following bank revetment projects on the concave side have been implemented between the nodes of Huangboshan and Xishan: Liuchuxian, Zhengjiawan, Panjiawan, and Yanglanduan (full length of 12.52 km). After years of construction (reinforcement and modification), the reinforced bank segments became stable. In general, the banks of the Huangzhou section are stable.

Recent changes of the river regime in the Huangzhou section are mainly as follows. The mainstream in the curvy segments has the tendency to move leftward but still follows the middle line during the high-water period and runs near the right bank during medium- and low-water seasons. The swinging of the mainstream in this section is influenced by the alternation of the primary and secondary branches in the upstream Tuanfeng section. Due to the impact of the swinging of the mainstream, the near-bank deep channel in curvy segments experienced sedimentation during high-water periods and scouring during medium- and low-water seasons. In the area of the Desheng sandbar-Huangzhou sandbank, sandbar cutting occurred during high-water periods, and sediment deposition occurred in low-lying areas to cause the merging of the sandbar with the sandbank during medium- and low-water seasons. Because of the effects of the swinging of the right branch toward the left bank in the upstream Tuanfeng section, the mainstream in the straight segment at the beginning of the Huangzhou section shifted to the left bank. The mainstream in the curvy segments swung to the concave bank, resulting in a stronger deflecting effect of the Xishan node on the flow, and the mainstream in the straight segment at the end of the Huangzhou section swung to the left. As a result, the Chihuqian sandbar lengthened, and the right branch shrank due to sediment deposition.

2.6.3.12 Daijiazhou Section

The approximately 21.6-km-long Daijiazhou section extends from Yanji to Huifengji with a mid-river sandbar – Daijia sandbar – and has a slightly curvy braided channel. The left and right banks have the following bedrock promontories: Fengji, Yanji, Guafuji, and Pingshanji. The right branch is the primary channel, and the left branch is the secondary one. The Xinyu sandbar is located at the exit of the right branch of this section. The mainstream of the section follows the right bank along the deep channel from Yanji to the right branch of the Daijia sandbar and starts to swing leftward at the end of the Yuzhou sandbar to Huifengji. The left branch of the Daijia sandbar continues its slow development trend, and the present diversion ratio has exceeded 40% under certain flow conditions. In recent years, new changes have taken place in the diversion area of the Daijia sandbar, and the sandbank on the left side in the diversion area has undergone changes because it has been subjected to both scouring and sediment deposition, making the deep channel of the left branch sway around. Floodlands on the left side at the beginning segment and in middle and lower segments of the right branch have kept altering as a result of scouring and sediment deposition, making the mainstream swing obviously and the deep channel disappear. Thus, the channel of the right branch has become a wide and shallow rectangular cross section, which is very unfavorable for the stability of the right branch. After the 1998 flood, a large bank revetment project was implemented on the bank of the left branch for the river segment downstream of the Ba River. The project has improved the stability of the bank of the left branch, maintained the left deep channel of the left branch to continue expanding through scouring. The stable banks have made the left channel more stable. The above evolutionary signs suggest that the upper tip of the Daijia sandbar may suffer further shrinking. Once this change occurs, the stability of the branches is in an extremely unfavorable condition. Although currently no substantial changes have occurred to the branches at the Daijia sandbar, the left branch will continue to expand, and the right branch will keep declining at the Daijia sandbar. In recent years, the dry season has been relatively longer and the water level relatively lower; measures should be taken in time to prevent changes that would adversely impact the navigable waterway.

2.6.3.13 Huangshi Section

The approximately 15.5-km-long Huangshi section extends from Huifengji on the left bank of the Yangtze River to Xisaishan on the right bank of the river and has a single slightly curvy channel. At the beginning and the end of the section, there are two bedrock promontories, Huifengji and Xisaishan, which have a strong controlling effect on the water flow. The right bank of the curvy section is a concave subject to the impact from the water flow and has the following bedrock promontories, Mao'er, Doucheng, Haiguan, and Huangshijiao, that have formed an arc and make the Huangshi section a stable curvy channel. The curvy section is about 900–1,200 m wide. The left bank has the broad Shanhua sandbank. The maximum width of the sandbank is about 700 m. The right bank consists of bedrock promontories, and the bank slope has been stable for years. The mainstream of the section starts at Huifengji on the left bank and gradually transitions to the right bank in Huangshigang District into the curvy Huangshi segment; and it follows the right bank of Huangshi City through Huangshi Port, Zhongyao sandbank (Shihuiyao dike section), and Daye Steel Plant Port (Huangsiwan dike section) to Xisaishan.

In the past decades, the river banks of the Huangshi section have been stable, and the deep channel and the mainstream have not changed much. While sandbars and sandbanks have experienced changes due to scouring and sediment deposition, the changes have not tilted to either single way. The implementation of bank revetment projects has made the banks of the section more stable. Moreover, because the water depth has been within a favorable range and there are no shallows, reefs, or other adverse conditions, Huangshi has become a robust natural harbor. However, both the deep channel and the dominant flow run adjacent to the bank, making the bank slopes relatively steeper and vulnerable to slope failure due to long-term scouring. Furthermore, there is no floodplain outside most of the dikes, which is a direct threat to the safety of the dikes. The dikes within Huangshi City are of the first class. However, lakes are present inside some sections of the dikes, and obstacles are present on the floodplain outside some sections of the dikes. As a result, the flood control situation is not optimistic.

2.6.3.14 Weiyuankou Section

The approximately 33.3-km-long Weiyuankou section extends from Xisaishan to Houerji with a typical slightly curvy braided channel. The section is about 800 m wide at the beginning near Xisaishan and about 920 m wide at the end of Houerji, but it is 1–3 km wide along its course with alternating wide and narrow segments. There are hills and natural bedrock promontories along the banks, such as Fengtouji on the left bank and Xisaishan Hill, Dahuoshan Hill, and Houerji on the right bank. The hills and natural bedrock promontories largely control the plan view of the river shape. The upper segment of the section from Xisaishan to Weiyuankou has a braided channel with the mid-river Guniu sandbar dividing the channel into two branches. The left branch is the primary channel, and the right branch is the secondary one. The middle segment of the section from Weiyuankou to Lijiazhou has a narrow river channel with widths ranging from 0.9 km to 1.1 km. The channel is deep with a rapid flow, and the river banks are stable. The lower segment of the section from Lijiazhou to Houerji is of a wide and shallow braided type with the mid-river Qizhou underwater sandbar. The mainstream of the section runs along the right bank through Xisaishan, and then, due to the impact of deflecting effects from nodes, the mainstream shifts to the left bank at Maoshan Gang and continues along the left bank through Qizhou Town until it swings back to the right bank at Huangsangkou and runs out of the section.

Historically, the river regime for the Weiyuankou section has been stable, and the mainstream has basically remained unchanged. The river banks have been relatively stable, and the plan view of the channel has had relatively small variations. In the past 20+ years, the deep channel of the section has been relatively stable with the exception that the deep channel in local transition or widening segments has swung at a relatively large range. Under the control of the natural nodes of hills and bedrock promontories, and bank revetment projects, the river banks of the Weiyuankou section have been relatively stable recently.

2.6.3.15 Tianjiazhen Section

The approximately 34.3-km-long Tianjiazhen section extends from Houerji to Matouzhen. The river from Houerji to Banbishan has a single channel with a length of 15.3 km. Both banks consist mostly of low mountains and hills. The following hills and bedrock promontories are on the banks: Laoyingzui and Niguanzui on the left bank and Houerji, Jianfeng Hill, and Yujia Hill on the right bank. The river section is narrow and deep with alternating wide and narrow segments and has an average width of only 1,000 m. The longitudinal profile of the riverbed has been relatively stable due to the special terrain and boundary conditions, and the riverbed can hardly widen because the river flows between bedrock banks through this section. Under the vertical vortex flow, the riverbed is cut deep downward, resulting in the elevation of the river bottom down to almost −100.0 m near Makou that is the deepest spot in the middle reaches of the Yangtze River. The segment from Banbishan to Matouzhen, called the Liyuzhou braided segment, is slightly curvy to the right and has a length of about 19 km. The segment is narrow at ends and wide between. At the beginning of the segment, Fengjiashan Hill is on the left bank and Banbishan Hill on the right bank with a channel width of about 680 m. In the middle of the segment, the maximum width reaches 2,300 m. At the end of the segment, Goutouji (a bedrock promontory) is on the right bank, and the river width is about 1,200 m. The Fu River, a tributary of the Yangtze River, converges from the right branch of the segment where a control sluice has been installed. The mainstream of the section follows the right bank through Jianfengshan Hill. Thereafter, due to the impact of the deflection from the hill, the mainstream transitions to the left bank at Makou. After Makou, due to the impact of Nieguanji, the mainstream gradually shifts to the right bank at Banbishan Hill. Owing to the deflection of Banbishan Hill, the water flow in the mainstream shifts to the left bank. After encountering the hill on the left bank, the water flow in the mainstream follows the left branch of the Liyu sandbar. After joining the right branch, the water flows toward downstream out of the section.

Many natural bedrock hills on both banks of the Tianjiazhen section have created very favorable boundary conditions for restricting the change of the river channel. Therefore, the shape of the riverbed in the section has historically undergone relatively small changes, and the evolution of the riverbed is mainly manifested in the formation and development of mid-river sandbars in wide segments. The shape of the channel in the Liyuzhou segment has remained basically the same in the last 100 years. The shape, size, and position of the Liyu sandbar have basically remained unchanged. The recent evolution shows that because of the control of many nodes on both banks of the Tianjiazhen section, the lateral swinging of the channel has been restricted. While the riverbed has undergone some changes due to scouring and sedimentation, the shape of the channel's cross section has basically experienced no changes, and the positions of the sandbars, floodlands, river banks, and the deep channel have largely remained stable.

2.6.3.16 Longping Section

The approximately 19-km-long Longping section extends from Matouzhen to Dashuxia and is a Class II regulation section of the middle and lower reaches of the Yangtze River. It has a goose-head-shaped plan view and is a braided river section with mid-river sandbars of Yaertan and Xinzhou. The channel of the section at Matouzhen is about 1.2 km wide and gradually widens to 2.22 km downstream of Goutouji. It reaches 2.41 km at the upper tip of the Xinzhou sandbar and narrows to 2.04 km at the lower tip of the sandbar. The floodplains between the normal water surface and the dikes are relatively narrow, but the shallows under the water surface have developed well and have low resistance against scouring. The segment upstream of Longping is the mid-river Yaerzhou segment, and the one downstream of Longping is the Xinzhou braided segment. The right branch of the Xinzhou segment is the primary channel that diverts about 75–89% of the river flow in the past years. The channel and floodplain of the Yaerzhou segment have remained relatively stable over the past years. In the Xinzhou segment, the left branch has remained as the secondary channel and the right branch the primary channel. However, the secondary channel has slightly declined, and the primary channel has slightly developed recently.

2.6.4 *Channel of the Lower Yangtze*

The lower reaches of the Yangtze River include 15 river sections: Jiujiang, Madang, Dongliu, Anqing, Taiziji, Guichi, Datong, Tongling, Heshazhou, Wuyu, Maanshan, Nanjing, Zhenyang, Yangzhong, and Chengtong.

2.6.4.1 Jiujiang Section

The Jiujiang section is from Dashuxia to Xiaogushan and approximately 90.7 km long. The section has a slightly curvy braided channel and consists of Renminzhou (also known as Danjiazhou or Bianyuzhou) braided, Jiujiang curvy, Zhangjiazhou braided, and Shangsanhaozhou-Xiasanhaozhou braided segments.

The Renminzhou segment extends from Dashuxia to Suojianglou and is divided into Renminzhou braided subsegment and Jiujiang curvy subsegment. The channel is slightly curvy with the right branch as the primary. The 2007 hydrological data indicates that the diversion ratio of the left branch of the Renminzhou braided subsegment was 2.48% in the dry season and 25.6% during the normal water period.

The curvy Jiujiang subsegment has a large curvature with the concave toward the south.

The Zhangjiangzhou braided segment is from Suojianglou to Balijiangr. The left branch is relatively curvy, while the right branch is relatively straight. Before the mid-1970s, the left branch had diverted slightly more water than the right branch, but recently the right branch has diverted slightly more than the left branch. Poyang Lake discharges into the lower part of the right branch. The flow path of the southern (or right) branch of the segment is relatively short. Due to the combined impact of the narrowing entrance, backwater from Poyang lake, and inflow and associated sediment contents, the diversion ratio between the two branches has remained relatively stable for a long time. After the 1998 and 1999 floods, the diversion ratio of the south branch increased to 50–60%. At present, the flowrate is large, and the diversion ratio for the two branches is quite equal during the flood season. However, during the normal- and low-water seasons, the diversion ratio of the south branch is about 60%.

The 14.5-km-long Shangsanhaozhou-Xiasanhaozhou braided segment extends from Balijian to Xiaogushan. The mainstream flows between the upper and lower Sanhao sandbars. The left branch of the upper Sanhao sandbar is curvy and secondary, while the right branch is straight and primary. The diversion ratio of the left branch has been decreasing and was about 5.5% in 1979. With the narrowing of the entrance of the branch, the diversion ratio was reduced to 2.4% in 1992 and about 0.24% in 1997. The left branch is basically dry during the dry season. However, the diversion ratio of the right branch has been increasing.

The left branch of the lower Sanhao sandbar is primary, and the right branch is secondary. The diversion ratio of the left branch has been increasing. Since the 1980s, due to the upward trend of the diversion ratio of the right branch, the tip and left edge of the lower Sanhao sandbar have been collapsing, and the entrance area of the left branch has been increasing. With improved flow conditions, the diversion ratio of the left branch was up to 86.8% in 2003.

2.6.4.2 Madang Section

The approximately 31.4-km-long Madang section extends from Xiaogushan to the mouth of Huayang River and has narrow beginning and end segments and a slightly curvy wide braided middle segment. The channel is divided into two branches by the Gepai sandbar with the left as the secondary channel and the right as the primary channel. The left branch is long and curvy, and water flows on the left bank year-round. The left branch is divided into two sub-branches by the Tiesha sandbar. With the diversion to the left branch decreasing, the Tiesha sandbar expands toward the right due to sedimentation and the friction to the flow increases along the branch, resulting in the gradual decline of the left branch. The right branch of the section starts at the tip of the Gepai sandbar and ends at the tail of the Guazihao sandbar with a total length of 16 km. It is straight to slightly curvy and has mid-river sandbars of Mianwai and Guazihao. Before 1995, the diversion ratio of the left branch of the Gepai sandbar had been relatively stable, basically remaining at about 27%, but it has been decreasing in recent years and was 13.6% in July 2007 and as low as less than 5.0% in January and October 2007 and March 2008.

2.6.4.3 Dongliu Section

The approximately 34-km-long Dongliu section extends from the mouth of the Huayang to Jiyangji and has a straight braided channel with multiple branches. Due to the frequently changing conditions of scouring and sedimentation, the mainstream swings and the navigable channel vary. The following mid-river sandbars are located in the section (from upstream to downstream) Laohutan, Tiansha, Yudai, and Mianhua, which divide the channel into the following three branches (from left to right): Lainhua Zhougang, Xigang, and Donggang. The Lainhua Zhougang branch is the primary channel during the flood season; the Xigang branch is the primary channel during the dry season; and the Donggang branch is always a secondary channel. The channel downstream of Hudongcun is narrow, deep, and slightly curvy where the banks are relatively stable.

2.6.4.4 Anqing Section

The 57-km-long Anqing section is located between Jiyangji to Qianjiangkou within Anhui Province and is listed on the Class I priority regulation. The section is divided into the following two segments by the convergence point of the Wan River at the West Gate of Anqing (from upstream to downstream), Guangzhou and Anqing, which are 32 km and 25 km long, respectively.

The Guangzhou and Anqing segments are both braided and connected with a single-channel segment. The Guangzhou segment is a goose-head braided type typical of the lower reaches of the Yangtze River. The beginning at Jiyangji is only 900+ m wide, and it widens gradually and reaches 8 km wide in the middle of the segment with two sandbars (Xinchang and Qingjie) and three branches (Dongjiang, Xinzhongcha, and Xijianjiang). The mainstream of the Dongjiang branch converges with Xinjiangcha and Nanjiajiang branches at the tail of the Guangzhou segment. The mainstream shifts from the left bank to the right bank at Yangjiatao and continues to Xiaozhamen. Then it shifts to the left bank and continues to the West Gate of Anqing where it runs along the river toward the downstream. Presently, the Dongjiang branch is the primary channel with an average diversion ratio over 80% for years, and the other two branches are the secondary ones.

The Jiangxinzhou branch (also known as the Emei branch) of the Anqing segment is a slightly curvy braided type with the Emei sandbar and the Jiangxin sandbar obliquely situated. The two sandbars are integrated into one during the normal and dry seasons. The left branch is straight and about 11 km long. There is a submerged sandbar at its entrance, and the deep channel is near the left bank. The left branch is the primary channel, and the approximately 15-km-long right branch is curvy and secondary with its deep channel near the right bank.

2.6.4.5 Taiziji Section

The Taiziji section extends from Qianjiangkou to Xinkaigou and has a curvy braided channel. The mainstream is 25.9 km long, and the secondary branch is 20 km long. The segment from Qianjiangkou to Lanjiangji has a slightly curvy single channel with the mainstream along the right bank. The segment from Lanjiangji to Sanjiangkou has a goose-head-shaped braided channel where the mid-river Tongban sandbar (consisting of the connected Tieban, Tongban, and Yuban sandbars) divides the channel into two branches. Currently, the right branch is the primary channel with a diversion ratio of about 85% for years. The right branch is divided into the east and west sub-branches by the Daochuang sandbar. The east sub-branch is deep and narrow, and the west sub-branch is wide and shallow. The former right branch of the Biandan sandbar from Sanjiangkou to Xinkaigou was filled with sediments in 1981. The general river regime of the section has been stable for years. The recent evolution of the section can be characterized mainly by the scouring of the right branch of the Tongban sandbar, alternating scouring and sedimentation process on the floodplain along the right bank, and the swinging of the deep main channel.

2.6.4.6 Guichi Section

The approximately 33-km-long Guichi section extends from Xinkaigou to Xiajiangkou and is braided with narrow ends and an approximately 9.5 km broad middle segment. The following sandbars are located within this section, Wanchuan, Fenghuang, Changsha, and Xinchang, all of which are connected into one sandbar during the dry season. The channel is divided into left, middle, and right branches. The left branch is curved to the left with a length of about 12 km, and the top of the curve is located at Yinjiagou. Near the mouth of the left branch is the Baidanzhahekou sandbank. The approximately 10 m long middle branch runs between Changsha and Fenghuang sandbars and is slightly curvy to straight. Presently, this branch is the mainstream. In recent decades, the diversion ratio of the middle branch has gradually increased and reached more than 66%. The right branch is curved to the right with a relatively small diversion ratio. At the upstream mouth of the right branch is the Niaoluo underwater sandbar which is an obstacle to navigation.

2.6.4.7 Datong Section

The approximately 21.8-km-long Datong section extends from Xiajiangkou to Yangshanji and has a slightly curvy braided channel. The section has a 9.5-km-long straight single-channel segment upstream of Meigeng and a slightly curvy braided segment with Heyue and Xinsha sandbars downstream of Meigeng. The diversion ratio of the straight left branch has remained about 89–94%. The straight deep channel runs close to the left side of the Heyue sandbar within the left branch.

2.6.4.8 Tongling Section

The Tongling section extends from Yangshanji to Digang Town, and its mainstream is about 60 km long. There are many mid-river sandbars within the goose-head-shaped braided section, including Chengde, Dingjia and Taiyang major mid-river sandbars. The section can be divided into Chengdezhou braided, Dingjiazhou braided, and Digang waterway segments. The Chengdezhou braided segment is from Yangshanji to Tuqiao and about 25 km long. The Chengde sandbar is located at the mouth of the left branch of the Dingjia sandbar, and the left branch of the Zhengde sandbar is the primary channel. It is straight with a length of approximately 17 km, and its diversion ratio is about 40%. The right branch of the Chengde sandbar is divided into two branches by the Dingjia sandbar at Xingou to form a three-branch segment until Qingcaogou. The diversion ratio of the middle branch between Chengde and Dingjia sandbars is about 35%.

The segment downstream of Qingjiagou is the Dingjiazhou braided segment with the left branch as the primary channel. The 27-km-long segment is curvy with a diversion ratio of approximately 95%. The 25-km-long right branch of the segment is the secondary channel. It is meandering and relatively narrow with a diversion ratio of about 5%. The channel of the left branch of the Dingjia sandbar is sharply curved and widened downstream of Liujiadu. The left side of the bend is the braided subsegment of Taibai and Taiyang sandbars. Its left branch is narrow, but it still diverts flow at a relatively small ratio. The branch between Taibai and Taiyang sandbars was blocked in 1992. The left branch of the Dingjiazhou segment diverts 97.9% of the water as measured in January 2004. The Digang waterway segment extends from the end of the segment between the Dingjia sandbar and Digang Town with a curvy single-channel length of 13.3 km.

2.6.4.9 Heishazhou Section

The Heishazhou section extends from Digang to the mouth of the Sanshan River and is about 33.8 km long (the left branch). The plan-view shape of the section has a curvy three-branch braided channel. Both ends of the section are narrow, and the middle is wide. The parallel mid-river Heisha and Tianran sandbars divide the channel into three branches. The right branch is slightly curvy, and the middle and left branches are relatively curvy. According to measurements collected in 2003, the diversion ratio was 41.2%, 0.9%, and 57.9% for the left, middle, and right branches, respectively.

Presently, the right branch is the main navigation channel. After water flows into the section, due to the deflecting effect from Banziji (a bedrock promontory) on the right bank, the water flow is bounced to the left bank near Nicha Town and then turns to the tip of the Heisha sandbar, causing the tip of the sandbar to collapse and move backward drastically. The deflecting effect of Banziji keeps the water flow far away from the tip of the Tianran sandbar, and, coupled with the scouring of the tip of the Heisha sandbar, the left edge of the Tianran sandbar has been enlarged significantly due to sedimentation. In order to control the river regime of the Heishazhou braided channel, a riprap revetment project was carried out from the tip to the right edge of the Tianran sandbar several years ago. There had been a slope failure on an approximately 5 km portion of the bank near the top of the Xiaojiangba bend in the left branch of the Heisha sandbar. In 1965, 899 m of the bank was revetted. In addition, after the confluence of the three branches, the confluent flow had caused local slope failures due to scouring around Gaoanwei of Fanchang County where riprap revetment projects have been implemented.

2.6.4.10 Wuyu Section

The Wuyu section extends from the mouth of the Sanshan River to East and West Liangshan, and its mainstream is about 49.8 km long. This section is divided into the upper and lower segments by a big bend. The upper segment, also called big bend segment, is about 25.3 km long. The lower segment, also called Yuxi braided segment, widens gradually and becomes braided. At the beginning of the big bend segment, due to the confluence of the Heishazhou branches, the deep channel runs near the right bank after passing Baodingwei where the deep channel gradually shifts to the left bank at Sanba. At Jiaoji, the river turns about 95 degrees with a turning radius of about 2 km, which is why it is called the big bend. At the bend, the west-east-flowing river turns to a south-north-running one. The protruding Shanxizui on the left bank forms the top of the bend.

The mainstream from the big bend to the tip of the Caogu sandbar in the Yuxi segment is about 10.1 km long, also called Wuhu subsegment. This subsegment is straight and gradually widens. From the tip of the Caogu sandbar to East Liangshan, the river is braided with mid-river Caogu and Chenjia sandbars located in series. The diversion ratio between the branches varies greatly. In general, the left branch is the secondary channel including Yuxikou and North Chenjiazhou waterways. The left branch is curvy with a length of about 12.8 km. The right branch is the primary channel, or the Xihua waterway, which is about 10.5 km long. A sandbar upstream and right of the Caogu sandbar has developed relatively rapidly recently, and the channel between the sandbar and the Caogu sandbar diverts about 10% of the river flow. The crossing channel between Caogu and Chenjia sandbars, called Caojie waterway, varies frequently due to scouring and sedimentation. Recently, the crossing channel has been experiencing sedimentation, and the current diversion ratio is less than 10% (of the mainstream flow). According to the hydrological measurement data collected in May 2003, the diversion ratio of the left branch of the Caogu sandbar was 21.6% (water level at 5.89 m), while the diversion ratio of the right branch was 78.4%. The data also indicates that the left branch of the Chenjia sandbar diverted about 15.6% of the flow and the Caojie waterway diverted about 6% of the flow.

2.6.4.11 Ma'anshan Section

The approximately 36-km-long Ma'anshan section extends from East and West Liangshan to Cimushan and is narrow at both ends with a wide straight braided middle segment. The section includes Jiangxinzhou and Xiaohuangzhou braided segments, and the following five sandbars (from upstream to downstream) are in the section: Pengxing, Taixing, Jiangxin, Hejia, and Xiaohuang.

The left branch of the Jiangxinzhou braided segment is the mainstream with a length of approximately 22 km and a width of about 2 km. The outline shape of the branch is straight, but the mainstream swings, causing sandbanks and the channel to exchange positions. The Niutun, Muxia, Taiyang, and Shiba rivers converge into the Yangtze River from the left bank of the segment. The secondary right branch of the Jiangxinzhou segment is curvy in the middle and is approximately 24 km long. The Guxi River joins from the right bank. The Pengxing sandbar at the mouth of the right branch forms a secondary braided segment. Since the 1950s, the diversion ratios of the left and right branches of the Jiangxinzhou segment have been relatively stable. The diversion ratio of the right branch ranges from 8% to 13%. The right branch of the Ziaohuangzhou segment is the main channel with a length of roughly 6.5 km and a diversion ratio of about 20%.

2.6.4.12 Nanjing Section

The approximately 95-km-long braided Nanjing section extends from Maozishan to Sanjiangkou and is composed of the Xinjizhou, Meizizhou, and Baguazhou braided segments and the Longtan curvy segment.

The Xinjizhou braided segment from Maozishan to Xiasanshan is the straight and approximately 25 km long with the mid-river Xinsheng and Xinji sandbars. The Zimu sandbar is located in the right branch of the Xinjizhou segment and is parallel to the Xinji sandbar. Between the tail of the Xinji sandbar and Qiba is the Xinqian sandbar resulting from sedimentation. The mainstream of the right branch runs through the Jiqian waterway and adjacent to Qiba into the Meizizhou braided segment.

The straight Meizizhou braided segment extends from Xiasanshan to Xiaguan. It connects the upstream Qiba segment and the downstream narrow Xiasanshan segment. The left branch is the mainstream, and the Qianzhou sandbar is in the branch. The diversion ratios are about 95% and 5% for the left and right branches, respectively. The Qianzhou braided subsegment is located in the lower left branch of the Meizizhou segment. The diversion ratio of the left branch of the subsegment is about 85%.

The Baguazhou segment extends from Xiaguan to Xiba and has a goose-head-shaped braided channel. Since the 1940s, the right branch has been the primary channel, and its diversion ratio has been increasing as the head of the Bagua sandbar has been retreating, and the left branch has been shrinking accordingly. According to the measurement data collected in January 2003, the diversion ratio of the left branch was 16.6%.

The Longtan curvy segment, originally called Xinglongzhou braided segment, extends from Xiba to Sanjiangkou. In 1985, a branch blocking project was implemented in the left branch. As a result, the segment has since become a single curvy channel.

2.6.4.13 Zhenyang Section

The 73.7-km-long Zhenyang section extends from Sanjiangkou to Wufeng Mountain and consists of (from upstream to downstream) the Yizheng curvy, Shiyezhou braided, Liuwei curvy, Hechangzhou braided, and Dagang curvy segments. The Doushan Mountain node is at the mouth and the Wufeng Mountain node at the exit of the section. The plan view of the section is a lotus-root shape.

The Yizheng curvy segment ranges from Sanjiangkou to Siyuangou and has a length of 10.5 km. The single-channel segment curves slightly toward the left.

The Shiyezhou braided segment ranges from Siyuangou to Guangzhou ferry with a length of 24.7 km. The right branch is the primary channel with a length of 15.8 km. The curvy segment has a moderate curvature. The straight left branch is the secondary channel with a length pf 13.5 km. Yingbangzi and Xumaozhou sandbanks are located within this segment. In June 2008, the diversion ratio of the left branch reached 37.1%.

The approximately 15.1-km-long Liuwei curvy segment extends from Guangzhou ferry to the convergent point of the Shatou River and has narrow ends and a wide middle portion that curves toward the left with a 7.5 km curve radius. The deep channel runs along the left side, and the Zhengzhou sandbank is on the right side.

The Hechangzhou braided segment extends from the convergent point of the Shatou River to the Qinglong Mountain in Dagong. The left and right branches are 10.9 km and 10.2 km long, respectively. After the first phase of a river regulation project was completed in the Zhenyang section, the development rate of the left branch was effectively controlled. However, because the subsequent bank revetment project was not implemented as planned and several large flood events occurred in the 1990s, the development of the left branch began to accelerate. In September 2002, the diversion ratio of the left branch reached 75.48%. Since the implementation of the regulation project at the mouth of the left branch in 2003, the diversion ratio of the left branch has basically remained at 72–73%.

The approximately 8-km-long Dagang curvy segment has a slightly curvy single channel, and the mainstream has run adjacent to the right bank for many years. Due to the hilly terrace terrain of the right bank, the channel of the segment is relatively stable.

2.6.4.14 Yangzhong Section

The Yangzhong section is a Class I key section in the middle and lower reaches of the Yangtze River. It extends from the Wufeng Mountain of the upstream Zhengyang section to Ebizui of Jiangyin of the downstream Chentong section. The mainstream of the section is 91.7 km long. The Yangzhong section can be divided into the upper and lower segments according to the morphological and flow characteristics. The segment upstream of Jiehekou is the Taipingzhou braided segment. The following mid-river sandbars are located in the segment: Taiping, Luocheng, Luan, and Paozi. The left branch of the segment diverts about 90% of the flow with annual variations less than 3.5%. The Jiangyin segment is located downstream of Jiehekou and has a straight to slightly curvy single channel.

2.6.4.15 Chentong Section

The approximately 97-km-long Chentong section extends from Ebizui of Jiangyin to Xuliujing of Changshu and consists of the Fujiangsha braided, Rugaosha, and Tongzhousha segments.

The approximately 22-km-long Fujiangsha braided segment is from Ebizui to Hucaogang and has the Ebizui and Paotaiwei controlling nodes with a width of about 1.40 km. The mainstream is near the south bank, and the channel downstream of Ebizui widens gradually. The Fujiangsha sandbar is located near the south bank in the vicinity of Cangshan Mountain and forms the braided channel. The north branch is the primary channel with a straight and wide waterway, and the curvy, narrow, and shallow south branch is the secondary waterway. The lower portion of the northern branch of the Fujiangsha segment is again divided into two sub-branches, Fuzhong and Fubei, by the Shuangjian sandbar. The Fubei sub-branch runs mainly through the north branch of the Shuangjian sandbar into the middle waterway of the Rugaosha segment. The Fuzhong sub-branch converges with the south branch of the Fujiangsha segment at the tail of the Fujiang sandbar and then runs through Hucao Port into the Liuhaisha waterway in the Rugaosha segment.

The approximately 23-km-long Rugaosha braided segment ranges from Hucaogang to Shierweigang and is wide with many sandbars, divergent waterways, and braided intersecting channels. The following relatively large sandbars are in the section (from upstream to downstream): Shuangjian, Youlai, Minzhu, Changqing, and Henggang. The head of the Shuangjian sandbar intrudes into the north branch of the Fujiang sandbar, and the tail of the Henggang sandbar extends into the east waterway of the Tongzhou sandbar. The Minzhu sandbar consists of Jiangxin, Youyi, Heping, and Shenglichang sandbars. Xue'an, Kaisha, Dongsha, and Hongbei sandbars are adjacent to the Changqing sandbar and connect to the Henggang sandbar. The Youlai sandbar was connected to the bank in 1991. The Liuhaisha waterway ranges from Zihucaogang to Shi'erwei with a length of approximately 22.4 km and has been the primary waterway for the Rugaosha braided segment since the last century with a water diversion ratio of about 70%. The central branch downstream of the convergence point runs adjacent to the right bank.

The 51.8-km-long Tongzhousha segment ranges from Shi'erwei to Xuliujing and is broad with many mid-river sandbars, shallows, and submerged sandbars. The Tongzhou sandbar divides the Yangtze River into east and west waterways.

2.6.5 *Yangtze Estuary*

The Yangtze Estuary extends from Xuliujing to Beacon 50 outside the mouth of the river and has a length of approximately 181.8 km. The section is fan shaped in a plan view and is braided in three levels. The first level of braiding by Chongming Island resulted in the North and South Branches. The second level of braiding by Changxing and Hengxing Islands led to the division of the South Branch into the South and North Channels. The third level of braiding occurred on the South Channel resulting from the Jiuduansha sandbar to form the South and North Passages. As a result, there are four waterways (North Branch, North Channel, North Passage, and South Passage) from the Yangtze Estuary to the sea, and the North Passage is the primary navigation waterway to the sea at the Port of Shanghai.

The South Branch extends from Xuliujing to Wusongkou and has a total length of 70.5 km. Qiyakou divides the branch into the upper and lower segments. The upper segment is approximately 35.0 km long and is braided. The lower segment, also known as the Sansha (three sandbars: Biandan, Xinliuhe, Zhongyang) segment, is 35.5 km long and has a multibranch braided channel. The North Branch extends from the head of Chongming Island in the west to Lianxinggang in the east, has a total length of approximately 83 km, and is curvy in a plan view with the top of the curve located between Dahonghe and Daxinhe. The channels upstream and downstream of the curve top are straight with an average width of about 2.3 km upstream of the curve top and about 6.2 km downstream of the top. The width of the outlet at Lianxinggang is about 12.0 km. The North Branch is dominant in tidal flow, and during a large tidal event, the salt water from the sea can be pushed by the tidal force backward through the North Branch into the South Branch.

Table 2.3 Deposition area distribution of sediments from Yangtze River

Sediment deposition area	Percentage (%)
Datong-Xulujing segment	10
North branch	6
Underwater in Delta area outside river mouth	31
Hangzhou Bay and offshore	40
Coastal waters in Zhejiang and Fujian provinces	11
Deep sea	Minimal

The sediments at the Yangtze Delta are mainly from the Yangtze River. According to estimates from monitoring data, the sediments from the Yangtze River are deposited in different areas with various percentages as summarized in Table 2.3.

The Yangtze Delta also receives sediments from the abandoned Yellow Delta which flowed into the sea from Jiangsu during 1128–1855, and large amounts of sediments from the ancient Yellow River were transported to the coastal area, which had led to the rapid expansion of the shore toward the sea. Part of the sediments from the Yellow River through Jiangsu into the sea diffused southward and converged with sediments from the Yangtze River. Since 1855, the Yellow River has moved northward and flowed into the Bohai Sea. The abandoned Yellow Delta gradually eroded back. Part of the eroding sediments is still diffusing southward into the Yangtze Estuary.

The change cycle (expansion or decline) for the North Branch of the Yangtze Estuary is mainly dependent on the supply of sediments and the balance of dynamic forces from the Yangtze River and the sea. The present North Branch has been in a decline process since the middle of the eighteenth century when the runoff of the Yangtze River began to mainly discharge through the South Branch. Since 1959, the North Branch has become in control of the tidal flow with increased sediments in the tidal flow. Consequently, the channel has gradually been filled with sediments, and the trend of decline appears to be irreversible. According to the life cycle of the ancient North Branch, under natural conditions, the North Branch will be filled out with sediments and disappear in a maximum of 500 years or a minimum of 200 years (Zhang and Meng 2009). Presently, the evolution of the Yangtze Estuary has the following characteristics:

2.6.5.1 North Branch to Continue Sedimentation and Decline

The North Branch has gradually developed into a dominant tidal flow channel. Under the existing river regime, it is difficult to effectively improve the inflow conditions of the upper mouth of the North Branch, and the present tidal-dominant channel characteristics will remain unchanged. Under natural conditions, the North Branch will continue to shrink due to sedimentation. Currently, the channel of the North Branch still has a volume of approximately 1 billion m^3 below the elevation of 0 m. The sediment deposition rate was about 50 million m^3/year during 958–1970 and was reduced to 17 million m^3/year after 1970. Since the impoundment of the TGR, the amounts of sediments into the Yangtze Estuary have further been reduced. Moreover, in recent years, the channel volume below the elevation of −2 m even appears to have increased. Therefore, even if no structural measures (such as installation of a control gate at the upper mouth of the North Branch) are taken, the natural decline process of the North Branch will undergo a longer period of time. The tidal flow convergent zone of the North and South Branches will remain stable in the upper portion of the North Branch for a relatively long time, and the upper portion of the North Branch will continue to undergo sedimentation. In recent years, scoring has occurred in the lower portion of the North Branch and appears to extend toward upstream. The causes for the scouring and the trend of the future development will need to be observed and studied.

2.6.5.2 Inadequate Control of Xuliujing Node

The formation of the Xuliujing node has greatly improved the control of the river regime. However, the width at the node is only 5.7 km, and the downstream channel is still wide with sandbanks along both banks that are prone to scouring. Therefore, the Xuliujing node does not have the adequate capability to control the river regime. The change of the upstream river regime has been affecting and will continue to affect the inflow to the North and South Waterways at Baimao sandbar and the stability of the diversion ratios between the North and South Waterways at the Baimao sandbar and between the downstream South and North Passages.

2.6.5.3 Changes in Baimaosha Braided Segment Unfavorable to Stability of River Regime

Because the head of the Baimao sandbar is still retreating and the channel that has carved into the sandbar at the southern edge of the sandbar's head is still developing, along with inadequate control of the upstream node and scouring on the sandbar by floodwater, etc., the north waterway in the Baimaosha segment has started to shrink slowly, and it is possible that the Baimao sandbar will experience scouring and dispersal. It is also possible that the dynamic axis of the North and South Waterways will swing at a relatively large range. All these changes are unfavorable to the stability of the river regime in the lower portion of the South Branch.

2.6.5.4 Qiyakou to Develop into New Node

Since the 1960s, the convergent area of the North and South Waterways at the Baimao sandbar has basically stabilized in the vicinity of Qiyakou. After the water flows from the North and South Waterways converge, the hydraulic energy of the joined water flow is concentrated and renders the channel to become narrower and deeper, causing Qiyakou to become a potential node. However, as the north bank boundary has not stabilized and the Biandan sandbank is low in elevation, it will take a long time for Qiyakou to develop into a relatively stable node through the natural evolution of the river channel.

2.6.5.5 Xinqiao Waterway Still Vital

Although the Xinqiao Waterway is not the mainstream of the large river regime, as the waterway is in the same direction as the tidal flow in the North Channel, the Xinqiao Waterway will gain enough channel volume to maintain a relatively good condition of water depth in its middle and lower portions in a relatively long foreseeable future because of the combined effects of the tidal flow from the North Channel and the water flow from the Biandan sandbank.

2.6.5.6 Unstable State of Diversion Inlets of North and South Channels

Because the lower portion of the South Branch is influenced by the wide river channel, numerous underwater sandbars, unstable boundaries of the North and South Channels, and fine sediments at the channel bottom prone to becoming suspended in the water, the diversion inlets of the North and South Channels will remain in an unstable state under the influence of the upstream change of the river regime, large flood events, and riverbed-making process associated with the rising and ceding tidal flow. After the 1998 large flood, the Zhongyang sandbar, Xinliuhe sandbar, and Xinliuhe sand mound all appear to have moved toward downstream, and the southerly expansion of the Xinliuhe sand mound has caused the upper channel of the Xinbaoshan waterway to become increasingly narrower. The North Waterway of the Xinbaoshan segment has continued developing and has presently become the main diversion waterway for the South Channel. The Nanshatou waterway has also undergone some development since 2001, and the tail of the Biandan sandbar downstream of the North Channel's mouth has experienced scouring to form the Xinqiao waterway. There are also signs for a new waterway to form at the mouth of the South Channel due to scouring and cutting into the middle part of the Biandan sandbar. The Xinqiao waterway now appears to be declining. Therefore, several waterways that divert water from of the North and South Channels appear in an unstable state under natural conditions. Thus, only the implementation of mitigation measures can the stability of the waterways be achieved.

2.6.5.7 South and North Channels to Remain in Relative Equilibrium for Considerably Long Time

The relative equilibrium among the waterways in the South and North Channels is controlled by the river regime of the North Channel, North Passage, and South Passage at the estuary. Although the diversion at Gate #121 in the South and North Channels is in an unstable state, the South and North Channels will inevitably coexist in a relative equilibrium state for a relatively long period of time due to the dynamic factors of the reciprocating powerful tidal flow and the four major alluvial landforms (East Beach of Chongming Island, Hengsha Island, Jiuduansha sandbar, and Nanhui sandbank) in the estuary.

2.6.5.8 South Channel to Maintain Primary Multiple Waterway Form Close to South Bank

With the estuary continuously extending toward the southeast and the impact from human land reclamation projects, the tidal dynamic force appears to have been weakened; the tidal fluctuation range appears to have gradually decreased; and the width of the South Channel appears to have narrowed further. However, the rising tidal waterway of Changxing Island is in the same direction as the rising tidal flow from the North Passage, and under natural conditions, it will be difficult for the Ruifeng shoal to join Changxing Island. Moreover, the implementation of the Dike Project along both sides of the North Passage has led to a stable rising tidal flow direction in the North Passage. At the same time, due to the Coriolis force, the flow pattern that the direction of the rising tidal flow is slightly northward and the falling tidal flow is slightly southward will remain unchanged for a long time. Therefore, under natural evolution, the Ruifeng shoal will maintain the landform of the riverbed in the sand mound area for a long time, and the South Channel will remain to be the mainstream along the south bank with multiple waterways for a long period of time. However, due to the impact of the lateral water flow over sandbanks, it is possible that the middle portion of the Ruifeng shoal will undergo scouring and cutting, which will threaten the stability of the Yuansha waterway at the mouth of the North Passage.

2.6.5.9 North Channel to Continue Changes Between Single Channel and Multiple Channels Under Natural Conditions

Owing to the relatively stable Xinqiao waterway to divert water from the North Channel in recent years, the North Channel appears to have developed into a single-channel waterway. If the upstream river regime does not change significantly, the North Channel will continue the trend to transform into a single-channel waterway, while the Qingcao sandbar will continue the trend to expand northward in the upper portion by sedimentation and to shrink in the lower portion due to scouring. Under natural conditions, due to the cyclical trend of changes in the diversion mouths and waterways of the South and North Channels, the North Channel will continue the cyclical evolution trend between a single channel and multiple channels in the future.

2.6.5.10 South Passage to Be Narrower and Deeper for Better Navigation Depth

Since the formation of the Jiuduansha sandbar, the total volume of sediment deposition has increased; the tail of the sandbar has extended further toward the sea, and the sandbar has risen due to sedimentation. In addition, the first stage, Diversion Inlet Stabilization, of the Navigable North Passage Deepening Project was implemented. All this has resulted in the connection of the Jiangyanan sandbar to the Jiuduansha sandbar to become a larger sandbar and stabilized the diversion inlets of the North and South Passages and the head of the Jiuduansha sandbar. With the implementation of the second and third stages of the Navigable North Passage Deepening Project, the southern dike along the North Passage will further extend toward downstream; the water and sand exchange between the South and North Passages will be weakened; and the tail of the Jiuduansha sandbar will receive more sediments and extend further toward the sea. At the same time, with more natural sedimentation and artificial sedimentation improvements on the Nanhui sandbank along the south bank, the borderline of the south bank will also extend northward. Therefore, the South Passage will gradually shrink in width, and the navigable depth condition of the passage will likely continue to be improved.

2.6.5.11 River Regime of North Passage to Be More Stable and Navigable Depth to Be Deeper

The implementation of the first stage, Diversion Inlet Stabilization, of the Navigable North Passage Deepening Project has stabilized the river regime of the division inlets of the North and South Passages and the inlet channel of the North Passage, stabilized the river banks of the North Passage and the pathway of the rising tidal flow along the North Passage, and weakened the water and sediment exchange between the North and South Passages. With the implementation of the second and third stages of the Navigable North Passage Deepening Project and the Land Reclamation Project on the East Beach of Hengsha Island, the East Beach of Hengsha Island will become an underwater island. The shallow underwater portion of Hengsha Island will be elevated due to sedimentation, and the river regime of the North Passage will become stable. In spite of the implementation of the first stage of the Navigable North Passage Deepening Project, the diversion ratio of the North Passage has decreased. However, due to the narrower channel of the North Passage and a unified waterway for the rising and falling tidal flow, as well as the diverting function of the navigable waterway dredging project, the water depth of the North

Passage is expected to be deepened. However, problems associated with sedimentation and navigable waterway maintenance outside the mouth of the dikes along both sides of the North Passage will hinder the effectiveness of the Navigable North Passage Deepening Project.

2.6.6 *Summary of River Channel Evolution*

On a short timescale (dozens to hundreds of years), the river channel of the upper Yangtze is basically stable and will be mainly affected by the construction and operation of cascade hydroelectric stations. The change of the river regime will be characterized mainly by sedimentation in the reservoirs and scouring of the river channel downstream of the reservoirs. For example, since the operation of the Gezhouba Hydroelectric Project in 1981, while there has been no significant change in runoff and suspended sediment contents in the flood season at the Yichang Station, the bedload of sand and gravel material has substantially decreased, and the downstream channel has experienced a certain degree of scouring. In the past 50 years, the development and utilization of the river banks in the middle and lower reaches of the Yangtze River have reached hundreds of km mainly for the construction of port terminals, bridges, intake and drainage structures, green belts, etc. Sixteen bridges across the river, 177 ports, and 1,900 intake structures have been or are being constructed. At the same time, large numbers of river channel regulation projects have been implemented in the mainstream of the river, including a total length of 1,200 km of bank revetment works, two meander cutoff projects in the lower Jingjiang River, five river branch blocking projects in the lower reaches of the Yangtze River, channel regulation projects in Nanjing, Zhenyang, and Jiepai river sections, etc. Although the impacts of these individual projects have been limited to local river sections, but the long-term and cumulative impacts on the river and the banks are significant, further restricting the range of the river's natural evolution.

The lower Jingjiang River has undergone the most intense evolution in the middle and lower reaches of the Yangtze River. Over the past 50 years, the following events have occurred: two natural meander cutoff events (at Nianziwan in 1949 and at Shatanzi); two artificial meander cutoff projects (at Zhongzhouzi in 1967 and at Shangchewan in 1969); two translocations between the primary and secondary branches in a curvy braided section (at the Wugui sandbar in Jianli in 1972 and 1995); and two large meander cutoffs at bends (Huangjiaguai bend downstream of Nianziwan bend in the 1960s and Shishouhewan bend in 1994). In 1983, the lower Jingjiang River Regime Control Project was implemented, and later the River Bank Revetment Projects at the Baxingwan and Shishuvan bends were completed. These projects have led to the basic control of the lower Jingjiang River's regime. Since the 1998 large flood in the Yangtze River basin and the 1999 large flood, the overall river regime in the lower Jingjiang River has still been largely stable, and no significant change of the river regime has occurred.

In the past 50 years, only two translocations have occurred between the primary and secondary branches: one at the Baguazhou braided segment of the Nanjing Section (1964) and the other at the Tianxingzhou braided segment in the Wuhan Section (1966). In addition, changes of the branch diversion ratio from less than 50% to slightly greater than 50% have occurred at the Xinshengzhou braided segment of the Nanjing Section and the Changzhou braided segment of the Zhenyang Section. The primary and secondary branches swinging alternately have occurred four times in Wuguizhou, Luxikou, Tuanfeng, and Guangzhou braided segments, respectively, in the Jianli Section of the lower Jingjiang River. One of the major causes for the alternation of the primary branches was that the straight segment upstream of the braided segment was relatively long and the mainstream swung, such as Tianxingzhou, Baguazhou, etc. In addition to the abovementioned primary branch alternation, changes in the diversion ratio of the primary and secondary branches and the local river regime have occurred in other 44 segments, but the overall river regime and the primary branch positions have remained unchanged.

The average amount of annual sediment transport at the Yichang Station is 516 million tonnes and 471 million tonnes at the Datong (Anhui) Station, indicating that after the Yangtze River runs into the hilly plains in the middle and lower reaches, although the Qing River, the Dongting Lake system, the Han River, the Poyang Lake system, and other tributaries converge into the river, their overall erosion impact on the river banks has not intensified, which can be characterized by erosion occurring in localized segments of the river and sedimentation taking place in the other sections of the river.

In the middle and lower reaches of the Yangtze River, lateral erosion has occurred in curvy segments, and longitudinal erosion has occurred only in localized segments, which has been manifested by the development of meandering and curvy segments from the shape of the river channel. However, local deep downward cutting has also been very severe, such as at Xisaishan in the Huangshi section, Xiaogushan in the Pengze (Jiangxi) section, Ebizui-Xishan segment in the Jiangyin section, etc. where the bottom of the riverbed has reached −40 to −60 m in elevation, far below the erosion datum of the Yangtze River. The severe downward erosion in these local areas was related to specific flow and geologic conditions. Dongting and Poyang Lakes have played a role in regulating the local discharge datum of the Yangtze River, especially Dongting Lake, which not only controls the formation of the mouths of the Xiang, Zi, Yuan, and Li Rivers but also plays a major role in controlling the erosion and sedimentation of the lower Jingjiang River.

Before the completion of the TGP, the middle reaches of the Yangtze River had been generally balanced in scouring and sedimentation with slightly more sedimentation. During 1975–1996, the total volume of sediment deposition in the river channel and floodlands was 179.3 million m^3, with an average annual amount of 8.54 million m^3. After the 1998 flood, the channel of the middle reaches of the Yangtze River started to experience scouring. During 1998–2002, the total scouring volume in the river section between Yichang and Hukou was approximately 547 million m^3 with an average annual scouring amount of 156.2 million m^3. After the impoundment of the TGR, from October 2002 to October 2010, the channel and floodlands between Yichang and Hukou generally experienced scouring with a total scouring volume of 979 million m^3 with an average annual scouring volume of 122 million m^3 and did not show any sign of increase. The reason for this is that the flood process in the middle reaches was evened after the regulation at the TGR. Moreover, the Yangtze River has not experienced any large flood over the last 10+ years. Even so, the clear water discharged from the TGR has caused some significant changes in the riverbed of the middle reaches of the Yangtze River. From the scouring pattern, deep downward cutting has been the major longitudinal scouring process, and the lateral deformation of the channel has been relatively small except for the lower Jingjiang River where the channel has been widened slightly during the dry season. From 2003 to 2011, the channel of the Yizhi section underwent downward cutting by an average of 3.5 m with the largest at 16.4 m (near Yidou); the channel of the upper Jingjiang River was cut by 1.71 m with the maximum at 10.5 m (at the tail of Wenziyuan); and the lower Jingjiang River underwent the same average scouring as average sedimentation, but the scouring was the main process with the largest scouring at 12.7 m near the Shishou curvy segment. From Chenglingji to Hankou, the channel underwent an average downward cutting by 0.65 m, and the scouring mainly occurred in the segment at and downstream of Luxikou with the largest cutting of 11.2 m near Chibi and cutting of 10.3 m at the end of the Paizhouwan section. From Hankou to Jiujiang, the channel was cut by an average of 0.71 m with the largest cutting of 14.6 m (near Yangluo), and scouring mainly occurred in the segment upstream of Huangzhou and the segment downstream of Longping.

In the future, the middle and lower reaches of the Yangtze River will be mainly affected by the operation of the regulating reservoirs and large flood processes in the upper Yangtze. The middle and lower reaches of the Yangtze River will undergo more than 100 years of scouring and sedimentation rebalancing process. Due to the significant decrease in sediment load, scouring will be the main process, and sedimentation will be the accompanying process. The process will develop with time and gradually shift from the middle reaches to the lower reaches, and eventually it will affect the evolution of the estuary. Since the impoundment of the TGR in 2003, not only the sediment load into the reservoir has reduced significantly, but the sediment amount out of the reservoir has also greatly reduced. The average amount of annual sediment transport during 2003–2010 at the Datong Station was only 152 million tonnes, less than one-third of that before the impoundment of the TGR. In the middle and lower reaches and at the estuary, there will be a sediment deficit. The annual sediment deficit in the Yangtze Estuary will reach 250 million to 300 million tonnes, which will have a long-term effect on the evolution of the river channel in the middle and lower reaches of the Yangtze River and the estuary.

References

Fan D, Li C (2007) Research progress on timing of the Yangtze initiation draining the Qinghai-Tibet Plateau throughout to the East China Sea. Mar Geol Quat Geol 27(2):121–130

Jia J, Zheng H, Yang S (2010) Spatiotemporal distribution of rock mass in the Yangtze River basin and tracing of clastic zircon source. J Tongji Univ (Nat Sci Ed) 1(9):1375–1380

Kong P (2009) When Jinsha River began to flow eastward. Chin J Geol 4(44):1256–1265

Mo X, Pan G (2006) From Tethys to the formation of the Qinghai-Tibet Plateau: constrained by tectono-magmatic events. Earth Sci Front 13(6):43–50

Su S (1992) Historical argumentation of genesis and evolution of Poyang Lake. J Lake Sci 4(1):40–47

William YBC, Liu J (1996) Formation and evolution of Tai Lake in past 11,000 years. Acta Palaeontol Sin 35(2):129–135

Xiang F, Zhu LD, Wang CS, Zhao XX, Chen HD, Yang WG (2007) Quaternary sediment in the Yichang area: implications for formation of the Three Gorges of the Yangtze River. Geomorphology 85(3–4):249–258

Yang D (1989) Evolution of the middle and lower Yangtze in past 5,000 years. J Nanjing Univ 25(3):167–173

Yang H, Tang R (1999) Study on evolution of Jingjiang River in the middle Yangtze. China Water & Power Press, Beijing

Yang SY, Li CX, Yokoyama K (2006) Elemental compositions and monazite age patterns to core sediments in the Changjiang Delta: implications for sediment provenance and development history of the Changjiang River. Earth Planet Sci Lett 245(3–4):762–776

Zhang Y (1995) Timing and significance of Yangtze channeling through Three Gorges. J Northwest Norm Univ (Nat Sci Ed) 31(2):52–57

Zhang J, Meng Y (2009) Formation and change of the North Branch at the Yangtze Estuary. Yangtze River 40(7):14–17

Zhang YF, Li CA, Wang Ql, Chen L, Ma YF, Kang CG (2008) Magnetism parameters characteristics of drilling deposits in Jianghan Plain and indication for forming of the Yangtze River Three Gorges. Chin Sci Bull 53(4):584–590

Zheng D (2003) Forming environment and development of the Qinghai-Tibet Plateau. Hebei Science & and Technology Press, Shijiazhuang

Zheng Y, Li Y (2009) Study on timing of initial formation of Three Gorges section in the Yangtze River system. J Sichuan Norm Univ (Nat Sci Ed) 32(6):808–811

Zhu RX, Potts R, Pan YX, Lu LQ, Yao HT, Deng CL, Qiu HF (2008) Paleomagnetism of the Yuanmou Basin near the southeastern margin of the Tibetan plateau and its constraints on late Neogene sedimentation and tectonic rotation. Earth Planet Sci Lett 272(1–2):97–104

Chapter 3
Hydrological Characteristics of the Yangtze River

Abstract The Yangtze River basin has a typical monsoon climate, with more rainfall during the summer and less precipitation during the winter. Floods and runoff in the Yangtze River result mainly from rainstorms and rainfalls. As 70% of the Yangtze River basin is located in mountainous and hilly areas with innumerous flash-flood gullies and serious soil erosion problems, the river is a sediment-rich water system. Its unique geomorphological environment and strong precipitation process are the major driving forces for the development of the Yangtze River water system and the hydrological processes. Other forces that accompanied these major forces included floods, sediment production, sediment transport, runoff, and other hydrological processes. This chapter introduces the basic characteristics of the Yangtze River water system and hydrology, such as of the Yangtze River's ranking among the world's largest rivers, the administrative regions involved, the basic features of the mainstream and major tributaries, the important lakes, and the complex river-lake relations. Additionally, this chapter analyzes the river basin's precipitation and runoff characteristics, material flux changes such as sediment yield and sediment transport, natural water quality characteristics, and water pollution conditions of the Yangtze River.

Keywords The Yangtze River · Changjiang River · Evolution of river system · Basin ecosystem · Water resources utilization · Floods and drought · Ecological and environmental protection · Basin management

3.1 Status of the Yangtze River

3.1.1 Ranking of the Yangtze in World Rivers

To measure the size of a river, there are many parameters, such as length, watershed area, amount of runoff, and the population living in the basin. The Yangtze River is ranked first in China in these four parameters. In the world river rankings, the Yangtze River is first in population living in the basin that is currently home to more

J. Chen, *Evolution and Water Resources Utilization of the Yangtze River*,
https://doi.org/10.1007/978-981-13-7872-0_3

than 430 million people. It is third in length, next to the Amazon and the Nile Rivers, and in amount of runoff, after the Amazon and Zaire Rivers. However, the Yangtze River is tenth in basin area, behind the Amazon, Zaire, Mississippi, Ob, Parana, Nile, Yenisei, Lena, and Niger Rivers. In addition, the Yangtze River is fourth in sediment load, following the Yellow, Brahmaputra, and Indus Rivers.

From the above parameters, the Yangtze River is third in the overall rankings of world rivers. Judging from the runoff generation per unit area of the river basin, the river is only behind a few rivers in South America and South Asia. The river is one of the world rivers with the highest flood and drought risks and most prominent issues.

The above rankings are based on average parameter values. Presently, some parameter values have changed. The length and watershed area of the river are the objective characteristics that are not impacted by human activities, but runoff and sediment load are susceptible to human activities. The amount of sediment transport in the Yangtze River has decreased markedly over the past 20 years. If the ranking is based on the amount of sediment transport in the last 20 years, the Yangtze River is seventh in the world, behind the Ganges, Amazon, Mississippi, etc. If the South-to-North Water Diversion Project is implemented, or the amount of water use and consumption outside the river basin increases, the flow to the sea will be reduced. Thus, the ranking of the Yangtze River will be surpassed by other rivers based on the amount of water discharging to the sea.

3.1.2 Geographic Location of Yangtze River

The Yangtze River is located in South Central China and originates from the snow-capped Geladandong Mountains in the central part of the Tanggula Mountains on the Qinghai-Tibet Plateau. On the north are the Kunlun Mountains, Bayan Har Mountains, Min Mountains, Qin Mountains, Funiu Mountains, Tongbai Mountain Range, Dabie Mountains, Huaiyang Hills, etc. and the drainage divides of the intermontane water system of the Qaidam Basin and the Yellow and Huai River basins; on the south are the Tanggula Mountains, Yun Range of the Hengduan Mountains, Mount Jizu in Dali, ridges of east-west-trending mountains in eastern Yunnan, Wumeng Mountain, Miaoling Mountains, Nanling Mountains and drainage divides of the intermontane water system in northern Tibet, and the Salween, Lancang, Yuan (Red), and Pearl River basins; on the west are the Kekeli Mountain, Ulan Wulashan Mountain, Xur Ken Wulashan Mountain, Gaqiadirugang Snow Mountain, and the drainage divide of the Jiangtang intermontane water system in northern Tibet; and on the southeast are the Wuyi Mountain, Shier Mountain, Yellow Mountain, Tianmu Mountain, and the drainage divides of water systems in Fujian and Zhejiang. In the Yangtze Delta area, the terrain to the north is flat, and there are many water systems that are difficult to differentiate. Usually, the north boundary of the Yangtze Delta area is considered to consist of the Jiangdou-Bingcha highway near the Tongyang Canal and the drainage divide of the Huai River basin; the south boundary consists of the hills south of the Hangzhou-Huzhou Plain and the drainage divide of the Qiantang River, and the east boundary is the East China Sea.

The upper reaches of the Yangtze River extend from the origin of the river to Yichang, Hubei, and is about 4520 km long with a catchment area of about 1 million km^2, making up 55.6% of the entire basin area of the river. The middle reaches of the river range from Yichang to Hukou, Jiangxi, and is approximately 955 km long with a catchment area of about 680,000 km^2. The river sections downstream of Hukou are the lower reaches of the river which have a length of about 938 km and encompass a catchment area of 120,000 km^2. Various sections of the Yangtze River have different names. The section of the true source of the river is called the Tuotuo River and is 346 km long. The section from the confluence of the Tuotuo River and the Dangqu River at Rangjibalong to the mouth of the Batang River at Yushu is called the Tongtian River with a total length of 828 km. The section from the mouth of the Batang River to Yibin is named the Jinsha River with a total length of 2290 km. All the sections downstream of Yibin are called the Yangtze River. However, the section from Yibin to Yichang is called the Chuan River with a total length of 1040 km. The section from Zhicheng to Chenglingji is called the Jingjiang River with a total length of about 340 km. The Jingjiang River is divided into the upper and lower Jingjiang River by Ouchikou. The lower Jingjiang River is a meander. In 1967 and 1969, artificial meander cutoff projects were implemented at Zhongzhou and Shangchewan in the lower Jingjiang River, respectively. A natural meander cutoff event occurred at Shatanzi in 1972. Moreover, natural meander cutoffs occurred at curvy segments of Wutozhou and Xiangjiazhou. All these events have shortened the channel by a total length of 83 km. The water of the Jingjiang River enters Dongting Lake through four outlet rivers – Songzi, Taiping, Ouchi, and Diaoxian (the Diaoxian outlet was blocked in 1959) from the south bank of the river. Dongting Lake now still receives water through three of the four outlet rivers and then discharges back into the Yangtze River at Chenglingji through the Xiang, Zi, Yuan, and Li Rivers. In the section between Hukou and Xulujing in the lower reaches of the Yangtze River, there are many mid-river sandbars and many braided segments. Moreover, the area consists of extensive lowland plains studded with lakes and is one of the regions with the most densely distributed water systems of China.

The river section downstream of Datong is tidal sensitive. The river's water level is affected by tides with periodic diurnal fluctuations. The section from Xuliujing to Beacon 50 at the mouth the Yangtze River at the sea is about 181.8 km long and is called the Yangtze Estuary. This tidal section has dual facies of land and sea with a moderate tidal intensity.

3.1.3 Yangtze River Basin and Administrative Regions

The mainstream of the Yangtze River runs from the west to the east through the following 11 provinces, autonomous regions, or municipalities – Qinghai, Tibet, Yunnan, Sichuan, Chongqing, Hubei, Hunan, Jiangxi, Anhui, Jiangsu, and

Shanghai – and into the East China Sea at Chongming Island, Shanghai, with a total length of approximately 6300 km. Its tributaries extend to the following eight provinces or autonomous regions: Gansu, Shaanxi, Henan, Guizhou, Guangxi, Guangdong, Fujian, and Zhejiang. The total area of the entire watershed is approximately 1.8 million km^2.

The Yangtze River basin involves 19 provinces, municipalities, or autonomous region and 127 prefecture-level cities as of 2000. Provinces (municipalities) with more than 95% of their total areas in the Yangtze River basin include Sichuan, Chongqing, Hubei, Hunan, Jiangxi, and Shanghai; with 50–75% of their total areas in the basin include Guizhou; with 25–50% of their areas in the basin include Shaanxi, Anhui, Jiangsu, and Yunnan; and with 10–25% of their areas in the basin include Qinghai, Zhejiang, Henan, Tibet, Gansu, Guangxi, Guangdong, and Fujian. Figure 3.1 shows the relationship between the Yangtze River basin and the provinces, municipalities, or autonomous regions.

3.2 Hydrologic Characteristics of the Yangtze River

3.2.1 Rivers

The Yangtze River water system is well developed with more than 7000 tributaries of all levels. Relatively large tributaries of the mainstream in the source area include the Dangqu and Chumar Rivers. Major tributaries of the upper reaches include the Songmai, Shuiluo, Yalong, Min, Tuo, and Jialing Rivers that converge from the left bank and the Longchuan, Pudu, Niulan, Heng, Chishui, and Wu Rivers that join from the right bank. In the middle reaches, the Juzhang and Han Rivers join from the left bank, and the Qing, Xiang, Zi, Yuan, and Li Rivers of the Dongting Lake water system and the Gan, Fu, Xin, Rao, and Xiu Rivers of the Poyang Lake water system converge from the right bank. In the lower reaches, joining from the left bank are the Yuan and Chu Rivers, and the Chao Lake water system, and from the right bank are the Qingyi, Shuiyang, Zhang, and Huangpu Rivers and the Tai Lake water system. In addition, the Huai River also discharges some water (mainstream of the Huai River) from the left bank at Sanjiangying of Yangzhou into the Yangtze River. The South-North Grand Canal crosses the Yangtze River between Yangzhou and Zhenjiang. All these tributaries from both the north and south banks of the Yangtze River, its mainstream, as well as the lakes, reservoirs, and the artificial canals within the Yangtze River basin have formed the Yangtze River water system.

In the Yangtze River basin, there are 437 tributary rivers each with a drainage area above 1000 km^2, 49 tributary rivers each with a catchment area greater than 10,000 km^2, and 8 tributary rivers each with a watershed area larger than 80,000 km^2 (four in the upper reaches, Yalong, Min, Jialing, and Wu Rivers, and four in the middle reaches, Yuan and Xiang Rivers of the Dongting Lake water system, the Gan River of the Poyang Lake water system, and the Han River from the north bank of the Yangtze River). The general information of tributary rivers each with a basin

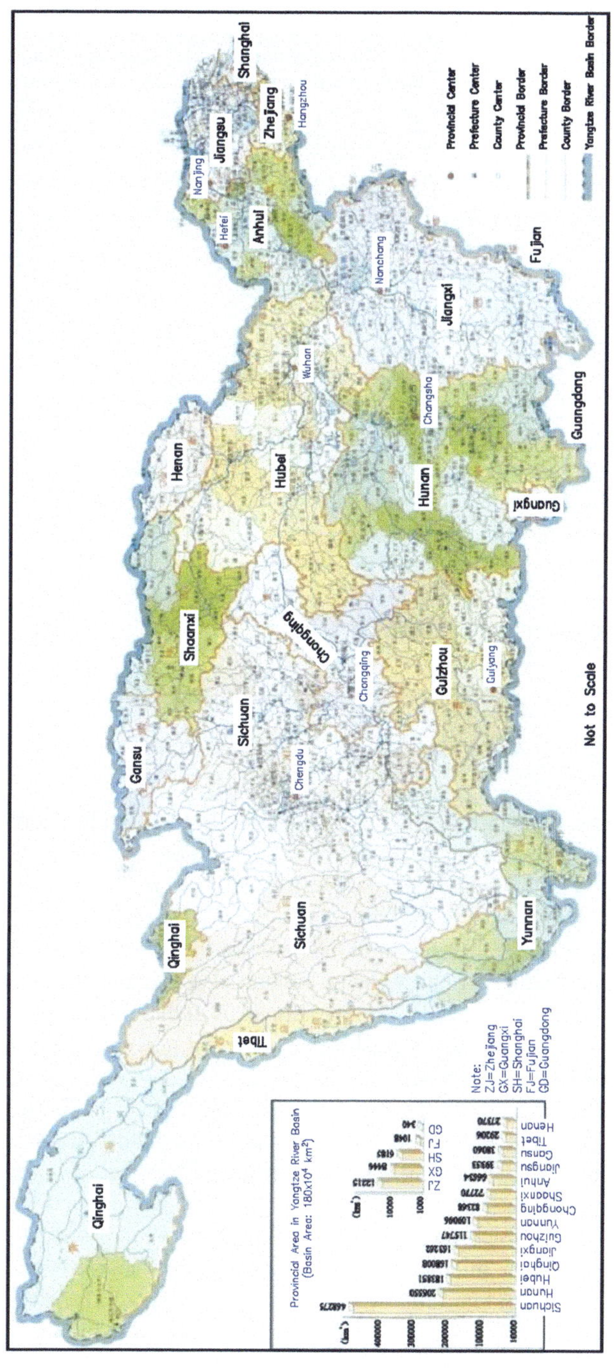

Fig. 3.1 Relationship between the Yangtze River basin and administrative regions

Table 3.1 General information of tributaries with basin area greater than 80,000 km^2

Water system	Tributary name	Basin area(km^2)	Average annual discharge (m^3/s)	Length (km)	Natural elevational relief (m)
Jinsha	Yalong	128,556	1914	1535	4420
Min	Min	135,000	2850	735	3560
Jialing	Jialing	159,000	2120	1120	2300
Wu	Wu	87,770	1690	1037	2124
Dongting Lake	Xiang	94,815	2070	856	756
	Yuan	88,451	2070	1033	1462
Han	Han	159,000	1640	1577	1962
Poyang Lake	Gan	80,948	2130	819	937

area larger than 80,000 km^2 are summarized in Table 3.1. The Jialing River boasts the largest basin area; the Min River has the largest discharge; and the Han River is the longest. The major tributary rivers listed in Table 3.1 possess a total basin area that makes up 51.9% of that of the entire Yangtze River basin and a total runoff amount that accounts for 57.4% of that at the Datong Station. The major tributaries encompass a significant portion in drainage area and contribute substantial amounts of water to the Yangtze River basin.

3.2.2 *Lakes and Wetlands of the Yangtze River*

There are many lakes in the Yangtze River basin. In addition to many alpine lakes with small surface areas in the source area, the middle and lower reaches of the Yangtze River are studded with lakes. The total area of lakes in the basin is approximately 20,000 km^2. Lakes in the source area are mostly saline or salty water lakes, accounting for about 4% of the total lake area in the Yangtze River basin. Alpine lakes in northern Guizhou and western Yunnan also account for about 4% of the total lake area in the basin. Lakes in the plain areas in the middle and lower reaches of the Yangtze River are mostly freshwater lakes and make up approximately 92% of the total lake area of the basin.

In the source area, there are ten relatively large lakes, including Quemocuo, Mazhangcuoqin, Hulu, etc. Famous alpine lakes in the middle reaches of the Jinsha River include Erhai, Napahai, Dianchi, Hugu, Chenghai, Ma, Qionghai, Xi (Qingshuihai), etc.

Most of China's freshwater lakes, including four of the five large freshwater lakes, are located in the middle and lower reaches of the Yangtze River. In 1949, the total lake area in the middle and lower reaches of the Yangtze River was 25,800 km^2, of which more than 17,200 km^2 belong to the lakes connected to the Yangtze River and located upstream of Datong Lake. By 1977, the area of the Yangtze-connected lakes had reduced to 10,500 km^2 in the middle and lower reaches, which was a 59.5% decrease when compared to that in 1949. The primary causes for the large reduction in area of the Yangtze-connected lakes were that large numbers of projects

Table 3.2 Major lakes in Yangtze River basin

Lake	Province	Elevation (above Wusong Datum) (m)	Area (km^2)	Storage capacity (billion m^3)	Average water depth (m)
Poyang	Jiangxi	22	3900	28.9	7.41
Dongting	Hunan	33.5	2623	16.7	6.37
Tai	Jiangsu	3.1	2338	4.87	2.08
Chao	Anhui	10	780	4.81	6.17
Hong	Hubei	25	344	0.659	1.92
Liangzi	Hubei	17	304	1.08	3.56
Dianchi	Yunnan	1887.5	311.9	1.59	5.11

for flood control, drainage improvements and blood-sucking parasite eradication had been built in the middle and lower reaches of the Yangtze River since 1950. Except for Dongting and Poyang Lakes that had remained connection to the river with a combined area of approximately 6000 km^2, the remaining large- and medium-sized lakes had all been separated from the river by control sluices by the late 1980s. Moreover, with sedimentation and continuous land colonization from beaches, islands, and shoals of lakes, the area of lakes had been shrinking, and the flood storage capacity had been decreasing with weeds rampant. Consequently, the lakes have been partially transforming into marshes, and most small lakes are diminishing.

There are various types of lakes in the Yangtze River basin, including tectonic lakes (such as Poyang Lake, Dianchi Lake, etc.), glacial lakes (such as the lakes in the glacial source area), barrier lakes (Dahaizi and Xiaohaizi Lakes in the Min River basin), dissolution lakes (such as Caohai Lake in Guizhou), lagoons (such as the West Lake in Hangzhou), oxbow lakes (such as the lakes in the Jianghan Plain), etc. Poyang, Dongting, Tai, and Chao Lakes in the middle and lower reaches of the Yangtze River are among the five large freshwater lakes of China. Table 3.2 summarizes the characteristics of major lakes in the Yangtze River basin.

Reservoirs are man-made lakes. By the end of 2009, 46,000 reservoirs of various types had been constructed in the Yangtze River basin with a total storage capacity of more than 250 billion m^3 and a total beneficial use capacity of more than 120 billion m^3. One hundred and sixty-six of the reservoirs are large reservoirs with a total storage capacity of 190.8 billion m^3 and a total beneficial use capacity of 98.3 billion m^3, accounting for 76% and 81% of the total reservoir storage capacity and total beneficial use capacity, respectively, of all reservoirs in the Yangtze River basin.

3.2.3 Complex River-Lake Relationship

3.2.3.1 Relationship Between the Jingjiang River and Dongting Lake

The Jingjiang River and Dongting Lake in the middle reaches of the Yangtze River are located in the most typical area with a complex river-lake relationship. Spatially, the water from the Jingjiang River used to enter Dongting Lake through the Songzi,

Taiping, Ouchi, and Diaoxian outles (the Diaoxian outlet was blocked in the winter of 1959 and has since no longer diverted water into Dongting Lake). Presently, the water of the Jingjiang River enters Dongting Lake through three diversion channels and three outlets. The three channels and outlets are the Songzi River and Songzi outlet, Hudu River and Taiping outlet, and Ouchi River and Ouchi outlet, respectively. The Songzi River is braided into two branches in Hubei, namely, East Branch (Dahukou River) and West Branch (Xinjiangkou River), and three branches in Hunan, namely, Central Branch (Zizhiju River), West Branch (Guanyuan River), and East Branch (Dahukou River). The Hudu River that is connected to the Taiping outlet joins the Songzi water system in the middle section, and a control sluice has been constructed in the lower section. Therefore, most of the water and sediment diverted through the Taiping outlet enter Dongting Lake through the Songzi outlet water system. The Ouchi River has the West Branch (Anxiang River), Central Branch (primary and secondary sub-branches of the Chenjialing River), and East Branch that is braided into two sub-branches in its upper section (Lianyuxu and Meitianhu Rivers). After the two sub-branches of the East Branch converge in its lower section, it is braided again into two sub-branches (the main sub-branch converges into the Chikou River and the other sub-branch joins the Tuo River) before entering the lake. Therefore, if all main channels and branches are counted, the three outlets connect to ten rivers before entering Dongting Lake. In addition, the Xiang, Zi, Yuan, and Li Rivers and two small rivers (Miluo and Xinqiang Rivers) also drain to Dongting Lake. Eventually, Dongting Lake drains to the Yangtze River at Chenglingji. The relationship among the water systems, water flow, and sediments is unusually complex. The diversion ratio of water flow and sediments for each river varies constantly due to the impact of natural inflow and sediment load and human activities. As of today, there is no deterministic quantitative relationship between water and sediment. Figure 3.2 shows the relationship of the Yangtze River with the four rivers south of the Jingjiang River and Dongting Lake.

In addition, Poyang Lake receives water from five rivers. The relationship between the outflow at the outlet of a lake and the water level and flowrate of the Yangtze River is also very complex. Although Tai Lake is connected to the Yangtze River through sluices and culverts at its main inlets, it outflows to the Huangpu River and eventually drains into the Yangtze River. The relationship between Tai Lake and the Yangtze River is not simple either.

3.2.3.2 Natural Factors for River-Lake Separation

River-lake separation resulted from the human construction of dikes and sluices under certain preexisting conditions created by natural evolution. Before the emergence of water pollution problems, river-lake separation had not had much impacts on human life, and had even been considered beneficial, because people had regulated water exchange between rivers and lakes based on the amounts of inflow and water demand, and helped people retain floodwater, drain excessive water, and provide irrigation water and water supply. However, after eutrophication and pollution

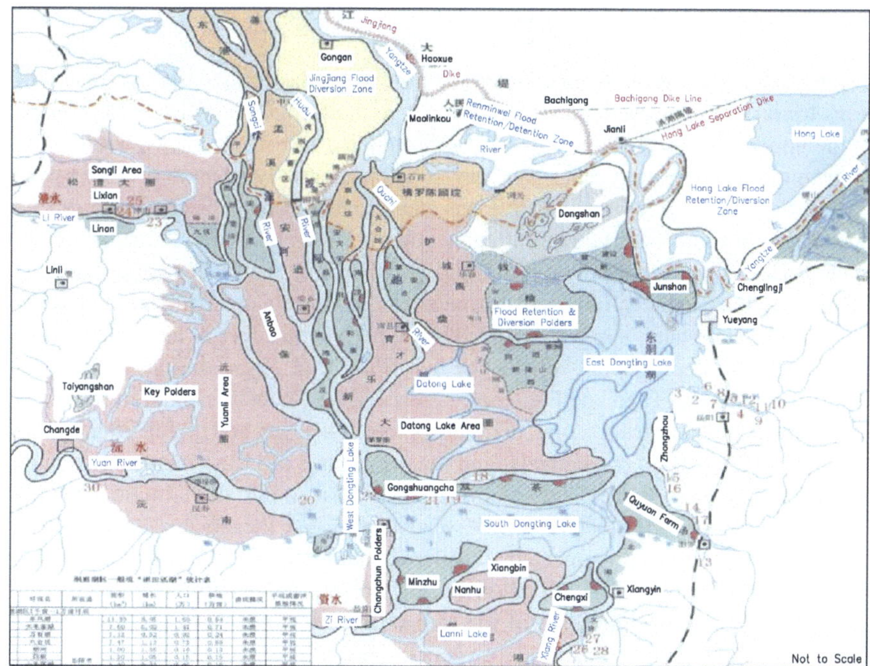

Fig. 3.2 Relationship between the Yangtze River and the water systems of Dongting Lake

of water bodies occurred, the river-lake separation increased the time required for a cycle of water exchange between rivers and lakes and decrease lakes' capacity of self-purification. Consequently, human sustainable use of water resources has eventually been impacted, and, of course, the aquatic organisms migrating between rivers and lakes have suffered the most adverse impacts. For example, from 1996 to 2005, the amounts of water diverted through the three outlets of the Jingjiang River were found to be clearly related to the fishing production of the FMCC downstream of the Gezhouba Dam. The general trend was that the more the amount of water diverted through the three outlets, the higher the yield of the FMCC and the correlation coefficient between the amount of water diversion, and the fishing yield was 0.88. After 1998, with the same amounts of water diverted through the three outlets, the yield of the FMCC was obviously lower than that before 1998. Since 1998, the production of the FMCC has gradually been decreasing, and especially since 2003, the yield has decreased even more. Therefore, the river-lake connectivity has an effect on the growth and yield of the FMCC. When the river-lake connectivity is reduced, the juveniles of the FMCC in the middle and lower reaches of the Yangtze River cannot enter lakes to feed or grow. Moreover, the juveniles are unable to get enough food from rivers where the flow velocity is high; water temperature is low; and food supplies are scarce. Consequently, the survival rate of juveniles is low. River-lake separation is the combined result of natural factors and human activities. The main natural factors include the following:

3.2.3.2.1 Neotectonic Movement

The subsidence resulting from neotectonic movement has formed the terrain conditions for lakes to hold water and is conducive to the formation of lakes. Most lakes in the upper reaches of the Yangtze River are faulted lacustrine basins, while lakes in the middle and lower reaches have developed from low-lying areas created by the crustal subsidence. Tectonic uplifting has caused lakes to become shallower or disappeared and weakened the natural connection between lakes and the surrounding water bodies. The changes in morphology and volume of lake basins in the Jianghan Plain have been greatly impacted by neotectonic movement. Hillocks and wavy plains in the Jianghan Plain were located in the peripheral zones of lacustrine basins and underwent depressional subsidence before the end of the middle Pleistocene and before the end of the late Pleistocene, respectively, which resulted in relatively thick fluvial and lacustrine sediment depositions of the lower and middle Pleistocene and upper Pleistocene, respectively, and the occurrence of lacustrine transgression. However, they were successively and intermittently rose into fluvial terraces at the end of the middle Pleistocene and the end of the late Pleistocene, respectively. Subsequently, they were subjected to different intensity of erosion and cutting to result in the formation of the hillocks and wavy plains and occurrence of lacustrine retreat. Naturally, the timescale of the impact from tectonic movement was relatively long and could be thousands of years to tens of thousands of years.

3.2.3.2.2 Climate Change

Climate change not only caused variations of precipitation but also resulted in changes of the inflow conditions of lakes. Moreover, drastic changes of temperature led to cyclic variations of ice ages and interglacial periods and might substantially changed the morphology of terrestrial waters including lakes. During interglacial periods, the Yangtze River and the lakes along both banks were often connected; rivers and lakes were not separated; and the river-lake connectivity was good. However, during the ice ages, due to lower sea level, lakes and river systems were separated; even lakes were drained dry; and only river networks remained. Due to the historical cyclical changes of climate, the extents of lakes expanded and shrank periodically, which led to the repeated changes in connectivity of water systems.

3.2.3.2.3 Flood and Sediment Deposition

Large floods generally carry large amounts of sediments into lakes and floodplains. Sedimentation often makes lakes shallower, and large lakes were dispersed into small ones or lakes and rivers became separated. The lakes in the Jianghan Plain and Dongting Lake are such typical lacustrine water bodies. The Quaternary intermittent subsidence of the Jianghan tectonic basin reached more than 300 m, which provided a favorable tectonic environment for sediment deposition. When sedimentation occurred in tributaries of lakes or tributary channels between lakes and the Yangtze River, the channels between river systems and lakes could be blocked, and the connectivity of water systems would be weakened.

Both Dongting Lake and the Yangtze River have abundant runoff without water shortage problems. The major issue regarding the water system connectivity is whether the connecting channels are unobstructed, which is mainly manifested by the changes of scouring and sedimentation in the channels of the three outlets from the Jingjiang River to Dongting Lake, the lake area, and the outlets from Dongting Lake to the Yangtze River. Due to sedimentation in the channels of the three outlets, the amount of water diverted through the outlets south of the Jingjiang River has been decreasing, and the diversion ratio was reduced from approximately 30% in the 1960s to about 15% in 2000, indicating that the diversion ratio reduced by about 50% in 40 years. In recent years, the inflow of the three outlets has decreased, especially at the Ouchi outlet where the no-flow phenomenon once occurred for a long time. Sedimentation in the channels of the three outlet rivers and the Donting Lake area has resulted in a smaller storage capacity and reduced the flood control function at Dongting Lake. Sedimentation at the Chenglingji outlet and the channel downstream thereof has also weakened the flow exchange between the Yangtze River and Dongting Lake in various seasons and deteriorated the two-way flow connectivity between Dongting Lake and the Yangtze River.

3.2.3.2.4 River Swinging

River swinging also had a great impact on the connectivity of the lakes along both banks of the river and was essentially caused by the changes of sedimentation and scouring. Over the past 5 ka, floodplains with various widths and a series of lakes have developed along river banks in the middle and lower reaches of the Yangtze River. The widening of the river channel led to the formation of sandbars and sandbanks. As a result, curvy channels and braided channels appeared, or channels changed their horizontal positions (e.g., the mainstream of the Jingjiang River moved southward). The swinging of the river caused the low-lying land in bends or branches to be isolated from the mainstream to form lakes, or lakes formed in the low-lying areas of floodplains and lost connection with the mainstream during dry seasons.

Although we have often blamed human to be responsible for the lake-river separation, actually most occurrences of the ancient lake shrinkages or river-lake separation were due to natural factors that first created the objective conditions for land reclamation. Sedimentation made lakes shallower or swampy (due to aquatic plants flourishing, lake areas gradually changed to land and terrestrial plant evolution occurred). Human took advantage of the conditions to construct berms for land colonization. A typical example was the disappearance of the Yunmeng Marsh.

3.2.3.3 Human Factors for River-Lake Separation

3.2.3.3.1 Land Reclamation

On a timescale of 10–100 years, land reclamation was one of the most important factors affecting the river-lake connectivity. Land reclamation began in Dongting Lake very early. In the 1950s to the 1980s, land reclamation in Dongting Lake reached its climax, resulting in substantial reduction of the lake area. Presently, the main body of Dongting Lake is limited by dikes, and large numbers of large and small polders are surrounded by berms. In the Jianghan Plain, land development had been carried out to a certain extent before the sixteenth century. Since the sixteenth century (middle and late Ming Dynasty), large-scale land reclamation from lakes has been rapidly carried out. Due to human reclamation activities of the lake area in the Jianghan Plain, the lake area decreased sharply from the 1950s to the 1970s, and the lake area in the 1970s was only 34.5% of that in the 1950s. While both the number and area of lakes decreased, the number of small lakes increased, and these small lakes were mostly due to the shrinking or disintegration of large lakes, which was the direct result of land reclamation. For example, Liangzi Lake had encompassed Ya'er, Bao'an, and Sanshan Lakes before land reclamation began, but due to land reclamation, it was divided into four independent disconnected lakes. The ancient Danyang Lake was also disintegrated into four independent lakes due to land reclamation, and then the ancient Danyang Lake disintegrated further to leave only Gucheng, Shijiu, and Nanyi Lakes alive. Affected by land reclamation, even if a large lake was not disintegrated into a number of smaller lakes, its area was greatly reduced, and the connectivity to the external water bodies was weakened.

3.2.3.3.2 Construction of Dikes and Sluices

The construction of dikes and sluices was the main cause for the isolation of the lakes that used to be connected to the Yangtze River. The lakes originally located along both banks of the Yangtze River had been largely connected to the river, but now only Dongting and Poyang Lakes have remained well connected to the river. Since the 1950s, a large number of dikes and sluices have been constructed along both banks of the Yangtze River, which has artificially cut off the natural connectivity between the river and the lakes and controlled the hydraulic connections of the Yangtze River and its tributaries with the lakes, thus converting the lakes originally connected to the river into barrier lakes. Since the dikes and sluices were constructed, lakes have been separated from the river, and the water levels in the isolated lakes have not been affected by the fluctuation of the water level in the river, which has reduced the risk of inundation by the floodwater of the river, accelerated land reclamation, and reduced the area of lakes greatly.

3.3 Characteristics of Precipitation and Runoff of the Yangtze River Basin

3.3.1 Historical Climate and Hydrologic Changes

China's last glacial period began 75 ka ago and ended 11 ka ago. A preliminary calculation indicates that 18 ka ago, it was the last glacial period with a cold and dry climate, and it is estimated that average annual temperature was 8–9 °C lower, precipitation 14% lower, and evaporation 15% smaller than the present time. At that time, sea level in China declined at the greatest amplitude, and the amount of annual precipitation in the Yangtze River basin (above the Datong Station) was about 682-1,278.7 billion m^3.

In the Holocene about 10 ka ago, it was a warm period globally. The temperature was in a rising period between 10 and 7.5 ka ago and was high in the early and middle Holocene period (7.5–5 ka ago) with the average annual temperature estimated to be about 3 °C higher than today, and the precipitation increased markedly. Consequently, large floods occurred around the world; the area of the global glacial permafrost shrank; and sea level rose significantly. The high-temperature period of the middle Holocene (7.5–2.5 ka ago) was the warmest and wettest time of the earth in 11 ka with the precipitation of 500–800 mm higher and the average annual temperature 2-3 °C higher than today. Large marine transgressions occurred on the eastern and southern coasts, and large floods rampaged in the Yangtze River, other major rivers, and lakes. At this time, human societies appeared in the Middle East, India, and China, and it was the period of time when the legend Yu the Great carried out his flood control efforts. It was also the period of time when the biblical legend about the great flood and the ark boat occurred.

In the late Holocene (5–0.1 ka ago), the global temperature began the cooling process until the industrialization occurred more than 100 years ago, and the global climate began to reverse its trend. In the recent 100 years (temperature measurements began to be recorded), the temperature has begun to rise with an average temperature increase of 0.6 °C per year, which is the global warming problem of the world's concern caused by massive emissions of greenhouse gases such as carbon dioxide.

The aforementioned climate changes of the recent antiquity have revealed the following:

① Before human activities affected nature, cyclic changes in the global climate and hydrology had occurred.

② The longer the timescale, the greater the range of changes. For example, on a timescale of thousands of years, the change in average annual temperature reached 2–3 °C, and the change in precipitation was in the range of 500–800 mm, which was considered normal. On a 100-year timescale, the change in temperature was less than 1 °C and that in average annual precipitation ranged from 100–200 mm.

③ On a relatively short timescale, it is very hard to measure the impacts of human activities on climate change, and even on a relatively long timescale, it is difficult to differentiate human impacts from the cyclic natural change.

Therefore, global warming caused by excessive emissions of greenhouse gases needs to be studied on a global scale and on a relatively long timescale. Of course, restricting emissions of greenhouse gases, conserving energy for emission reduction, and advocating low-carbon lifestyle are of great importance and necessity for human sustainable development.

3.3.2 *Characteristics of Precipitation*

Most of the Yangtze River basin is located in temperate and subtropical monsoon regions with an average annual precipitation of 1087 mm. The precipitation is uneven in distribution and shows the general trend to decrease from southeast to northwest. The water system in the source area is complex with long river channels, accounting for 19% of the total length of the Yangtze River, but the precipitation and runoff are relatively small. The annual runoff amount per unit area (runoff modulus) is only 93,000 m^3/km^2; the annual sediment yield per unit area (sediment modulus) is 50–200 tonnes/km^2; and erosion occurs primarily in freezing-thawing and hydraulic types. Because the source area is high in elevation and is affected by the Himalaya, not much vapor comes to the area, and thus, the annual precipitation is less than 400 mm. Although the annual precipitation in the source area is small, the evaporation is not high due to low temperatures. Moreover, the glaciers, permafrost, and marshes can store large amounts of water, and perennial ice and snow can reach 100.4 km^3, equivalent to 88.75 billion m^3 of water or five times the annual amount of runoff at the Zhidamen Station. Furthermore, the area is sparsely populated, and the regional water resources are relatively rich. Therefore, the area is called China's water tower.

Except for the source area, the annual precipitation in most regions of the Yangtze River basin ranges from 800 to 1600 mm, and the regions with an annual precipitation greater than 1600 mm are mainly in the western edge of the Sichuan Basin and parts of Jiangxi and Hunan Provinces. The regions with an annual precipitation greater than 2000 mm are in the mountainous regions with a relatively small area. For example, the annual precipitation at the Jinshan Station in Xingjing County of Sichuan Province reaches 2518 mm, the highest in the entire Yangtze River basin. The average annual precipitation of the major water systems or tributaries in the Yangtze River basin is summarized in Table 3.3.

The Yangtze River basin is a typical monsoon region, and the precipitation is significantly affected by the monsoon activities with an uneven distribution throughout the basin. The general pattern of the precipitation can be characterized by the following:

Table 3.3 Annual precipitation in major water systems or tributaries in the Yangtze River basin

Water system or tributary	Annual precipitation (mm)
Jinsha River	715
Min River	1089
Tuo River	1014
Jialing River	935
Wu River	1151
Dongting Lake water system	1431
Han River	904
Poyang Lake water system	1648
Tai Lake water system	1177

① In March every year, the wet season begins in the upper reaches of the Xiang and Gan Rivers where the wet season starts first in the Yangtze River basin.
② In April, except for the Jinsha River, the regions in the upper reaches and north of the Yangtze River and the upper and middle reaches of the Han River basin, all the other regions in the Yangtze River basin are in the rainy season.
③ In May, the rainy season is mainly limited in the water systems of the Xiang and Gan Rivers.
④ From mid-June to early July, the middle and lower reaches of the Yangtze River are in the rainy season, and the rainy belt hovers across the mainstream of the Yangtze River.
⑤ From mid-July to August, the rainy zone moves to the region in the upper reaches and north of the Yangtze River and the middle and upper reaches of the Han River. At this time, the middle and lower reaches of the Yangtze River and eastern Sichuan are susceptible to high temperature and drought hazards because the area is often controlled by the subtropical high pressure.
⑥ In September, the rainy zone moves southward to the middle and upper reaches of the Yangtze River. The rainy area moves from western Sichuan to eastern Sichuan and the Han River. Autumn rains often occur in the Yalong River, portions of the Dadu River, the upper mainstream of the Yangtze River, Jialing River, most parts of the Wu and Qing Rivers, and the upper and middle reaches of the Han River. Meanwhile, the autumn rain also falls to the Yangtze Delta.
⑦ During October–November, the rainy season of the entire basin comes to an end.

There are several typical rainstorm zones in the Yangtze River basin. From Yibin to Yichang in the upper reaches, the rainfall in the western Sichuan rainstorm zone is the main source of floodwater for the Min River, and the rainfall in the Daba Mountains' rainstorm zone is the primary source of floodwater for the Jialing River. From Yichang to Luoshan in the middle reaches are southwestern Hubei and northwestern Hunan rainstorm zones, through which the Qing River and Dongting Lake water system flow, respectively. The Han River converges to the Yangtze River somewhere between Luoshan and Hankou, and the Dabie Mountains' rainstorm zone is located downstream of Hankou. In the lower reaches, the Jiangxi rainstorm zone and the rainstorm zone across the Jiuling Mountains in Jiangxi and Mount

Huangshan in Anhui are the principal sources of floodwater for Poyang Lake and the lower reaches of the Yangtze River.

Due to the impact of the global climate change, the average annual precipitation in the upper reaches of the Yangtze River basin has been generally trending lower in the past 50 years. The decreasing trend of the inflow from the tributaries on the left bank is more prominent than that from the tributaries on the right bank. There is no obvious change in precipitation in other areas. However, the lower reaches of the Yangtze River have showed an insignificant increasing trend. Specifically, the annual precipitation in the middle and lower reaches of the Jinsha River, the Qu River which is a left tributary of the Jialing River, and most part of Dongting Lake and Poyang Lake basins is trending up with the highest increase in the Poyang Lake water system. However, the annual precipitation in most areas of the Min, Tuo, Jialing, and Han Rivers is trending lower with the central area of decreasing in the lower reaches of the Min and Tuo Rivers.

3.3.3 Glaciers in the Yangtze River Basin

According to *the Glacier Inventory of China*, 1332 glaciers of various sizes are distributed in the source area and the watersheds of the Jinsha, Yalong, Min, and Jialing Rivers in the Yangtze River basin. The glaciers have a total surface area of 1894.98 km^2 and a total ice reserve of 24.752 billion m^3. The eastern Jialing River basin is relatively high in temperature and boasts China's highest longitudinal Mount Xuebaoding Glacier and lowest latitudinal Yulongxue Mountain Glacier in the rolling Hengduan Mountains. The glaciers of the Yangtze River basin consist of both continental and marine ones. The continental glaciers are mainly in the source area of the Qinghai-Tibet Plateau, and all the others are marine glaciers. As for the types of glaciers, there are 822 hanging and cirque glaciers, accounting for 61.7% of the total number of glaciers. In addition, there are other different types of glaciers, such as cirque-hanging glaciers, cirque-valley glaciers, valley glaciers, canyon glaciers, piedmont glaciers, ice caps, etc.

There are 662 glaciers each with an area of less than 0.5 km^2 in the Yangtze River basin, accounting for 49.7% of the total number of glaciers in the basin. The glaciers with the largest area and highest ice reserve range in area from 2.01 to 5.0 km^2. Only one glacier has an area greater than 50.01 km^2, and 31 glaciers each has an area larger than 10.00 km^2, accounting for only 2.3% of the total number of glaciers, with a total area of 627.79 km^2 and a total ice reserve of 16.704 billion m^3, accounting for 33.1% of the total glacier area and 70.8% of the total ice reserve in the Yangtze River basin, respectively. All this shows that most of the glaciers in the Yangtze River basin are small.

It is noteworthy that the area of the glaciers has been declining markedly with global warming in the past 30+ years. From 1977 to 1999, the shrinkage rate was high with an average annual reduction of 4.6 km^2. From 1977 to 2009, the total area of glaciers was down by 126.33 km^2 or a 11.84% reduction in 32 years. Meanwhile, the area of the alpine lakes has gradually expanded, which has a certain relationship with the change of precipitation pattern. In recent 10 years, the amount of snowfall has decreased, but the quantity of rainfall has increased.

3.3.4 Characteristics of Runoff in the Yangtze River

3.3.4.1 Runoff Characteristics

The runoff of the Yangtze River basin is mainly replenished by precipitation. In the source area, alpine snowmelt and glacier runoff provide a very small portion of replenishment. The regional distribution of runoff is largely consistent with precipitation. Due to the combined influence of climate, precipitation, topography, and geological conditions, the runoff has both zonal and vertical changes, as well as special changes in local areas. If annual runoff depth contours of 800, 200, 50, and 10 mm are used to divide the water-rich, wet, transition, little-water, and dry zones, the Yangtze River basin is mostly in the water-rich zone (more than 800 mm) and wet zone (200–800 mm) with a very small portion in the transition zone (50–200 mm). The source area has the lowest annual runoff depth that is below 50 mm. The annual runoff depth is mostly below 200 mm in the catchment area of the Tongtian River upstream of Yushu, only 100–200 mm in the area west of the Zhimenda Station, and 600–1200 mm in most of the middle and lower reaches of the Yangtze River. The areas with an annual runoff depth greater than 1200 mm are in the water-rich zones of the western edge of the Sichuan Basin, the southern Daba Mountains, and the water systems of Dongting and Poyang Lakes. The areas with an annual depth above 1400 mm are mainly in the windward slopes of the water-rich mountainous regions. The western edge of the Sichuan Basin is special. The northern foot of the Daxiangling Mountain Range, including Mount Emei and Erlang Mountain, is considered as the center of the maximum runoff depth value. This region is in the climate zone of the "Ya'an Leak of Heaven," and the annual runoff depths are up to 1606 and 1580 mm at the Xingjing and Tianquan Stations, respectively, southwest of Ya'an. Figure 3.3 shows the distribution of the average annual runoff depth for the Yangtze River basin.

At the Zhimenda Station, the average annual discharge is 408 m^3/s, and the average annual runoff volume is 12.87 billion m^3, accounting for 1.3% of the surface runoff volume of the Yangtze River. The surface runoff in the source area contributes very little to the runoff of the Yangtze River, but the total area of the glaciers in the source area is 1247 km^2 with relatively large amounts of water frozen in the permafrost and an annual melting amount of about 989 million m^3, which accounts for 7.7% of the runoff volume at the Zhimenda Station. During 2000–2009, the average discharge was 472 m^3/s that is 15.7% higher than the average. This may be attributed to the higher precipitation and accelerated melting from glaciers and permafrost, but the former was the main reason.

Fig. 3.3 Average annual runoff depth distribution of the Yangtze River basin

The average annual runoff volumes at main control stations, Pingshan, Yichang, Hankou, and Datong, on the mainstream of the Yangtze River are summarized in Table 3.4.

Of the runoff sources to the Yichang Station, the Jinsha River contributes one-third. Although the precipitation is small in the Jinsha River basin, groundwater replenishment is abundant due to its large catchment area. The Min River basin is a famous rainstorm area with a large amount of water and is the largest tributary in amount of runoff to the Yangtze River, and the amount of runoff at the Gaoyang Station is 19.9% of that at the Yichang Station, which is greater than its area ratio. The ratio of the runoff amount at the Beibei Station on the Jialing River to the Yichang Station is 15.3% which is similar to the area ratio. The catchment area of the Wulong Station on the Wu River is 8.3% of that at the Yichang Station, but the amount of runoff accounts for 11.4% that is greater than the area ratio. The runoff volume at the Hankou Station is mainly from the runoff upstream of Yichang. The catchment area above Yichang accounts for 67.6% of that at Hankou, but the average amount of annual runoff is only 60.8% of that at Hankou. The second largest

Table 3.4 Long-term average annual runoff at main control stations on mainstream of Yangtze River

	Catchment area				Average annual runoff			
	Area	Relative to Yichang	Relative to Hankou	Relative to Datong	Relative to Yichang	Relative to Hankou	Relative to Datong	Relative to Yichang
Hydrologic station	(km^2)	(%)	(%)	(%)	(billion m^3)	(%)	(%)	(%)
Pingshan	458,592	45.6	30.8	26.9	144.6	33.4	20.3	16.1
Yichang	1,005,501	100.0	67.6	59.0	433.1	100.0	60.8	48.2
Hankou	1,488,036		100.0	87.3	712.1		100.0	79.2
Datong	1,705,383			100.0	899.2			100.0

contributor of runoff volume to the Yangtze River is the four rivers in the Dongting Lake water system. Due to flowing through the rainstorm zone in northwestern Hunan, the amount of runoff from the four rivers is 23.7% of that at Hankou although their catchment area is only 14% of that at Hankou, indicating the runoff volume ratio is much larger than the area ratio. The catchment area of the Han River is 9.5% of that at Hankou, but the runoff amount of the Han River is only 6.7% of that at Hankou, lower than the area ratio, indicating that the Han River is a tributary with a relatively low volume of runoff. Of the runoff sources to the Datong Station, the catchment area upstream of Yichang is 59% of that upstream of Datong, but the average amount of annual runoff at Yichang accounts for only 48.2% of that at Datong; the catchment area upstream of Hankou accounts for 87.3% of that upstream of Datong, but the average amount of annual runoff at Hankou accounts for about 79.2% of that at Datong. The catchment areas of the four rivers of the Dongting Lake water system in the western Hunan rainstorm zone and the Poyang Lake water system in the Jiangxi rainstorm zone account for 12.2% and 9.5%, respectively, of that of Datong, and their runoff amounts are 18.8% and 16.8%, respectively, of that of Datong, larger than their respective area proportions. The Dongting and Poyang Lake water systems are important sources of runoff to the Datong Station.

The abovementioned statistical data of runoff represents average values. Because the Yangtze River basin traverses multiple climatic and geographical regions, the runoff range of each river section or tributary is very uneven, and the runoff in every river has a relatively large variation both interannually and annually, especially when a rare flood event or dry event occurs, which has a great impact on human life and work safety and deserves attention.

3.3.4.2 Trend of Runoff Change in the Yangtze River in Recent Years

With socioeconomic development, the interference of human activities on the runoff process has increased gradually. For instance, the increased consumption of water resources as a result of more effective irrigated area, coupled with the impact of climate change and the hydrological cycle, has resulted in significant changes in amounts of runoff of some tributaries. According to an analysis of the 1950–2007 hydrological data (Wang and Yu 2009), the annual amount of runoff at the Gaoyang Station on the Min River decreased significantly, and two sharp reductions occurred in 1969 and 1992, respectively. The annual amount of runoff at the Beibei Station on the Jialing River was markedly reduced, and a sharp decrease occurred in 1993. Table 3.5 indicates that the amount of runoff from the northern tributaries, including the Han River, of the Yangtze River has been trending down since the 1990s.

According to the statistics of flood and drought disasters in recent millennium, there were irregular cyclical changes in climate and hydrology in the Yangtze River basin, with a large cycle of about 120 years, a medium cycle of 39 years, and a small cycle of 10–20 years. There were small cycles in a large cycle. For example, during the 120-year drought cycle, there might be three 40-year small cycles, namely, drought, flood, and drought. Most of the Yangtze River basin is currently in a new

Table 3.5 Ratio between decade-average runoff and average runoff in secondary river sections

Reaches	1950s	1960s	1970s	1980s	1990s
Upstream of Shigu on Jinsha River	0.899	1.048	0.940	1.039	1.012
Downstream of Shigu on Jinsha River	0.967	1.061	0.953	0.935	1.058
Min and Tuo Rivers	0.982	1.045	0.949	1.033	0.982
Jialing River	0.963	1.109	0.919	1.117	0.881
Wu River	0.874	0.989	1.028	0.969	1.059
Yibin to Yichang	0.971	1.007	0.993	1.050	0.964
Dongting Lake water system	0.882	0.949	0.999	0.952	1.134
Han River	1.012	1.067	0.934	1.178	0.834
Poyang Lake water system	0.859	0.863	1.018	1.006	1.154
Yichang to Hukou	0.943	0.900	0.927	1.110	1.078
Mainstream downstream of Hukou	1.011	0.865	0.955	1.087	1.080
Tai Lake water system	1.120	0.838	0.802	1.088	1.204
Yangtze River	0.932	0.979	0.971	1.020	1.052

cycle of large drought that has started with a small drought period. Therefore, in the past 10+ years, the precipitation has decreased with fewer large flood processes, relatively more dry years and higher probability for drought than flood or waterlogging in many areas of the Yangtze River basin. As to which, climate change or hydrological cycle, has had more impact, or which is the main factor affecting the runoff of the watershed, it is now not possible to make a determination. It will take a relatively long time, and more statistical data will be required to make such a determination.

3.4 Sediments of the Yangtze River

3.4.1 Soil Erosion in the Yangtze River Basin

According to the data obtained from the second remote sensing survey for the entire country, the area of soil erosion in the Yangtze River basin was 622,200 km^2 or 34.6% of the total land area of the basin, and the annual amount of eroded soil is 2.4 billion tonnes. The hydraulic erosion area was 569,700 km^2, accounting for 91.6% of the total area of soil erosion. The area of soil erosion in the watershed was mainly in the upper and middle reaches, accounting for about 98% of that in the entire basin, especially in the area of the upper reaches. The area of soil erosion in the upper reaches was 352,000 km^2, and the average annual amount of eroded soil was 1.6 billion tonnes, accounting for 62.3% and 66.7%, respectively, of those in the entire basin. Soil erosion areas were mainly located in the lower reaches of the Jinsha River, Jialing River basin, Tuo River, middle reaches of the Min River, upper reaches of the Wu River, Chishui River, the TGR area, and the upper reaches of the Han River.

Soil erosion in the Yangtze River basin is dominated by hydraulic erosion and followed by gravity erosion. The major sources of soil erosion are natural factors such as landslide, debris flow, etc. and human activities, of which the latter contribute much more. According to field monitoring data, the main material of soil erosion originates from arable land, accounting for 60–78% of sediments in the river. Over the past 2 ka, continuous conversion of hilly woodlands and grasslands into arable land in the Yangtze River basin has been the main cause of soil erosion. Meanwhile, mining, road construction, quarrying, development and construction projects, etc. are also one of the important causes for soil erosion. More than 1000 km^2 of new erosion area per year is generated in the Yangtze River basin due to mining, road construction, urban construction activities, etc., resulting in eroded soil of more than 150 million tonnes per year. In recent years, as the precipitation in the source area of the Yangtze River has increased and the rodent problem has been severe, coupled with widespread disorderly sand quarrying on the banks and in the channel of the Tongtian River, soil erosion in the source area of the river has been aggravating. Soil erosion and slope failures have become prominent problems due to the construction of traffic roads and railroads, power grids, communication lines, and cascade hydroelectric stations in mountainous areas in the upper and middle reaches of the Yangtze River.

3.4.2 Basic Sediment Conditions of the Yangtze River

Sediment, like flood, is the main material source and dynamic factor to shape the river, rive-lake relation, and estuarine landform. The change of the sediment-carrying capacity of the water flow in the Yangtze River directly affects the evolution of the Yangtze River and the habitats for aquatic organisms on a medium to short timescale. Before the TGR impoundment, the Yangtze River had carried more than 400 million tonnes of sediments into the sea every year. The flood season is the primary time for sediment transport. The percentage of sediment transport amount during the maximum 3 months of the flood season versus that in an entire year has been 72.5% at the Yichang Station and 56.7% at the Datong Station. However, the percentage of sediment transport amount during the minimum 3 months of the dry season versus that in an entire year has been 0.4% at the Yichang Station and 2.5% at the Datong Station. Therefore, the flood season is the primary time for sediment transport, and the floodwater is the main driving force for sediment transport.

Most of the sediment in the Yangtze River is from the Jinsha, Jialing, Min, Tuo, and Wu Rivers in the upper reaches, of which the Jinsha and Jialing Rivers are the most sediment-producing areas. Therefore, the average annual amount of sediment transport at the Yichang Station is the highest among the hydrologic stations on the mainstream of the Yangtze River, even higher than that at the Datong Station in the lower reaches, indicating that sedimentation is dominant in the middle and lower reaches.

Up to 2006, the average annual amount of sediment transport at the Pingshan Station on the Jinsha River and the Beibei Station on the Jialing River was 245 million tonnes and 110 million tonnes, respectively. The sum of the sediment transport from the two rivers accounted for 86% of that at the Cuntan Station, but the combined average annual runoff volume of the two rivers was 209.3 billion m^3 or 61% of that at the Cuntan Station. Other tributaries of the upper reaches of the Yangtze River had contributed less amount of sediments. For example, the average annual amount of sediment transport at the Gaoyang Station on the Min River was 47.6 million tonnes or 12% of that at the Cuntan Station. Meanwhile, the average annual amount of sediment transport at the Wulong Station on the Wu River was 26 million tonnes. Since the 1990s, the amount of sediment transport of most tributaries in the upper reaches of the Yangtze River was markedly reduced compared with that before 1990 due to the combined effects from sediment interception, changes in precipitation conditions, soil conservation measures, etc.

According to the monitoring data collected during the 1980s, the average annual amount of soil erosion in the upper reaches of the Yangtze River was 1.57 billion tonnes, and during the same period of time, the amount of suspended sediments out of the Yichang Station was 549 million tonnes per year on average. As a result, the sum of suspended sediment transport and sediment deposition at reservoirs accounted for about 45% of the amount of soil erosion. In other words, more than one half of the eroded soils were deposited in creeks, gullies, and rivers. Most of the suspended sediments were hydraulically transported to the outside of the basin, but not all the suspended sediments resulting from landslides, debris flow, etc. were transported to the outside of the basin. Some of the suspended sediments were transported with the flow from one place to another, which was only a spatial position change over time. Due to large differences in natural conditions among various regions of the basin, the soil erosion modulus varies greatly. In the upper reaches, 1.4 billion tonnes of eroded soils were mainly from 352,000 km^2 of the basin area. In other words, 35% of the basin area generated 89% of the total soil erosion amount of the entire basin. In terms of amounts of eroded soil and sediment transport among the water systems, the Jinsha River ranked first, the Jialing River second, and the Min River third. With the construction of cascade reservoirs in the upper reaches of the Yangtze River, sediment deposition in the reservoirs will gradually increase. Moreover, sand quarrying on river channels will also reduce sediments. As a result, the amount of sediment transport to the middle and lower reaches of the Yangtze River will be significantly reduced in the future decades to 100 years.

Up to 2006, the average annual amounts of sediment transport at major stations on the mainstream of the middle and lower reaches of the Yangtze River were as follows: 462 million tonnes at the Yichang Station (1950–2006), 386 million tonnes at the Luoshan Station (1954–2006), 378 million tonnes at the Hankou Station (1954–2006), and 408 million tonnes at the Datong Station (1953–2006). Between Yichang and Luoshan, the Songzi, Taiping, and Ouchi outlets of the Jingjiang River diverted 112 million tonnes of sediments into Dongting Lake annually on average. Other tributaries to Dongting Lake contributed relatively small amounts of sediments. For example, the annual average amount of sediment transport at the

Changyang Station on the Qing River was 8.2 million tonnes. The average annual amounts of sediment transport at the Xiangtan, Taojiang, Taoyuan, and Shimen Stations on the four rivers of the Dongting Lake water system were 10.5 million tonnes, 2.3 million tonnes, 12.5 million tonnes, and 6.3 million tonnes, respectively. Tributaries to the mainstream of the Yangtze River between Yichang and Luoshan contributed very small amounts of sediments, most of which were deposited in Dongting Lake and on floodplains of the mainstream of the Yangtze River. Between Luoshan and Datong, tributaries contributed relatively small amounts of sediments. The average annual amount of sediment transport was 36.8 million tonnes at the Xiantao Station (1955–2006) on the Han River and was 10.3 million tonnes at the Hukou Station (1952–2006) of Poyang Lake.

The sediment transport in the middle and lower reaches of the Yangtze River is mainly suspended material with a very small amount of bedload. Based on years of monitoring, the ratio between the bedload and the total sediment transport was 1.59%, 1.33%, 0.47%, and 0.037% at the Xinchang, Luoshan, Hankou, and Datong Stations, respectively, with a decreasing trend toward the downstream. The sediment transport in the middle and lower reaches of the Yangtze River is affected by the regulating function of Dongting and Poyang Lakes, and the effect of sediment transport in the middle and lower reaches can be characterized by transporting sediments from the concave bank to the convex bank and from the upstream to the downstream.

3.4.3 Sediment Characteristics in the Upper Yangtze

3.4.3.1 Characteristics of Sediment Generation and Transport in the Upper Yangtze

The sediment of the Yangtze River is mainly from overland soil erosion and channel scouring. The major characteristics of the major sediment-producing and sediment-transporting areas in the upper reaches of the Yangtze River are as follows:

1. Intense Sediment-Producing Area Is Not Large But Rather Relatively Concentrated

The soil erosion area in the upper reaches of the Yangtze River can be divided into slight, mild, moderate, and intense erosion zones, of which the intense erosion zone includes the rocky mountain areas in the middle and lower reaches of the Xihan and Bailong Rivers in the Jialing River system, the lower reaches of the Anning River in the Yalong River system, and the high canyons in the section between Dukou (now called Panzhihua City) to Pingshan in the lower reaches of the Jinsha River. The total area of this zone is 112,000 km^2, accounting for 11% of the total area in the upper reaches of the Yangtze River basin, and the average annual amount of soil erosion in the zone is 383.52 million tonnes, accounting for 24.5% of that in the upper reaches of the Yangtze River.

2. Ratio Between Sediment Transport and Yield Well Below Unity

Based on the survey conducted in the 1980s, the ratio between the sediment transport and yield was 0.338 at the Yichang Station, and 0.1–0.5 in the tributaries, all well below 1. One of the main reasons for this is that eroded soil was usually deposited as piedmonts, in low-lying areas, as alluvial fans, etc., and certain amounts of eroded soil were deposited in a short distance. The riverbeds of the tributaries in the Yangtze River basin consisted mostly of rocky soils, gravels, and coarse sands with limited supply along the river, but the eroded overland soils contained lots of coarse-grained material that is not readily transported through water to a long distance. Moreover, reservoirs, ponds, and agricultural irrigation facilities intercepted certain amounts of sediments.

3. Sediment Yield Well Related to Quantity of Water Flow and to Intensity and Location of Precipitation

The amount of sediment yield is dependent mainly on the amount of runoff, but the location and intensity of precipitation also have an obvious impact on the amount of sediment transport. When the center of a rainstorm falls in the main sediment-producing area, the amount of sediment transport in the river would be large or otherwise it would be small.

4. Sediment from Overland Erosion into the Mainstream Yangtze Is Mainly Suspended Sediment with Very Small Portion of Bedload

At the Cuntan Station, for example, the average annual amount of suspended sediments was 439 million tonnes (1953–2000), and the bedload of pebbles, gravel, and sand was 510,000 tonnes that is about 1.2% of the total amount of sediment transport.

5. Suspended Sediment in the Upper Yangtze Is Primarily from the Jinsha and Jialing Rivers

The amount of sediment from the Jinsha River was 50.9% of the total amount of sediment in the upper reaches of the Yangtze River (at the Yichang Station), and the Jialing River contributed 24%. The combined average annual amount of sediment transport from the Jinsha and Jialing Rivers made up 74.9% of the total amount of sediment transport in the upper reaches of the Yangtze River, which could be as high as 90% in individual years. Since the 1990s, the amount of sediment transport from the Jialing River has decreased significantly, and since 2002, the amount of sediment transport in the Jinsha River has also begun to decrease. With the construction of Xiluodu and Xiangjiaba hydroelectric stations, the amount of sediment transport from the Jinsha River will be substantially reduced in the future.

6. Characteristics of Pebble Bedload in the Upper Yangtze

The annual amount of pebble bedload at each station was only hundreds of thousands of tonnes. The average annual bedload was 269,000, 220,000, 341,000, and 247,000 tonnes (1966–2002) at the Zhutuo, Cuntan, Wanxian, and Fengjie Stations, respectively, in the mainstream of the upper reaches of the Yangtze River. The amount of the pebble bedload at each station was affected by the pebble quarrying activities for building materials in the river section. The average annual pebble bedload reduced to 173,000 tonnes at the Zhutuo Station during 1991–2001 from 348,000 tonnes during 1981–1990 and 162,000 tonnes at the Cuntan Station during 1991–2001 from 284,000 tonnes during 1971–1980. Since the 1980s, thanks to the rapid economic development, large-scale pebble mining has been one of the principal causes for the decrease of annual pebble bedload.

3.4.3.2 Bedload in the Upper Yangtze

Although the sediment of the Yangtze River is mainly suspended material, bedload is an important factor affecting the long-term operation of reservoirs, to which scholars, and designer and operators of reservoirs have been paying more attention. The movement of bedload has been studied for centuries, but very accurate quantitative method has yet to be developed. Because the movement of bedload does not synchronize with the water flow, it is difficult to perform mathematical simulation or calculation. Due to the problem associated with the similarity relationship, it is also very difficult to use physical model experiments to simulate. Therefore, predictions are generally made based on field measurements and empirical equations. Presently, quantitative methods that have been used are mainly: ① Establish hydrologic stations on the river and take bedload measurements. ② Perform geologic and geomorphologic investigation and lithologic mineral analysis. ③ Use equations of bedload rate to perform calculations. ④ Carry out bedload transport tests. ⑤ Monitor the deposition of bedload in reservoirs. In the above listed methods, if the method of monitoring the bedload deposition in reservoirs provides a high precision, the estimated quantity of bedload is relatively reliable. The quantity estimated using equations or bedload tests generally represents the upper limit value and therefore is higher than the actual amount.

Materials resulting from gravity erosion under the action of water flow have become the partial source of bedload in modern rivers. The average annual amount of gravity erosion in the upper reaches of the Yangtze River is approximately 120 million tonnes, making up about 8% of the total erosion amount, of which more than 60% is from the Jinsha River. From a reserve analysis, the upper reaches of the Yangtze River are abundant in reserve of bedload, which has been a key problem that experts have been worried about to potentially impact the effective storage capacity of large reservoirs in the future.

Although the upper reaches of the Yangtze River have the sufficient condition to supply bedload of pebbles and gravels, to make these loose materials become bedload will require the necessary dynamic condition. The necessary condition is the ability of water flow to move solid runoff or the hydraulic factor. According to the statistics of five gullies of debris flow, including the Jiangjiagou Gully on the Xiao River that is a tributary of the Jinsha River, the solid runoff reserve reaches 1.015 billion m^3 in the 137-km^2 watershed area, and 6.8–9.3 million m^3 of solid runoff flows out of the gullies per year. Even if the solid runoff moves as debris flow, the reserve is still much larger than the amount of solid runoff.

The medium-sized and small rivers in the mountainous area east of the Sichuan Basin (Jialing River), such as the Long River, Dahong River, etc., have shallow and thin overburden riverbeds, and bedrock is often exposed. Even stone ridges appear to form waterfalls and rapids. Accumulation of pebbles and gravels occurs only in areas where the water surface is relatively wide, and the flow velocity is relatively small, which indicates that the sediment-carrying capacity of the rivers exceeds their supply capacity. The Qu River is similar to the Long and Dahong Rivers.

The Sichuan Basin is located in the middle and lower reaches of the Jialing, Fu, Tuo and Min Rivers. The fluvial deposits below the elevation of the "Huangjiaoshu Terrace" that formed in the late Pleistocene (10,000–10,300 years ago) were generated in the Holocene and formed relatively large mid-river sandbars. Moreover, pebbles and gravels have been washed as if they had been coated with oil or called "oily stone," and bedrock is exposed on riverbed in individual sections. The riverbed overburden material consists of pebbles and gravel with sand of the middle and upper Holocene, and the thickness is generally less than 20 m. For example, the overburden in the Cangxi-Hechuan section after the Jialing River runs out of the canyon is generally 5–20 m thick with the known maximum thickness of 28.6 m. The thickness of the riverbed overburden is 10–16 m at Dujiangyan after the Min River runs out of the mountainous area and 6–14 m after the Fu River flows out of the mountainous area.

The mountainous rivers in western Sichuan (west of the Min River), such as the upper reaches of the Min River, the upper and middle reaches of the Qingyi and Dadu Rivers, the Jinsha River, etc., have 20-m-thick riverbed overburdens consisting of the Holocene sediments according to the data obtained from large numbers of boreholes.

Major rivers that run from mountains to the Sichuan Basin include the Jialing, Fu, Min, Qingyi, Dadu, and Jinsha Rivers, and the positions where they run out of mountainous areas are Zhaohua, Wudou, Zipingpu, Caoyutan, Gongzui, and Xinshi Town, respectively. The six rivers converge to the Yangtze River at Chongqing and Yibin. As the rivers flow from mountainous areas to the Sichuan Basin, the gradient and flow velocity are reduced sharply, and supplies of bedload within the Sichua Basin are minimal. As a result, the bedload is unbalanced and decreases along the rivers. In other words, the bedload at the Chongqing section should be much less than the total bedloads of the abovementioned six rivers.

Under natural conditions, the total bedload at the locations where the Min, Qingyi, and Dadu Rivers flow out the mountainous areas is 2.05 million tonnes. At the mouth of the Min River, the bedload is only 440,000 tonnes, a reduction of about four-fifths. The average distance is 287 km from the locations where the rivers flow out of the mountainous areas to the mouths of the rivers where they join the Yangtze River, and the average reduction of bedload is 5600 tonnes or about 2800 m^3 per km. If the width of the rivers is assumed to be 400 m and all bedload reduction is due to deposition, the average thickness of deposition is about 7 mm/year and only up to 0.7 m in a century. However, due to loss resulting from wear, the actual average deposition per year is less than the estimated value. On average, the thickness of riverbed deposition is in the order of magnitude of "mm" and will be in the order of magnitude of "m" in hundreds of years. The deposition is considered "very slight."

The six rivers flowing into the Sichuan Basin bring a total bedload amount of less than 5 million tonnes per year. The bedload of pebbles and gravels in the Yangtze River are mainly from the Jinsha and Min Rivers. The total amount of bedload at the confluence of the two rivers is about 2.3 million tonnes per year, slightly more than that from the Min, Qingyi, and Dadu Rivers at the locations where they run out of the mountainous areas. The approximately 400-km-long Yibin-Chongqing section of the Yangtze River is one half in longitudinal gradient of the lower reaches of the Min River but is thrice wider than the latter. In between, the Tuo and Jialing Rivers and small tributaries from the hilly area in southern Sichuan converge to the Yangtze River and provide small amounts of bedload as well as 0.5- to 1-mm grain-sized sand bedload transformed from suspended sediments. Without considering the sand bedload, the amount of pebbles and gravels entering the Yangtze River through the Yibin cross section is less than 2.5 million tonnes per year. If the bedload is balanced, the amount of pebble and gravel from the Chongqing section would be in the range of 2–3 million tonnes per year. However, if the bedload is not balanced, similar to the Min River, the amount of pebble and gravel bedload in the Chongqing section should be in the magnitude of hundreds of thousands of tonnes per year but no more than 1 million tonnes per year. If large-scale sand quarrying from riverbed and deposition in reservoirs for 20 years are considered, the amount of bedload will be even lower. During 2003–2011 after the impoundment at the TGR, the average annual bedload of sand and pebble at the Cuntan Station was 17,000 tonnes and 43,000 tonnes, respectively, or a decrease by 95% and 70%, respectively, when compared to that during 1991–2002. In fact, great changes have taken place in the upper reaches of the Yangtze River regarding bedload transport. The main reasons for the changes are not only the decrease of sediment yield resulting from soil conservation but also primarily to the deposition at cascade reservoirs and sand quarrying in river channels.

3.4.3.3 Characteristics of Sediment Yield and Transport of the Jinsha River

Because the Jinsha River has a large gradient, its water flow has a relatively higher sediment transport capacity. Therefore, it is the most typical river that has the highest sediment yield and transport capacity in the upper reaches of the Yangtze River. Even so, sedimentation has occurred on the riverbed with a thick overburden layer. There are 379 rapids in the 793-km long channel section between Xinshuangjie of Panzhihua to Yibin in the Jinsha River. The average density is 0.48 rapid/km, and the average elevation difference is about 1.3 m per rapid. Eighty-eight percent of the rapids resulted from debris flows and deposition of pebble bedload and 12% from bedrock obstacles. Debris flows from creeks and gullies have forced large amounts of meter-sized stones into the river, or large boulders from banks have collapsed into the river, or mountain slope failures have caused the blocking of the river. The river water could not carry these large boulders with it, and thus submerged dams or rapids are formed. The submerged dams have changed the base of erosion surface and raised the upstream water level, which has caused sedimentation and elevated the riverbed, and as this has continued, the whole riverbed would continue to rise due to sedimentation.

The Jinsha River is in a tectonically active zone of the southeastern Qinghai-Tibet Plateau with many alpine canyons. The area has the highest frequency for occurrences of landslides and debris flows in the Yangtze River basin. Frequent occurrences of landslides and debris flows have often caused river channels to be blocked and mid-river sandbars to change greatly. For example, an origianl Class III sandbank at Baihetan changed to a Special Class sandbank as a result of a debris flow that occurred at Mengzigou in 1924. The debris flow event caused the riverbed to rise by 15–25 m and 7- to 13-m thick boulders to be deposited at the upstream Baihetan dam site. Laojuntan (known as the "king of sandbanks") used to be a small "flowering sandbank" before debris flow events occurred several times at Baishatao and caused the channel to be blocked during 1932–1933. Consequently, the riverbed was raised by approximately 40 m with roughly 20 million m^3 of deposits. On the night of December 22, 1935, a massive landslide occurred at a mountainside in Xintian, Sichuan, causing the Jinsha River to be blocked for 3 days, and later a Class I sandbank formed. As a result, the riverbed was elevated by about 10–15 m. On the night of April 22, 1954, a debris flow occurred at Shibantao and a Special Class sandbank formed overnight by 15- to 25-m thick material that filled on the riverbed. On the night of September 4, 1965, a rainstorm poured down overnight (Zhu, 1999). On September 5, 1954, Zhu Jianyuan and others, during their fieldwork, witnessed a debris flow occurring in an unnamed indiscernible little gully to block two-thirds of the Jinsha River in width. The blocking material was later washed open to form a Class III sandbank, and the riverbed was elevated by about 10 m. On the night of November 22, 1965, a massive landslide occurred at Zehei of Luqin County, Yunnan, about 1.8 km from the Jinsha River (170 million m^3 of collapsed materials, which was really a rare event in the world), damming a small river to form a barrier lake. In 1967, the Xiaohe Dam broke, and debris flows ran into the Jinsha River, raising the riverbed by 20–30 m, forming a small sandbank on which the length of backwater reached approximately 20 km.

The riverbeds of the Jinsha River's tributaries have also undergone sedimentation. The most typical one is the Xiao River that has developed along the Xiao River fault zone and has been recognized as a world-known "museum" of debris flows. According to the *Records of Dongchuan Prefecture*, during the Kangxi years, the Xiao River was only 13–17 m wide, and clear water was used to irrigate the fields along the river, resulting in high yields of rice and sorghum. However, a complete destruction of vegetative cover resulting from large-scale copper mining and logging for smelting, coupled with poor geological conditions, led to an eventual catastrophe. There are currently 107 large and small gullies of debris flows in a 105-km river section and contribute an annual sediment amount of 30–40 million tonnes to the Xiao River. However, an average annual amount of only 6.2 million tonnes of suspended sediments flows out of the Xiao River with the remaining 80% of the sediments settled on the riverbed. As a result, a landscape has developed where "there is a trumpetlike exit for each gully to the river and you would step on stones every step if you take a walk on the floodland of the river." In the early 1960, the Kunming Institute of Hydroelectric Survey and Design under the Ministry of Electric Power analyzed the data collected from drilling and found that over the previous 200 years, the 42-km lower reaches of the Xiao River channel had been raised by about 134 m, and the 5-km river mouth section had risen by about 80 m due to sedimentation. Consequently, the slope of the riverbed in the lower reaches had evolved from the original 6.6‰ to 9.7‰. The drilling data collected from the Dongjiang area by the Ministry of Railways in the 1950s revealed that over the previous 200 years, the riverbed of the Xiao River had elevated by 54–144 m. The railroad, which had been constructed along the Xiao River, was buried in the early 1990s and had to be rebuilt later. According to the investigative data collected from the Longchuan River, more than 30 small streams in the Longchun River basin have generally been elevated by 0.5–2 m due to sedimentation over the recent 30 years. The characteristics of sediment generation and transport in the Jinsha River basin indicate that the area is prone to frequent landslides and debris flows due to natural conditions and the impact of human activities.

According to investigations and field observations, historical documents have recorded changes of the river channels resulting from scouring and sedimentation over hundreds of years, and the drilling data have recorded changes of the riverbeds due to scouring and sedimentation over thousands of years. It can be inferred from the drilling data that during the geological time before the Holocene, the terrain of the Qinghai-Tibet Plateau had undergone uplifting movements and the water flow of the Jinsha River and others (or glacier) had cut downward to bedrock. During the 10,000-year Holocene, due to warm climate, precipitation and runoff increased, and sediment deposition on the riverbed was greater than scouring, which finally led to the formation of the overburden in the riverbed consisting dominantly of the Holocene sediments.

The upper reaches of the Min River and the Dadu and Qingyi Rivers in the mountainous areas of western Sichuan and the upper reaches of the Jailing River are similar to the Jinsha River. There are large numbers of gullies of debris flows and boulders deposited in the riverbeds. Consequently, rapids developed, and wandering channels formed in broad valleys. For example, wandering channels have formed in broad valleys at Baishui in the Bailong River, Hanyuan in the Dadu River, and Qiaojia in the Jinsha River. The riverbed in Wudou County, which is located along the riverside of the Bailong River, has been higher than the ground, resulting in a situation in which "the water is higher than the county". Consequently, the entire county has relied on dikes for blocking the water of the river.

3.5 Water Quality of the Yangtze River

The quality of the aquatic environment (hereinafter referred to as water quality) not only determines the health of the ecosystem of the Yangtze River water system but is also directly related to human life, production, and sustainable use. The water quality of a water system has a broad meaning and can be generally divided into two aspects. The first aspect refers to the physical, chemical, and biological characteristics of water, and the second is the quality of water or the contamination condition of the water body. The former refers mainly to the natural properties of water, and the latter reflects the effects of human activities, especially the impact of pollutant discharges, which is the social characteristic of water quality. Water is vital for all kinds of life. All human beings and organisms need water, and various water users need different properties and quality of water. Scholars from various fields have different focuses on water quality. Geophysical chemists are concerned about the relationship of the natural conditions, such as the source of solutes, the distribution of ions, regional geology, etc., of water bodies, e.g., rivers, lakes, and groundwater. Physical geographers pay more attention to the physical and chemical rate and spatial transport distribution pattern of land denudation. Climatologists and environmentalists are interested in the circulation of nutrients such as carbon, nitrogen, phosphorus, and silicon in terrestrial and aquatic organisms. Oceanographers need to understand the contribution and impact of materials from rivers on the ocean. Aquatic biologists and ecologists are interested in the temporal and spatial changes in water temperature, dissolved oxygen, ions, nutrients, organic carbons, pH values, etc. Workers of water conservancy, sanitation, urban construction, agriculture, and environmental protection are concerned about whether water quality meets the standards of human life, production use, etc. Therefore, water quality is the common concern of the whole society.

3.5.1 Physical and Chemical Characteristics of the Yangtze River Water

3.5.1.1 Chemical Characteristics of Major Ions

Due to its large amount of runoff, the Yangtze River has a high water-dilution capability. Although the total salinity of the water is not high, the concentration of total dissolved solids (TDS) of the Yangtze River water is thrice the average TDS concentration of world's rivers. The TDS concentration in the Yangtze River water has temporal and spatial differences more than one order of magnitude with the minimum concentration of 49.7 milligrams per liter (mg/L), the maximum level of 518.1 mg/L, and a median value of 205.9 mg/L. The TDS concentration of the Yangtze River water decreases gradually with the increase of water quantity from the upper reaches to the lower reaches. The highest concentration occurs in the marginal area of the Qinghai-Tibet Plateau mainly because the upper reaches of the Jinsha River are not only relatively lower in precipitation than the middle and lower reaches but also intense in soil and rock erosion in the mountainous areas. The TDS levels in the water of the rivers in the mountainous areas in the Yunnan-Guizhou Plateau and the upper reaches of the Jialing River are higher than 250 mg/L. The TDS concentration is below 150 mg/L in the water of the mainstream and tributaries of the Yangtze River downstream of Wuhan. The Gan River, a tributary in the middle reaches of the Yangtze River, has the lowest TDS concentration that is below 100 mg/L (Chen 2006).

The dominant anion in the Yangtze River water is bicarbonate (HCO_3^-), and the dominant cation is calcium ions (Ca^{2+}). Generally speaking, HCO_3^- makes up 64% of the total ions, and Ca^{2+} only 16% of the total ions. The water of the Yangtze River has very low concentrations of sulfate ions (SO_2^{4-}) and chloride ions (Cl^-), accounting for 8% and 3%, respectively, of the total ions, but the SO_2^{4-} content is greater than 100 mg/L at a few monitoring stations. The concentration of Cl^- is relatively high at the Shawan Station in the Min River, and no more than 43 mg/L in all the other stations.

According to the seasonal variation of the ionic composition at two important stations (Hankou and Datong Stations) in the mainstream of the Yangtze River, while the seasonal variation in the discharge of the Yangtze River is very large, the seasonal change in the concentration of TDS and major ions is very limited with the ratio between the highest and the lowest levels not exceeding 2. This is very different from the rest of the world's rivers, such as the Amazon and Lena Rivers that have a large variation in TDS concentrations during the year. The decrease in the TDS concentration of the Yangtze River water during the flood season is not large because there are other sources of TDS at the same time, which offsets the diluting effects of the floodwater. One of the potential reasons is that limestone, which is widely distributed in the Yangtze River basin, has increased dissolution under high-temperature and rainy summer conditions. The second potential reason is that the clastic calcite contained in large amounts of suspended sediments (as much as 6–55 times the amounts during the dry season) resulting from soil erosion in the flood period increases the dissolution of solids, which offsets the diluting effects from the increased water quantity.

Similar to the major rivers in the northern hemisphere, the Yangtze River is a typical carbonate river and has the highest relative proportion of HCO_3^- of the rivers in the northern hemisphere because carbonate rocks (limestone) are widely distributed in the upper and middle reaches of the Yangtze River, and the dissolution of the carbonate rocks plays a leading role in supplying ions to the river water.

The ionic composition in the water of the Yangtze tributaries is also mainly controlled by the weathering of carbonate and evaporite rocks but is less affected by the weathering of silicate or alumina silicate. However, the degree of influence of carbonate and evaporite rocks on various rivers differs. For example, the concentrations of Na^+, K^+, SO_2^{4-}, and Cl^- are relatively high in the water of the Pudu, Fu, and Xin Rivers, indicating that these rivers are more affected by the weathering of evaporite rocks, and Ca^+ and HCO_3^- are dominant in the water of the Qing, Juzhang, and

Qiupu Rivers, and the chemical compositions of these rivers are mainly controlled by the weathering of carbonate rocks (Xia et al. 2008). According to the data of average annual runoff and ion contents, Dongting and Poyang Lakes are the two largest sources of ions to the Yangtze River. The Xiang River is the largest source of all ions, but Mg^{2+} and dissolved SO_2^{4-} to Dongting Lake and the inflow from the three outlets from the Yangtze River to the lake are the largest sources of Mg^{2+} and dissolved SO_2^{4-}. The Gan River is the largest source of each ion and dissolved SO_2^{4-} to Poyang Lake. In addition to Dongting and Poyang Lakes, the Min River, among the other major tributaries of the Yangtze River, is the largest source of Na^+, K^+, Ca^{2+}, Mg^{2+}, F^-, and HCO_3^- to the Yangtze River, and the Jialing River is the largest source of SO_2^{4-} and dissolved SiO_2. The Ju River is the largest source of Cl^-, followed by the Jialing and Min Rivers. In short, the ionic composition and concentration in the water of each section of the Yangtze River are related to the geological conditions, soil erosion, mining, and other factors in the catchment area.

3.5.1.2 General Evaluation of Natural Water in the Yangtze River

The average concentration of salinity in the Yangtze River ranges from 50 to 500 mg/L, generally trending down from the upper reaches to the lower reaches. The source area upstream of the Tongtian River is high in salinity, indicating that soil erosion in the area has a major impact. The eastern and southeastern portions and the upper reaches of the Han River are low in salinity with a concentration of less than 200 mg/L. The Maguo and Jinxi Rivers, which are tributaries of the Jinsha River, and the source area in Qinghai, are high in salinity with a concentration greater than 500 mg/L. Jiangsu and Shanghai in the Yangtze Estuary area are affected by tides and seawater, and the salinity is slightly higher with concentrations of 300–500 mg/L (Fan and Zhang 2008).

The total hardness of the water in the Yangtze River is below 250 mg/l, which is in the soft water range and of good quality. The total hardness of the river water increases with the increase of salinity, and the zonal distribution of geochemical regions is the same as that of salinity. The eastern and southeastern regions and the source area of the Yuan and Xiang Rivers have a low total hardness, less than 55 mg/L. The Fuxi River in the Sichuan Province and the source area of the Yangtze River in Qinghai Province have a high total hardness that is greater than 250 mg/L.

The natural chemical conditions of the surface water in the Yangtze River basin are relatively good with moderate pH values, salinity, and total hardness. The water use for human production depends primarily on the surface water of the Yangtze River system. The atomic structures of chemical elements and their distributions on the earth's crust have determined that the major dissolved solids in natural rivers are chlorides, sulfates, and carbonate salts of alkali and alkaline earth metals. In the chemical composition of the water in the watershed, the dissolution and equilibrium of carbonate are the dominant factors in controlling the chemical stability of surface water and groundwater. Because the dominant anion is $HC0_3^-$ and the dominant cation is Ca^{2+}, the chemical stability in the water body is generally good with a relatively strong buffer capacity and a robust aquatic environmental capacity and carrying capacity.

Geochemists have divided the mineralization process of natural water bodies into four stages. Silicate-carbonate water is the first stage; sulfate-carbonate water is the second stage; chloride-carbonate water is the third stage; and sulfate-chloride water is the fourth stage. The overall quality of the surface water in the Yangtze River basin is still in the first stage of the mineralization process of good water quality with high contents of carbonates, Ca^{2+} and Na^{+}. The chemical type, total hardness, salinity, and pH value of the natural water in the Yangtze River basin are consistent with the trend of the regional zonal distribution that decreases gradually from the upper reaches to the middle and lower reaches (but increases in the estuary area due to tidal effects).

3.5.2 Present Condition of Aquatic Environmental Quality in the Yangtze River

3.5.2.1 Discharge of Sewage

According to the wastewater discharge data for the period of 1998–2008, the discharge of wastewater in the Yangtze River basin (excluding the discharge of the spent cooling water from thermal power plants and drainage water from mining pits) increased significantly. In 2008, the total amount of wastewater discharge reached 32.511 billion tonnes, 60.5% higher than that in 1998, of which industrial wastewater increased by 67.7% and domestic sewage rose by 47.6%. In recent years, the sewage discharge mainly concentrated in the Tai Lake water system, Dongting Lake water system, mainstream of Yangtze River downstream of Yibin, Poyang Lake water system and Han River system, making up approximately 80% of the amount of wastewater discharged into the Yangtze River.

The amount of wastewater and sewage discharge in the Yangtze River basin showed the following characteristics:

① The total annual amount is trending up. The industrial wastewater discharge increased rapidly, and the domestic sewage increased steadily.

② The total amount showed a rapid growth in the early period, a moderate growth in the middle period and a steady state in the later period. During 1999–2000 and 2001–2003, the annual growth rate was up to 10%–13%; during 2004–2007, the growth rate fell but still reached 3%–5%, and the growth rate for 2008 was the lowest which was only 1.4% when compared with 2007.

③ The amount of industrial wastewater discharge accounted for two-thirds of the total amount of wastewater release in the basin, and the trend of change showed the same characteristics as the total amount of wastewater. During 1998–2000 and 2001–2003, the industrial wastewater growth was fast with an annual rate up to 19%; during 2004–2008, the growth of the wastewater discharge fell to an annual rate of less than 4%.

④ The amount of domestic sewage discharge accounted for one-third of the total amount of wastewater discharge in the basin. During 1998–2003, the growth was relatively slow with an annual rate of less than 5%. The growth rate of the domestic sewage during 2004–2007 increased to and remained at 5–8%.

⑤ After 2004, the annual growth rate of the domestic sewage was higher than that of the industrial wastewater, and the proportion of the domestic sewage to the total waste water discharge slightly increased (Yu et al. 2011).

3.5.2.2 Change of Water Quality in Rivers

An overall evaluation of the water in the Yangtze River basin performed during 1998–2008 indicated that the proportion of Class II or Class III water was the highest in length of rivers, and more than 65% of the river length in the entire watershed had water quality better than Class III.

The 1998–2008 statistical data indicated that the average proportion of Class II water in terms of the length of rivers was the highest at 37%, and the second highest proportion of river length was the Class III water at 31%. In most years, the mainstream of the Yangtze River had the highest proportion of Class III river length with an average of 42%, and the second highest was the Class II river length with an average proportion up to 30%. Most of the tributaries of the Yangtze River had the longest Class II river length with an average proportion of 41%, and the second longest was Class III with an average proportion of 26%.

During 1998–2008, the proportion of the river length in the Yangtze River basin with the water quality better than Class III was generally 65–88%. It was 65.1–88.4% for the entire year, 65.1–88.0% during the flood season, and 67.3–85.7% during the non-flood season, indicating that the difference in the proportion of river length with the quality better than Class III was not significant among the flood season, non-flood season, and the entire year.

During 2004–2008, the proportion of the river length in the mainstream of the Yangtze River with the water quality better than Class III was generally 52–91%. It was 66.5–75.6% for the entire year, 52.7–71.8% during the flood season, and 70.7–90.8% during the non-flood season, indicating that the water quality was much better during the non-flood season than the flood season.

The water quality of the mainstream of the Yangtze River and the provincial border rivers of Jialing, Wu, and Han was relatively better. The sections with relatively serious pollution mainly included the tributaries of the Jinsha River, the tributaries in the upper reaches of the Yangtze River, the Dongting Lake water system and the tributaries in the lower reaches of the Yangtze River, and the total length of the sections with water quality exceeding the standard was over 60% of the total length of the rivers. The provincial border river sections with relatively severe water pollution and the water quality poorer than Class V included the Baodoushen section of the Linglang River along the Yunnan-Sichuan border, the Wongdong section of the Qingshui River along the Guizhou-Hunan border, the section from Huanggaihe Farm of Huanggai Lake to the mouth of the Ouchi River along the Hubei and Hunan border, the section from Xindianpu of the Bai River to Huangquhezhen of the Huangqu River along the Henan-Hubei borer, and the section from the State Highway 104 Bridge of the Qu River to Lai'an of the Qingliu River along the Anhui-Jiangsu border.

Because of large amounts of runoff, the mainstream of the Yangtze River and its major tributaries have high capacities of self-purification. Although the amount of wastewater discharge has increased in recent 10 years, the overall water quality has remained stable. The aforementioned monitoring results were used to evaluate the overall conditions of the river sections. As a matter of fact, the water quality of the river sections in the vicinity of large- and medium-sized cities has been poor, and especially the drinking water sources on riparian zones and other water-intake areas have often been threatened with pollution. Some parameter values often exceeded their standards. In addition, the Yangtze River system has been a well-developed busy navigation system; emergent incidents of water pollution have occurred frequently either from ships or from roads along the banks. Consequently, the water quality of urban drinking water sources has often been affected. Moreover, in some small- and medium-sized reservoirs and the bays of large reservoirs, eutrophication has occurred frequently, and water quality has also become a problem. Therefore, it is still a daunting task to protect water resources.

3.5.2.3 Change of Water Quality in Lakes and Reservoirs

Comparisons of water quality evaluations for more than 20 key reservoirs in the Yangtze River basin during 1999–2008 indicate that the proportion of Class II reservoirs was the highest during the 10-year evaluation period with an average of 46.4%, followed by that of the Class III reservoirs with an average of 26.6%, and the reservoirs with other water quality classes each were less than 10% in proportion. The proportion of the major reservoirs that were better than Class III for the entire year in the Yangtze River basin was 58.6–89.7% during 2002–2008, 62.1–92.1% during the flood season, and 65.5–89.7% during the non-flood season, indicating that no significant difference exists in the proportion of reservoirs with the quality better than Class III among the flood season, non-flood season, and the entire year.

During 1999–2008, the major lakes evaluated in the Yangtze River basin were dominantly eutrophic, and the reservoirs assessed were primarily mesotrophic. A eutrophication evaluation was performed on the data collected from ten major lakes in the Yangtze River basin during 1999–2008. These lakes were Dianshan, Tai, West, Chao, Gantang, Poyang, Qionghai, Dianchi, Lugu, and Chenghai. The results of the evaluation showed that the major lakes were dominantly eutrophic with an annual proportion of 56–75%. The proportion of the mesotrophic lakes was 20–33% and that of oligotrophic lakes was about 10%. Lugu Lake had remained oligotrophic for many years, and Qionghai and Poyang Lakes were mesotrophic for most of the years. Chenghai Lake had been mesotrophic before 2001 and began to change to

Fig. 3.4 Shore waters of the Eutrophic Chao Lake

slightly eutrophic in 2002. Gantang and Dianshan Lakes had remained moderately eutrophic in recent years. West Lake was slightly eutrophic, and Dianchi Lake was moderately eutrophic. The eastern half of Chao Lake was mesotrophic to slightly eutrophic, while the western half of the lake was moderately eutrophic (see Fig. 3.4). Tai Lake was moderately eutrophic.

During 2003–2008, a eutrophication evaluation was performed on nearly 30 major reservoirs in the Yangtze River basin, and the evaluated reservoirs were all in the range of being mesotrophic to eutrophic. Most of them were mesotrophic with an annual proportion of 83–93%; the second most were eutrophic with an annual proportion of 7–17%; and no reservoirs evaluated were oligotrophic. The Qingshan, Wujiangdu, and Hongfenghu Reservoirs had been eutrophic for a long time.

Although the water quality at the TGR had conformed to Class III water standard (except for the fecal coliform group) in the whole year, the reservoir bay areas where the Xiangxi River, Daning River, and other tributaries converge had high concentrations of nitrogen, phosphorus, and other nutrients, which had caused different degrees of algal blooms. While the Danjiangkou Reservoir had been assessed to be stable in water quality at Class II (except for total nitrogen) for years and met the water quality requirements for the South-to-North Water Diversion Project, the Shending and Laoguan Rivers and other tributaries discharging to the reservoir had suffered serious pollution with water quality poorer than Class V.

The abovementioned evaluations were mainly performed at large reservoirs and lakes. Generally, the water quality of small- and medium-sized lakes and reservoirs was worse than that of large lakes or reservoirs, especially the lakes near urban areas, most of which were in the state of serious pollution or eutrophication. With the development of industries and the improvement of people's living standards, the discharge of wastewater has continued to increase. While the proportion of treated urban sewage effluents has increased faster than the growth of sewage effluents, the water quality of most lakes and small- and medium-sized reservoirs continued to deteriorate due to cumulative pollution in the past 30+ years, coupled with increasingly serious agricultural nonpoint source pollution.

3.5.2.4 Water Quality Change in Water Function Zones

During 2006–2008, evaluations were performed on water function zones of surface waters in the Yangtze River basin. Approximately, 200 major water function zones were evaluated during 2006–2008, and about 1200 water function zones were assessed in 2007 and 2008. Based on the results of the evaluations, the water function zones in the Yangtze River basin were not high in meeting the targeted standards of water quality. During 2007–2008, less than 60% of the water function zones in the basin, in terms of river lengths, met the targeted standards of water quality. Both the number and area for the water function zones to meet targeted standards were less than 50%. However, for major water function zones, the proportion of the river length and number to meet targeted standards was obviously higher than that of the entire basin. The proportions of river lengths and number that met the targeted standards were above 70% and 65%, respectively. Overall, the proportion of the water function zones meeting targeted standards in terms of river lengths was higher than that of the number, and the proportion of the number meeting targeted standards was higher than that of the area.

The proportion of key water function zones that met the standard is quite different in various secondary water resource regions of the Yangtze River basin. Both the Jialing River and Poyang Lake have over 80% of their water function zones in compliance with water standards. On a 3-year average, the fraction of water function zones in compliance with water standards in terms of number was below 55% in the Min and Tuo Rivers, and the mainstream at and downstream of Hukou was in the range of 60.0–80.0% in other water resource regions. In addition, the ratio of water function zones in compliance with water standards in number varied greatly interannually in the Han River, the mainstream at and downstream of Hukou, and the Min and Tuo Rivers where the water quality conditions were unstable.

A comparative analysis of the proportion of key water function zones that met standards in various water seasons indicated that the fraction in river length of water fuction zones in compliance with standards was higher during the non-flood season than the whole year and higher for the whole year than that in the flood season. The ratio in number of water function zones in compliance with standards during non-flood season was similar to that during the flood season. The proportion in area of water funtion zones in compliance with standards during non-flood season was similar to that during the whole year and the flood season.

3.5.2.5 Water Quality of Municipal Drinking Water Sources

During 2003–2008, water quality monitoring was performed at approximately 40 primary water sources for 16 major cities in the Yangtze River basin, including Kunming, Guiyang, Chengdu, Chongqing, Xiangfan, Yichang, Wuhan, etc. During 2007–2008, water quality monitoring was carried out for more than 200 municipal drinking water sources in the river basin. The results of the evaluations indicated that the proportion of the sources that met the drinking water standard for entire years in the Yangtze River basin was not high and 40% of the water sources still did not meet the standard. During 2003–2008, the quality of drinking water in the 16 major cities had a relatively high percentage of meeting the standard for the entire year which was over 80%. The parameters at water sources that did meet the standard were mainly total phosphorus, ammonia nitrogen, fecal coliform bacteria, iron, manganese, etc. Other unsafe factors, such as heavy metals, microorganisms, and nutrients, were also found at the water sources. Failures to meet standards at water sources mainly occurred in the first and third quarters, and more failures occurred during the dry season (January–March and November) and the flood season (July–August).

The principal problems of the municipal drinking water sources in the Yangtze River basin are as follows: ① A municipal system had only one single source, and there was no backup, or the backup source was of a poor water quality.

② Most municipal drinking water sources were surface water from rivers that flow through or nearby lakes. Therefore, they were vulnerable to releases of contaminants or accidental water pollution incidents that occurred upstream.

③ Water source management was not up to requirements since it was difficult to eradicate nearby pollution sources.

④ Online water quality monitoring and notification systems were not up to standard.

⑤ The contingency plan and measures for emergent water pollution events were not in compliance.

3.5.3 *Future Trend of Water Quality in the Yangtze River*

3.5.3.1 Characteristics of the Yangtze River Water

In recent 30–40 years, except for the concentration of SO_2^{4-} that has increased significantly, the concentration of Ca^{2+} at some sites in tributaries that has increased, and hardness and alkalinity that have increased remarkedly, the long-term trend of water quality changes in the Yangtze River has been toward an apparent acidification. In the upper reaches of the Yangtze River, Sichuan and Guizhou Provinces are among the areas with most serious acid deposition in China where the rainwater has a pH value usually between 4.0 and 5.0 and is of a sulfuric acid type. Limestone and carbonate-rich Triassic sandstone and shale are dominant in Sichuan and Guizhou. Under such conditions, severe acidic rains are bound to accelerate the dissolution of carbonate minerals, resulting in higher concentrations of SO_2^{4-} and Ca^{2+} in surface water and lower pH values; accordingly, the ratio of total hardness over the total alkalinity of the river water increases. The acidification of the water in the Yangtze River is not only due to the severe acid deposition process caused by sulfur emission from burning coal in the middle and upper reaches, but also, to a certain extent, the oxidation process of nitrogen fertilizers lost from farmlands.

The construction of a large number of cascade reservoirs in the future will gradually result in deposition of large amounts of sediments in the reservoirs. The reduced amounts of sediments discharged from the reservoirs and scouring occurring on the downstream channels will cause a redistribution of ions in the Yangtze River water system due to the change of sediment distribution, as well as a temporal variation in the distribution of ions in the water bodies of the Yangtze River basin due to the change in time of incoming water.

3.5.3.2 Trend of Change in Quality of the Aquatic Environment

From the total amount of sewage effluents, the value in the Yangtze River basin reached 33.9 billion tonnes in 2010 that was an increase of 1.4 billion tonnes from 2008, and the total amount is still increasing but at a slower rate. Based on the water quality of the evaluated sections of the Yangtze River, the proportion of the river length with better quality than Class III was 67.4% in 2010, which was lower than the previous 10 years. In the Yangtze River basin, 89.2% of major reservoirs were better than Class III in 2010, slightly better than 10 years earlier (58.6–89.7%). In 2010, 60.8% of major lakes in the Yangtze River basin were still dominantly eutrophic, comparable to 10 years earlier (56–75%). As for the water quality in the water function zones, 64.6% of the total evaluated water function zones met the standard in 2010 and 74.2% of the total evaluated river lengths met the standard, slightly better than the previous 2 years. However, only 41.5% of the total area of evaluated lakes and reservoirs were in compliance with standard. In 2010, only 64.2% of the total evaluated source areas met the drinking water standard for the entire year, indicating the safety issue of drinking water was still prominent.

References

Chen J (2006) Principle of river water quality and water quality of China's Rivers. Science Press, Beijing

Fan K, Zhang J (2008) Analysis of evolutionary trend of surface water quality in the Yangtze River Basin. Yangtze River 39(17):82–84

Wang M, Yu Y (2009) Analysis of runoff change trend in the Upper Yangtze. Yangtze River 40(19):68–69

Xia X, Yang Z, Wang Y (2008) Chemical characteristics of major ions of the Yangtze River system. Earth Sci Front 15(5):194–203
Yangtze River Water Conservancy Committee (2010) Bulletin of water resources of the Yangtze River Basin and Southwestern Water Systems. Changjiang Publishing House, Wuhan
Yu M, Fan W, Wang R et al (2011) Analysis on change of water quality in the Yangtze River Basin in recent years and recommended countermeasures. Yangtze River 42(2):75–78
Zhu J (1999) Study on changes of riverbed sand and transport of bedload gravel in the Upper Yangtze. J Hydroelectric Eng 3:86–102

Chapter 4
Ecosystem of the Yangtze River Basin

Abstract The Yangtze River traverses the three great steps of the Chinese geomorphological "staircase." The climate, geography, and geomorphology in the upper, middle, and lower reaches are very diverse, resulting in rich biodiversity, a large number of rare and endemic species, and a close relationship between the terrestrial and aquatic ecosystems. This chapter describes the characteristics of the ecosystem in the Yangtze River basin through the lens of ecology and environmental science/management. This chapter introduces the major problems of the watershed, which are water pollution, habitat fragmentation and functional decline, overharvesting of wild fauna and flora resources, and invasion of nonnative species. This chapter preliminarily analyzes the history of change and existing conditions of forest cover, the evolution and existing conditions of major lacustrine wetlands, the rare and endemic animals and fish species, and the fishery resources in the Yangtze River basin. Then everything is concluded with a discussion on the structural and integral protection of the Yangtze River ecosystem.

Keywords The Yangtze River · Changjiang River · Evolution of river system · Basin ecosystem · Water resources utilization · Floods and drought · Ecological and environmental protection · Basin management

4.1 Characteristics of Ecosystem of the Yangtze River Basin

The Yangtze River basin is a large river watershed. How to understand the ecosystem of the watershed with the water cycle as its link is a new discipline worthy of further study. In order to fully comprehend the characteristics of the ecosystem of the Yangtze River basin, it is necessary to define the watershed ecosystem.

J. Chen, *Evolution and Water Resources Utilization of the Yangtze River*,
https://doi.org/10.1007/978-981-13-7872-0_4

4.1.1 Definition of Watershed Ecosystem

4.1.1.1 Ecosystem

An ecosystem can be defined as a physical system in which the biome interacts with its environment. The system is generally composed of four components: inorganic environment, biological producers (such as green plants), consumers (herbivores and carnivores), and decomposers (saprophytic microorganisms). An inorganic environment is the nonliving component of an ecosystem, including sunlight, oxygen, and all other substances and conditions that form the basis of an ecosystem, such as water, inorganic salts, substrates (soil, rock), air, habitats, climate, and hydrological process. The inorganic environment is the foundation of an ecosystem, and its conditions directly determine the complexity of the ecosystem and the richness of the biome. For example, in the North and South Poles of the earth and China's Qinghai-Tibet Plateau, due to cold climate, there are relatively fewer species and biomass, while in the rainforest where the climate is wet and warm, there are abundant biomasses such as forests. The conditions of the environment determine the status of the biodiversity and stability of the ecosystem. Similarly, the biome can counteract with the inorganic environment. For instance, after death, organisms can turn into fossils or deposit after decomposition and then turn into inorganic matters after consolidation. The biome in the ecosystem not only needs to adapt to the environment but also to change the surrounding environment. Various basic matters link the biome with the inorganic environment closely together, so that various components in the ecosystem are closely linked and interdependent and become certain ecological elements of the organic whole.

4.1.1.2 Structure of Ecosystem

A good ecosystem has not only large numbers of organisms but also has a complete structure. The structure of an ecosystem refers to the components of the ecosystem and the ratios of their quantities, the temporal and spatial distribution of each component, and the communication methods and transmission relationship of energy, matter, and information among the components. Only the system with a complete structure is a stable and healthy ecosystem. An artificial forest may have as much biomass as a natural forest. However, due to the lack of biodiversity, adaptivity to extreme weather, and resistance to various types of bacteria and insects, a large area of biomass may die in the event of a natural disaster or plant disease resulting from attacks by pests, or a devastating destruction may occur in the event of a forest fire. Therefore, the ecosystem of an artificial forest is very fragile. In a natural forest, the amounts of biomass may not necessarily be large, and trees may not look neat. However, because of its robust biodiversity, size variety, and coexistence of herbs, shrubs, and trees and their survival through long-term natural competition, a natural forest has the capability to resist various natural disasters or plant diseases resulting from pest attacks. Thus, a single plant disease resulting from pest attacks does not endanger the entire forest ecosystem, and even in the event of a forest fire, it will revive soon. Moreover, some tree species even need a forest fire to break their shells for germination. Therefore, the more complete the structure of an ecosystem, the more stable the system and the stronger the function of the ecosystem.

Ecosystems include natural ecosystems and artificial ecosystems. Natural ecosystems mainly include forests, grasslands, wetlands, and oceans. The first three types appear on land, and their occurrences are determined by specific climate and geographical environments. The greatest advantage of a natural ecosystem is its complete structure and rich biodiversity. The artificial ecosystems are mainly distributed in farmlands and urban areas and can be characterized by scarce species of plants and animals, being apparently influenced by human activities, an incomplete structure, and dependence on and interference to the natural ecosystem. Due to the growingly larger extent of human activities, the area of China's natural ecosystems has become increasingly smaller, while that of the artificial ecosystems has become progressively larger.

4.1.1.3 Biodiversity and Ecological Function

Biodiversity includes diversity in three aspects: species, genetics, and habitat. The species diversity constitutes the basic requirement for the biological food chain where plants survive mainly through competition for sunlight, water, and nutrients; herbivores live in competition for plants and water; carnivores persist in competition for herbivores and other small animals; and microorganisms stay alive through competing for organisms such as dead animals and plants, which forms a food chain that depends on one another. In an ecosystem, only when the food chain is complete and each species has a certain redundancy can the system be maintained stable and function normally ecologically. Therefore, it should be realized that all the species that survive in nature play a certain biological role in the ecosystem and deserve conserving theoretically, no matter how large or small in quantity, whether their forms are ugly or pretty, and whether they are beneficial or not for mankind. The diversity of the environment or habitats is the basis of biodiversity; biodiversity is the basis of genetic diversity; and only when the three diversities coexist is there a real biodiversity.

4.1.1.4 Watershed Ecosystem

A traditional watershed is mainly described in the category of hydrology that refers to the region surrounded by dividing ridges (or lines) of surface water or groundwater or a catchment area of a water system such as a river, lake, etc. In fact, a watershed not only has the connotation of hydrology but also has the meaning of a certain multilevel structure and overall functional ecology. Therefore, it is a comprehensive ecosystem that includes not only water resources but also the natural elements such as the atmosphere, earth and life, and the human socioeconomic factors. Therefore, the basic characteristic of a watershed ecosystem is the water circulation that connects all the other components. Under the interaction of climate, hydrology, geology, and biology, the ecosystem includes not only the interaction of water with rock, soil, biology, and human activities but is also manifested in their transformation relationship on various spatial-temporal scales.

The greatest difference between a watershed ecosystem and other terrestrial ecosystems can be characterized by large difference on geometric scale and the flowing water and sediments. The longitudinal (in the upstream-downstream direction) dimension of a river is several orders of magnitude larger than the dimension in the lateral direction and the depth. A river is a narrow and long ecological area or a narrow ecological corridor that may flow through various geological, geomorphic, and climatic zones and may also run through various terrestrial ecological regions and lacustrine wetland ecological areas. Therefore, a watershed ecosystem extends an expansive area, is influenced by different ecological zones along its course, and is the most open ecological area. The greatest challenge to a watershed ecosystem is the dramatic changes in the river's hydrological process and habitat environment caused by human activities.

A watershed ecosystem has the following characteristics: ① On a medium to short timescale (below one millennium), the upstream ecosystem of a river basin directly affects the downstream and estuarine ecosystems, and the tributary ecosystems have a direct effect on the mainstream ecosystem. ② On a long timescale (over a millennium), the downstream environmental changes can also affect the upstream environmental variations. For example, when sea level rises, the total gradient of the river will be reduced; sediment deposition will increase; and lacustrine wetlands will expand. However, when sea level falls, the headward erosion in the source area of the river will intensify, and the lacustrine wetland area will decrease. ③ The biomass per unit area in a watershed system is less than that of a lacustrine wetland or a forest due to the impact of the change of the incoming water to the river between flood and dry seasons, rapid flows, etc. ④ Aquatic organisms, such as migratory fish species, are very dependent on a river system's connectivity of four dimensions (three geometrical dimensions and one temporal dimension). Therefore, the continuity of a river is vital for the watershed ecosystem.

4.1.2 Characteristics of the Yangtze Ecosystem

The ecosystem of the Yangtze River basin is a complex and huge system. Based on the horizontal dimension, the system can be defined as a river water system in a narrow sense or a basin in a general term. The former is mainly for rivers, lakes, and other transition areas between land and water, and the latter includes additional land areas where water, sediments, and nutrients are generated and water areas. Because sediments, nutrients, mineral ions, organic matters, and pollutants in water are mostly from land, the watershed ecosystem in a general sense has a more practical significance.

Table 4.1 Change in spatial-temporal scale of watershed environment and organisms

Habitat scale	Habitat characteristics	Biological evolution
Geometric scale of 1 μm to 100 mm of sediments in microhabitat	Gravel, sand, clay, and fine stone particles, deposition, and transport time from 1 s to 1 year	Algae and other microorganisms, life cycle from days to several weeks, the range of activities within 1 m^2
Small habitat or bio-speckle 0.1 m–1000 m	Pebbles, boulders, sandbank, islets, driftwood, and cobbles. Evolution time from 1 day to several years	Benthos, insects, etc. Life span: several months to 1 year. Range of activities: hundreds of m^2
Engineering river section (several km to dozens of km) generally within potential impact area of a hydraulic project	Including river bends, braided sections or rapids, and slow-flowing sections. Evolution: several years to 100 years	Grasslands, shrubs, amphibians, etc. Life span: several months to several years
Scenic river section (dozens of km to hundreds of km), located in land ecological area with river crossing	With similar geological and geomorphic features and vegetation types. Evolution from 100 years to tens of thousands of years	Large aquatic plants, etc., several months to several years, range of activities hundreds of m^2; fish species, reptiles, and birds, several years to more than 10 years; riparian vegetation, decades
Full river length, macrohabitat (dozens of km to thousands of km)	Changes of river and river-lake relationship, hundreds of years to tens of thousands of years	Aquatic and riparian vegetation succession, thousands of years to tens of thousands of years; and animal evolution
Watershed level (from water to land) 10 km^2–10^6 km^2	Tens of thousands of years to hundreds of thousands of years	Terrestrial vegetation succession, thousands of years to hundreds of thousands of years
Continental regional level: above 10^6 km^2	Interbasin, inter-region, and overlap 10^4–10^7 years	Species succession: tens of thousands of years to millions of years

From the view of landscape ecology, a river ecosystem has multiple levels in scale. On the spatial-temporal scale of a river ecosystem, it can be divided into seven levels from the microcosmic to macroscopic range. Table 4.1 shows the change in the spatial-temporal scale of a watershed ecosystem.

There is a close interrelationship among various scales of biological habitats, and the habitat change has a direct relationship with biological evolution. The smallest mesoscopic habitat is a grain of sediment, which can be attached to nutrients and pollutants and can also inhabit bacteria, algae, and other microorganisms. Although it is small, it flows in a very large range and may flow from the upper reaches to the lower reaches and even to the coastal waters. The habitat, one level higher than a mesoscopic habitat, is called a bio-speckle that is as small as a pebble to as large as thousands of meters of a sandbar or an islet in the middle of a river channel. Bio-speckles in a river are plants, benthos, insects, amphibians, and reptile habitats. The more such speckles in a river, the more biological habitats. The more variety of speckles, the more abundant the habitat diversity and the more biological species. A river section, one level higher than a bio-speckle, can be called an engineering section (generally within the potential impact area of a hydraulic engineering project) or an ecological corridor, ranging from a few km to dozens of km. The engineering section generally includes a typical river landform, such as a deep main channel, slow-flowing zones, bends, even braided channels, etc. These river landforms are not only habitats for many biological species but are also spawning sites and migratory pathways for fish species and activity zones for static-water-loving fish species. A scenic river section may have a length ranging from dozens of km to hundreds of km, including the river sections with the banks having the same type of geologic or geomorphic features, such as the middle reaches of the Jinsha River, the Three Gorges section, the upper Jingjiang River, etc. The land corresponding to the scenic section can be a typical terrestrial ecological area. From the characteristics of the river, a scenic river section has basically the same characteristics of geology, landform, river shape, soil, and vegetation, and the riparian vegetation types and habitat conditions are similar, which forms the medium-scale habitats of the river. The study of the biological relationship between terrestrial organisms and river organisms, or the study of the habitat relationship under the river-lake connection conditions, can be carried out on the scale of the scenic section. Another level higher is a large river or a major tributary of a large river. It is generally divided into the source area; upper, middle, and lower reaches; and the estuary. A medium-sized or small river can be divided into the upper and lower reaches and can constitute a separate watershed ecological area. If the whole including the water system and land is a watershed level, the Yangtze River system is the scale of a watershed ecological zone. Above the watershed level is a large regional or continental plate level scale, such as the Eurasia, the African continent, and the Australian continent.

The scale characteristics of a biological habitat determine the spatial and population features of biological life. Because a water body has the flowing and open nature, organisms can live in a relatively large space in a river ecosystem. The most typical example is the migratory fish species and dolphins whose living space may extend from the upper reaches to the lower reaches of the Yangtze River and even to the estuary and coastal waters. They need the continuity and connectivity of the river's aquatic system, and their range of activities may extend up to thousands of km.

As the Yangtze River flows through the three-step topographic staircase in China and across a number of climatic, geographical, and ecological zones, the water system and river-lake relationship are complex, ranging from microhabitats to biological habitats with rich biodiversity in all types of landscape ecosystems. The water system not only includes forests, wetlands, and artificial ecosystems (such as reservoirs) but also connects these ecosystems either in series or parallel. The flows of information, genetics, material, and energy in all types of ecosystems are manifested in the flow of water, which not only shapes the freshwater ecosystem of large rivers and great lakes but also provides the estuary and coastal waters with abundant nutrients and biological resources.

The ecosystem of the Yangtze River basin is a coupling of terrestrial ecosystems and aquatic ecosystems. The terrestrial ecosystems are nourished by the water ecosystems, while water ecosystems directly and indirectly reflect the characteristics of the terrestrial ecosystems through which the Yangtze River flows. For example, the mainstream of the Yangtze River can be divided into 13 scenic sections and dozens of engineering sections, as shown in Table 4.2.

Table 4.2 Scenic and engineering reaches of mainstream of Yangtze River

Yangtze	Reaches	Scenic section	Characteristic description	Engineering section
Mainstream	Estuary	Downstream of Jiangyin	Confluence of seawater and river water	Slightly
	Lower reaches	Jiangyin–Datong	Plain braided tidal section	Slightly
		Datong–Hukou	Hilly meandering braided section	Slightly
	Middle reaches	Hukou–Wuhan	Hilly meandering section	Slightly
		Wuhan–Chenglingji	River-lake intercrossing section	Wuhan urban section
				Engineering section at Baishazhou Bridge
				Engineering section at Junshan Bridge
				Paizhouwan section
				Section at Jiayu County seat
				Chibi (Jiayu–Honghu) section
				Honghu urban section
				Luoshan section
				Chenglingji lake-river intercrossing section
		Chenglingji–Zhicheng	Plain wandering braided section	Slightly
		Zhicheng–Yichang	Hilly meandering section	Slightly
	Lower reaches	Yichang–Fengjie (Three Gorges section)	Three Gorges deep-cut valley section	Slightly
		Fengjie–Jiangjin (Chuan River)	Parallel ridge valley section	Slightly
		Jiangjin–Yibin (Chuan River)	Basin edge river section	Slightly
		Yibin–Shigu (Lower Jinsha River)	Alpine deep-cut valley section	Slightly
		Shigu–Yushu (Upper Jinsha River)	Plateau valley section	Slightly
	Source area	Upstream of mouth of Chumar River	Plateau wandering section with shallows	Slightly

4.1.3 Ecosystemic Problems in the Yangtze River Basin

Due to the impacts of climate change and human activities, the ecosystem of the Yangtze River basin has faced enormous challenges, such as water pollution, habitat fragmentation or functional degradation, overexploitation, and invasive nonnative species.

4.1.3.1 Increasingly Serious Water Pollution

The water bodies in the Yangtze River basin receive more than 34 billion m^3 of waste effluents per year and a similar amount of agricultural return water (approximately 30 billion m^3) that contains residual fertilizers and pesticides. Although municipal wastewater treatment plants have been built and gradually put into operation, the average capacity for large- and medium-sized cities to collect and actually treat wastewater is less than 50% of required needs, and small cities and towns basically have no sewage treatment facilities, or even though they have treatment plants, the treatment results are not satisfactory. Consequently, large amounts of wastewater have been discharged directly into rivers, lakes, and other water bodies without treatment. The capacities for wastewater treatment have been below demands for more than 30 years. Moreover, air pollution, mining, urban ground hardening, and unregulated storage of solid waste and toxic and harmful substances have caused large amounts of contaminated rainwater to flow into the water system in every storm event. Presently, the amounts of contaminants carried by the first rainstorm event from nonpoint pollution sources are no less than the amounts of industrial and domestic wastewater discharge. Furthermore, with many petrochemical enterprises located along the Yangtze River and a large number of freight vessels sailing in the river, accidental water pollution incidents have occurred frequently. Therefore, the current water pollution conditions in most lakes, river sections near urban areas, and riparian zones in the Yangtze River basin have been increasingly serious, which has not only seriously affected the river ecosystem but also the human health and quality of life.

4.1.3.2 Fragmentation and Degradation of Habitats

Due to agricultural production and the construction of hydraulic projects, bridges, roads, communications facilities, etc., the ecological zones or habitats of the watershed have been seriously fragmented. There are only a few terrestrial ecological zones, such as primitive and continuous natural rivers and large natural forests, and the function of the whole ecological zone of the basin or scenic sections has been seriously degraded. For example, industrial and urban development and agricultural activities in floodplains, construction of dikes and control sluices along rivers, and installation of cascade reservoirs on the river channels have seriously affected the continuity of the rivers, the change process between flood and dry seasons, and the connectivity of ecological corridors. The construction and operation of large cascade reservoirs have markedly changed the natural hydrological process, and the distribution process of water and sediments in the Yangtze River, and affected the function of ecological corridors of continuous channels. If all the hydroelectric projects that have currently been planned are implemented, the mainstream in the upper and middle reaches of the Yangtze River and a considerable number of its tributaries will be partially or completely channelized, and the natural rivers will be transformed into seminatural or artificially controlled streams, which will have long-term and far-reaching impacts on the ecosystem and environment of the Yangtze River. The decrease of natural forests, the increase of artificial ones, and the construction of highways, railways, and oil and natural gas pipelines have caused seminatural and semiartificial forests to be fragmentated, area of farmlands and urbanization to increase, area of wildlife habitats to decrease drastically, especially habitats for high-grade and large wildlife to sharply decrease, and the animal food chain to be broken. Consequently, the balance of the watershed and terrestrial ecosystems has become growingly problematic.

4.1.3.3 Excessive Exploitation and Utilization of Resources

Excessive exploitation and utilization of natural resources such as land, forests, mines, water, and fisheries in the watershed have rendered the watershed ecosystem unsustainable. Since natural resources are limited and their generation and renewal tend to be rather slow. If the rate of development and utilization is much higher than the rate of their renewal, their carrying capacity will be exceeded, resulting in their depletion, and the human socioeconomic development will not sustain. Fisheries production, for example, does not take into account the fish production cycle, and the intensive use of dense nets and trawls to catch all sizes of fish has resulted in severe depletion of wild fish resources. Presently, there are tens of thousands of professional fishermen who rely on fishing for a living in the Yangtze River basin. Moreover, eating wild fish has become one of the recreational activities ubiquitously. Consequently, large numbers of wild fish resources are close to exhaustion. If no longer-time (several years) fishing ban is implemented and no solutions are developed to provide fishermen with alternatives for a living after their abandoning of fishing, the wild and rare fish species of the Yangtze River, such as the Yangtze (or baiji) dolphin and Reeve's shad, will become extinct.

4.1.3.4 Invasive Nonnative Species

Ordinary people may not notice the invasive nonnative species, but biologists are very worried. Because globalization brings about the rapid flow of materials and people, biological invasion has become very serious. Once nonnative exotic species adapt to the environment, they will spread rapidly, thanks to the lack of natural enemies. Consequently, the local native species cannot compete with them and start degrading or even become extinct. The introduction of water hyacinth is an example.

The fact that the Yangtze River is abundant in water has resulted in an impression that the river is still very vital. However, when the water quality is prudently monitored, and aquatic organisms and riparian vegetation are vigilantly observed, an increasingly serious problem will be identified. As large cascade hydroelectric stations are being constructed in major tributaries in the upper Yangtze, continuity of the rivers has been significantly affected. Both sides of the middle and lower reaches of the Yangtze River are studded with ports and bridges; banks have been intensively used; and vegetation on bank slopes has been replaced with hard protective surfaces. The Class II water in the Yangtze River is mainly in the deep channel; the water near banks is mostly Class III; and the water near the banks in some urban areas has even degraded to Class V. Moreover, concentrations of nitrogen and phosphorus in the Yangtze River water have reached the levels of Class III. When the river water enters reservoirs, lakes, or the ocean, the water is assessed to be only Class IV or even Class V in accordance with the water quality criteria for lakes and oceans. If the lake water has become eutrophic, an algal bloom can easily develop. If the water of this quality flows into the ocean, a red tide will result. Presently, rare wild fish species, such as the baiji dolphin, Chinese sturgeon, Yangtze River sturgeon, white sturgeon, and Reeve's shad, are on the verge of extinction.

4.2 Forests in the Yangtze River Basin

The forest ecosystem is one of the three major ecosystems of the earth, and it is also the ecosystem of richest biodiversity. The larger the forest area in the basin, the stronger the water storage and holding capacities. The forest can not only slow down the intensity of the flood process but can also make water slowly flow to the river in the dry season, which functions as a natural and risk-free reservoir. If the river basin has good forest vegetation, less soil and water loss will result; lower sediment yield and transport will occur; clearer water will appear in rivers and lakes; and increased aquatic biomass will emerge. Therefore, research and understanding of a watershed ecosystem inevitably involve the forest ecosystem.

4.2.1 Historic Evolution of Forest Cover in the Yangtze River Basin

In the geological period before the emergence of human beings, China's land forest cover had been generally over 80%. In the middle Holocene that is about 8–3 ka ago and corresponds approximately to the Daxi, Qujia, and Hemudu cultures representative of the Neolithic age in terms of archeology of the Yangtze River basin, the forest vegetation in the Yangtze River basin had been distributed very extensively. Except for the steppes in the source area, exposed rock in alpine areas, low-lying lacustrine marshes, and coastal saline-alkali areas, the Yangtze River basin had been almost all covered with dense subtropical (including some temperate and tropical) evergreen broad-leaved forests, coniferous forests, and deciduous forests whether they were plains, hilly terrains, or mountainous areas. At the beginning of the middle Holocene, the estimated forest cover should be around 80% (Zhou 1999). During this period, human activities were scarce; food supplies were mainly from forests by picking and hunting; and large-scale farming had yet to start.

The disappearance of primeval forests in the Yangtze River basin was closely related to the emergence and development of rice farming. The process of plain forests' disappearance can be divided into three stages, namely, Neolithic age, Shang and Zhou Dynasties and Spring and Autumn Period, and Qin and Han Dynasties.

First Stage: Neolithic Age The most common farming practices of our ancestors were "slash and burn." Specifically, after trees and weeds were cut, they were burned to ashes, and then crops were planted. Although the destructiveness to natural vegetation by farming in a different area using the "slash-and-burn" farming approach was quite severe, the destructed forest had sufficient time to recover due to the scarcity of population and the long fallow period after cultivation. Therefore, by the late Neolithic age, the forest area in the plains of the Yangtze River basin is estimated to have been still large.

Second Stage: Shang and Zhou Dynasties and Spring and Autumn Period The prevailing era of bronze farming tools in the Yangtze River basin began in the middle of the Shang Dynasty and extended to the early Warring States Period, while the iron farming implements started to appear in the late Spring and Autumn Period. Thus, it can be inferred that metal farming tools, especially the extensive use of iron implements, had led to relatively large-scale development of hilly areas in the Yangtze River basin. With frequent wars between various troops in the northern Yellow River basin, a large number of northern populations migrated into the Yangtze River basin. As a result, farming tools had a qualitative leap, and copper and iron metal tools gradually had replaced the stone tools. Compared with stone tools, metal tools were not only strong and sharp, but also could reduce labor needs and improve efficiency, which had undoubtedly helped advance the development of rice farming in the Yangtze River basin. Based on recent archaeological excavations, metal tools were unearthed in the Sichuan Plain in the upper reaches of the Yangtze River, plains of Dongting and Poyang Lakes in the middle reaches, the Tai Lake plain in the lower reaches, etc. The number of unearthed metal tools and their varieties was quite different among various places. The number of metal implements unearthed in the Tai Lake plain was relatively small.

Third Stage: Qin and Han Dynasties The agricultural productivity of the people in the Yangtze River basin was higher than that in the previous generations, but the agricultural activities were still concentrated in plains and valleys. Therefore, by the Eastern Han Dynasty, the hilly areas of the Yangtze River basin had still maintained flourishing forest vegetation, and by the end of the second century, the forest cover had still been nearly 70%.

From the third to the fourteenth centuries, the emergence of drought-tolerant crops began to deplete the hilly forests. During the approximately 1 ka from the Three Kingdoms and Jin Dynasty to the Song and Yuan Dynasties, the destruction of forests in the hilly areas (including some low mountains) in the Yangtze River basin mainly came from the expansion of dryland farming: (1) expansion of dryland areas for cultivating crops, (2) prevalence of tea planting, and (3) spread of Champa rice. According to legend, the Champa rice was from the Champa state and was one of the indica rice varieties. It was one of the many high-quality seeds introduced by the Song Dynasty from overseas that could be cultivated in the widest range of lands. The greatest advantage of the Champa rice was drought-resistant and suitable to be cultivated on infertile lands. Therefore, once introduced to Fujian in the early Song Dynasty, it was soon spread to the hilly area of the Yangtze River basin. By the middle of the fourteenth century, the natural forest cover in the Yangtze River basin was below 40%.

During the fifteenth to twentieth centuries, the emergence of corn and other American crops resulted in the shrinkage of mid-alpine forests. Large-scale destruction of mountainous forests in the Yangtze River basin is estimated to have begun in the Ming Dynasty, intensified in the Qing Dynasty, and continued to the modern times. In the mountainous region of Qinba, especially in the mountainous region of Daba, the forests were nearly completely destroyed in less than 100 years during the Qing Dynasty. The reason for this was the heavy population pressure. Without the introduction of corn, potato, sweet potato, or other American crops, the destruction of forests would definitely not have been so fast. The generally accepted reason for the American crops, such as corn, potato, etc., to have replaced the traditional dryland crops and become the main crops in mountainous regions was that corn, potato, and other American crops were drought-resistant, suitable for infertile lands, highly productive, and cold-resistant. Generally speaking, the higher the altitude, the lower the temperature; accordingly, the soil temperature was reduced; and the accumulated temperature decreased. Consequently, the crop-growing season naturally shortened, as was the case in the mountainous area in the middle and lower reaches of the Yangtze River. The traditional dryland crops such as wheat and millet are suitable to grow at an altitude no more than 1000 m, but corn and potato are suitable to grow at an altitude of about 1500 m. By 1949, the natural forest cover in the Yangtze River basin was only 5–6%, and if other secondary natural forests and artificial forests are considered, the forest cover was around 16%.

4.3 Existing Forest Conditions in the Yangtze River Basin

Since the founding of the PRC, China's forest resources and their structure have undergone great changes. During 1950–1981, the overall cover of forests was in a period of fluctuation, but the area of natural forests was declining and that for artificial forests was gradually increasing. Conversely, during 1981–1993, China's forest cover was in a period of slow growth; the decreasing trend of natural forests was controlled and then reversed to an increasing trend; and the area of man-made forests continued to increase steadily. During 1994–2003, the forest cover was in a period of a relatively fast growth. After the Yangtze River large flood of 1998, the Central Government implemented protective policies for natural forests. As a result, both the natural and man-made forests underwent an accelerated growth, and the growth rate of natural forests began to exceed that of the man-made forests. As of 2008, the 7th Forest Census indicated that the forest cover was at 20.36%, which was a significant improvement over 1949. However, the forest cover was still well below the world average of 31%, and the per capita forest area was only one-fourth of the world average.

Before the 1980s, forest logging in China had been dominated by clearcutting. Although the cutting quantity was relatively low, the forest cover was still at a negative growth rate. After the 1980s, selective cutting gradually became dominant, but multiple selective cuttings made the traditional rotational clearcutting seem to have become dispensable. The lower the stocking per unit area, the more frequently the selective cutting occurred. This vicious cycle was the crucial cause for the difficulties in recovery of the forest quality in China.

As the human land has been used intensively, the increase in the area of natural forests has been limited. As a result, man-made forests are the only way to increase forest cover and the total amount of forest resources. After more than 50 years of planting, protection, and utilization, China's forest area and stocking increased significantly. The forest area increased by approximately 111.74%, but the stocking increased by only about 17.4% when compared to the early PRC (Zhang 2006, 2007).

In the past 60 years, great changes have taken place in the forest structure of China. The main characteristics are as follows: ① The proportions of the area and stocking of man-made forests have increased gradually. The fraction of the area increased to 33.77% in 2003 from 4.49% in 1964. While the proportions of the area and stocking of natural forests have decreased noticeably, the ratio of the stocking of natural forests still accounted for 89.65% in 2003. ② The proportion of the area and stocking of coniferous forests have been decreasing gradually from the original 70%. Presently, the ratio between coniferous forests and broadleaf forests is close to 1:1, and the dominant stocking has also undergone a great change. For example, the high-quality timber forests such as Korean pines, Scots pine, and Chinese white pines are almost completely depleted. ③ The forest structure has become young, and the stocking per unit area has been reduced. Consequently, harvestable resources have been reduced. ④ The structure of forest types has also changed significantly. The shelter forests have increased noticeably, and the timber forests have decreased greatly. The area and stocking of the two forest types are nearly equal. ⑤ The management of the shelter forests is in serious trouble, and the management of the forest resources is still in the early stage of the process that is gradually achieving a sustainable operation of forests consisting of multi-resources and multifunctional uses.

After the 1980s, the forest resources began to recover gradually with an upbeat trend. However, in 2003, the quality of forest resources still did not match the 1962 level. Newly man-made forests cannot be accounted into the forest cover until they have grown into dense woods, and the time required for trees to grow into dense woods is roughly 2–4 years in southern China and about 3–5 years in northern China.

Although the area and stocking of natural forests in the Yangtze River basin had been decreasing, the basin's forest cover has still increased because large-scale afforestation efforts have been carried out in the basin since the 1950s. According to the annual survey and statistical analysis of forest resources conducted by the Ministry of Forestry during 1977–1981, the forest cover in the basin was about 18%.

In 1998, pilot forest protection projects began, and in 2000, the Natural Forest Resources Protection Project (simply referred to as Natural Protection Project) and the Returning-Farmland-to-Forest Project were formally implemented, which were the largest ecological projects implemented in China in the early twenty-first century. The implementation of the Natural Protection Project was a timely response to the large floods that occurred in the Yangtze and Nen River basins in 1998. The implementation of the Natural Protection Project has completely stopped commercial logging of natural forests in the upper reaches of the Yangtze River and in the upper and middle reaches of the Yellow River. As a result, the forest resources in China have been recovering rapidly in the past 10+ years, and the accumulated carbon sinks have increased greatly.

The Yangtze River is in the middle latitudes, with a total length of more than 6300 km, of which more than 3600 km flows through subtropical humid areas with relatively rich water and soil resources. The Yangtze River basin has a total land area of 180 million ha, of which 30.8 million ha are presently farmland with 18.1 million ha of paddy field and 12.7 million ha of dryland. The ratio between paddy field and dryland is 3:2, and the area of farmland accounts for 17% of the total land area in the basin. The cultivated area per capita is 0.072 ha which is below the national average. Therefore, the conflict of a vast population with less arable land in the basin is prominent. Nonetheless, the combination of concentrated fertile land and excellent agricultural climate in the Yangtze River basin has made it boast the vast majority of the agrobiological resources in China. Moreover, the basin encompasses abundant grassy hills and slopes and freshwater, oceanic, and beach resources.

As most of the Yangtze River basin is in a typical subtropical climate zone, forest plants can grow throughout the year. As a result, it has become China's primary source of forest products. With a forest area of 36 million ha, the current forest cover is 20%, and the timber stocking accounts for about one-fourth of the country. Historically, the Yangtze River basin has been the traditional commercial base for fir, bamboo, camellia oil, tung, tea, lacquer, etc.

The forests in the river basin are mostly concentrated in the upper reaches of the Jinsha, Min, Dadu, and Jialing Rivers and are an important part of the Southwest Forest Region of China. The main species are Yunnan pine, spruce, and precious species such as camphor, zhennan, teak, rosewood, etc. The hilly areas in the middle and lower reaches boast relatively more forests, such as the upper reaches of the five rivers (Qing, Xiang, Yuan, Qingyi, and Shuiyang Rivers) in the Poyang Lake system and the Shennongjia area in Hubei. Besides pine and fir, there are camellia oil, tung, Chinese tallow tree, lacquer, privet, bamboo, tea, and other economic trees.

In addition, as the forests possess the rolling mountainous terrain, towering trees, exotic flowers, and rare plants, they provide suitable habitats for the survival and breeding of a variety of wildlife species. There are many world's rare and endemic species that are among the major species protected by the government of China, such as giant panda, golden monkey, white-lipped deer, wild donkey, black-necked crane, takin, macaque, tufted deer, hoatzin, etc.

4.4 Wetland Ecosystem of the Yangtze River

Under the Ramsar International Wetland Convention Treaty (*Ramsar Convention on Wetlands of International Importance especially as Waterfowl Habitat*), "wetlands are areas of marsh, fen, peatland or water, whether natural or artificial, permanent or temporary, with water that is static or flowing, fresh, brackish or salt, including areas of marine water the depth of which at low tide does not exceed six meters." Wetlands are one of the earth's three major ecosystems. As the importance of wetlands to the ecosystem had not been recognized in the past, more than one-half of the world's wetlands had disappeared. However, presently, wetlands have been recognized as very important to the ecosystem and human beings. Wetlands not only have great ecological values but also provide human beings with comprehensive values such as important water sources, flood control, food, materials, environment, and culture.

4.4.1 Distribution and Characteristics of Wetlands in the Yangtze River Basin

4.4.1.1 Regional Distribution of Wetlands

As the Yangtze River basin extends across three major regions, namely, eastern, central, and western China, and through plains, basins, mountainous areas, and plateaus, the geological and geomorphic conditions in the basin are very complex, and the physical geographical environment varies markedly. Moreover, hydrothermal conditions vary regionally and geomorphologically. Therefore, wetland types are diverse, and the distribution of wetlands is also extensive and unbalanced.

The source area of the Yangtze River is located in a plateau. Due to abundant water replenishment from the melting ice in glaciers and the characteristics of the plateau's landform, drainage conditions are poor, which have helped the development of large marshes, lakes, and wandering rivers on the plateau. The source area is China's largest swampy area. Moreover, the region is sparsely populated with little impact from human activities. As the Yunnan-Guizhou Plateau in the upper reaches of the Yangtze River has been dominated by the deep downward cutting by streams, lakes and palustrine wetlands are sporadic and scarce. However, due to the influence of the collision between the Indian Plate and the Tibet Plate, many tectonic lakes have developed, and most of the marshes have resulted from lakes. The middle and lower reaches of the Yangtze River, where several of China's largest freshwater lakes are located, are the most concentrated area of freshwater lakes in China. Because of the high density of population, the area has been intensively influenced by human activities. The delta region of the Yangtze Estuary, which is located in the junction of an alluvial river network, the sea, and adjacent land, boasts a complicated river network of plains and large areas of coastal tidal-beach wetlands.

The distribution of wetlands is very uneven among the provinces, municipalities, and autonomous regions in the Yangtze River basin. Table 4.3 shows the area of natural and artificial wetlands in each administrative region of the Yangtze River basin.

Table 4.3 indicates that Qinghai, Hunan, and Jiangxi have relatively more areas of wetlands, while Guizhou and Yunnan have relatively less. Gansu and Henan have only sporadically distributed wetlands. As Tibet, Guangxi, Guangdong, and Fujian have very small portions in the Yangtze River basin and the areas for natural and artificial wetlands are very small, they are not included in the table. There are no statistical data available for riverine wetlands, and therefore, they are not listed in this table either. The statistical data for artificial wetlands mainly includes areas of paddy fields and reservoirs.

The spatial distribution of wetlands in the Yangtze River basin can be characterized as being widespread and uneven. Marshes are mainly distributed in the Dangqu River basin in the source area and the plain areas of Dongting and Poyang Lakes in the middle and lower reaches. However, in Sichuan and Chongqing where human development and utilization began earlier, there are very few natural palustrine wetlands. In the Yunnan-Guizhou Plateau, because deep downward cutting by rivers was not conducive to wetland development, there are only sporadic lakes and marshes. The Jianghan Plain in the middle reaches used to be the original Yunmeng Marsh area that has evolved into a fragmented low-lying region with sporadically distributed lakes due to the natural process and the impact of human activities.

4.4.1.2 Characteristics of Major Wetlands

The major wetlands in the Yangtze River basin have their own characteristics. The wetlands in the Qinghai-Tibet Plateau consist of alpine meadows or marshes in the subregion of the Tibetan wormwood-lichen marshes in southeastern Qinghai and

Table 4.3 Wetland area in each administrative district of Yangtze River basin

Province, municipality, or autonomous region	Marsh area (km^2)	Lake beach area (km^2)	Natural wetland area (km^2)	Paddy field area (km^2)	Reservoir area (km^2)	Wetland area (km^2)
Qinghai	9153	650	9803	–	–	9803
Tibet	–	–	–	–	–	0
Yunnan	467.44	445.52	912.96	2242.1	84.6	3239.7
Sichuan (Chongqing)	191	55.6	246.6	30,656.1	368.8	31,271.5
Guizhou	25	13	38	5797.5	156.9	5992.4
Gansu	–	0.8	0.8	42.7	17.1	60.6
Hubei	478.6	2383.5	2862.1	17,589.5	1538.5	21,990.1
Hunan	860	3577.3	4437.3	24,936.9	885.6	30,259.8
Jiangxi	640	3627.9	4267.9	17,381.7	800.4	22,450
Shaanxi		–	–	2387.4	31.7	2419.1
Henan	12	–	12	625.5	77.2	714.7
Guangxi	–	–	–	623.3	7.3	630.6
Guangdong	–	–	–	14.1	–	14.1
Anhui	155.3	2330.1	2485.4	6745.9	293.4	9524.7
Jiangsu	181.4	2517.4	2698.8	11,605.9	47	14,351.7
Shanghai	2660.1	68	2728.1	2573.9	27.8	5329.8
Zhejiang	–	–	–	3400.3	–	3400.3
Fujian	–	–	–	34.6	–	34.6
Total (rounded)	14,824	15,669	30,493	126,658	4336	161,487

western Sichuan plateaus. The wetlands in the Yunnan-Guizhou Plateau and hilly areas in Central China are located in the subregion of shallow water and peatland moss-marsh wetlands and vegetative wetlands. The wetlands in the middle and lower reaches of the Yangtze River are situated in the subregion of reedy wetlands in plains and shallow-water vegetative wetlands.

In the Qinghai-Tibet Plateau in the source area of the Yangtze River, large areas of palustrine and lacustrine wetlands have developed, for example, Cuorendejia and Duoergaicuo Lakes. Marshes in the source area of the Yangtze River, such as Dangqu Marsh, cover a total area of more than 6700 km^2. The marshes are located in the hinterland of the Qinghai-Tibet Plateau and have an average elevation of approximately 5000 m. There are many lakes and lacustrine palustrine wetlands in the source area of the Chumar River which is a tributary of the Tongtian River. Many marshes are distributed in the Beilu and Yawu Rivers, which are also tributaries of the Tongtian River. These marshes are not only wormwood-marsh wetlands endemic to the alpine region of northern Tibet but also the main pastoral areas of Qinghai Province. The genesis types of the lakes in the source area of the Yangtze River are complex and diverse, but most of them are located in bases between parallel mountains or large valleys. Large- and medium-sized lakes, such as Lake Nam, are formed by tectonic movements. The lakes are very deep with the deepest at

150 m, and the slopes of the lake banks are very steep. Many lakes are in the hinterland of the plateau, and most of them are in the tails and backwater zones of rivers. Some outflowing freshwater lakes resulted from the headward erosion and deep downward cutting in the source area of the Yangtze River, such as Duoergaicuo and Yaxingcuo Lakes. The marsh zone in the Qinghai-Tibet Plateau is unique in the world, such as the wormwood and moss marshes in northern Tibet that were mainly transformed from lakes, rivers, and meadows. Although the area of the marshes is large, the types of them are simple. Most of the marshes are eutrophic, and the wetlands are mainly replenished by snowmelt.

There are alpine lakes and marshes in the Yunnan-Guizhou Plateau and western Guizhou Plateau areas of the upper Yangtze. For example, the Caohai Lake Marsh is the largest marsh in Guizhou with very rich biodiversity, and it is home to many rare wildlife species and one of the wintering habitats for black-necked cranes that are on the priority list for protection in China. Marshes mainly developed in lacustrine beaches and basins, and most of them were transformed from lakes. The type of the marshes is mainly mossy, but some are reedy or other types. Due to tectonic plate movements, large numbers of tectonic lakes have developed in the Yunnan Plateau, such as Dianchi, Jianhu, Chenghai, and Lugu Lakes.

The middle and lower reaches of the Yangtze River boast broad plains studded with lakes and are areas most concentrated with lakes in China. Four of the famous five freshwater lakes in China are in this area: Dongting, Poyang, Chao, and Tai Lakes. The middle and lower reaches of the Yangtze River and the Yangtze Delta region possess crisscrossing water networks and many lakes throughout the areas. The genesis of the lakes was mostly related to the rise of sea level and the evolution of the river systems at the end of the late Quaternary. For example, the lake system in the Jianghan Plain and Dongting Lake in the middle reaches of the Yangtze River resulted from the interaction from the mainstream of the Yangtze River and its following tributaries: Han, Xiang, Zi, Yuan, and Li Rivers. Longgan, Huangda, and Bo Lakes in the middle and lower reaches of the Yangtze River formed due to the southward movement and swinging of the mainstream of the Yangtze River. As many lake systems are shallow, coupled with favorable sunlight and thermal conditions, aquatic plants have flourished with diversified shallow-water plant species. The formation of marshes was mainly through the transformation of lakes, supplemented by the conversion of rivers. Due to thousands of years of dike construction and land reclamation, large numbers of natural marshes have been converted into artificial wetlands in the primary form of paddy fields, and natural marshes are only distributed in some lake areas and floodplains. The basic types of these marshes are freshwater reedy swamps, followed by freshwater mossy marshes with poorly developed peat layers. These reedy swamps boast many related plant varieties and are mainly replenished by water from lakes and surface runoff, followed by groundwater and precipitation. The lacustrine marshes vary greatly with the change of seasons and are mainly affected by the flood process during the wet season and large rivers, such as the Yangtze River, and the hydrological process of lakes with rivers flowing

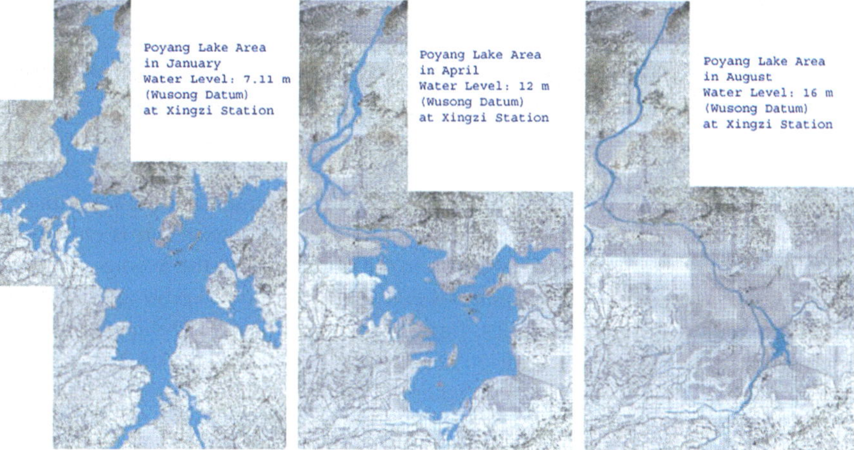

Fig. 4.1 Seasonal change in surface area with various water level at Poyang Lake

through. The plain wetlands in the middle and lower reaches of the Yangtze River are rich in plants and forage. As a result, wildlife resources are abundant, and many fish and bird species live here. For example, Dongting and Poyang Lakes provide all bird species with habitats for reproduction, wintering, and migratory destination. Figure 4.1 shows various areas of the water surface of Poyang Lake at different water levels and indicates that Poyang Lake is the most typical lake with rivers flowing through, which is manifested by the fact that it appears a linear feature during the dry season and shows a large surface area during the flood period.

The estuarine wetlands at the Yangtze Estuary include the coastal wetlands along the river estuary and the wetlands on sandbars and islands at the mouth of the river. The wetland resources are rich, and the biodiversity in the wetlands is very high. There are commonly 157 bird species, 112 fish species, 58 large benthic animal species, 128 zooplankton species, and 136 wetland plant species in the Yangtze Estuary.

4.4.2 Artificial Wetlands

Artificial wetlands mainly include reservoirs, paddy fields, and other man-made wetlands. The ecological function of the artificial wetland is not as good as the natural wetland. However, the area of artificial wetlands has been increasing, while that of natural wetlands has been decreasing, resulting in a structural change of wetlands. Especially in the densely populated area, as the area of natural wetlands has greatly reduced, the ecological function of artificial wetlands has become prominent, and some aquatic organisms and birds can survive and even massively reproduce in the artificial wetlands.

4.4.2.1 Reservoirs

A reservoir is a man-made lake constructed on the original channel of a river. Compared with a common natural wetland, the artificial wetland has the following characteristics: ① The body of water is deep. As most reservoirs are constructed in mountainous areas, they are formed by dams that make the water levels of the rivers higher. For larger reservoirs with high dams, the temperature of the water in the reservoirs is stratified with depth. ② The water surface of a reservoir is elongated and irregular. For example, the TGR is up to 600 km long, but the average width is only 1–2 km. Moreover, many tributaries converge into the reservoir area, and the reservoir extends into the tributaries, forming reservoir bays or archipelagos. ③ A reservoir often has a large fluctuation zone where the water level of the reservoir varies greatly with seasons and the gradient of environmental changes in the fluctuation zone is large. As a result, the fluctuation zone becomes a seasonal wetland. ④ As the water in a reservoir is deep, the quantities of large aquatic plants (aquatic grass) or benthic wildlife are relatively small, resulting in a small biomass per unit volume of water.

The reservoir area in Table 4.3 does not include newly constructed large reservoirs, such as the TGR. In the area of recently constructed reservoirs, the reservoir area in the Yangtze River basin should exceed 5000 km^2. For example, Danjiangkou Reservoir and TGR have very large areas of water surface and fluctuation zones, and their wetland function cannot be ignored. As for the runoff hydroelectric stations or avionics reservoirs in the plain areas, although they increase the area of the water surface for the entire year, the increase is largely at the price of encroaching the wetlands on floodplains. The ecological function of the increased area of the water surface is much lower than that of the natural wetlands on the floodplains.

4.4.2.2 Paddy Fields

A paddy field is a typical artificial wetland where rice grows fast. The longest mature time for a rice crop is 1 year, and the fastest mature time is 3–4 months through the process of germination, flowering, and grain production. Paddy fields are inundated by water for at least 3–4 months during most of the rice-growing season and are home to many amphibians, insects, and birds. Moreover, paddy fields have certain wetland functions.

The rice-growing area in China can be divided into 6 regions and 16 subregions. The rice-growing area in the three south regions makes up 93.6% of the total area in China. Bordered by the East China Sea on the east, the western edge of the Chengdu Plain on the west, the Nanling Mountains on the south, and the Qinling Mountains and the Huai River on the north, the Central China Double- and Single-Crop Paddy Rice Region has a subtropical warm humid monsoon climate and consists of all or majority of Jiangsu, Shanghai, Zhejiang, Anhui, Hunan, Hubei, and Sichuan Provinces and Chongqing Municipality, as well as southern Shaanxi and Henan Provinces. The year-long rice-growing area in the region is about 18.3 million ha, accounting for 61% of rice-growing area of China and more than 50% of the natural wetland area in the Yangtze River basin, and most of the paddy fields were transformed from prior natural wetlands.

4.4.2.3 Artificial Wetlands

In the past 20 years, with people's increased awareness of the importance of wetlands, many cities have constructed large numbers of small wetlands in parks, riversides, lake shores, and residential subdivisions. Although the area of these individual wetlands is small, the total number and cumulative area have been gradually increasing. These wetlands are important ecological speckles in urbans areas, as well as important recreational places for urban residents. Moreover, the wetlands have not only beautified the urban environment but have also provided habitats for some aquatic organisms and bird species.

4.5 Rare Fish Species and Fish Resources in the Yangtze River

Fish are the top-level organisms of the biological chain living in the water, and their survival conditions can reflect the health of the water ecosystem, just as the survival conditions of tigers and lions are the most reflective of the biological conditions of forests and grasslands. If rare and endemic fish species can have a normal, stable living environment, the aquatic ecosystem can generally be considered as healthy, and its water quality can generally meet the needs of human beings. Therefore, the flagship fish species is often considered as an indicator species in evaluating the health of a river, lake, and wetland ecosystem.

4.5.1 Fish Resources in the Yangtze River

The water system-lake relationship in the Yangtze River basin is complex with large areas of aquatic habitats of various diversities and rich fish resources. According to statistical data of surveys, the Yangtze River water system has 370 fish species, accounting for 48% of the total 772 freshwater fish species in China and is ranked first in the water systems of Asia (Wang and Wang 2004). Among the 370 fish species, there are 294 freshwater species, 22 brackish species, 9 species migrating between freshwater and seawater, and 45 marine species. In terms of endemic species, there are 228 non-endemic species and 142 endemic species. In terms of the regional distribution of endemic species, there are 112 endemic species in the upper reaches, 21 in the middle and lower reaches, and 9 distributed in the entire river basin. As indicated, the upper reaches boast most endemic fish species, and there are 188 fish species distributed solely in the "National Nature Reserve of Rare and Endemic Fish Species in the Upper Reaches of the Yangtze River." In addition to the protected species of Chinese paddlefish, Dabry's sturgeon, and Chinese sucker, 63 other fish species live in the National Nature Reserve, most of which are adapted to the rapid flow environment and live by feeding on algae or benthic invertebrates on the ground floor or crawling onto gravel, indicating that the water system in the upper reaches of the Yangtze River possesses unique aquatic habitats and a unique hydrological and hydrodynamic environment.

Of the fish species in the Yangtze River, nine species have been on the List of Animals under Special State Protection, of which three species receive Level I protection, Chinese sturgeon, Dabry's sturgeon, and Chinese paddlefish, and six species receive Level II protection, Sichuan taimen, Chinese sucker, golden-line barbel, Dali schizothoracin, giant mottled eel, and roughskin sculpin (Liu and Cao 1992).

In addition to the very important biological genes, species, and population diversity, the fishery in the Yangtze River is vital in China's freshwater fishery. Fish are an important source of protein in human food for thousands of years, and it can be claimed that China's freshwater fishery would not be ranked as the world leader without the fishery in the Yangtze River basin. Taking the annual production in 2000 as an example, Shanghai, Jiangsu, Anhui, Jiangxi, Hubei, Hunan, Chongqing, Sichuan, Guizhou, and Yunnan Provinces or Municipalities in the Yangtze River basin had a combined area of freshwater aquaculture of 2.81 million ha, making up 53.2% of China's total area of 5.278 million ha; the freshwater fishery production was 9.839 million tonnes, accounting for 56.5% of China's 17.403 million tonnes; and the revenue from the freshwater fishery was 52.67 billion yuan, accounting for 56.5% of China's 93.25 billion yuan. Of the top ten provincial or municipal freshwater fishery producers in China, seven are in the Yangtze River basin, of which Hubei is ranked second nationally and first in the Yangtze River basin, whose freshwater fishery production accounts for one-fourth of the Yangtze River basin. The Yangtze River is home to the FMCC, Chinese mitten crab, Reeve's shad, and Japanese eel. The fish quality in the Yangtze River is the best in China. The Yangtze River is not replaceable by other water systems or artificial facilities due to its precious genetic resources.

In the past 30+ years, due to the combined impact of water pollution, overfishing, habitat loss, and barriers resulting from hydraulic projects, of the renowned "three delicacies" (Reeve's shad, pufferfish, and swordfish), presently only swordfish have the value for development. The Reeve's shad has been prohibited from harvesting in the Yangtze River since 1988, but to no avail. Only several pufferfish can be caught each ship a day. Although the yield of swordfish can still reach a certain amount, there are signs of decline. If not protected, the swordfish will potentially follow the fate of the Reeve's shad. The number of natural fries of the FMCC has been greatly reduced.

Research on the biological characteristics of typical fish species can not only help protect representative flagship species but also provide the important technical basis for watershed and reservoir managers to carry out the ecological regulation of reservoirs and discharge the ecological flow from reservoirs.

4.5.2 *Bronze Gudgeon (*Coreius heterodon*) and Coreius Zeni (*Coreius guichenoti*)*

Bronze gudgeons and coreius zeni are common Chinese names for the ray-finned fish species in the family of Cyprinidae, subfamily of Gobioninae, and genus *Coreius*. They are endemic species of the Yangtze River and typical river fish species, which prefer to live in a certain water flow velocity zone in the river.

The biological characteristics of coreius zeni (see Fig. 4.2) are the most representative of fish species in the upper reaches of the Yangtze River. The necessary aquatic environmental conditions can be used as one of the indicator species to determine the ecologic flow and implement the ecological regulation of reservoirs. However, bronze gudgeons are distributed in both the mainstream and tributaries of the Yangtze River, mostly in the middle and lower reaches.

Coreius zeni are a bottom-feeder and are commonly seen in the lower reaches of the Jinsha River, the mainstream in the upper reaches of the Yangtze River, middle and lower reaches of the Jialing River, the Tuo River, the lower reaches of the Min and Wu Rivers, etc. Adult coreius zeni live in rapids and backwater zones of the mainstream, seldom migrate into tributaries, and generally do not migrate to small tributaries where the water flow is slow. Coreius zeni are omnivorous and love to feed at the bottom of rapids in flowing rivers. Their food composition consists mainly of freshwater mussels, clams, snails, and mollusks, followed by higher plant debris and some diatoms. The spawning period of the fish in the Yangtze River is approximately 3 months from the beginning of April to the beginning of July. Due to the difference in water temperatures, the annual spawning activities in the Pingshan section of the Jinsha River begin in early April, peak in mid-April to early May, slow down in late-May, and basically end in July.

Coreius zeni spawn in turbulent flow sections and their spawning sites are distributed in the upper reaches of the Yangtze River. Their eggs need to drift while being hatched; some fries drift to the sections downstream of the middle reaches; and they migrate gradually upriver during the growing process. In addition to their intrinsic physiological factors, coreius zeni require certain external hydrological conditions when they lay eggs. There is a very close relationship between the ovi-

Fig. 4.2 Coreius zeni

position and the water level. The rise in water level in the river includes a series of hydrological, hydrodynamic, and water quality changes, such as increased flowrate, higher flow velocity, reduced water transparency, and turbulent flow. According to the data of the fish eggs collected from four river sections over the past years, the appropriate change in flow velocity is one of the main external conditions to prompt coreius zeni to spawn. When the water temperature of the spawning section reaches above 17 °C, the rise in water level (increase in flowrate) can play the role of stimulating bronze gudgeons and coreius zeni to spawn (Liu 1990).

*4.5.3 Chinese Paddlefish (**Psephurus gladius**)*

The Chinese paddlefish (see Fig. 4.3) is a pure freshwater fish species and has been designated as a rare animal species under the state Level I protection. The current number of Chinese paddlefish is very low, and they mainly live in the upper reaches of the Yangtze River. In the past, the fish species used to be distributed in the entire mainstream and the lower reaches of major tributaries of the Yangtze River. Presently, the fish spawning sites can only be found in the mainstream section from the lower reaches of the Jinsha River to Chongqing. The past annual yield of the fish is estimated to be approximately 25,000 kg. In October 1972, the Luzhou Fishery Society of Sichuan caught one Chinese paddlefish that weighed 400 kg. Due to the impact of human activities, the current number of the fish has been greatly reduced. The Xiangjiaba Hydraulic Project in the Jinsha River that is under construction now will have most of the spawning sites of the fish submerged in the reservoir.

Chinese paddlefish are a semi-anadromous fish species, and they occur in the middle and lower water layers of the mainstream of the Yangtze River, but sometimes migrate into large lakes along the river. Large individuals mostly inhabit the deepwater channel of the mainstream, but the juveniles often feed to tributaries, ports, river mouths and associated waters, and even the brackish zone of the Yangtze Estuary. Chinese paddlefish are bottom-feeding carnivorous animals with a fero-

Fig. 4.3 Chinese paddlefish

cious nature. Adult and juvenile Chinese paddlefish feed mainly on fishes and shrimps and vary with seasons and habitats. Gudgeons and bronze gudgeons are their main food in the upper reaches of the Yangtze River during the spring, and shrimps are the primary food during the fall and winter. The first sexual mature age is 7–8 years old; and an individual is usually about 2 m in length and above 25 kg in weight. After their sexual maturity, they migrate upriver to spawn in the Yibin section of the Yangtze River (Zhu 1987; Ma and Cai 1996). One 30 kg female can carry about 200,000 eggs, and an 87 kg parent fish can produce up to 740,000 eggs.

The breeding season of the Chinese paddlefish is during March–April when the spring water comes. Eggs sink and are laid on the rocky riverbed. The spawning temperature is generally above 17 °C. The spawning river section is required to have a width of approximately 360 m, a water depth of about 10 m, a flow velocity of 0.49 m/s, and a rocky or gravel riverbed (Li and Xi 1997). The eggs are oval-shaped, expand rapidly by absorbing water in a few days after spawned, float from the rocky or gravel riverbed, drift with the flowing water, hatch slowly, and grow. The fertilized eggs hatch into fries in 45 days. The newly hatched translucent fries are 40 mm long.

*4.5.4 Dabry's Sturgeon (*Acipenser dabryanus*)*

The Dabry's sturgeon is a freshwater fish species. It has been designated as a rare animal species under state Level II protection. The fish are mainly distributed in the lower reaches of the Jinsha River and the upper reaches of the Yangtze River. According to a survey conducted in the 1970s, the fish occurred in every major tributary in the upper reaches of the Yangtze River, but they have disappeared from the major tributaries in recent years. Only a small number of juveniles and sexually mature individuals are distributed in the river section between the Nanxi District of Yibin and Hejiang County of Luzhou. The spawning parent fish are dispersed and sporadic, and the spawning sites are not concentrated.

Dabry's sturgeons often live in the substrate of the river but also migrate into large lakes. The fish especially prefer to live in the slow-flowing river bend or deep pools where humus and benthic organisms are rich over substrates of sand and gravel barriers. The fish grow fast; the reproductive males reach sexual maturity at 4–7 years old and females at 5–8 years old. The body of a mature fish is generally 0.8–1.0 m in length and 5–10 kg in weight. They spawn during March and April of the spring and sometimes in October and November of the fall. During the breeding season, mature individuals migrate upriver toward the upper reaches and spawn in the river's mainstream. The eggs sink and stick to the riverbed gravel for development at the spawning sites. The fish feed mainly on benthic invertebrates and on aquatic plants, algae, and humus as well. The juveniles feed on aquatic zooplankton and small fish, and larger juveniles and adults feed principally on humus and benthic invertebrates (Huang 1980).

4.5.5 *Chinese Sucker (*Myxocyprinus asiaticus*)*

The Chinese sucker has been designated as a wild animal species under state Level II protection. Presently the fish only occur in the Yangtze River system, mostly in the upper reaches, and their spawning sites are mainly distributed in the Jinsha, Min, and Jialing Rivers (Zhang and Zhao 2001). Chinese suckers feed mainly on benthic invertebrates and organic matters contained in the substrates of mud and on some of the higher plant debris and algae as well. Generally sexual maturity is reached at 5–6 years old. In every mid-February (around the Chinese rainwater seasonal point), the parent fish whose gonads are close to mature migrate upriver toward the upper reaches. Most parent fish spawn in fast-flowing shallow gravel riverbed during March and April when the water temperature rises to above 13 °C and other outside conditions are appropriate. The most suitable temperature for spawning is 15–18 °C, but spawning does not have any noticeable relationship with the river's water level, flowrate, flow velocity, or sediment content. As long as there is a suitable substrate and water temperature, spawning can occur. The eggs are slightly sticky and adhere to the gravel for development. The developmental time from fertilization to initial food ingestion is approximately 20 days (Yu et al. 1988). After they are hatched, the juveniles drift to the middle and lower reaches for growing. The fries and juveniles often live in groups among gravels where the water flows slowly, and they are mostly active in the upper water layer and swim slowly. The half-grownup fish are accustomed to living in lakes and the middle and lower reaches of the Yangtze River. They are active in the middle and lower layers of the water and move slowly. The adult fish live in the upper reaches of the river, and they are active in the middle and lower layers of the water (Jiangsu Provincial Freshwater Aquatic Products Institute 1987; East China Sea Aquatic Products Institute of China Institute of Water Resources and Hydropower Research 1990). A 10 to 15 kg female fish can carry about 0.1–0.2 million eggs, and the eggs stick to the riverbed stone or algae after they are laid. However, the larvae drift.

4.5.6 *Chinese Sturgeon (*Acipenser sinensis*)*

4.5.6.1 Basic Characteristics

The Chinese sturgeon (see Fig. 4.4) is a large anadromous fish species. It is one of the three endemic sturgeon species in China and has been designated as an animal species under state Level I protection. The anadromous fish migrate upriver toward the upper reaches from the estuary during May–August every year and spawn in an appropriate section between the lower Jinsha and the upper Yangtze during October–November of the next year. More than 16 spawning sites have been reported (Zhang et al. 2008). The breeding season is between the early October and the early November. The Chinese sturgeon is of a one-time ovulation type, and an individual

Fig. 4.4 Chinese sturgeon

releases 300,000 to 1.3 million large eggs with a density of 1.72–4.45 eggs/gram. The eggs sink and stick to gravel for hatching. The Chinese sturgeon is carnivorous and stops feeding in the spawning period. During its life in the sea, it feeds intensively mainly on sole, followed by yellow croaker, crucian carp, black carp, cutlassfish, Indian anchovy, conger eel, shrimp, and crab. In the estuary area, a juvenile has a level 3–4 stomach feeding intensity and feeds on the inshore benthic fishes. In the middle and lower reaches of the Yangtze River, it feeds on shrimp and crab and sometimes on some tribonema and aquatic microtubule.

The Chinese sturgeon used to spawn in the 800 km river section between Pingshan and Mudong in the upper Yangtze. Because the Gezhouba Dam has blocked its migratory channel, the Chinese sturgeon cannot reach the spawning sites in the upper reaches, and the fish has been constrained to spawn in a 7 km section downstream of the Gezhouba Dam, 38 km from the TGD.

4.5.6.2 Hydrological Process Required for Spawning

The Chinese sturgeon has requirements for both the flowrate and water level before it spawns, and the optimal flowrate and water level can provide a better spawning environment. An analysis of the hydrologic data and the oviposition records collected in the river section between Pingshan and Yibin during the oviposition period between 1961 and 1980 indicates that the Chinese sturgeon spawned in the river section between Pingshan and Yibin mainly in every October and November when the water level was in decline, which was just opposite of the hydrological conditions for the FMCC. During the spawning season, the characteristic flowrate and water level were 3512–7079 m^3/s and 282.0–286.2 m, respectively, at the Pingshan spawning sites and 3628–7451 m^3/s and 262.0–265.1 m, respectively, at the Yibin spawning sites. The water temperature in the spawning period was 17–21 °C, and the sediment content was 0.23–1.46 kg/m^3. The fish spawned mainly on the substrates consisting primarily of gravel when the sediment content in the water was in

Table 4.4 Hydrological parameter values at spawning sites both upstream and downstream of Gezhouba Dam during spawning season

Parameter	Value
Water depth (m)	9–28
Water level decline rate (m/day)	0.034–0.209
Flowrate decline rate ($m^3/s/h$)	2.8–9.7
Velocity (m/s)	1.0–2.0
Water temperature (°C)	17–21
Water sediment content (kg/m^3)	0.1–1.46

Table 4.5 Current hydrological parameter values at spawning sites downstream of Gezhouba Dam during spawning season

Parameter	Value
Flowrate (m^3/s)	8813–16,723
Flowrate decline rate ($m^3/s/h$)	3.5–9.7
Water level (m)	41.7–45.1
Upper and lower limits (m)	49.8 & 40.0
Water level decline rate (m/day)	0.034–0.209
Water temperature (°C)	17.2–20.8
Average velocity (m/s)	1.00–1.66
Water sediment content (kg/m^3)	0.1–1.32

a declining process. Moreover, the fish generally spawned in the river bends where there were deep pools and the flow condition was very complex with whirlpools.

A comparison of the hydrological factors between the original and current spawning sites both upstream and downstream of the Gezhouba Dam based on comprehensive investigations has revealed the common ranges of the hydrological parameter values as summarized in Table 4.4. The Gezhouba Dam is a sluice dam and possesses a runoff hydroelectric station and basically did not change the hydrology or water temperature. Therefore, the Chinese sturgeon did not change its spawning time before and after the dam was constructed, and both occurred from late-October to mid-November, corresponding to the water level declining cycle of the last flow peak of the year. Moreover, the topography and substrates at the original and current spawning sites are very similar.

The present hydrological parameter values at the spawning sites downstream of the Gezhouba Dam are summarized in Table 4.5.

The Chinese sturgeon has a long life and becomes mature late. Maturity is generally reached when the fish is more than 10 years old. The fish migrates upriver from the Yangtze Estuary toward the upper Yangtze to spawn, and then the parent fish must return through the dam to the coastal area for recovery. The fish repeats this migratory process once every few years. Based on the lifecycle of Chinese sturgeons, the general fish facilities not only cannot help Chinese sturgeons migrate upriver over the dam but also cannot assist the postpartum parent fish swim down the dam. The construction of fish facilities cannot solve the problem regarding the fish passing the dam. A 150 to 250 kg spawning fish cannot go up, and the postpartum parent and her juveniles cannot get down. Therefore, no fish facilities were considered when the Gezhouba Dam was constructed, and since then the spawning sites and breeding space of Chinese sturgeons have been reduced greatly.

4.5.6.3 Existing Problems

4.5.6.3.1 Shorter Spawning River Section

The original spawning sites for Chinese sturgeons were distributed in an 800-km-long river section in the upper Yangtze with 16 recorded spawning sites. Presently, the spawning sites are limited to a mere 7-km-long river section downstream of the Gezhouba Dam. Because of the small and narrow spawning sites, 48.9% of the parent fish had degraded gonads, and 85% of the mature female fish suffered gonadal degeneration. A spawning site can accommodate an average of 38 female fish for spawning at a time, and the smallest site can only accommodate 13 female fish (Chang and Cao 1999). A Chinese sturgeon reaches sexual mature late. The male first spawns at 8–18 years old and the female at 13–26 years old. After the construction of the Gezhouba Dam, the Chinese sturgeon born at the spawning sites downstream of the dam could not migrate and join the breeding population until more than 10 years later. The breeding population of the Chinese sturgeon was about 2000 (Chang and Cao 1999) before the dam was constructed. Since the construction of the Gezhouba Dam, the spawning sites have been small and narrow, the natural reproduction has been smaller in scale, and the replenishment has been insufficient. Therefore, when the individuals that were born after the dam was built began to join the breeding population, the breeding population started to decline because the population of the individuals that had been born before the dam was constructed commenced to decrease (Ji et al. 1999). The survey data collected during 1997–2002 showed that the breeding population had fallen to 300–500, while the survey data gathered during 2003–2006 indicated that the breeding population had decreased to 100–300. Although currently hundreds of breeding parent fish are migrating to the spawning sites downstream of the dam, the breeding resources have continued to decline because the individual fish that had been born before the dam was constructed are gradually and slowly retiring from the breeding population. As the spawning scale is so small, replenishment to the breeding population is bound to be insufficient, and the resources will inevitably decline markedly.

4.5.6.3.2 Little Effect of Replenishment on Breeding Population Through Artificial Reproduction and Release

The Chinese sturgeon can be artificially reproduced. However, the population of juvenile Chinese sturgeons in the Yangtze Estuary consist mainly of naturally reproduced ones, and artificially reproduced juvenile fish only accounts for 2.94–17.64%, which is not enough to change the genetic diversity of the entire population (Liu et al. 2008). Biomarker tests conducted in recent years indicate that the contribution rate of the released 10,000 10–15-cm-long artificially reproduced juvenile fish to the total juvenile population of the Chinese sturgeons in the Yangtze Estuary area was less than 5% (Zeng et al. 2007), which suggests that the current natural reproduction of Chinese sturgeons at the spawning sites downstream of the Gezhouba Dam still plays a leading role in the propagation of the Chinese sturgeon species.

4.5.6.3.3 Impact of the TGP

Under the preliminary design specifications, the TGR would begin impoundment in October every year after the construction of the TGP. During the impoundment period, the average water discharge would be reduced by about one-third. This period happens to be the breeding season of the Chinese sturgeon. The impoundment at the TGR would change the hydrological conditions in the downstream river section and likely impact the natural reproduction of the Chinese sturgeon. In every October, due to the need for the hydropower generation from the reservoir, the average monthly flowrate in the Yichang section would be reduced from the past 18,980 m^3/s to 11,090 m^3/s, which is a reduction of 41.6%. It is the spawning season from October to November for the Chinese sturgeon, and when the flowrate is reduced, the water surface of the spawning sites is decreased. The variations would consequently result in the change of the corresponding hydrological elements, such as water level, water temperature, flowrate, flow velocity, and sediment content, in the river section downstream of the Gezhouba Dam and the hydrological environment at the spawning sites that the Chinese sturgeons have long been adapted to. These changes may potentially have adverse effects on the propagation of the Chinese sturgeon.

Since the TGR began to impound in 2003, the inflow to the Gezhouba Reservoir has decreased; the water level has been lowered; and the hydrological conditions have changed markedly. As a result, the initial spawning time of the Chinese sturgeon has significantly been later than before the impoundment; the number of spawning has reduced to once; and the number of spawning fish has reduced to less than 10. If the annual spawning number is only 3–4 female fish, the number of the reproductive population resources may be only 40 fish in decades. Generally speaking, the impoundment operation of the TGR may have reduced the suitable spawning area for the Chinese sturgeon; the lower flowrate during the spawning period may not meet the ecological flow required for the Chinese sturgeon to spawn, reduced the size of the spawning population, and decreased the resources of the breeding population. Therefore, effective protective measures should be taken to preserve the resources of the Chinese sturgeon species.

4.5.6.3.4 Natural Enemies and Human Impact

During the autumn breeding season of the Chinese sturgeon, the eggs of the Chinese sturgeon are the main food for the bottom-feeders such as the coreius zeni, bronze gudgeon, *Pseudobagrus vachelli*, longnose catfish, *Saurogobio dabryi*, *Gobiobotia filifer*, *Xenophysogobio nudicorpa*, and *Leiocassis longirostris*, of which the coreius zeni, bronze gudgeon, and *Pseudobagrus vachelli* swallow most quantities of the eggs of the Chinese sturgeon. On their upriver migratory route, the Chinese sturgeons are often captured by fishermen or killed due to vessel collisions. The death of the migratory parent fish will very seriously affect the number of eggs spawned.

*4.5.7 Rock Carp (***Procypris rabaudi***)*

Rock carps are distributed in the middle and lower reaches of the Jinsha River and the mainstream and major tributaries in the upper Yangtze. The fish live in the substrate of gravel. During the daytime, they often inhabit the slow-flowing water in stone caves and prefer to feed in clusters on the relatively dark river bottom where the flow is slow. They are omnivorous bottom-feeders. In the winter, they are more concentrated in the rocky area in the deep river waters, stop feeding, and reduce activity significantly. After the spring starts, they migrate upriver to the upper reaches or tributaries of the Yangtze River. They start to increase food intake during March–April and peak ingestion during July–August. Their eggs are sticky and pale yellow in color. The spawned eggs stick to rocks for development. Their spawning peaks during February–March (water temperature above 10–12 °C). According to reports from fishermen, spawning sometimes also occurs during August–September. The spawning sites are generally distributed in the rapid-flowing sections with substrates of gravel in the tributaries of the Yangtze River. The flow velocity at spawning sites is about 1 m/s. The fish feeds mainly on aquatic insects, snails, freshwater mussels, and other mollusks that have grown in voids among stones in rapids and also on small amounts of phytoplankton, zooplankton, and epiphytic algae.

*4.5.8 Prenant's Schizothoracin (***Schizothorax prenanti***)*

Prenant's schizothoracin have been designated as a precious species protected by Sichuan Province and belong to the schizothoracin branch of the carp species. The fish are widely distributed in the upper Yangtze and the lower reaches of the Jinsha, Min, and Wu Rivers. The largest individual can reach more than 8 kg. The fish are China's endemic species of important economic values. Prenant's schizothoracin are omnivorous fish with plant-feeding habits, and their natural food consists mainly of algae, such as epiphytic green algae and filamentous green algae. In addition, the fish also eat benthic animals such as oligochaetes, chironomid juveniles, and mayfly juveniles. Their suitable breeding temperature ranges from 9 to 19 °C, and their spawning season is during March–April. The slightly sticky eggs are mainly laid on gravel shallows where the flow is rapid. The eggs sink to the bottom and often flow with water to voids among gravels where they will continue to develop. The suitable growth temperature for juveniles is 5–27 °C, and within the range, the temperature is positively correlated to their feeding and growth. During September and October, the postpartum parent fish migrate to deepwater zone or underwater cave for wintering.

4.5.9 *Longnose Catfish (***Leiocassis longirostris***)*

Longnose catfish are distributed in the Jinsha River; the mainstream in the upper Yangtze; the middle and lower reaches of the Yalong, Min, Tuo, and Jialing Rivers; and the lower reaches of the Wu River. The benthic fish like to live in irregular stony substrates and fast-flowing rapids. During the breeding season, the fish spawn in clusters, and other times they are usually dispersed. The fish are mainly carnivorous and feed primarily on small fish and aquatic insects. The type and characteristics of the spawned eggs are similar to those of bronze gudgeons. The eggs drift, and the fertilized eggs develop with the flowing water. The adult fish population begin to mature every March–April, and they migrate upriver to the fast-flowing zone with substrates of gravel. They spawn during April–June and start to migrate downriver around August. The spawning sites in the Yangtze River are mainly concentrated in the upper reaches of the Tuo River.

Longnose catfish have a long spawning period from March to July, and the spawning sites are mostly in the sections with substrates consisting of sand and gravel and at the tails of fast-flowing rapids in the upper reaches of rivers. The number of female and male spawning populations is very close. They spawn in batches, and the sticky eggs expand after absorbing water. The colorless eggs are transparent and elastic. When the water temperature is 21.5–23 °C, fries emerge after 47 h hatching time. Each fish produces 16,800–107,600 eggs with an average of 44,000. The number of eggs a fish can produce increases with fish's body size. The carnivorous fish feed mainly on crustaceans. As their bodies grow, their diet changes. Larvae feed on other young fish species, and juveniles devour gammarus, nipponense, and juveniles of various aquatic insects with a relatively small proportion of fish. Adults feed mainly on fish and sometimes on small fish and shrimp.

4.5.10 *Varicorhinus Angustormatus (***Onychostoma angustistomata***)*

Varicorhinus angustormatus are distributed in the Jinsha River and the mainstream of the Yangtze River; the middle and lower reaches of the Min, Dadu, and Qingyi Rivers; the Jialing, Fu, and Qu Rivers; and the lower reaches of Yalong and Wu Rivers. The fish are bottom-feeders that like to live in the river section with a high sand content. The fish feed primarily on diatom plants and edible vegetable detritus and completely stop feeding during the wintering period. During April–May, the fish spawn often on rapid shallows, and their sticky eggs often adhere to gravel for incubation.

4.5.11 *Elongate Loach (***Leptobotia elongata***)*

Elongate loaches are distributed in the upper reaches of the Yangtze River and sometimes occur in the middle and lower reaches of the river. Elongate loaches are bottom-feeders and live in voids among gravels near riversides where the water flows slowly. The fish like a clear miniature aquatic environment are photophobic and adapt to a wide range of temperatures. The ferocious elongate loaches are carnivorous and feed mainly on small fish and shrimp, and other fish eggs, also on phytoplankton and zooplankton, and sometimes on plant detritus, filamentous algae, etc. In their various developmental stages, the fish have different food compositions, but with no mixing or gradual transition (not from phytoplankton to zooplankton or from herbivorous to carnivorous). Elongate loaches have the habit to migrate upriver to spawn in the fast-flowing zone during April–June. The sticky eggs adhere to stones where they go through the incubation process. The fertilized eggs become fries in 34 h of hatching when the water temperature is 22.0–23.5 °C.

4.5.12 *Four Major Chinese Carps (FMCC)*

4.5.12.1 General Information

Grass, black, silver, and bighead carps are four well-known fish species of economic importance in China (see Fig. 4.5) and are excellent breeds, known as "four major Chinese carps." They mainly live and spawn in the Yangtze River, and their spawning sites are distributed in the mainstream of the Yangtze River and its tributaries such as Xiang, Han, and Gan Rivers (Yi et al. 1988). Raising the FMCC began in the Tang Dynasty more than 1 ka ago. Fries were harvested from the Yangtze

Fig. 4.5 Four major Chinese carps

River and then released to the nearby ponds (Gui 2003), which is why they have been called "house fish," but that is not accurate. Although they can be raised in an artificial feeding environment, the fries mainly originate from the natural environment such as the mainstream and major tributaries of the Yangtze River. The carps need the fast-flowing environment to spawn, and the eggs are hatched during their drifting process.

4.5.12.1.1 Silver Carp (*Hypophthalmichthys molitrix*)

Silver carps inhabit the upper and middle water layers and are very active. When startled, they jump out of water. They feed mainly on phytoplankton. Sexual maturity can be reached when they are 3 years old, and parent fish mainly spawn in the rapid-flowing water with whirlpools during April–June when the water temperature reaches above 18 °C and the water in the river rises or the flow velocity increases. The juveniles independently migrate to river bends or lakes for foraging, and the postpartum parent fish migrate to forage-rich lakes to recover. During the winter when the water level in lakes fall, adult individuals migrate back to the deep riverbed of the mainstream for wintering. The majority of immature individuals migrate to nearby deepwater body to winter. The fish are not so active in the winter.

4.5.12.1.2 Bighead Carp (*Aristichthys nobilis*)

Bighead carps prefer to live in the upper and middle layers of static water, are slow in movement, and do not like to jump. The fish feed mainly on zooplankton and eat some algae as well. Sexual maturity is reached when they are 4–5 years old, and the parent fish spawn in the rapid-flowing water with whirlpools during May–July when the water temperature reaches 20–27 °C. The juveniles generally grow in lakes along the river and return to the river for spawning when they are sexually mature. The postpartum parent fish return to forage-rich lakes to recover. During the winter they stay in the riverbed and deep rocky pit for wintering.

4.5.12.1.3 Black Carp (*Mylopharyngodon piceus*)

Black carps generally live in the middle and lower layers of water zone where there are abundant of snails. The fish feed mainly on snails and a variety of bivalves and eat shrimp and insect juveniles as well. The fish grow fast and can reach up to 3–5 kg through 2–3 winters. The largest individual can grow up to 70 kg, and the common individuals in the Yangtze River weigh about 15–20 kg. Sexual maturity is reached when they are 4–5 years old. The fish spawn in the mainstream of rivers where the water velocity is relatively high during April–July. Postpartum fish are generally concentrated in river bends and lakes that connected to rivers to recover and stay in deepwater zone for wintering.

4.5.12.1.4 Grass Carp (*Ctenopharyngodon idellus*)

Grass carp generally inhabit near-bank grassy areas in the middle and lower layers of water bodies such as rivers, lakes, etc. The fish have the habit to migrate between rivers and lakes. Sexually mature fish spawn in the flowing water in rivers. The postpartum parent fish and juveniles migrate to tributaries and lakes that connected to rivers and usually feed in grasslands of shallows and floodplains, as well as in major tributaries connected to water bodies (grassy areas in lakes, streams, small drainage ditches, etc.). During the winter, the fish stay in the mainstream of the Yangtze River or deepwater zones of lakes for wintering. Grass carps are very active, swim fast, and often feed in clusters. Grass carp are the typical greedy herbivorous fish. The reproductive habit of the grass carp is similar to that of the other FMCC. Under natural conditions, the fish cannot spawn in the static water. The fish generally choose the spawning sites at the confluences of tributaries and the mainstream of the Yangtze River, deep channels of river bends, and narrow river sections where the banks sharply approach to each other. Grass carps generally spawn during April–July but primarily in May. Generally, when the water level of the river rises early and rapidly, and the water temperature can be stabilized at about 18 °C, grass carp spawn a great deal.

All the FMCC species live in natural-food-rich lakes connected to the Yangtze River, while they migrate to the water-flowing mainstream of the Yangtze River to spawn during the breeding period. As far as the mainstream of the Yangtze River is concerned, their spawning sites are widely distributed in the upper, middle, and lower reaches. Based on years of field investigations, there are 36 large, medium-sized, and small spawning areas on the mainstream of the Yangtze River, of which 9 sites are distributed in the Yichang section (including Yichang) and 27 sites are distributed in the section downstream of Yichang until Pengze in Jiangxi Province.

Grass, black, silver, and bighead carps are freshwater fish species, can only migrate between rivers and lakes for a short distance, and thus are called semi-migratory fish. The FMCC don't have high swimming capabilities when compared to the true migratory fish species, which renders them unable to pass through fish facilities such as fishing ladders. Opening the sluices that separate the river and lakes to allow fries to migrate from the river to the lakes is one of the measures to restore the river-lake connection specifically designed for the FMCC species.

4.5.12.2 Hydrological Conditions Required for Spawning

Due to the similarity in lifestyle of the four carps, scholars often treat them as a whole in their studies, commonly known as "four major Chinese carp species." Overall, the four species breed from late April to early or middle July every year when the average water temperature of the river is 18–27 °C. The fish spawn when the average temperature steadily rises to above 18 °C, and they stop spawning when the water temperature is below 18 °C. Their spawning peaks when the temperature is 21–24 °C. The FMCC's spawning activity basically occurs when the water level rises (during a medium-sized flood process). The water-level-rising process includes the change process of a series of hydrological elements, such as rising water level, increasing flowrate, higher flow velocity, decreasing water transparency, and increased turbulence. The continuous rise in water level of the river generally lasts 4–7 days, and the rate of daily rise is generally around 0.3 m/day, which are suitable conditions for the FMCC to spawn. Whenever the Yangtze River's water temperature rises to above 18 °C, the sexually mature parent fish at the spawning sites are able to spawn under the stimulation of the rising water. When the water level decreases and the water flow slows down, most spawning activities cease. The produced eggs need to be carried by water and hatched into fries while drifting in the middle of the flowing water. The fries continue to drift in the water until they have developed into larvae that can swim into lakes and other water bodies where they continue to grow. The eggs generally drift 80–100 km and take 1–2 days to be hatched. The fries continue to drift for 5–7 days, and they cannot survive if they are blocked. This breeding habit was the result of the interaction between the fish and the water flow of the Yangtze River and the adaptation and compromise of the fishes for eons (Li 2001). The FMCC are not only very important in economic values, but they also characteristically represent all fish species in reproductive characteristics in the middle and lower reaches of the Yangtze River. Therefore, they can be used as the targeted fish species to be protected in the biological regulation of control reservoirs such as the TGR.

4.5.12.3 Existing Problems

According to the monitoring data of the FMCC during the breeding season in recent years, the number of fries of the FMCC in the Yangtze River has been decreasing. The number of FMCC fries in the Yangtze River through Jianli during May–June was 3.587 billion in 1997, 2.747 billion in 1998, and 2.154 billion 1999, respectively, which was 53.53%, 41%, and 32.15%, respectively, of the number during the same months in 1981. The causes for the decline of the FMCC resources include hydraulic projects (construction of dams and sluices), land reclamation from lakes, overfishing, and the deterioration of the aquatic environment.

4.5.13 Yangtze River Dolphin (**Lipotes vexillifer**)

The Yangtze River dolphin or baiji (see Fig. 4.6) is a dolphin-like freshwater mammal that lives in rivers and lakes and is a small- to medium-sized member of the whale family. It has a Chinese pinyin of baiji and belongs to the Lipotidae family. A baiji has a spindle-shaped body with bare skin and long lips. Fossil records suggest that the baiji first appeared 25 Ma and migrated from the Pacific Ocean to the Yangtze River when the Qinghai-Tibet Plateau had yet to form, and they had lived in the middle and lower reaches of the Yangtze River before the Yangtze River channeled through. In the ancient Chinese dictionary *Near to Correctness* compiled more than 2 ka ago, the baiji were described as Goddess of the Yangtze River. The baiji had been widely distributed in the Dongting Lake and Poyang Lake areas of the Yangtze River basin and once distributed in the Yangtze River up to the Three Gorges area 35 km upstream of Gezhouba and down to the Yangtze Estuary near Shanghai. The species is estimated to have historically reached 5000 in number and only about 300 20 years ago when it was near extinction.

The China Ministry of Agriculture has established a nature reserve for the baiji in the Xinluo section of the Yangtze River that extends from Luoshan to Xintankou with a total length of 135 km. The channel of the river section is broad and meandering with a water depth of about 25 m and a flow velocity about 0.3–0.8 m/s. The section is also studded with shallows and mid-channel sandbars and is the water zone where the baiji are most densely distributed. Immediately upstream of the section is connected to the exit mouth of Dongting Lake. Moreover, Hong, Huanggai, Xiliang, and Wu Lakes and the Lushui Reservoir converge into the section.

Fig. 4.6 Yangtze River dolphin

Furthermore, bedrock promontories such as Huanji, Sanmaoji, and Chibiji are present along the banks of the river to naturally protect the banks and control the direction and pattern of the water flow. From November 6 to December 13, 2006, nearly 40 scientists from China, the United States, Great Britain, Japan, Germany, and Switzerland inspected the 1700-km-long Yichang-Shanghai mainstream section in the middle and lower reaches of the Yangtze and did not identify any baiji. After their joint inspection concluded, the disappointed foreign scientists declared: "the Yangtze River dolphin may have already perished."

Hua Yuanyu and others made 47 expeditions to observe the baiji's behavior in the middle and lower reaches of the Yangtze River during 1979–1990, and each expedition was 15–45 days long. They found that the baiji fed on fish and preferred to occur in a group with a small group of 2–3 or a large group of 9–16. The mammal likes to stay in the fast-flowing deepwater zones. The baiji is a lung-breathing aquatic mammal, and every time when it breathes, the top of its head and the breathing hole are first exposed above the water surface, and then its back and lower triangular dorsal fin are exposed. It breathes for about 1–2 s and dives into the water for about 20 s and up to 200 s. A mature individual has a maximum length of 2.5 m for a female and 2.3 m for a male, a weight of 100–150 kg, and a constant body temperature of 36 °C. The baiji's audiovisual organ has been severely degraded, but the sonar system is particularly sensitive and can detect and identify objects in the water.

The group activities of the baiji are closely related to the habitat environment. During the dry season, the group activity is relatively stable, and generally the group only migrates for a short distance. During the water level rising season, the group often makes a long-distance migration. The baiji feeds mainly on fish. Analysis of its stomach content indicated that the baiji preys on fish both in the upper water zone and the benthic zone as well. The baiji seeks forage mainly in slow-flowing water zones, the water zones formed by small weirs, or deepwater pools where the water flow is stable. According to the observation of Qiqi, a human-raised male baiji, its estrus is during March–May every year, and it is thus estimated that the courtship season of the baiji in the Yangtze River is during March–May.

The baiji has been most significantly affected by human activities. According to incomplete statistical data collected from 1973 to 1985, 22 baiji were killed by roll-hook fishing, and 12 were killed by propellers. The baiji is also sensitive and nervous to the noise from ships' dynamic systems. When the noise is weak and far away, baiji dive to a great depth to stay away from the source of noise, and when the noise is strong and close, they dive to the deep water, turn around rapidly, and try to find their way to escape.

The baiji is a rare aquatic mammal endemic to China and is distributed in the middle and lower reaches of the Yangtze River. Because of its rare occurrence, it is near extinction. Due to its beautiful shape, it is called the "Goddess of the Yangtze River." The baiji is one of the most severely endangered mammal species in China and is also one of the world's most endangered animals. To some extent, the baiji is more precious than the panda. "Qiqi," one of the "goddesses," was captured by a fisherman at the mouth of Dongting Lake in January 1980 and sent to the Wuhan

Aquatic Research Institute (WARI) of the Chinese Academy of Sciences for artificial rearing, the world's only one artificially raised baiji. Qiqi only ate live freshwater fish. It ate mainly silver carps, about 6 kg a day, from spring to autumn, and fed primarily on common carps and crucian carps, 7.5 kg per day, from autumn to winter. Due to living in a small space for a long time, it suffered endocrine disorders, hepatitis, and diabetes. Its injured liver caused high blood lipids and hyperglycemia. Its GPT value was up to 1500 units, 150 times the normal value; its triglyceride value was up to 11,000 mg/dL, 100 times the normal value; and its blood sugar level was as high as 900 mg/dL, 9 times the normal value. On July 14, 2002 at about 6:00 AM, researchers at the WARI inspected the baiji museum as usual and Qiqi was seen swimming in the pool and nothing unusual was noticed. They also reviewed the previous day's records that indicated that Qiqi was fed at night as normal. However, at 8:00 AM when the staff entered the "Baiji Museum," Qiqi was seen lying in the water without any movement. After intensive rescue efforts, Qiqi never revived. Thus, the world's only artificial-reared baiji died. Qiqi was about 25 years old, equivalent to a human age of 70+ years old.

4.5.14 *Yangtze Finless Porpoise (***Neophocaena phocaenoides asiaeorientalis***)*

There are two species of toothed whale animals living in the Yangtze River: the Yangtze River dolphin (*Lipotes vexillifer*) and the finless porpoise (*Neophocaena phocaenoides asiaeorientalis*; see Fig. 4.7). The finless porpoise belongs to the Cetacea infraorder, Odontoceti parvorder, Phocoenidae family, and *Porpoise* genus. Measurement studies on the shape and skeleton of the finless porpoise indicate that the finless porpoise has three subspecies: Yellow Sea finless porpoise (*Neophocaena phocaenoides sunameri*), South China Sea finless porpoise (*Neophocaena phocaenoides phocaenoides*), and Yangtze finless porpoise. It is estimated that these three subspecies originated from their ancestors in the sea 0.72–1.08 Ma. The Yangtze finless porpoise is the only species of the finless porpoise in the world known to live in freshwater and is only distributed in inland waters such as the middle and lower reaches of the Yangtze River, Dongting Lake, and Poyang Lake. The Yangtze finless porpoise feeds mainly on small freshwater fish and shrimp. It has been designated as an endangered species by the International Union for Conservation of Nature and Natural Resources (IUCN), and it has also been listed in the Convention on International Trade in Endangered Species of Wild Fauna and Flora (CITES) (Jiang 2010). The Yangtze finless porpoise has been designated as an animal species under state Level II protection and will be promoted to receive state Level I protection.

According to the results of surveys conducted before 1991, it was estimated that the population was about 2700 at the time, and the results of surveys performed after 1991 showed that the population was significantly reduced to only 1200–1400 (Jiang 2010) and the species was estimated to become extinct in 24–97 years.

Fig. 4.7 Yangtze finless porpoise

The finless porpoises often form a basic unit or "core unit" of 2–3, and the core unit is usually composed of one mother and one newborn, one mother and one juvenile, or one male and one female. Their mating period is mainly from March to June. Years of observations on their mating behavior in the Tianezhou Nature Reserve for Yangtze Finless Porpoises indicate that their mating mainly occurred during March–June, and it is estimated that their pregnancy may be 10 months. The companionship between the juvenile and their mother may last for more than 2 years, but the actual weaning time was earlier. It is speculated that such a long period of maternal-child companionship may also be a kind of maternal protection behavior or the process of juvenile learning.

In order to prevent the repeat of the baiji's fate, China's whale scholars have been paying attention to the conservation of the Yangtze finless porpoise, and governments of various levels have established the following dolphins nature reserves since the 1980s: three national dolphins nature reserves along the Yangtze River (Tianezhou National Baiji Nature Reserve in Shishou, Yangtze Xinluo Section National Baiji Nature Reserve in Hubei Section of the Yangtze River, and Tongling Freshwater Dolphin National Nature Reserve in Anhui) and three provincial dolphins nature reserves (East Dongting Lake Provincial Dolphins Nature Reserve of Hunan, Poyang Lake Provincial Dolphins Nature Reserve of Jiangxi, and Zhengjiang Provincial Yangtze Finless Porpoises Nature Reserve of Jiangsu).

The protection efforts through relocation of Yangtze finless porpoises in the open water areas in Tongling and Shishou have been successful. However, since the semi-natural water zones of Jiajiang in the Tongling Nature Reserve and of Gudao in the

Shishou Nature Reserve are controlled by sluices, there is no hydrological characteristic of seasonal change or water exchange with the river, which has led to the eutrophication of the water bodies and the decline of the biological biomass of aquatic plants. This has further severely influenced the productivity of fish species in the water bodies and resulted in the decrease of natural forage for Yangtze finless porpoises, especially in the Tongling Nature Reserve. In recent years, the proportion of artificial feeding has markedly increased. In addition, extreme weather changes also have adverse effects on the protection efforts through relocation. For example, surface water freezing and death of porpoises in the Tongling and Shishou reserves occurred during the winter snowstorm events in early 2008.

Since the baiji was functionally extinct, scholars and the society have begun to take the protection of finless porpoises as a priority mission. Although a small population of different porpoise species remain in the mainstream of the Yangtze River and Dongting and Poyang Lakes, the channels for genetic exchanges among various populations of finless porpoises in different water zones will be significantly affected if sluices are constructed at the two lakes. Thus, it is urgent to establish more and safer protective water zones for the relocation of the porpoises, or otherwise the porpoises will face the fate of the baiji.

*4.5.15 Reeve's Shad (*Macrura reevesii*)*

The Reeve's shad (see Fig. 4.8) got its Chinese name because it migrates into the Yangtze River at a fixed time annually. The fish belongs to the class of Actinopterygii, order of Clupeiformes, family of Clupeidae, and genus of *Macrura*. It has an elliptic-shaped body, a flat side view, a medium-sized head, pointed lips, and a large mouth with no tooth. The lower jaw is slightly longer, and there is a distinguished central notch in the upper jaw. The rear ends of the upper jaw reach the lower edges of the eyes. The sawtooth-shaped scales are large, sharp, and thin with fine grains but no lateral lines. The gill rakers are fine and dense, and the adipose eyelids are well-developed. The fish has a dark gray body and head, a slightly bluish-green

Fig. 4.8 Reeve's shad

upper side, and silver-white lower side and abdomen. The silver scales are as shiny as jade. The fish are tender, fatty, scaly, and delicious and are well-known in China and abroad due to their deliciousness. An individual is generally 1–1.5 kg in weight, but a larger one can reach 3–3.5 kg.

The Reeve's shad is an anadromous fish species. Adults live in the coastal area. Sexual maturity is reached when they are 3–4 years old, and the ratio of male and female is 1:1. A female individual can produce 1 million to 3.3 million eggs. A ripe egg is an oil-ball-shaped and floating. The shad breeds in the Yangtze River basin during June–July each year when the water level has a relatively high fluctuation to provide certain stimulation from flowing water. The suitable water temperature for spawning is 24.5–32.0 °C. Eggs become fries in 17–18 h of hatching. In late April every year, mature individuals migrate in groups upriver through Jiangsu, Anhui, and Jiangxi into Poyang Lake in Jiangxi and continue up to the Gan River to spawn in the Xingan-Ji'an section of the Gan River, of which the Xinyu-Xiajiang section is the main spawning area for the shad in the Yangtze River. The spawning season is from June to July every year. The postpartum parent fish migrate back into the sea. Juveniles migrate from the spawning area down the Gan River into Poyang Lake and feed in southern Poyang Lake until they migrate into the Yangtze River through the mouth of the lake in the autumn when the water temperature starts to fall. During the winter they migrate from the Yangtze River to the sea to grow.

The Reeve's shad had been the fishing target of the important fishery industry in the Yangtze River before the 1970s. Before 1962, the annual production of the Reeve's shad had been steady at 300–500 tonnes with the highest annual output of 580 tonnes. During the 8 years between 1968 and 1975, the total output of the shad fluctuated from 385 tonnes in 1968 down to 70 tonnes in 1971, and then rebounded to a historical high of 1575 tonnes in 1974, and then began to decline significantly to only 345 tonnes in 1975. In the late 1970s, the yield began to decrease continuously. In the 1980s, the annual output was 12–192 tonnes. In 1986, the production was only 12 tonnes, which cannot be considered as a fishing season. In the 1990s, the Reeve's shad were hardly seen.

In May 2006 and May 2007, 25-day investigative fishing outings were conducted in the river sections at Ma'anshan, Wuhu, and Tongling of Anhui. A total of 124 boats were used with a cumulative fishing time of 462 h, and fishing outings occurred 310 times following the swordfish harvesting boats. However, no Reeve's shads were captured. During the 2 years, 230 senior fishermen were interviewed, and it was found that no Reeve's shad had been captured from the river sections in Anhui since 1994.

The Reeve's shad has become an endangered species mainly due to overfishing of the parent shad and associated loss of their juveniles, the aquatic environmental pollution, and changes of the spawning sites and hydrological conditions resulting from dams.

Loss of Juveniles of the Reeve's Shad Poyang Lake in Jiangxi Province is a good feeding site for larvae and juveniles of the Reeve's shad. During June–July every year, the larvae from the Gan River are concentrated for growth in the southern lake area where the water is shallow and the flow is slow. Fishermen in the Poyang Lake area have the habit to densely distribute huge fishing nets to harvest the whitebaits and Spanish mackerels. During the operation using the nets, a large number of the juvenile Reeve's shads were captured, resulting in great damages to the shad resources. During the peak year, the fishermen in Hukou County alone caught 7.74 tonnes of juvenile Reeve's shads. During the autumn and winter, the juvenile shads would migrate from the Yangtze River to the sea for growth. They were caught on their way through Jiangxi, Anhui, Jiangsu, and the Yangtze Estuary area in Shanghai. The operations in which these juvenile shads were caught were not specifically for harvesting the juvenile shads, but rather for harvesting juvenile swordfish, eels, white shrimps, etc. Because small-opening nets were used in the operations, the juvenile Reeve's shads were also incidentally captured, and damages to and killing of the juvenile Reeve's shads were quite serious. During August to October of 1970, the Bayu Aquaculture Station in Jingjiang, Jiangsu, bought a total of 56.7 tonnes of all kinds of juvenile fishes, 40% of which were about 3-cm-long juvenile Reeve's shads, or 22.7 tonnes. As a result, the damage to the juvenile Reeve's shad had serious impacts on the replenishing population.

Aquatic Environmental Pollution With industrial development and intensified human activities, water pollution of the Yangtze River has become increasingly serious. According to incomplete statistical data, the daily amounts of industrial wastewater and domestic effluents released from Jiangsu, Anhui, and Shanghai into the Yangtze River were more than 8 million tonnes from 153 major sources of pollution. As the wastewater containing all kinds of harmful substances has been discharged into rivers and lakes, serious effects have resulted to the normal migration, spawning, hatching, and larval and juvenile growth of the Reeve's shad, leading to the decline of the Reeve's shad population.

The Xiajiang section of the Gan River in Jiangxi is the main spawning area of the Reeve's shad. The rising water process and frequent occurrences of flood peaks in the Gan River during June–July provide very favorable conditions for Reeve's shads to spawn. The dimension of the spawning sites is closely related to the frequency of the rising water level and the intensity of flood peaks in the Gan River during June–July. Some of the cascade hydroelectric stations (especially Wanan and Xiajiang Hydroelectric Stations) constructed in the middle and upper reaches of the Gan River, on the one hand, have blocked the passage for Reeve's shads to migrate upriver for spawning and, on the other hand, have caused the changes of the hydrological conditions and the aquatic ecological environment, such as the slower flow velocity; reduced water transparency, which are not conducive to the breeding of Reeve's shads; and have directly led to the further decline of the Reeve's shad resources.

4.6 Birds and Other Wildlife Species in the Yangtze River Basin

Most birds live in riparian and lacustrine wetlands. The living conditions of the birds can indirectly reflect the health status of the waters. According to surveys, 762 bird species are distributed in the Yangtze River basin (Wu et al. 2004) and belong to 20 different orders, 66 families, and 291 genera, accounting for 61.2% of all bird species in China. The bird species also include 72 endemic and important species of China, 26 species designated to be under state Level I protection, and 92 under state Level II protection. Various bird species inhabit the living conditions suitable to them or they have adapted to. Therefore, the habitats in various regions of the Yangtze River basin are always very suitable to certain species, thus resulting in the distributional differences of bird species.

4.6.1 Bird Species

4.6.1.1 Yangtze Source Area

The source area of the Yangtze River has the characteristics of high altitudes and a long distance to the ocean and is in an alpine arid climate. The Yangtze River water flows slowly on the plateau toward the east. The river channel is relatively wide with many marshes. Dark coniferous forests, alpine shrubs, alpine meadows, and snow and ice caps are the main vegetative types in this area. The region is the richest in endemic bird species, and the representative species distributed in the area include spotted-tailed hazel chicken (*Tetrastes sewerzowi*), blood pheasant (*Ithaginis cruentus*), green-tailed monal (*Lophophorus lhuysii*), white-eared grouse (*Crossoptilon crossoptilon*), blue-eared grouse (*Crossoplilon auritum*), Lady Amherst's pheasant (*Chrysolophus amherstiae*), black-necked crane (*Grus nigricollis*), Lady Derby's parakeet (*Psittacula derbiana*), Tibetan lark (*Melanocorypha maxima*), Mongolian accentor (*Prunella koslowi*), ground tit (*Pseudopodoces humilis*), Sichuan jay (*Perisoreus Internigrans*), Kessler's thrush (*Turdus kessleri*), Chinese fulvetta (*Fulvetta striaticollis*), white-rumped snowfinch (*Montifringilla taczanowskii*), pink-rumped rosefinch (*Carpodacus eos*), etc.

In the Yangtze source area, birds are the most noticeable wildlife species, but mainly sparrows and other small birds. There are a few predatory raptors such as hawks. As a result, the rat problem has been serious in the source area.

4.6.1.2 Upper Yangtze

In the upper reaches of the Jinsha River, the rapid water flow has cut the originally relatively flat plateau into alpine canyons and created high topographic reliefs between ridges and valleys. As a result, the natural vertical change in mountainous area is unusually prominent. From foothills to peaks, the vegetation types include tropical, subtropical, highland warm temperate, and alpine sub-frigid vegetation species. The intricate natural conditions make the ancient northern and oriental species crisscross each other in the region. The birds of the ancient northern species are mainly distributed in meadows of the alpine shrubs and the alpine dark coniferous forest in the highland warm temperate climate zone, and their representative species include snow partridge (*Lerwa lerwa*), Tibetan partridge (*Perdix hodgsoniae*), common raven (*Corvus corax*), alpine chough (*Pyrrhocorax graculus*), horned lark (*Eremophila alpestris*), streaked rosefinch (*Carpodacus rubicilloides*), Himalayan beautiful rosefinch (*Carpodacus pulcherrimus*), snow bunting (*Montifringilla* spp.), Brandt's mountain finch (*Leucosticte brandti*), Tibetan sandgrouse (*Syrrhaptes tibetanus*), little owl (*Athene noctua*), alpine accentor (*Prunella collaris*), etc. The oriental bird species are mainly distributed in the evergreen broad-leaved forest of relatively lower elevations, and the representative species include crested goshawk (*Accipiter trivirgatus*), mountain bamboo partridge (*Bambusicola fytchii*), Chinese bamboo partridge (*Bambusicola thoracicus*), golden pheasant (*Chrysolophus pictus*), silver pheasant (*Lophura nycthemera*), Mrs. Hume's pheasant (*Syrmaticus humiae*), pin-tailed green pigeon (*Treron apicauda*), slaty-headed parakeet (*Psittacula himalayana*), brown wood owl (*Strix leptogrammica*), red-headed trogon (*Harpactes erythrocephalus*), striated bulbul (*Pycnonotus striatus*), white-winged magpie (*Urocissa whiteheadi*), laughingthrush (*Garrulax* spp.), shrike-babbler (*Pteruthius* spp.), Nana wren-babbler (*Spelaeornis* spp.), minla (*Minla* spp.), alcippe (*Alcippe* spp.), white-collared yuhina (*Yuhina* spp.), heterophasia (*Heterophasia* spp.), slaty-billed tesia (*Tesia* spp.), prinia (*Prinia* spp.), niltava (*Niltava* spp.), verditer flycatcher (*Muscicapa thalassina*), etc.

4.6.1.3 Lakes in Middle and Lower Yangtze

In the middle and lower reaches of the Yangtze River, due to the lowly undulating terrain, there are many crisscrossing rivers, many lakes connected to rivers, large areas of marshes, and thick aquatic plants that provide excellent habitats for all kinds of waterfowls and waders. Representative species include little egrets (*Egretta garzetta*), grey heron (*Ardea cinerea*), mallard (*Anas platyrhynchos*), Indian spot-billed duck (*Anas Poecilorhyncha*), swan (*Cygnus* spp.), Siberian crane (*Grus leucogeranus*), plumbeous water redstart (*Rhyacomis fuliginosus*), white-crowned forktail (*Enicurus leschenaulti*), vinous-throated parrotbill (*Paradoxornis webbianus*), etc. A series of lakes, especially Poyang Lake, are important distribution areas for waterfowls. The lake areas boast a suitable climate, abundant water, and a long frost-free season where aquatic plants flourish. The water levels in lakes are low during the autumn and winter, forming large swampy shallows. In different seasons, large numbers of waterfowls of different varieties are attracted to the lake areas for forage, habitat, breeding, and wintering. Because cranes need a large area of swampy shallows to peck the underground rhizome of aquatic plants, Poyang Lake becomes the main wintering habitat for cranes. Every year, large numbers of white-naped cranes (*Grus vipio*) and Siberian cranes are distributed in Poyang Lake for wintering. For example, during the winter of 1995, the wintering cranes in Poyang Lake included 3716 white-naped cranes, 2896 Siberian cranes, and nearly 100 common cranes (*Grus grus*) and hooded cranes (*Grus monacha*). Rare bird species, such as oriental storks (*Ciconia boyciana*), black storks (*Ciconia nigra*), tundra swans (*Cygnus columbianus*), and scaly-sided mergansers (*Mergus squamatus*), and a variety of geese and ducks swarmed to Poyang Lake for wintering. Figure 4.9 shows rare and typical bird species that rely on the waters in the middle and lower reaches of the Yangtze River.

4.6.1.4 Yangtze Estuary

The Yangtze Estuary is a magnanimous and vast plain with a broad river channel and a slow-moving flow of water where the river water intermingles with the seawater. The spacious intertidal zone, high-yield crustaceans, and aquatic plants provide a suitable habitat for waterfowls and waders and attract swarms of seagulls, sandpipers, herons, and ducks to live, winter, and breed. Especially for some migratory species, the vast coastal intertidal area provides an important resting place. Typical species distributed in this area include the little grebe (*Tachybaptus ruficollis*), great cormorant (*Phalacrocorax carbo*), Chinese egret (*Egretta eulophotes*), little gull (*Larus minutus*), bar-tailed godwit (*Limosa lapponica*), little curlew (*Numenius minutus*), whimbrel (*Numenius phaeopus*), Far Eastern curlew (*Numenius madagascariensis*), ruddy turnstone (*Arenaria interpres*), sharp-tailed sandpiper (*Calidris acuminata*), red knot (*Calidris canutus*), great knot (*Calidris tenuirostris*), grey wagtail (*Motacilla cinerea*), Kentish plover (*Charadrius alexandrinus*), common moorhen (*Gallinula chloropus*), Baillon's crake (*Porzana pusilla*), tawny fish owl (*Ketupa flavipes*), Indian scops owl (*Otus bakkamoena*), oriental scops owl (*Otus sunia*), osprey (*Pandion haliaetus*), etc. The birds that breed in this area include the striated heron (*Butorides striatus*), intermediate egret (*Egretta intermedia*), yellow bittern (*Ixobrychus sinensis*), whiskered tern (*Chlidonias hybrida*), dunlin (*Calidris alpina*), little ringed plover (*Charadrius dubius*), pheasant-tailed jacana (*Hydrophasianus chirurgus*), watercock (*Gallicrex cinerea*), etc. The wintering waterfowls include the Mandarin duck (*Aix galericulata*), Eurasian teal (*Anas crecca*), ruddy shelduck (*Tadorna ferruginea*), mallard, tundra swan, northern lapwing (*Vanellus vanellus*), swan goose (*Anser cygnoides*), greylag geese (*Anser anser*), long-billed plover (*Charadrius placidus*), seagull (*Larus canus*), common greenshank (*Tringa nebularia*), white-naped crane, common crane, hooded crane, red-crowned crane (*Grus japonensis*), etc.

Fig. 4.9 Rare and typical bird species in lacustrine wetlands in the middle and lower Yangtze

4.6.1.5 Bird Habitat Protection

Birds are flying poultry. They choose habitat mainly based on the environmental factors such as food source, temperature, landing place, concealment, and natural enemies. Most bird species migrate due to changes of temperature and adequacy of food on ground (or water). Also, they need good concealment for their breeding sites, less interference, and a suitable habitat environment. Water zones and transition zones between water and land such as riparian zones and lacustrine wetlands are the most suitable habitat for birds because there are rich aquatic life and adequate food; the terrain is open; the water body acts as a barrier; and there are not only fewer natural enemies but also less human interference.

The diversity of bird distribution is closely related to beneficial factors (first factor is food source and hidden place) that habitats can provide. The adaptability of birds to habitats is the result of the long-term evolution of the species. When compared to the distribution of bird species in other habitats, the similarity of the bird species' distribution among rivers, lakes, and marshes is the lowest. Because of various types and sources of food for different birds, the structures of their beaks have evolved to different types (see Fig. 4.10). Some birds' foods come from water and even fish and weeds from various depths, while some birds' foods are the benthic animals or grassroots from mudflats, as well as insects and grasslands on uplands. Special attention must be paid to the uniqueness of the habitat needs of bird species when implementing protection measures to conserve habitats of bird species. Attention also needs to be paid to the habitat that is merely a migratory resting or wintering place for birds. Figure 4.11 is an illustration of vegetation-elevation relationship in a typical land-water transition zone of Poyang Lake, including shallow water, mudflats, and various types of marshes, which are the ideal habitat and food source for all kinds of birds. While it is difficult for human beings to set foot in this open land-water transition zone, it is a paradise to birds and needs special protection.

The key bird species that are particularly dependent on the Yangtze River basin, such as the white-naped crane, Siberian crane, etc., which are special species under

Fig. 4.10 Various bird species have different beak structures due to different food sources

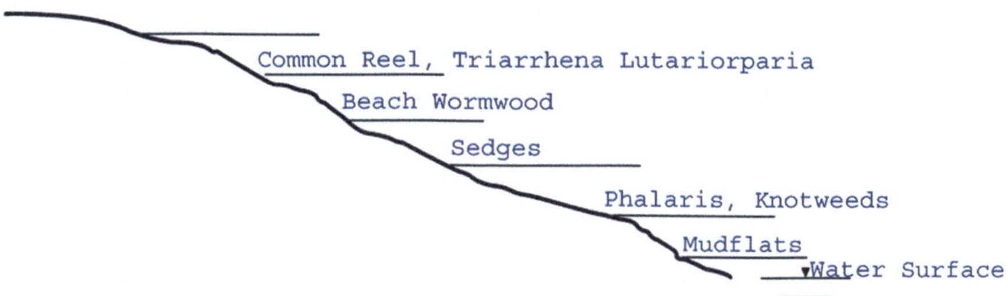

Fig. 4.11 Illustration of vegetation-elevation relation in Poyang Lake's land-water transition zone

the state Level I protection, have adapted to the seasonally changing land-water transition zones of the lacustrine wetlands in the middle and lower reaches of the Yangtze River. Although the practice to control the water level during the dry season through sluices at the mouth of the lakes is beneficial to human beings, it is not conducive to the habitat environment of the bird species because the change of the hydrological and hydrodynamic conditions between the river and lakes would cause the natural ecological environment of wetlands to change. Moreover, the number of bird species and the number of all birds in the Yangtze River basin are also influenced by the water quality. As industrial wastewater is discharged from the urban areas along the Yangtze River into river and lakes, some toxic substances are released into the waters without effective treatment, leading to the destruction of the living environment for the bird species in the Yangtze River basin, especially for the bird species dependent on the water bodies. Therefore, attention should be paid to the environmental protection of the Yangtze River basin during economic development. Measures should be proactively implemented, and wastewater should be properly treated and comprehensively reused to eliminate harm resulting from pollution.

Figure 4.12 shows a large tract of poplar trees planted in the Dongting Lake area. Although it looks beautiful and has economic values (such as paper-making materials), it is no good for many migratory birds since it has occupied their habitats.

4.6.2 Other Wildlife Species

4.6.2.1 Amphibians

According to records, 145 species of amphibians occur in the Yangtze River basin and belong to 2 classes, 10 families, and 30 genera, including 23 salamander species, more than one-half of those in China, and 122 are koala species, close to one-half of those in China. There are 49 endemic amphibian species in the Yangtze River basin, 33.79% of all the species in the river basin, and 118 amphibian species in the Yangtze River basin that are native to China, accounting for 81.38% of all the endemic amphibian species in China. The endemic amphibian species in the Yangtze

Fig. 4.12 A large tract of poplar trees planted on the beach of Dongting Lake

River basin are distributed more in the upper reaches than in the middle and lower reaches of the river and gradually decrease in number of the species with descending elevations. The endemic amphibian species in China are also distributed more in the upper reaches than in the middle and lower reaches, but the variation range with the change of altitudes is very small. Five amphibian species (only seven in China) are on the "List of Wild Animals under State Priority Conservation," including the giant salamander, Guizhou knobby newt, black knobby newt, and Asian bullfrog, and 69 threatened species are on the *China Species Red List*.

Except for the Yangtze source area and the upper and middle reaches of the Jinsha river where there are not many amphibian species because they are located in the plateau, an average of 48 amphibian species occur in the lower reaches of the Jinsha River and tributaries and mainstream in the upper reaches of the Yangtze River, and 30 and 27 amphibian species are located in the middle and lower reaches, respectively, of the Yangtze River, of which less than 20 are in the Poyang Lake water system.

Amphibian species are more dependent on temperature and humidity. The middle and lower reaches of the Yangtze River basin used to be the ideal habitat. However, due to economic development, natural vegetation has been seriously destructed, and the habitat has been fragmented, resulting in the decline of the species. The upper region can be characterized by a vertical distribution from the subtropical zone in valleys to the permanent ice cap in highlands and boasts a variety of habitat environments, abundant vegetation, relatively few human activities, and, therefore, relatively more species.

4.6.2.2 Water Buffalo and Other Wading Animals

Water buffalo were domesticated about 4000 BC and are now used for plowing land or transportation. Presently, there are no wild water buffalo, but there are large numbers of human-raised water buffalo used for land plowing and meat. They like to live by water bodies and often swim in water ponds or wallow in mud to prevent insect bites. Some birds (such as little egrets) like to rely on and help water buffalo and feed on parasites on the bodies of water buffalo. Water buffalo and birds can live by water bodies together in harmony. Many poultry, birds, and reptiles need or depend on water bodies to live.

4.7 Structure and Integrity of Yangtze River's Ecosystem

4.7.1 Environment of the Yangtze River Basin

The nonliving system in an ecosystem can be called an environment. When viewed in a large perspective, an environment includes climate, atmosphere, hydrology, geology, geomorphology, soil, and aquatic environment. These environmental elements are closely related and affect one another. Climate determines the air temperature, soil temperature, humidity, precipitation, and precipitation forms. Atmospheric circulation not only affects the climate but also affects the transformation among the three water elements (atmospheric water, soil water, and surface runoff). Topography, geology, and geomorphology affect not only climate and precipitation but also water generation, sediment yield, and water's physical and chemical properties. Precipitation, floods, and aquatic environment can also impact geomorphic changes and local climate.

The large karst landforms in the upper Yangtze were the result of the interaction between water and geology, and debris flows and landslides, which were not caused by human beings, along river banks in the upper Yangtze were due to the interaction between precipitation, and soil and rock. On the geological timescale, erosion effects such as soil and water erosion can flatten a high mountain and can also cut a mountain into a canyon. In the alpine area such as the Qinghai-Tibet Plateau and the Jinsha River, the climate has vertical zoning characteristics because of high altitudes. At the same time or in the same season, different seasonal characteristics are present on the mountain peak, mountainside, and canyon. It is winter up in the mountain, and it is spring and autumn at the foot of the mountain. The Yangtze River basin has a typical monsoon climate where water and heat are in synchronization. During the summer, the temperature is high; it rains more; and plants grow fast. In the mountainous area, especially in the alpine area, the precipitation on the windward side is quite different from that on the leeward side, which is the result of the influence of the topography on climate.

From the view of the aquatic environment, mountain waterfalls, deeply submerged sandbanks, rapids and slow-flowing zones, and the static water in the lacustrine wetlands are all the manifestation of the diversity of the aquatic environment. Deepwater zones, shallow-water zones, sandbanks, mid-channel sandbars (stones), and fallen trees are all bio-speckles or habitats. Curvy river sections, braided rivers, and river-lake connections are all the external environments needed by the aquatic organisms of the Yangtze River.

There are many rare and endemic fish species in the mainstream and tributaries in the upper reaches of the Yangtze River, which are related to the rapid-flowing turbulent water and large amounts of gravels and boulders in the substrate. Some sticky eggs (such as Dabry's sturgeon) can adhere to gravels, while the eggs of other fishes (such as coreius zeni) need a rapid water flow to drift downriver for a period of time and distance to hatch into fries. If the water flows too slowly, the eggs will sink to bottom and become dead. Therefore, the rapid-flowing environment is the necessary environment for these endemic fish species to survive. The FMCC in the middle and lower reaches of the Yangtze River need the water temperature to reach above 18 °C to spawn in the process of a rising water level, and the eggs drift a certain distance to become fries. The juveniles can only grow up in lakes with a slow water flow and rich aquatic life, relying greatly on the river-lake connectivity. Large areas of mudflats and marshes in Poyang Lake are exposed every winter, and large areas of the water fluctuation zone in the lake are the wintering habitat for migratory birds such as Siberian cranes. Therefore, the Yangtze River's endemic bird species are the result of the aquatic environment of the Yangtze River system, and the change of the environment will alter the biological population and structure.

4.7.2 Ecological Structure of the Yangtze River Basin

4.7.2.1 Structure of the Ecosystem

From the biological function in the ecosystem, each species can be categorized as a decomposer, a producer, or a consumer. A species can be a consumer of a lower-level species but a producer of a higher-level species. Lower-level producers survive through competing for energy (such as solar energy) and nutrients; consumers gain energy and nutrients by competing for food; the decomposing organisms restore the dead to nutrients or energy; and the three categories of species form the cycle of life, nutrition, and energy. From the biological scale or level, all species can be divided into primary producers (such as plankton), secondary producers (such as higher plants), primary consumers (such as zooplankton), and secondary consumers (invertebrates, fish, birds, amphibians, reptiles, mammals, etc.), and omnivorous species. A normal ecosystem is composed of a multilevel structure in which individual organism species have their defined functions and all species are interdependent. From the ecological point of view, each biological species has the function of maintaining the succession, evolution, and balance of the ecosystem, which makes the structural integrity and balance of the ecosystem most important. Therefore, it is more important to preserve the integrity of the ecosystem than to protect a single rare species.

Although the Yangtze River basin is large in size, it lacks large, continuous natural forests and undisturbed wetland environment. The areas that are really reserved in natural state are limited to the Yangtze source areas and main tributaries and a few nature reserves. Therefore, both the number of high-level wild animal species and their counts are low. For the ecosystem of the river system, the integrity of habitats has been significantly affected by constructed dikes and dams, and the rare large migratory fishes and dolphins are very sparse. Thus, it is now very difficult to discuss the structural integrity of the ecosystem from the perspective of the entire Yangtze River basin, but it is more appropriate to discuss it only in a local area or a local river section.

As mentioned previously, the forest area in the Yangtze River basin is increasing slowly, and the stocking quantities of producers such as plants have maintained stable. However, the main problem is that the increase in the number of natural forests is limited and the structure of forest plants is irrational or needs to be improved. In the waters, due to extensive use of chemical fertilizers for agricultural production and the use of large quantities of laundry products in human life, rivers and lakes have become eutrophic with excessive primary organisms such as algae. Both on land and in waters, high-level animals are very scarce and rare, and they are the priority objects that need to be protected.

4.7.2.2 Utilization of Producers' Energy and Nutrients

From the point of view of circulation of energy and matters, all creatures mainly do two things: one is energy conversion and the other is circulating matters. Most of the energy on the earth comes from the sun and solar energy, of which about 33% is deflected back to the space by clouds and the earth's surface; 20% is used to transform water into vapor that forms clouds, rain, and snow; and the other 47% is used to drive the water cycle and used by organisms. Plants derive energy from sunlight through photosynthesis; animals obtain energy and nutrients by eating food (including plants and animals); and microbes and bacteria survive through decomposing (eating) dead animals and plants, each species of which plays a different role in its own material cycle.

The growth of organic matters needs elements such as nitrogen, phosphorus, carbon, silicon, iron, etc., all of which can be called nutrients. The level of nutrients in waters determines, to a large extent, the level of primary producers. After the nutrients in waters are absorbed and used by organisms, new organic cells will develop. Various organisms need different nutrients and varying proportions of nutrients. For example, the growth of algae normally needs a 106:6:1 ratio of nutrients for carbon/nitrogen/phosphorus; animal's growth of lipids needs to eat matters rich in phosphorus and carbon; and the biological growth of amino acids and proteins requires large amounts of element nitrogen. Each organism needs nourishment to survive, grow, and reproduce. Most primary producers and many bacteria can use inorganic nutrients, and they are called "self-nutrition" organisms, while most animals and non-"self-nutrition" organic matters require organic nutrients. Biological absorption and transformation of nutrients require outside energy (such as sunlight). Due to the influence of human activities, when nutrients become excessively rich, crops or flowers may grow well, but the residual nutrients will be transported to waters through water and soil. The nutrients will make the waters, especially the relatively static waters, become eutrophic; algae become a predominant species; other species have difficulties surviving; and the structure and balance of aquatic ecosystems be destroyed.

Vegetation can absorb hydrogen and oxygen from water through photosynthesis or directly use carbon from carbon dioxide. Other nutrients come mainly from rock and soil on the surface of the earth and are transported into rivers and lakes through surface runoff and underground water. Some nutrients are dissolved in waters, and some are deposited to the bottom of the waters. Aquatic plants can extract nutrients not only from waters but also through their root systems from soil. According to many studies, the main nutrients controlling primary producers are nitrogen and phosphorus, and the role of phosphorus is more prominent.

Aquatic organisms can also produce elements of inorganic nutrients from the microbial decomposition of organisms, especially the decomposition of dead organisms, and through physical and chemical actions. This transformation is an important part of the nutritional replenishment of waters. Therefore, nutrients can come from outside (atmosphere, rainfall, runoff, etc.), but also be produced through the biological decomposition of substances inside waters and sediments, which is the manifestation of the laws of energy and mass conservation.

4.7.2.3 Level of Nutrients and Primary Producers

The nutritional level of a water body can be determined by natural factors and human activities. An oligotrophic water body generally has less algae or low-level producers, a high transparency, and a great penetration depth by sunlight. This water body generally has few aquatic plants and is usually a safe source of drinking water if mineral composition meets requirements. Presently, such water bodies only exist in the remote areas of minimal human activities and the source areas of the rivers. In a eutrophic water body, primary producers such as phytoplankton and algae flourish, and it is usually difficult for fish species to survive because of oxygen deficiency. Therefore, a mesotrophic water body is generally the richest in aquatic organisms. The nutritional level of a water body can be measured by the concentration of suspended algae, parasitic algae, and nutrients in the water body. Once a water body becomes eutrophic, especially hypereutrophic, not only the water stinks and affects the landscape of the water body, but also algae generate toxin and cause oxygen deficiency in the water, which so seriously impacts the aquatic biological system that the restoration of the system becomes very difficult.

Under natural conditions, lakes generally accumulate sediments and nutrients, transform gradually from oligotrophic waters toward mesotrophic waters, then become marshes, and eventually disappear. However, this process is generally relatively long, ranging from hundreds of years to more than thousands of years. The evolution of some large tectonic lakes varies with the geological time. For example, although Lake Baikal in Russia is considered the deepest lake in the world and has thousands of meters of sediments, it has a long span of life. Under natural conditions, a water body needs a certain level of nutrition to maintain the diversity of aquatic organisms and the number of species. Through the competition among organisms and material circulation, a balance can be achieved in a long period of time. In the absence of human interference, the change process of the aquatic environment and aquatic species is slow.

Especially since the industrialization, eutrophication in rivers and lakes, which is often a rapid process, has resulted from human activities, such as the large-scale use of chemical fertilizers and pesticides on farmlands, aquaculture and poultry farming, soil erosion associated with land reclamation and deforestation, mining, discharge of nutrient-rich wastewater, etc.

4.7.2.4 Food Chain and Competition

Creatures need to pillage food or nutrients to survive, grow, and reproduce, and they themselves form the food chain of other creatures. Primary producers compete primarily for nutrients and energy (such as sunlight). In some water bodies, aquatic organisms can intercept more than 90% of sunlight (although they cannot fully utilize it), and emergent plants can use the sunlight from water and above the water surface. Flourishing plants indicate that they have taken full advantage of sunlight and nutrients; they then also provide food for herbivores; and herbivores provide food for carnivores. When plants and animals complete their life cycles, they also provide food for microbes (through decomposition and absorption). While microorganisms absorb organic matters, they decompose organic matters into inorganic nutrients to participate in the nutrient cycle. As every creature is a part of the food chain, the lower-level organisms have to rely on quantities or scale to prevail in order to maintain their own circulation and provide food for the higher-level lives. At the same time, to leave sufficient quantities of species for the next generation to reproduce, the lower-level organisms generally need considerable redundancy to maintain their circulation. Thus, a rational biological structure is a pyramid shape. At the bottom of the structure, the biological population is large. In the meantime, however, the environmental condition for the higher-level biological lives is harsher, or they face more intense competition; their quantities are lower. When implementing the ecological protection, human beings should not think that some species do not deserve conservation because their quantities are large. In fact, some organisms rely on their quantities to maintain the cycle of their species, or otherwise if the quantities are not large enough, the stability and integrity of the entire biological system will be impacted.

4.7.3 Integrity of the Yangtze Ecosystem

Although the Yangtze River basin runs through various regions and has different habitats and environments, they are all closely related to one another. Geological conditions, the type of rock, and the characteristics of soil erosion in the Yangtze River basin determine the background conditions of water bodies' physical and chemical properties; forest vegetation cover governs the characteristics of runoff generation and sediment yield and the content of organics in the waters of the basin; and the continuity of the river dictates the living conditions of aquatic organisms such as migratory fishes. The hydrological connection of Dongting and Poyang Lakes to the Yangtze River and the seasonal variation of the wetland environment define the species of migratory birds and the environment of wintering habitats. The transport of sediments and nutrients in the Yangtze River has shaped the estuarine wetlands and the offshore ecosystem. Therefore, the ecosystem of the Yangtze River basin is a system of interdependent integrity, and the functional degradation in one link is bound to cause the other functions to adjust and degrade, and even a "butterfly effect," resulting in a great change of other species in the ecosystem and ultimately affecting the survival of mankind.

The integrity of an ecosystem varies with the size of the watershed. In a medium-sized and small river section, such as microscopic, mesoscopic, and engineering sections, environmental elements change rapidly, on which organisms used to rely to evolve. Therefore, they can generally adapt to the changing environment. However, on a relatively large space and timescale, the natural evolution is slow, and the relatively great impact of human activities often obscures the evolution of natural habitats. Most species live in a limited space, and it is very difficult for them to adapt to large and rapid environmental changes, and when faced with sudden environmental changes, many species have no time to complete their normal succession or evolution before they become extinct. The author is not intended to condemn the development and utilization of the river during the human economic development, but instead to discuss the integrity of the Yangtze River's natural ecosystem where the problem is indeed getting worse, as is the case with the Yangtze River dolphin (or baiji) and the Reeve's shad. A baiji has a huge body, is highly sensitive to the environment, and needs a relatively large range of waters, a relatively quiet environment, and adequate fish forage. The barriers between the river and lakes and large numbers of motorized vessels sailing on the river have blocked and occupied their migratory routes. Moreover, overfishing has occurred in the river. Consequently, the baiji did not have the environmental conditions for their survival. The Reeve's shad used to spawn in the upper reaches of the Gan River of the Poyang Lake water system. However, with the construction of dams such as the Wanan Dam, their main spawning sites have been occupied. Moreover, the Reeve's shad has high economic values since people love to eat them, resulting in overharvest of them. Eventually, the species has become extinct before it could find a new spawning site or maintain a stable number of a basic population.

Although an ecosystem without large wildlife is incomplete, the watershed's ecosystem needs to be maintained or evolved. If the intensity of human disturbance is lowered, a new ecosystem will gradually develop, and even new species will appear to fill the lost large wildlife, which requires comprehensive conservation measures to reduce the increasingly intensified activities of interference to forests, rivers, and lakes and provide space and time for the ecosystem to repair, restore, and evolve. For example, for the conservation of fish species, a 3-month ban on fishing is obviously inadequate, which only bans fishing during the spawning period. However, the fish need more time to complete their growth and change. Moreover, they need additional time to adapt to the changes and to complete succession in the new environment. As a result, some experts have recommended a 10-year ban on fishing so that the fish can "self-repair and recover." According to the experts involved in the exchange of fishery resources between China and the United States, they and American fishery scientists conducted research-related fishing in the Mississippi River and caught several tonnes of fish with one net in one try. When compared to the fish density in the Mississippi River, the Yangtze River is equivalent to possessing no fish. This may be attributed to the fact that Americans do not like to eat freshwater fishes, but it is more related to the intensity the United States has used in the conservation of rivers and fish resources. From a healthy cycle of the Yangtze River's ecosystem, the Yangtze River needs low-intensity interference and a period of recuperation time to restore the structure and integrity of the ecosystem.

References

Chang J, Cao W (1999) History and future of Chinese Sturgeon species conservation. Acta Hydrobiol Sin 23(6):712–720

Fish Species Section of Hubei Provincial Aquatic Organism Institute (1976) Fish species of the Yangtze River. Science Press, Beijing

Fishery Bureau of Sichuan Province (2002) Comprehensive investigation report of rare fish species in state nature reserve in Yibin-Luzhou Section of the Upper Yangtze

Gui J (2003) History and current conditions of releasing parent "Four Major Chinese Carps" to the Yangtze River. China Fish 1:11–12

Huang D (1980) Preliminary observation on development of digestive system and early feeding habit of Dabry's Sturgeon's fries. J Fish China 4(3):285–293

Institute of Hydroecology, MWR & CAS (2007) Research report on identification of ecological and environmental sensitive areas in the Yangtze River Basin

Jiang W (2010) Overview of protective relocation of Yangtze finless porpoise in the Yangtze River. J Anhui Univ (Natural Science Edition) 1(4):104–108

Jiang G (2008) Large hydraulic projects in the Yangtze River Basin and rescue of fish resources. Yangtze River 39(23):62–64

Li S (2001) Study of biodiversity and conservation of important fish species in the Yangtze River. Shanghai Scientific & Technical Publishers, Shanghai

Li Y, Xi X (1997) Investigation on shape development of Chinese Paddlefish's Juveniles and spawning grounds in the Upper Yangtze. J Southwest Agric Univ 19(5):447–450

Liu L (1990) Breeding ecology of swordfish and *Coreius Guichenoti* in mainstream of the Yangtze River after construction of Gezhouba hydroelectric project. Acta Hydrobiol Sin 14(3):205–215
Liu J, Cao W (1992) Fishery resources in the Yangtze River Basin and protective strategies. Resour Environ Yangtze River Basin 1(1):17–22
Liu H, Cheng Y, Li D, et al (2008) Technical report on hydrological mechanism of effect of three gorges reservoir's operation on spawning ecology of Chinese Sturgeon and protective countermeasures
Ma J, Cai M (1996) Preliminary study on age identification and growth of Chinese Paddlefish. Acta Hydrobiol Sin 20(2):150–159
Sichuan Provincial Yangtze Aquatic Resources Investigation Group. (1988) Studies on biology and artificial propagation of Sturgeon resources in the Yangtze River. Sichuan Science and Technology Press, Chendu
Wang L, Wang D (2004) Fishery and wetland conservation in the Yangtze River Basin. Yangtze River 35(5):37–39
Wu Y, Zhong Y, Yangong X (2004) Preliminary analysis of bird species in the Yangtze River Basin. Chin J Zool 39(4):81–84
Yi B, Yu Z, Liang T et al (1988) Gezhouba hydroelectric project and four major Chinese carps of the Yangtze River. Hubei Science & Technology Press, Wuhan
Yi J, Chang J, Tang D et al (1999) Preliminary study on current situation of breeding population of Chinese Sturgeon in the Yangtze River. Acta Hydrobiol Sin 23(6):554–559
Yu Z, Deng Z, Cai M et al (1988) Preliminary report of reproductive biology and artificial propagation of Chinese Sucker Downstream of Gezhouba. Acta Hydrobiol Sin 12(1):87–89
Zhang Y (2006) Analysis of structural change of China's forestry resources during 1950–2003. J Beijing For Univ 28(6):80–87
Zhang Y (2007) Analysis of contribution from afforestation to forest cover. J Northeast Univ 35(3):76–78
Zhang C, Zhao Y (2001) Migration of Chinese sucker and impacts of hydraulic projects on fish resources. Acta Theriol Sin 47(5):518–521
Zhang K, Zhang Y, Zheng J et al (2007) Distribution and evaluation of Zooplankton in Chongqing section in three gorges reservoir area. Guizhou Agric Sci 35(1):57–59
Zhang H, Weiou Q, Hao D (2008) Relationship between natural breeding behavior and meteorological conditions of Chinese Sturgeon. Sci Technol Rev 26(17):42–48
Zhong X, He L, Liu S, et al. (1998) Study on improvements of Shelter forests along banks of the Middle and Upper Yangtze. Chengdu University of Science and Technology Press, Chengdu
Zhou H (1999) Historic investigation of forest changes in the Yangtze River Basin. Agric His China 18(4):3–14
Zhu C (1987) Preliminary study on shape, growth and feeding habits of Juvenile Chinese Paddlefish at the Yangtze Estuary. Acta Hydrobiol Sin 11(4):289–296

Chapter 5
Water Resources and Flood and Drought Hazards in the Yangtze River Basin

Abstract The Yangtze River basin has carried China's socioeconomic development for many thousands of years. Because there are more mountains, fewer plains, more population, and less land, the watershed has experienced periodic large floods and severe droughts; the socioeconomic development of the basin is reflected in the history of fighting flood and drought disasters everywhere as the construction of large water conservancy projects occurred after each large flood or drought disaster. This chapter describes the spatial and temporal distributional characteristics of water resources in the Yangtze River basin. Most of the water resources in the Yangtze River basin are surface water with little groundwater resources. The mainstream and tributaries in the upper Yangtze have abundant hydraulic energy resources, and the middle and lower reaches have good navigable channels and water transportation resources. This chapter overviews the historical droughts, typical water-scarce areas, and existing water quality-related shortage problems of the Yangtze River. This chapter also analyzes flood characteristics of the Yangtze River basin, reviews major historical flood processes and associated disasters, and discusses drought characteristics of the Yangtze River and representative drought years and associated damages. Finally, there is a discussion of extreme hydrological events resulting from climate change.

Keywords The Yangtze River · Changjiang River · Evolution of river system · Basin ecosystem · Water resources utilization · Floods and drought · Ecological and environmental protection · Basin management

5.1 Characteristics of Yangtze River Water Resources

Water is the source of life and the foundation of ecology. Water provides not only basic physical conditions for the earth's ecosystems but also basic resources for human production and life, including the water supply for human life and industrial production, as well as the water source for agricultural irrigation. Hydraulic energy can provide us with tremendous electric power, and river channels facilitate

J. Chen, *Evolution and Water Resources Utilization of the Yangtze River*,
https://doi.org/10.1007/978-981-13-7872-0_5

navigation and transportation for human beings. Water zones are the most important cultural and recreational resources. There is a need to understand the characteristics and distribution of water resources so as to scientifically and rationally develop, utilize, and protect the resources.

5.1.1 Shortage of Water Resources in the Yangtze River Basin

The Yangtze and the Pearl Rivers are located in the forefront of the southern and southeastern monsoon regions of China where the precipitation is relatively high but the temporal distribution within the year is very uneven with about 70% falling in the wet season. In China's southeast coastal areas, the rainfall resulting from a few short-time typhoons may make up about 25% of the annual precipitation. After the wet season, the precipitation decreases, and seasonal droughts and water shortages to varying degrees occur in most of the Yangtze River basin. As relatively large portions of the Yangtze River basin are located in mountainous and hilly areas and relatively small portions of the basin are on plains, even if precipitation occurs in large quantities, the resulting runoff quickly flows down from mountains and hills to the middle and lower Yangtze River and eventually into the ocean due to the lack of water storage facilities (natural lakes, wetlands, and regulating reservoirs). As a result, physical water shortages occur in considerable mountainous and hilly areas. In the middle and lower Yangtze plains, large areas of lakes, wetlands, and ponds are present, and the water resources are relatively rich. However, due to the discharge of urban domestic sewage and industrial wastewater, and pollution from agricultural non-point sources, there are large numbers of riparian pollution zones where the river flows through and by urban areas, and lakes in urban areas suffer serious pollution. Consequently, there is a general water shortage problem due to water quality issues. Therefore, although the Yangtze River basin is relatively rich in water resources, occurrences of various types of droughts and water scarcities are not less frequent than in north China. Moreover, the main crops in south China are rice and other high water-consumptive products which are highly reliant on the availability of water and more vulnerable to droughts.

Based on the water use information, although the Yangtze River basin boasts large quantities of water resources, they are mainly surface water. Groundwater resources, if not accounted repeatedly, makes up only about 1% of the total water resources. Most urban water supplies rely mainly on the surface water which is greatly influenced by the changes of runoff. Due to the lack of reliable groundwater and backup water sources, water supplies may be interrupted once the flow of the surface water in rivers is low or an accident of water pollution occurs. Therefore, the safety of water supplies is not highly guaranteed. Moreover, most of China's high water-consumptive industries and pollutant-generating enterprises are located in south China. For example, more than 8000 chemical enterprises are located along major tributaries of the Yangtze River, posing a high risk of water pollution. As a result, the safety of the aquatic environment is not optimistic. The amount of water in the Yangtze River accounts for one-third of that of China, but the amount of industrial and domestic wastewater discharged to the Yangtze River is close to one-half that of China. Consequently, the water pollution problem is increasingly serious, and the associated water scarcity is widespread.

Due to the impact of climate change and human activities, droughts and associated hazards have occurred more frequently in the Yangtze River basin. The continuous drought that occurred in the Yunnan and Guizhou area during 2009–2012 resulted mainly from low precipitation, but it also was closely related to the special geography, landform, socioeconomic development, and the uneven distribution of water resources in the region. For example, in Yunnan Province, there are six water systems, the Yangtze, Pearl, Lancang, Hong, Nu, and Irrawaddy Rivers, and the total amounts of annual water resources are 222.2 billion m^3, ranked third in China. Both water resources per capita and per unit area are more than twice the national average. However, because more area of the province is in mountainous areas, mountains are high, the available water is low and far, and the current utilization rate of water resources is only 7.1%. Moreover, the spatial and temporal distributions of water resources are very uneven. Only 6% of the land area of the province is located in Baqu County which, however, boasts two-thirds of the population, one-third of the arable land, and only 5% of the province's water resources. There are many mountains in Yunnan Province, but there are not many flatlands. There are dividing mountains for the six water systems or large karst areas. The natural conditions are not favorable for the construction of storage reservoirs. Consequently, the costs for construction, operation, and management of hydraulic projects are high. Therefore, there exist typical physical water scarcity problems. Furthermore, water pollution has occurred in alpine lakes, such as Dianchi Lake. Thus, the water scarcity resulting from water pollution is also prominent.

In some traditionally water-rich areas in the Yangtze River basin, the problem of water scarcity has also been very conspicuous in the past decade. For example, in the four river areas south of the Jingjiang River in Hubei Province and the Dongting Lake area in Hunan Province, due to the impacts of comprehensive factors such as sedimentation, meander cutoffs of the Jingjiang River, and scouring due to clear water discharge from the TGP, the inflow of the Yangtze River has decreased, the inflow time has been shorter, and three of the four rivers south have not had inflow from the Yangtze River for up to 10 months in a year. As a result, the rivers have become static waters; their environmental carrying capacity has reduced; and, consequently, water scarcity resulting from water pollution has been very pronounced. Presently, in the traditionally water-rich areas, such as the Dongting Lake and Poyang Lake areas, seasonal water shortages and water pollution-related water shortages have become more prominent than flood threats in most years (when there is no large flood). Moreover, it has become an increasingly more pronounced challenge to secure the safety of water supply in the Yangtze Estuary area such as Shanghai due to salt tides.

Although many reservoirs have been constructed in the Yangtze River basin, there are not many regulating reservoirs that boast relatively large beneficial storage capacities. There are at most 1–2 regulating reservoirs on each river, and they have limited regulating capacities for solving problems associated with the downstream droughts or low-flow processes, or too much revenues would be lost as a result of the loss of excessive hydropower generation due to the use of the regulating capacities. Nonetheless, the effect would be no good as for water replenishment.

Therefore, although the Yangtze River basin is rich in the total amount of water resources, the utilizable quantities are not as much as expected. Moreover, there are not enough large reservoirs and associated water conservancy works that can be used for storage and regulation. Furthermore, water pollution is serious in the economically developed areas that are excessively dependent on the upstream water for water supply. Consequently, the reliability and safety of water supply have become a problem.

5.1.2 Available Water Resources

Compared with the river basins in north China, the Yangtze River basin is relatively rich in water resources. The average annual amount of water yield and amount of water flowing into the ocean have been nearly one trillion m^3 with not much water consumption, and nearly one-half of the amount of water use has returned to rivers and lakes. Therefore, no obvious reduction in total water quantities flowing into the ocean has occurred, although the total amount of water use in the Yangtze River basin has increased gradually in the past 30 years. However, there are three points worth noting. First, large amounts of water stored in rivers have not been counted. The actual quantity of water use has been very large, and the water flowing into the sea has been used by cascade hydropower stations or thermal power plants for cooling purposes for many times. Second, the water that has been used outside the river has become poor in quality; the function of the returned water has been greatly reduced; and returned water from agricultural use has also been subjected to non-point source pollution. Third, large amounts of freshwater are needed in the Yangtze River channel and Estuary for purposes of navigation and the ecological environment. Therefore, the Yangtze River does not have much water available for use outside the river.

The available quantity of water resources is the amount of water resources that can be used for human socioeconomic activities outside the river channel, or it is the total amount of water resources minus the basic quantities needed to maintain the function of the ecological environment in rivers and lakes and the quantities (mainly floodwater) that are difficult to control and utilize in a certain time period due to technical and economic factors. The key problem herein is that it is very difficult to determine the water demand for maintaining the function of the ecological environment. It is not only unclear how much water is needed to maintain the ecosystem in a good condition, but also the water demand for maintaining the function of the ecological environment depends on subjective factors such as the attitude and natural view of the people and society toward the ecological environment. If a better ecological environment is desired, more water should be maintained in the river, or otherwise less water is left in the river, and thus more water resources are available for use. According to the *National Comprehensive Plan of Water Resources* recently approved by the State Council, in evaluating the water use for the ecological environment, the main consideration is to meet the requirements of water quality and water function zones. However, no detailed consideration was given to how much water is indeed needed for the ecosystem of the river.

Table 5.1 is a comparison of available quantities of water resources in several primary water resources regions of China. The table indicates that the ratio between the available amount of water resources for utilization over the total amount of water resources (thereafter referred to as utilization rate) in the rivers in south China is much smaller than that of the rivers in north China. On the one hand, the amount of floodwater in the southern rivers accounts for a large percentage of the total quantity of water, and most of the floodwater is not available for use. On the other hand, large amounts of water are required for the ecological environment in the southern rivers. Table 5.2 shows the utilization rate of the secondary rivers in the Yangtze River basin and indicates that the Han River has the highest utilization rate. If taking the future Han River-Wei River Water Diversion Project and the Second Stage of the South-to-North Water Diversion Project into consideration, the Han River's utilization rate will reach 50%, which is the highest by a tributary in the Yangtze River basin. In addition, the utilization rates are above 40% in local areas or river sections such as the Tai Lake basin, the upper Min River in Sichuan and the Yuanshui River in the lower reaches of the Gan River. Similarly, the utilization rate is over 25% in the Jinsha, Min, and Tuo Rivers of the upper Yangtze. As the utilization rate in the upper Yangtze is going up, the discharge of wastewater to the river will continue to increase at the same pace. Thus, the future pressure to protect the water resources in the middle and lower Yangtze River will increase.

5.1.3 *Groundwater in the Yangtze River Basin*

The distribution of groundwater resources is affected by the factors such as hydrogeology, topography, surface water, and the development of water resources. The overall distribution of the groundwater resources per unit area (or groundwater modulus) in the Yangtze River basin can be characterized by the following: it is

Table 5.1 Comparison of available water resources in representative southern and northern rivers (unit in billion m³)

Primary river region	Surface water	Groundwater	Total resources	Available surface water resources	Withdrawable groundwater in plains	Available water resources	Available utilization rate (%)
Hai	21.6	23.5	37.0	11.0	15.2	23.7	64.1
Yellow	60.7	37.6	71.9	31.5	11.9	39.6	55.1
Huai	67.7	39.7	91.1	33.0	19.9	51.2	56.2
Yangtze	985.6	249.2	995.8	282.7	15.0	282.7	28.4
Pearl	470.8	116.3	472.2	123.5	4.7	123.5	26.2

Table 5.2 Available amount of water resources of secondary rivers/regions in Yangtze River basin

Primary water resource region	Drainage area (km^2)	Mean annual surface water resources (billion m^3)	Mean annual total water resources (billion m^3)	Available surface water resources (billion m^3)	Available utilization rate of surface water (%)	Available utilization rate of water resources (%)
Jinsha River	473,974	156.517	156.517	56.633	36.18	36.18
Min/Tuo Rivers	163,042	106.497	106.610	26.808	25.17	25.15
Jialing River	159,357	69.878	69.881	11.747	16.81	16.81
Wu	87,773	55.113	55.113	10.223	18.55	18.55
Upstream of Yichang	984,192	451.489	451.604	120.196	26.62	26.62
Dongting Lake Water System	262,289	207.757	208.592	57.038	27.45	27.34
Han River	154,804	55.471	57.318	22.950	41.37	40.04
Poyang Lake	162,065	151.302	153.251	36.624	24.21	23.90
Upstream of Datong	1,682,805	940.486		243.866	25.93	

Table 5.3 Summary of average annual groundwater per unit area in various river sections

Yangtze river section	Average annual groundwater per unit area (m^3/km^2)
South bank tributaries	198,400
North bank tributaries	120,100
Upper reaches	116,000
Middle reaches	181,500
Lower reaches	156,300
Dongting Lake, Poyang Lake, and Tai Lake	207,300
Non-lake areas	121,400

higher in the south bank than the north bank; higher in the middle and lower reaches than the upper reaches; higher in the three lake areas than the non-lake areas; and higher in plains than mountainous and hilly areas. The average groundwater moduli for various areas are summarized in Table 5.3.

The groundwater modulus in the plain area is greater than that in the hilly area. Table 5.4 summarizes the information for different areas.

The groundwater recharge in the Yangtze River basin is summarized in Table 5.5.

The quantity of groundwater is not only related to precipitation and runoff but also to the annual distribution of precipitation and regional topography, landform, vegetation, and geology. Moreover, the quantity of groundwater in a region is basically consistent with the regional (high or low) values of groundwater modulus and

Table 5.4 Summary of average annual groundwater per unit area in different area

Yangtze river area	Average annual groundwater per unit area (m^3/km^2)
Upper plains (after deduction of hilly areas)	426,100 (362,300)
Upper mountainous and hilly area	115,100
Middle plains (after deduction of hilly areas)	196,900 (189,400)
Middle mountainous and hilly area	180,600
Lower plains (after deduction of hilly areas)	192,500 (187,300)
Lower mountainous and hilly area	135,300

Table 5.5 Proportion of groundwater recharge sources

Area	Precipitation recharge (%)	Surface water recharge (%)
Yangtze River basin	97.7	2.3
Mountainous and hilly area	100	
Plain area	76.8	23.2[a]
Upper plain area	47.2	52.8[a]
Middle plain area	80.2	19.8[a]
Lower plain area	80.0	20.0

Notes: [a]Including seepage from surrounding mountains

the regional (high or low) values of runoff and precipitation. As for groundwater discharge, in mountainous and hilly areas in the Yangtze River basin, most (more than 99%) groundwater discharges to rivers, and very little is drained through extraction, seepage from mountainsides and evaporation of resurfaced groundwater in valley plains. Of the total discharge in plain areas, 56% is through river channels and 44% through evaporation and actual extraction.

Groundwater in the Yangtze River basin is the main source of domestic water supply for many rural areas, and some factories and mines have also relied on the withdrawal of groundwater. In recent 20 years, groundwater has begun to be withdrawn for air-conditioning use in some large municipal office buildings and residential areas. Groundwater problems are mainly concentrated in the Yangtze Delta region. In the Suzhou-Wuxi-Changzhou area, groundwater withdrawal started in the 1980s and 80 m deep depression cones that cover an area of 5483 km^2 have gradually developed with the centers of the depression cones located in Suzhou, Wuxi, and Changzhou Cities. In the Hangzhou-Jiaxing-Huzhou area, eight depression cones have formed as a result of groundwater withdrawal. In Shanghai, wells are as deep as about 200 m (Cheng et al. 2003). In recent 10 years, Shanghai and other cities have taken stringent measures to control mining of groundwater, which has resulted in the preliminary control of groundwater overdraft. In addition, due to traditional habits and the lack of strict management measures, large amounts of solid hazardous substances have been randomly stockpiled; seepage control measures have not been in place at urban landfills, and pesticides have been extensively used in rural areas. Consequently, groundwater contamination in the Yangtze River basin has become increasingly serious, and the potential risk of groundwater quality can no longer be ignored. The protection of groundwater has never been taken seriously in the past. Moreover, it is more difficult to protect and restore groundwater than surface water. In the future, groundwater will become backup water sources in many urban areas. Therefore, from now on, strict protective measures must be implemented to prevent groundwater from further contamination.

5.1.4 Hydropower Resources

The Yangtze River basin consists of 70% water-rich mountainous and hilly areas. The natural fall of the Yangtze mainstream is about 5400 m. The upper tributaries such as the Yalong, Dadu, Min, Jialing, and Wu Rivers have a fall of up to 2000–4000 m. The rich runoff along with the large fall in elevation means high potential of hydropower resources. The upper Yangtze is the key base for China's current and future hydropower development.

The theoretical potential of hydropower resources in the Yangtze River basin is 305,000 MW (including the 10 MW theoretical potential in rivers with single stations of 0.1–0.5 MW installed capacities), accounting for 40% of China's hydropower potential and an annual theoretical hydroelectric potential of 2.67 trillion kWh. However, the technically explorable potential is 281,000 MW of installed capacity with an annual hydroelectric potential of 1.32 trillion kWh, accounting for 47.3% and 48%, respectively, of China's technically explorable hydropower resources. The hydropower resources in the Yangtze River basin are mainly concentrated in its upper mainstream and tributaries and the technically explorable potential and annual hydroelectric potential account for about 87% and 90%, respectively, of the entire basin. The technically explorable hydropower potential in the mainstream is concentrated in the Jinsha and Chuan Rivers upstream of Yichang, and that in tributaries is concentrated in the Yalong, Dadu, Wu, and Yuan Rivers and the upper and middle reaches of the Min, Jialing, Han and Gan Rivers. Hydropower resources are especially concentrated in the Jinsha River, Three Gorges section of the mainstream of the Yangtze River, and the following seven tributaries: the Yalong, Dadu, Wu, Jialing, Min, Yuan, and Han Rivers, where the technically and economically explorable hydropower resources consist mainly of large and medium-sized hydroelectric stations, accounting for about 68% of the total. By 2007, 2458 hydroelectric stations that had been or were being constructed had had an installed capacity of about 120,000 MW, accounting for 42% of the technically explorable hydropower potential in the entire Yangtze River basin and including 42 large hydropower stations with a combined installed capacity of 90,000 MW, of which 16 large hydropower stations each have an installed capacity of over 1000 MW.

5.1.5 Navigation Resources

The Yangtze River, known as the "Golden Waterway", is China's most important navigable waterway and most developed navigable inland river and is the primary navigable channel that links Eastern, Central, and Western China. The Yangtze River basin boasts a broad hinterland, abundant water, a mainstream that flows across the land from west to east, and tributaries that extend toward the north and the south. Both the mainstream and its tributaries are linked to lakes and are frost-free year-round. As a result, they have a great navigation potential. There are more than 3600 navigable rivers in the Yangtze River system, and the main navigable rivers include the mainstream of the Yangtze River; the Min, Chishui, Jialing, Wu, and Han Rivers; and the lake systems of Dongting, Poyang, Chao, and Tai. Moreover, the Yangtze River connects to the Huai River system through the Beijing-Hangzhou Grand Canal. All these constitute China's most important inland waterway navigation system. The successful navigation through the permanent ship lock at the TGP has greatly improved the navigable conditions, and all ships that fit the ship lock can sail all the way to the East China Sea. By the end of 2007, the total navigable length had reached more than 71,000 km in the Yangtze River basin, accounting for 55% of China's total inland navigable length, of which primary navigable channels accounted for more than 85% of that of China's total. Furthermore, the Yangtze River has valuable bank line resources. The Yangtze River has a 7000 km total length of all kinds of bank lines along its mainstream downstream of Yibin and is peerless of China's rivers.

By the end of 2010, the mainstream of the Yangtze River had achieved passages of 1000 tonne ships to Yibin year-round, 10,000 tonne fleets to Chenglingji year-round and to Chongqing seasonally, 5000 tonne ships to Anqing year-round and to Yichang seasonally, 8000 tonne ships to Wuhan seasonally, and 50,000 tonne ships to Taicang year-round and to Nanjing seasonally.

In 2010, 1.502 billion tonnes of freight were transported through the mainstream of the Yangtze River, an increase of 12% over 2009; the freight through sizable ports reached 1.38 billion tonnes, a surge of 22.1% over 2009; the freight of foreign trade reached 169 million tonnes, up by 16.5% from 2009; and handled containers were 9.079 million twenty-foot equivalent units (TEU), a 26.2% rise from 2009. The average load per ship was increased from 600 tonnes during the "Eleventh Five-Year Planning Period" to the current 850 tonnes in the mainstream of the Yangtze River; ports with handling capacity of 10,000 tonne ships along the Yangtze River reached 298; and there were 4100 million tonne handling capacity ports in Nanjing, Zhenjiang, Suzhou, and Nantong.

Of course, the development of the shipping industry needs to continuously use the river banks for dredging the river and constructing ports, which has negative effects on the river's ecosystem. Moreover, the navigable main channel is also the migratory pathway for the Chinese sturgeon and the Yangtze finless porpoise on which the navigation of ships has the obvious impact.

5.1.6 *Ecological Flowrate*

5.1.6.1 Definition of Ecological Flowrate

The concept for the water demand of the ecological environment is relatively extensive, including the water use on land and in an aquatic ecological environment. However, this book focuses on the water demand of an aquatic ecological environment such as rivers and lakes, hereinafter referred to as the water use of an ecological environment or ecological flowrate. The ecological flowrate can be defined as a basic flow regime consistent with the ecological and environmental needs of a river. This definition limits the scope and content to make the concept clearer and easier to determine. The ecological flowrate mainly involves rivers and associated lakes but does not cover the water needs for the terrestrial ecological environment such as forests and steppes. The special ecological flowrate mainly refers to the water need of a natural aquatic ecosystem but does not consider the downstream water need by human production or living, of which a considerable portion overlaps each other though. If the downstream human water demand is considered at the same time, it can be called the minimum discharge, which is the general ecological flowrate. Although it is easy to adopt the general ecological flowrate in real management, the requirement for human water use is different from the regime required by the ecological environment. It is easy to ignore the characteristics of water use of the ecological environment; as a result, the water use of the ecological environment is replaced with the human water use.

The water use of a river's ecological environment and the human water use are not only different in quantity and quality but also quite different in the use process. The water use process of industrial production and living is relatively uniform and stable, but the agricultural water use is dependent primarily on seasonal needs of the crop growth cycle and the natural precipitation. However, the ecological and environmental water use requires a relatively natural hydrological process and rhythm with certain requirements for chemical and physical properties of the water body such as the flow velocity, water level, water temperature, dissolved oxygen, turbidity, and nutrient levels. Of course, the management of the ecological flowrate is not only to satisfy the natural ecology and environment, but also for human beings to ultimately sustain the use of the rivers and associated water resources. To maintain the ecological flowrate is not to restrict human from development and utilization of rivers but to mainly minimize adverse effects on the ecology and environment. Therefore, the ecological flowrate can only be "basically" satisfied.

The ecological flowrate required by a river's ecosystem has its specific requirements and includes the following aspects: ① **A river's low-flow, flood, and periodical hydrogeological processes**. The low-flow and flood processes have positive ecological and environmental functions. During a river's low-flow process, the water body has a low sediment content and high transparency, and large numbers of floodlands and slow-flowing areas emerge near banks of the river, which is conducive not only to the growth of aquatic plants, zooplankton, benthic animals, amphibians, and birds but also to the decomposition of contaminated substrates in the fluctuation zone. The flood process can enhance the transport of sediment and nutrient, stimulate fish species to spawn and migrate, eliminate dominant species, and promote the diversity of aquatic organisms. By reaching the floodplain, floodwater can promote the lateral connectivity of the river (including river-lake connectivity) and stability of the riparian wetland ecosystem. ② **Hydrological and hydrodynamic factors, such as water quantity, water quality, flow velocity, and water level's fluctuation range**. The ecological flowrate not only refers to the flowrate itself only but also includes the hydrodynamic parameters such as flow velocity and water level's fluctuation range. The Yangtze River's FMCC will spawn only when the water level rises sharply, and the water temperature increases above 18 °C, but invertebrates require a low flow velocity to grow safely. ③ **Water quantity, water quality, physical and chemical properties, and nutrient transport of natural water body**. Aquatic organisms not only need water but also have requirements for physical and chemical properties and nutrient levels in the water body. Salinity, turbidity, pH value, sediment content, dissolved oxygen, water temperature, nutrient level, and organic contaminant level are all very important to the river's ecosystem. ④ **Various rivers or river sections with different landmark objectives for ecological and environmental protection**. The difference of a river's environment determines the variance of the river's ecosystem. In some river sections, rare and endemic species need to be protected, but in some other sections, the spawning sites of economic fishes need to be preserved. For biodiversity-rich waters or ecologically sensitive waters, the entire habitat may need to be protected. Of course, it is relatively difficult for river managers to determine or secure the ecological flowrate when the objects of protection are not determined, but it will be a long-term goal to quantitatively understand the ecosystem. We should not wait to carry out protection work for the ecosystem until after the scientific research is completed. Therefore, a relatively feasible way is to allow hydraulic engineers and ecologists to negotiate and manage the ecologic flowrate qualitatively and quantitatively.

5.1.6.2 Problems Associated with Ecological Flowrate of the Yangtze River

Due to the complexity of the Yangtze River system, the natural conditions and development approaches in the upper, middle, and lower reaches and the estuary of the river are quite different, and the required ecological flowrate for different river sections also varies. Based on the topographic location and reaches of the Yangtze River, the source area or main tributaries in the upper Yangtze do not have many development and utilization activities of water resources and are the water holding zone or the fragile zone of the ecological environment and, therefore, should be protected. If no hydropower development had been carried out in the area, the natural river state could have been largely maintained. However, the upper Yangtze is rich in hydropower potential, and, thus, the conflict between the protection of rare fishes and hydropower development is prominent. The main problem associated with the ecological flowrate in the middle and lower reaches of the Yangtze River is the inconsistency between the regulation of the control reservoirs in the upper reaches and the hydrological rhythms for the reproduction of the FMCC, migratory fish species and animals, followed by the conflicts among the river-lake connectivity, conservation of wetland's biodiversity and lake openness, and lake control projects. The primary problem in the Yangtze Estuary is the conflicts among the development and utilization of water resources in upper, middle, and lower reaches, salt water intrusion, water pollution and sediment reduction, and the protection of the ecological environment. In some tributaries in the upper, middle, and lower reaches of the Yangtze River where hydropower stations have been developed using the water diversion approach, or undertake the peak-regulating task, many losing, or even dry, river sections have emerged, which is the most conspicuous problem for the ecological flowrate. Therefore, there are different problems associated with the ecological flowrate in various reaches of the Yangtze River.

In terms of the water flow process and water quality required by aquatic organisms, it is not enough to only discharge a minimum ecological flowrate. After the future control reservoirs on the mainstream and major tributaries of the Yangtze River are constructed, the regulating capacity of the reservoir group will be greatly increased, the flood peak will be reduced, and the river flowrate and water level will be increased during the dry season, but the aquatic organisms will have a difficult time adjusting their rhythm or need a long time to adapt to the changes. The flood process can stimulate fish species to spawn and the migratory fishes to migrate. During the dry season, the floodplain is exposed, which is beneficial to the survival of aquatic plants, amphibians, and birds. The reservoirs of high dams often cause the stratification of water temperatures, and the low-temperature water is discharged to cause the fish to delay their spawning time and even not to spawn at all because the opportunity is missed. The discharge of floodwater from high dam reservoirs is generally conducted through aeration to dissipate energy and reduce scouring. As a result, the water is saturated with air, and after they swallow such water, juvenile fish suffer from bubble disease and die. Presently, the ecological flowrate is discharged through hydro-turbines or penstock from most hydropower stations, and it is difficult for biologic organisms to survive after migrating through the dam. As a considerable number of hydropower stations provide the peak load, the flow of the downstream channel fluctuates drastically. For a daily regulating hydroelectric power station, there is a period of time for the reservoir to store water every day, during which time the downstream channel is completely cutoff of water. These phenomena seriously affect the function of the ecological flowrate.

From the perspective of a river's ecological protection and management, because no long-term ecological monitoring system has been established on the Yangtze River, even ecologists cannot determine which species in a typical section need protection. Therefore, it is difficult to scientifically determine the required ecological flowrate for protected species, and it is also difficult to set ecological protection as one of the reservoir's multiple objectives of regulation. Presently, many methods for calculating the ecological flowrate are introduced from abroad, and they are mainly divided into hydrological, hydraulic, habitat simulation and holistic methods, and the integrated procedure. Some of the methods were developed from scientific observations and research in small streams and are not applicable to large rivers such as the mainstream and the major tributaries of the Yangtze River. A considerable number of methods require large numbers of detailed ecological monitoring data. Due to the lack of the basic information, the relatively simple hydrological method is currently used. Presently, the builders and managers of hydropower stations often use the low limits calculated from the method as the control standard of the ecological flowrate and do not pay much attention to the process or the physical and chemical properties of the water flow. Therefore, the control standard is very low. Because of the uncertainty associated with the protection objective and the calculation method, the managers of the river basin and reservoirs have difficulties in the ecological regulating. Only when the ecological regulating, the protection of the water function, the ecological protection objective, and the water demand of the downstream human production and living are comprehensively considered can an effective management of the ecological flowrate be carried out.

Even through an ecological flowrate is provided through a scientific method, it is still difficult to implement the ecological flowrate. The basic requirement for the minimum ecological flowrate is that the flow is never cut off. However, as some hydroelectric stations have only one outlet, or the bottom elevation of the outlet is too high, during the maintenance period or low water level, even the minimum base flow cannot be guaranteed, let alone to ensure the discharge of the continued ecological flowrate. The ecological discharge from most hydropower stations is secured through a turbine unit so that the discharge of the ecological flowrate is closely related to the function and regulating operation of the hydropower stations in the power grid. However, it is still difficult to make both demands meet. In case of an extremely dry time, there exist prominent conflicts in water use between the Yangtze River and upper and lower reaches of tributaries and between human being and the ecological environment, and it is even more difficult to coordinate the interests of all parties.

5.2 Floods and Their Characteristics of the Yangtze River

Since the beginning of the Quaternary, the monsoon climate has prevailed in China, especially in the Yangtze River basin. Precipitation varies as the season changes, and there is more precipitation as the temperature rises. During the summer, rainstorms and floods occur in the basin. The basic feature of the monsoon climate is that the process of floods and low flows are the natural phenomena of the monsoon climate. Floods are the main driving force for the changes of the landform and the river-lake relationship. Only since the formation of the human society have the floods had impacts on human life and production safety and thus become hazards. As a result, flood and waterlogging disasters have occurred. Therefore, floods have had social attributes. In ancient times, because of the low level of human productivity and weak ability to resist floods, the impacts of flood disasters had been threatening and endangering the safety of human lives and agricultural production. The history of the human society in China has always been accompanied by the process of fighting floods. Since the birth of mankind, legendary and historical records have shown a large number of floods and stories of fighting floods. For example, the legendary story of "Yu the Great Controlled Water" that occurred more than 4 ka ago has been widely spread in Chinese history. The earliest formal record of floods on the Yangtze River can be found in the *History of Han • Annals of Empress Lǔ Zhi*. In the third year of the empress' reign (185 AD), "the Yangtze and Han Rivers overflowed, which made more than 4,000 families homeless in the summer". This record has been in history for more than 2 ka. The characteristics of floods and associated disasters of the Yangtze River are not only closely related to the natural conditions of the Yangtze River basin, such as climate, geography, and water system conditions, but also have a strong correlation with the population distribution, the level of socioeconomic development, and the construction of flood control projects.

5.2.1 Characteristics of Floods in the Yangtze River

5.2.1.1 Characteristics of Rainstorm in the Yangtze River Basin

The characteristics of floods are related largely to the climate and geography of the river. The floods in the Yangtze River are mainly caused by rainstorms, and the river is a typical rain-flood type where the time and location of floods are basically consistent with spatial-temporal variation of rainstorms. Based on the major rainstorm distribution in the Yangtze River basin, the northern watershed divide of the Yibin-Yichang section in the upper Yangtze is the monsoon-ward side of the Bayan Har, Min, and Qin Mountains where there are two rainstorm areas of West Sichuan and the Daba Mountains, and the Min and Jialing Rivers flow through the two rainstorm areas, respectively. The Yichang-Luoshan section in the middle Yangtze has the windward side of mountains in western Hubei and western Hunan where the rainstorm areas of western Hubei and western Hunan are located and the Qing River and the Dongting Lake water system flow through the two rainstorm areas, respectively. The Han River joins the Yangtze River within the Luoshan-Hankou section. The section downstream of Hankou in the middle Yangtze has the rainstorm area of the Dabie Mountains. In the lower Yangtze, there is the rainstorm area of Jiangxi, or the rainstorm area from the Jiuling Mountains in Jiangxi to Mount Huangshan in Anhui.

From the temporal distribution of rainstorm occurrences, rainstorms in the Yangtze River basin usually occur earlier in the middle and lower reaches than the upper reaches and earlier in the areas south of the river than the areas north of the river. This indicates that the southern and southeastern monsoons are the main factor of heavy rainfall in the Yangtze River, and the rainstorms first occur in the most frontier mountain area (middle and lower reaches of the Yangtze River and areas south of the river). The general rule of the rainfall distribution is as follows: in May, the rain zone is mainly distributed in the two lake (Dongting and Poyang) water systems in Hunan and Jiangxi; in mid-June to mid-July, the rain zone hovers over the southern and northern areas of the mainstream of the Yangtze River; a rainy season occurs in the middle and lower reaches of the Yangtze River; and the rain zone in the upper Yangtze has an east-west distribution, and the rainfall is greater in the area south of the river than north of the river. From mid-July to early August, the rain zone moves to Sichuan and the Han River basin; in the upper area except the Wu River basin where the precipitation is slightly reduced, the other areas have an increased precipitation; and the main rainfall area is in western Sichuan with the northeast-southwest belt-shaped distribution. From middle August to late August, the rain zone moves northward to the Yellow and Huai River basins, and the Yangtze River basin is under the influence of the subtropical high pressure and often experiences hot weather and droughts. In September, the rain zone moves southward to the middle and lower reaches of the Yangtze River, such as the upper reaches of the Han River, and the precipitation center moves from western Sichuan to eastern Sichuan, and the precipitation in western Sichuan is reduced greatly.

5.2.1.2 Characteristics of Floods in the Yangtze River

As mentioned earlier, the time and location of floods are basically consistent with spatial-temporal variation of rainstorms in the Yangtze River basin. The main times of occurrences are as follows. Floods generally occur in the Poyang Lake water system and the Xiang, Zi, and Yuan Rivers in the Dongting Lake system in the middle and lower reaches during April–July; in the Li and Qing Rivers of the Dongting Lake system and the Wu River that is the upper Yangtze's south bank tributary during May–August; in the Jinsha River and the Yangtze's north bank tributaries during June–September; and in the Han River that is a north bank tributary of the middle Yangtze during June–October. It should be noted that these characteristics are only the rule for the statistical average. In the event of an abnormal climate change, the rule for the rainstorm and flood will vary.

Floods in the Yangtze River system can be categorized into three types: watershed, regional, and local (flash) floods.

A watershed flood is caused by a series of large-scale rainstorms and occurs in the upper Yangtze, middle and lower reaches. Moreover, when all the floods in the mainstream and tributaries of the upper, middle, and lower reaches of the Yangtze River run into one another, a large flood forms in the middle and lower Yangtze with a high volume of floodwater, a high peak flowrate, and a long duration. The floods that occurred in 1931, 1954, and 1998 are of a typical watershed type. The key to the formation of a large watershed flood is the coincidental occurrence of a flood associated with the rainy season in the middle and lower Yangtze and a flood in the upper Yangtze. Under normal circumstances, floods in the Yangtze River occur mostly regionally or locally. Only when a weather anomaly causes an unfavorable flood combination may a watershed flood occur. Therefore, a watershed flood is generally a large flood or an extremely large flood.

A regional flood is caused by a large-scale heavy rainfall and occurs in some sections of the mainstream or tributaries of the Yangtze River. The flood can be characterized by a high peak flowrate, a short duration, a large volume of floodwater, and a short flood process. A regional flood may occur in the upper, middle, or lower reaches of the Yangtze River. The 1981 flood in the upper Yangtze; the 1935, 1969, 1995, and 1996 floods in the middle Yangtze; and the 1981 flood in the lower Yangtze are typical regional events.

A local flood (flash flood) occurs in a tributary of a river basin and is caused by a rainstorm that occurs in a local area. A flash flood caused by a short-duration and small-scale rainstorm and a flood associated with a tidal surge caused by a typhoon storm in the lower Yangtze can also be included in this category. A local flood can be characterized by locality and instantaneity. Because the flood is small in area and is not large in volume of floodwater, it does not cause a large flood in the mainstream of the Yangtze River, and its influence is much smaller than a regional flood. However, as local floods (flash floods) may occur in many more areas and more randomly, it is difficult to take preventative measures. Moreover, such floods may occur frequently within a relatively large watershed and can often result in human deaths. Therefore, the cumulative losses resulting from local floods are substantial. Presently, the cumulative casualties resulting from local floods in China and the Yangtze River basin have exceeded the losses caused by large floods. A flood that occurs in a local area or a river section may be a 100-year event locally, and such a flood may occur every year in the hundreds of the flash flood-prone areas in the Yangtze River basin, which would give the public the impression that the flood has a very high frequency.

The magnitude of a flood in the Yangtze River is closely related to the extent, intensity, and the moving path of the rainstorm. During the flood season, the classification of floods is a dynamic process. In general, before the type of a flood is determined, it is necessary to ascertain the yearly highest water levels observed or predicted at representative hydrological stations in the middle and lower mainstream of the Yangtze River and then calculate the long-time volume of floodwater and the duration of the high water level. Then the type and magnitude of the flood can be determined. As the flood season progresses, it is necessary to reassess the flood based on the new changes. Therefore, it is a dynamic and cumulative process. The evaluation of the scale and damage resulting from a flood is generally performed after the flood season is over. However, flood control and rescue operations must be carried out in time, which would result in some uncertainties to flood prevention and rescue operations.

5.2.1.3 Frequency of Floods in the Yangtze River

As the middle and lower mainstream of the Yangtze River receives inflow from the upper Yangtze and tributaries of the middle and lower reaches and has many sources of floodwater, the flood composition is complex. Therefore, the frequency of a same flood at various control stations is inconsistent. Even for a peak flowrate of a same flood at a same station, the volume of floodwater for various durations of time and their frequencies also often vary greatly. For example, let's compare the floods that occurred in 1954 and 1998. At the Yichang Station during the 1954 flood, the largest peak flowrate was 66,800 m^3/s, and its recurrence period was about only 10 years. However, the 7-day volume of floodwater had a recurrence period of about 30 years; the 15-day volume of floodwater had a recurrence period close to 85 years; and the 30-day volume of floodwater was up to 138.6 billion m^3 that was equivalent to a 100-year recurrence period. In that year, the 30-day volume of floodwater at the Luoshan and Hankou Stations in the middle Yangtze River had a recurrence period of 200 years. At the Yichang Station during the 1998 flood, the peak discharge was 63,300 m^3/s, and the recurrence period was only 6–8 years, but the 30-day volume of floodwater was 137.9 billion m^3, and its recurrence period was close to 100 years.

In the Yangtze River system, especially the middle and lower reaches, due to the complex river-lake relationship, large amounts of sediment and great variations of the river channel's topography and river's cross sections, the water level-flow curve and channel's storage capacities of major cross sections often change. Therefore, it is difficult to accurately describe the magnitude or recurrence period of a flood solely based on a single hydrological element. Meanwhile, because of the broad distribution of rainstorms in the Yangtze River and extremely complex combination, it is difficult to formulate a watershed design storm or probable maximum precipitation (PMP). Moreover, due to the complexity of runoff and concentration, it is even more difficult to estimate the design flood of typical river sections in the mainstream of the Yangtze River. Therefore, the design flood cannot be simply expressed by an occurrence per specific years of a flood event. The parameters that are used to describe the magnitude of a flood include the maximum flood peak flow; 7-day, 15-day, 30-day, and 60-day volumes of floodwater, etc.; and the parameters such as the extent and duration for the occurrence of the highest water level and super high alarming water level are used to describe the risk level of a flood. Thus, over years, in the flood control planning and emergency scheduling program for the Yangtze River basin, a typical major flood that had really occurred in the past has been used as a standard of flood control. Although such a definitive flood process cannot be fully reproduced, using the flood that had actually occurred as the target for the flood control planning can achieve the overall goal of flood control and safety. Moreover, this gives a relatively clear and straightforward defensive target, which is suitable for the complex river-lake relationship in the middle and lower Yangtze River. Hence, in the development of a flood control plan and regulating program, it is generally necessary to study the typical flood processes that had occurred in history, rather than simply using a single recurrence period or a single parameter of frequency.

5.2.1.4 Flood Characteristics of Typical Control Sections in the Mainstream Yangtze

The flood characteristics in typically poorly controlled sections of the Yangtze River are described as follows.

1. Characteristics of Floodwater Composition at the Yichang Station

The control area of the Pingshan Hydrological Station at the mouth of the Jinsha River is about one-half of that of the Yichang Station, but the average annual amount of passing water in the flood season (May–October) is one-third of that at the Yichang Station. Because the flood process is relatively flat and the interannual change is small at the Pingshan Station, the flood process at the Pingshan Station is the base portion of the flood process at the Yichang Station. As the Min and Jialing Rivers flow through the rainstorm zones of western Sichuan and the Daba Mountains, respectively, the volumes of floodwater from the two rivers are very large. The control areas of the Gaoyang and Beibei Stations are 13.5% and 15.6%, respectively, of that at the Yichang Station, while the average annual volume of water passing the stations during the flood season are 20.2% and 16.6%, respectively, and the combined volume passing the two stations is about 37% of that at the Yichang Station, indicating they are the main sources of the Yichang Station. In addition, the floodwater from the upper mainstream cannot be ignored. The Cuntan-Yichang section is also a major rainstorm area in the upper Yangtze, and the control area is 5.6% of that at the Yichang Station, but the average annual volume of water from the section during the flood season is about 8% of that at the Yichang Station, and in some years, it is even more than 20% of that at the Yichang Station (such as 1982). Therefore, the floodwater of the upper Yangtze basin consists mainly of the abovementioned four portions. For different floods, the proportion of floodwater from each of the four portions may not be the same. If floods occur in all the four areas at the same time, an extremely large flood process will occur in the upper Yangtze.

2. Characteristics of Floodwater Composition at the Hankou Station

The control area upstream of the Yichang Station accounts for 67.6% of that at the Hankou Station, and the average annual volume of water during the flood season accounts for two-thirds of that at the Hankou Station. Therefore, the floodwater at the Hankou Station is mainly from the area upstream of the Yichang Station. In the Yichang-Hankou section, the Qing River and the Dongting Lake system are right-bank tributaries. The Dongting Lake system includes the rainstorm zones of northwest Hunan and southwest Hubei which generate relatively large amounts of floodwater during the flood season. The control area at the Changyang Station on the Qing River and that of the four rivers of the Dongting Lake system account for 1% and 14%, respectively, of that at the Hankou Station, but the average annual volumes of water during the flood season account for 1.9% and 20.7%, respectively, of that at the Hankou Station, which are greater than their areal proportions. The large contribution from the four rivers in the Dongting Lake system is a major composition of the floodwater at the Hankou Station. The average annual volume of water during the flood season at the Huangzhuang Station on the left-bank tributary Han River accounts for 6.7% of that at the Hankou Station, which is also one of the major sources of the floodwater at the Hankou Station.

3. Characteristics of Floodwater Composition at the Datong Station

The floodwater at the Datong Station is mainly from the river sections upstream of the Hankou Station. The catchment area of the Hankou Station accounts for 87.3% of that at the Datong Station, and the average annual volume of water at the Hankou Station during the flood season accounts for 81.8% of that at the Datong Station. The Poyang Lake system is a major contributor to the Datong Station and has the Jiangxi rainstorm zone where rainstorms occur in a large area many times in a year. The five rivers in the system contribute large amounts of floodwater. The catchment area of the Poyang Lake system is only 9.5% of that at the Datong Station, but the average annual volume of water in the flood season accounts for 14.8% of that at the Datong Station.

5.2.2 Representative Historical Large Floods

The first hydrological survey of a river in the world was conducted in the eighteenth century, only 300 years ago. In China, the hydrological survey has a history of only 150 years, which is obviously too short for measured flood series in large rivers and provides too few data points. Therefore, it is very important for hydrologists to investigate historical flood marks and search for records from historical literature for the extension of hydrological series. Before the Qin Dynasty, there were rare records about floods in the Yangtze River basin. This may be mainly attributed to the fact that the watershed was sparsely populated; nobody lived on floodplains; and the flooding hazards were low. Earlier records about floods and associated disasters in China were concentrated in the Yellow River basin. Since the Han Dynasty, as the population in the Yangtze River basin gradually increased, history books have begun to record flood events regularly.

According to the statistics of records in historical literature, 214 flood events occurred during the 2117 years from 206 BC (Western Han) to 1911 AD (late Qing), averaging about one flood event every 10 years. During the Tang Dynasty, flood events occurred once every 18 years on average. During the Song and Yuan Dynasties, floods occurred once every 5 years on average. During the Ming and Qing Dynasties, the average was once every 4 years. The statistical results indicate that floods appeared to have occurred more frequently in modern times. In fact, historical literature has mainly recorded flood disasters and has not necessarily accurately reflected the frequency of floods. More recently, more people have lived in the watershed, especially in floodplains. Consequently, floods caused greater losses and could be readily recorded in historical literature. Moreover, the more recent, the more details of flood events have been recorded. Accordingly, the statistics of the records show that floods occurred in the Yangtze River more frequently.

The early flood records were obtained from history books. Because there was no reliable test data or gauging marks, it is now difficult to calculate the magnitude or impact area of those floods. Historical records were generally an account of disasters. Therefore, the records are of limited value for the quantitative evaluation needed in the modern flood control planning, and only the frequency of flood occurrences is of certain statistical significance.

Continuous water level records at a hydrological station on the Yangtze River officially began in the 4th year of Emperor Tongzhi in the Qing Dynasty (1865). The years for the installation of hydrological stations on the mainstream Yangtze are Chongqing in 1892, Wanxian in 1917, Yichang in 1877, Shashi in 1903, Chenglingji in 1904, Hankou in 1865, Jiujiang in 1904, Wuhu in 1900, Nanjing in 1912, Zhenjiang in 1904, Shanghai in 1890, and Datong in the 1920s. The Hankou Station is the one that had the earliest continuous records of the water level of the Yangtze River and can now be used to make quantitative estimates of floodwater from the data recorded in 1870. In the mid-nineteenth century, two consecutive extremely large flood events occurred in the Yangtze River basin in 1860 and 1870, respectively. During the twentieth century, four extremely large flood events occurred in the Yangtze River in 1931, 1935, 1954, and 1998, respectively. These major flood events caused significant damages to properties and economic losses, and they are now the major targets for the study of the flood history of the Yangtze River.

5.2.2.1 1870 Flood

The 1870 catastrophe occurred in the upper and middle Yangtze. During the time, the Hankou Station had begun to record water levels. Because the 1870 flood was a catastrophe in several hundred years, it had left many marks of the event. More than 90 engraved water level marks were left on the river banks in the Chongqing-Yichang section, and about 250 marks of the flood were confirmed. The records for the water levels of the maximum flood peak flowrate are mostly reliable. According to these records, it can be inferred that the 1870 flood was mainly caused by rainstorms in the Three Gorges section and in the middle and lower Jialing River, coupled with the inflow of floodwater from the middle Qing River. According to a comprehensive analysis of historical data, the maximum peak discharge was estimated to be 57,300 m^3/s at the Beibei Station on the Jialing River and 100,000 m^3/s at the Cuntan Station, 105,000 m^3/s at the Yichang Station, and 110,000 m^3/s at the Zhicheng Station on the mainstream of the Yangtze River. The maximum peak flowrate of more than 50,000 m^3/s in a tributary of the Yangtze River was the indication of the high-intensity and menacing impacts of the flood. A preliminary estimate indicates that the 3-day volume of floodwater passing the Yichang Station reached 26.5 billion m^3, more than the current flood control capacity of the TGR, 7-day volume of floodwater reached 53.7 billion m^3, and 30-day volume of floodwater was 165 billion m^3. The flood process was the largest of all flood events recorded in the upper Yangtze and exceeded the process for a 1000-year flood event.

The 1870 flood was the primary reference standard for the structural design of the TGD and flood control of the reservoir. The TGD was designed based on a 1000-year flood and checked using a 10,000-year flood event plus 10%. The 1870 flood inundated an area of 30,000 km^2 in the Jianghan Plain, caused the south bank of the Jingjiang River to break at Pangjiawan and Huangjiapu of Songzi County, created the Songzi River to flow into Dongting Lake, and submerged an area of 40,000 km^2 in plains in the middle and lower Yangtze. Sichuan, Hubei, and Hunan suffered catastrophic damages from the flood. If this type of disastrous flood is to occur in the future, the flood control capacity of the TGR will reach the limit. Consequently, to ensure the safety of the dam and the reservoir, the TGR cannot reduce any flood effect on the middle and lower Yangtze River where flood control will have to rely mainly on the flood diversion/retention facilities in the area and the loss of the flood will still be large. However, with the improvements on the modern flood forecasting and pre-warning capabilities, services from other control reservoirs in the upper reaches and advancement of communications technologies, the government's ability to mobilize personnel to help evacuate residents will not be a big problem, and the resulting loss of human lives will not be too many. Presently, the diversion/retention facilities planned in the middle and lower Yangtze River have a much higher development and utilization rate than the past. However, because the safety improvements are behind, and many fixed assets cannot be relocated, the receding time will be several months long in the event of a catastrophic flood, and the resulting loss cannot be ignored.

Because the level of productivity, standard of the dikes, and capability of flood control were all low at that time, it was hard to avoid dikes from being broken or new floodways from being created by the rushing floodwater. If such a large flood process occurs today, because the capability of flood control has been greatly improved, the floodwater can be expected to be stored in rivers, reservoirs, and flood diversion/retention facilities. However, the regulating risk is still high. For example, if the flood retention facilities are not used, a lot more pressure will be put on reservoirs and dikes for flood control, and the flood will be more dangerous to the reservoirs and dikes.

5.2.2.2 1931 Flood

The 1931 flood was a large disaster that afflicted the entire Yangtze River basin and was caused by the concurrence of floods in the upper, middle, and lower Yangtze. The maximum peak flowrate was 40,800 m^3/s at the Gaoyang Station on the Min River, 50,000 m^3/s at Danjiangkou on the Han River, 30,300 m^3/s at the mouth of the Li River, 57,900 m^3/s at the Chenglingji Station (exit of Dongting Lake), and 59,900 m^3/s at the Hankou Station where the high water level was 28.28 m and the maximum 30-day volume of floodwater was 192.2 billion m^3. Areas most seriously affected by the flood included the provinces located in the middle and lower Han River and the Dongting Lake area such as Hubei and Hunan where the death toll was 145,000. During the flood, all the three districts of Wuhan City were inundated

Fig. 5.1 A Hankou Street during the 1931 Flood

for nearly 3 months, and dike failures occurred at 1600 places in Hunan. The flood caused the most casualties since such disasters began to be recorded. Figure 5.1 shows a street in Hankou during the 1931 flood.

5.2.2.3 1954 Flood

The 1954 flood was a 4-month-long catastrophic process in the Yangtze River basin and was caused by nearly 20 rainstorms. At the Yichang Station during the flood, the 30- and 60-day volumes of floodwater reached 138.6 billion m^3 and 244.8 billion m^3, respectively. At the Hankou Station, the recurrence period of the average peak daily flow was 1400 years; the highest water level was 29.73 m; the maximum peak flowrate was 76,100 m^3/s; and the 30-day maximum volume of floodwater was 218.2 billion m^3. The floodwater inundated 3.17 million ha of farmland in the middle and lower Yangtze, inflicted a population of 18.88 million, and caused death of 33,000. The Beijing-Guangzhou Railroad did not resume normal operation until 100 days after the flood was over. The 1954 flood caused 23,000 km^2 of dike area to fail and 102.3 billion m^3 of floodwater to be diverted to the area beyond the river channel.

The 1954 flood has the following characteristics: ① The volume of floodwater was large, and the water level was high during the peak flowrate. The frequency of the floodwater volumes at the Yichang, Hankou, and Datong Stations during the flood period (May–October) was equivalent to once every 100–200 years. ② The shape of the flood process curve was wide and high, and the flood duration was long. For example, the flowrate that exceeded 40,000 m^3/s lasted for 45 days at the

Yichang Station. ③ Floods concurrently occurred in the upper, middle, and lower Yangtze River. ④ The volume of floodwater diverted or outflowed through failed dike sections in the middle and lower Yangtze was tremendous, and the maximum volume of floodwater stored in the river channel between dikes reached 102.3 billion m^3.

The 1954 flood was one of the main targets for flood control at the TGR. In the event of such a flood, although the control reservoirs including the TGR in the upper Yangtze will play a great role in flood regulation, there will be 30–40 billion m^3 of floodwater that will still need to be diverted to the planned retention facilities in the middle and lower Yangtze, because the reservoirs in the upper Yangtze will have a combined effective capacity for flood control of only 50+ billion m^3 and cannot be used at the same time. Thus, the excessive amount of floodwater will need to be diverted to retention facilities. Therefore, the planned retention zones in the middle and lower Yangtze must be reserved for a long term.

5.2.2.4 1998 Flood

At the Yichang Station during the 1998 flood, the maximum peak flowrate was measured at 63,300 m^3/s, the 30- and 60-day volumes of floodwater were 137.9 billion m^3 and 254.5 billion m^3, respectively. Although the peak flowrate was about a 7-year event, the 30-day volume of floodwater was a nearly 100-year event, and the 60-day volume of floodwater had a recurrence period of more than 100 years. At the Hankou Station, the maximum water level reached 29.43 m; the maximum peak flowrate was 71,100 m^3/s; and the 30-day maximum volume of floodwater was 188.5 billion m^3 that is close to that in the 1935 flood, second to that in the 1931 flood, and fourth in rankings since 1865 with a recurrence period of about 30 years.

Although the flood was generally smaller than the 1954 flood, because no active diversion measures were taken, only few dike breaks occurred; the floodwater was mainly contained in the river channel; the water levels in many river sections were higher than those in the 1954 flood. During the more than 3-month period of the flood event, 240,000 ha of farmland was inundated; 2.13 million houses collapsed; and 2.32 million people were affected, including 1526 deaths.

The 1998 flood of the Yangtze River had the following characteristics: ① The flooding area was large; the volume of floodwater was tremendous; the high water level lasted a long time; and the flooding condition was devastating. As a result, the water level in the Zhicheng-Datong section in the middle and lower Yangtze remained high for a long time. The water levels at the Zhicheng, Hankou, Huangshi, Anqing, and Datong Stations were the second highest of the historical record, and the water levels at the Shashi, Shishou, Jianli, Lianhuatang, Luoshan, Wuxue, and Jiujiang were all the highest in history. Moreover, the time for the water level above the warning level at each station was mostly between 57 and 96 days. ② At the two outlet control stations for Dongting and Poyang Lakes, the water levels and discharges at the Chenglingji and Hukou Stations were measured at historical highs. In the five rivers of the Poyang Lake system, the water levels and discharges at the Lijiadu Station on the Fu River and at the Dufengkeng Station on the Chang River were all measured at historical highs, and the water level and discharge at the Meigang Station on the Xin River were the historical high and the second highest, respectively. In the four rivers of the Dongting Lake system, the measured water level and flowrate at the Shimen Station on the Li River also broke the historical record, and the measured water level at the Taoyuan Station of the Yuan River was the second highest in history. ③ The water levels at the three outlet control stations on the Jingjing River were all historical highs. The flowrate at the Shiquan Station on the Han River was the highest measured discharge, and the water level at the Hanchuan Station on the Han River was the historical high. The record-breaking high water levels were mainly caused by temporarily raised dikes but not using the flood diversion/retention facilities. Although the loss resulting from inundation was small, the cost from extra manpower, rescue operation, and emergency management was high.

Figure 5.2 shows a rescue site during the 1998 flood.

5.2.2.5 Comparison Between 1954 and 1998 Floods

During the 1954 flood, the Yangtze River basin did not have any large reservoir with a storage capacity of more than 100 million m^3, while during the 1998 flood, the basin boasted about 100 large reservoirs. During the 1954 flood, the volume of floodwater from dike breaks in the middle and lower Yangtze was up to 102.3 billion m^3, and the other excessive floodwater went to low-lying floodplains. However,

Fig. 5.2 Flood-fighting and rescue operation during the 1998 flood

during the 1998 flood, only about 10 billion m^3 of floodwater was stored in the river channel between dikes, and the floodwater was mainly stored in the river channel and reservoirs. Therefore, at the same flowrate, the water levels in many river sections in 1998 were higher than those in 1954. Moreover, the change of the river-lake relationship and the evolution of the river channel rendered the water levels and flood processes recorded at control hydrological stations on the mainstream of the Yangtze River incomparable during large floods. In order to compare the floods of 1954 and 1998, it is necessary to regenerate the floods. For example, let's compare and analyze the regenerated annual high water levels at the Shashi, Luoshan, and Hankou Stations:

1. **Regenerate 1954 Flood**. Under the assumption that no dike breaks had occurred, what would have been the high water level? The volume of floodwater from the dike breaks in 1954 was 102.3 billion m^3, of which 70 billion m^3 occurred in the river section upstream of the Luoshan Station. Moreover, the Jingjiang Diversion Project was used three times in 1954; the largest diversion flowrate was near 8000 m^3/s, resulting in the largest water level decrease of about 0.96 m at the Shashi Station; and large numbers of overflowed dikes or dike breaks occurred in the middle and lower Han River. After taking into account the impact of the backwater from the regenerated water level of the Han River, the regenerated water level in 1954 would have been about 46.10 m at the Shashi Station, about 36.10–36.30 m at the Luoshan Station, and 31.60–31.90 m at the Hankou Station.
2. **Regenerate 1954 and 1998 Incoming Water to the River-Lake Conditions in 1998**. During the 1954 flood, the regenerated water level would have been 46.10–46.30 m at the Shashi Station, 36.80–37.00 m at the Luoshan Station, and 32.00–32.30 m at the Hankou Station. During the 1998 flood, the regenerated water level would have been 45.60–45.80 m, 35.50–35.70 m, and 30.00–30.30 m at the Shashi, Luoshan, and Hankou Stations, respectively.

The above preliminary analysis indicates that if the water levels during the 1954 and 1998 floods had been regenerated to the 1998 river-lake conditions, a comparison of the highest regenerated water levels in the 2 years would have yielded: the water levels at the Shashi, Luoshan, and Hankou Stations in 1954 would have been higher than those in 1998 by 0.5 m, 1.3 m, and 1.7 m, respectively. Even the regenerated 1954 water levels to the 1954 conditions would have still been higher than those as expected for 1998.

Through the comparison of the two catastrophic floods in the twentieth century, the following conclusions can be made. ① With the change of the natural conditions and the impact of human activities, the overall scale of the catastrophic floods may not change much, but with the change of time, the flood storage capacity of rivers, reservoirs, and flood diversion/retention facilities may change substantially, and the distribution of the excessive floodwater in rivers, reservoirs, and flood diversion/retention facilities may vary greatly. ② Because of the rapid socioeconomic development inside the dikes, even in the flood diversion/retention zones, once floodwater needs to be diverted and retained, the loss from inundation will not be small, and the pressure of future flood control will be mainly transferred to reservoirs and dikes, especially reservoirs. As a result, the requirements for optimizing the flood regulating operation of reservoir groups are becoming increasingly higher. ③ In the event of a large flood, the loss may be small when floodwater is not diverted, but the water level within the dikes will remain high, and the pressure to protect the dikes will be tremendous. Once a dike break occurs, the loss will be even greater. Therefore, even if larger reservoirs such as the TGR have been constructed, improvements on the diversion/retention facilities will still need to be continued to defend against catastrophic floods. If multiple aspects such as the economy, society, and ecological environment are comprehensively considered, considerably large capacities of flood diversion/retention facilities should be maintained and utilized when needed. This will not only make the flood diversion/retention facilities be worthy of their names but more importantly can ease the pressure on reservoirs and dikes and prevent even larger losses resulting from dam and/or dike breaks. ④ Due to the impact of human activities, especially encroachment activities, such as river-lake barriers, utilization of rivers' sandbars and floodlands, and land reclamation from lakes, will not only reduce the storage capacities of river channels and lakes but also the available areas of flood diversion/retention zones. Moreover, river-lake relations and water level-flowrate curves will also be changed significantly. As a result, there will be many uncertainties in the use of floods for typical recurrent years in the design for flood control. Thus, the strategy for flood control should be adjusted in time to adapt to the changing river-lake relationship, and the regulating operation and flood control should be carried out scientifically.

5.2.3 *Flash Flood Hazards*

The state and society have paid great attention to flood control on large rivers, and the construction of flood control systems has also focused on large rivers all along. However, a large flood is a rare event and, generally, only occurs once in decades, but local floods or flash floods occur every year. Although the frequency of occurrence in a particular area is not high, for such a large basin as the Yangtze River, flash floods occur at different places every year. For example, if a 100-year flash flood event occurs at one location and there are 100 such locations, one such flood will occur once every year in the entire basin with a probability of 100%, and, of course, they occur in different parts of the basin. According to incomplete statistical data, debris flows and landslides resulting from flash floods caused the death toll of 800–1000 in China every year. The direct economic loss was 2 billion to 4 billion yuan, and 6000 ha of farmland was wiped out, of which about 600 ha could not be restored. In Sichuan Province alone, about 300 disastrous landslides and debris flows have occurred annually on average since 2000, endangered more than 100 villages, and caused the destruction of large areas of farmland and large numbers of human injuries or deaths.

Debris flows and landslides are widely distributed in Western China. They occur frequently in fluvial valleys in southeastern Tibet, lower Jinsha River valley, Yanlong

River valley, and Min River valley, which are one of the areas with most serious landslide and debris flow disasters in the upper Yangtze. According to incomplete statistical data, tens of thousands of landslides and more than 10,000 landslide gullies, of which 7561 gullies and 2435 landslides have been accurately located, have been recorded in the upper Yangtze and southwestern region. Debris flows and landslides have widely developed in the areas with severe downward cutting by rivers, relatively high topographic reliefs, and tectonically active structures. For example, in the areas of the Yalong, Anning, and Dadu Rivers, the lower Jinsha River, the upper Min River, the upper Jialing River, and the Bailong River that have developed along tectonic structures in the direction of longitudinal lines, debris flows and landslides follow fault zones; rivers and highways are densely distributed in a linear pattern; and seismic zones are distributed in clusters. For instance, the lower Jinsha River is only 138 km long, and its basin area is 3043 km^2, but 172 debris flow gullies have developed along both sides of the river, of which the Jiangjiagou debris flow gully is the largest in the world and an average of 15 debris flows occur annually with the maximum of 28 occurrences in 1 year. More than 20,000 landslides, slope failures, and debris flows are densely distributed in the Wenchuan earthquake area of "May 12, 2008". Figure 5.3 shows a debris flow disaster site caused by a catastrophic flash flood in Zhouqu in 2010.

The formation of a debris flow is mainly controlled by topography, loose soil mass, and soil moisture condition. Many mountainous areas in the upper Yangtze have the conditions of topography and solid materials necessary for the occurrence of debris flows. Rainstorms and flash floods have become the main factors for debris flows to occur in the areas. Based on various water sources for the formation of debris flows, debris flows can be categorized into three major types: rainfall type, outburst type, and ice melt type. Generally speaking, debris flows and landslides occur suddenly and last for a short time. The process of a debris flow or landslide lasts only minutes to tens of minutes from the beginning to the end. As the sudden occurrence of debris flow or landslide makes it difficult to forecast accurately, the evacuation time is short before the disaster occurs, and often large amounts of moving rock masses with strong kinetic energy swamp to houses, roads, and bridges. Consequently, rivers are blocked, lakes filled, farmlands buried with muds, and forests destroyed, resulting in disasters to the mountainous areas. For example, on July 14, 1987, due to a glacier motion, about 360,000 m^3 of an ice tongue at the Bomimi gully slid from the glacier into a glacial lake, and the surface level of the lake rose by 1.4 m on average, resulting in a sudden failure of the moraine dam and an outburst flood. As the outburst flood carried loose solids along its way, it transformed into a debris flow. The debris flow wiped out the Midui Village in the gully and destroyed large areas of farmlands and 27 km of road embankment of the Sichuan-Tibet Highway.

Over the recent 100 years, the following typical large flash floods and geological disasters have occurred in the Yangtze River basin. ① In 1933, a landslide was induced by the Diexi earthquake. The landslide blocked the Min River and formed a barrier lake. A week later, the debris dam broke, resulting in a flood in the Min River. The flood directly impacted about 200 km river section from the break point

Fig. 5.3 Zhouqu debris flow disaster caused by catastrophic flash flood

down to the proximity of Dujiangyan, causing 2500 deaths. ② In November 1979, debris flows occurred concurrently at Luwang and Ganxi gullies in the suburb of Ya'an City. The debris flows caused damage to 17 villages, 4 factories and some streets, 164 deaths, and industrial and agricultural losses of 24.6 million yuan. ③ In 1997, a debris flow that was transformed from a landslide in Meigu County of Sichuan Province wiped out a village, caused 151 deaths and missing. ④ On July 9, 1981, a debris flow in Liziyida gully destroyed a railway and bridges and caused a train to be overturned, 360 deaths and 15 days of interruption of traffic. ⑤ On May 30, 1984, a debris flow occurred at Heishui gully within the Yinmin mining area of Dongchuan City, Yunnan, and struck Yinmin Town, resulting in 121 deaths, more than 30 injured, more than 1000 people affected, 50,000 m^2 of buildings and a large number of production and living facilities destroyed, mining production suspended for half a month, and direct economic loss of 11 million yuan. ⑥ On July 11, 2003, at 23:05–23:30, a debris flow occurred at Qiongshan gully in Badi Township of Danba County, Sichuan, and led to the disappearance and death of 51 people and destruction of the popular scenic tourist site. ⑦ On August 13, 2005, a debris flow at the scenic Hailuogou tourist site in Sichuan washed away roads and all power stations, affected more than 1200 tourists, and caused the tourism business to be suspended for 3 months. ⑧ On August 7, 2010, a sudden heavy rainstorm occurred in Zhouqu County of Gannan Tibetan Autonomous Prefecture, followed by debris flows at Luojiayu and Sanyanxia north of the county seat. The debris flows swamped southward toward the county seat, destroyed the houses along the Bailong River, blocked the Bailong River, formed a barrier lake, and caused 1434 deaths and 331 missing.

5.3 Droughts in the Yangtze River Basin and Their Characteristics

5.3.1 Droughts and Associated Disasters

A drought is a relatively long period of abnormal weather cycle in a given watershed or region where precipitation is well below average and has caused a significant decrease in soil moisture content and river runoff. Based on progression and impacted aspects, droughts can be divided into meteorological, hydrological, and agricultural droughts. A meteorological drought happens mainly when the atmospheric and soil moisture content has decreased due to abnormally low precipitation, affecting the normal growth of plants and crops. A hydrological drought occurs mainly when low water supply has become evident, especially in rivers and lakes where the water level is low because of abnormally low runoff resulting from decreased rainfall. An agricultural drought occurs mainly when crops have become affected due to low precipitation or scarce water supply from rivers, lakes, and reservoirs, resulting in a low production, or even no production at all.

Drought and drought disaster are different concepts. A drought is a natural meteorological and hydrological phenomenon. Generally, plants, crops, and aquatic organisms can adapt to the general process of a drought, and periodic droughts are required to inhibit the development of dominant species. However, a drought disaster occurs when a shortage of water supply for human and livestock use is evident because relatively large areas are inflicted with a relatively long time period of droughts and, consequently, the cost of water supply increases or the growth of crops is affected, resulting in a low yield of crops and economic losses. When a drought disaster is compared with a flood, the former is a "chronic disease," and the latter is an "acute illness." An "acute" flood readily causes harm to the human body, so the state, governments of all levels, and the society pay special attention to it. However, it takes a certain time for a "chronic" drought disaster to start and develop until it becomes evident. It also takes time to evaluate its impact or damage. Therefore, a drought disaster develops gradually, and it is difficult to forecast it and evaluate the associated loss. The longer the duration of a drought and the greater the affected area, the higher the damage and loss, or the severity of a drought is proportional to the duration and the affected area of occurrence.

Droughts are classified into different levels due to differences in the developmental spatial and temporal extent, severity of impacts and loss. A common drought does not necessarily cause disastrous loss because crops can endure droughts to a certain extent. For example, some flowers can grow well if they are watered once for a few days; aquatic organisms, such as aquatic weeds and fish species, can also endure droughts to a certain degree. After rain comes, the function of the ecosystem will be gradually restored. The awful thing to do is that a lot of fish are harvested, while it rains little, which will have devastating impacts on aquatic organisms and fish species. Various crops have different levels of drought tolerance. Wheat and corn are drought-tolerant crops. In north China where it is dry due to low precipitation, drought-tolerant crops are generally planted. However, in the Yangtze River basin and south China where rainfall is high, rice and other high water-consumptive crops are planted. Therefore, once a severe drought occurs, economic loss will naturally result.

In the past, droughts and drought disasters mainly affected rural life and agricultural production. A severe drought caused no crop yield at all. Consequently, widespread famine and human death due to starvation occurred, and the starved were forced to leave their home for survival. With industrialization and urbanization, the population is highly concentrated. A severe drought will cause a shortage of municipal water supply, which will not only affect urban residents' normal life and production but also cause social stability issues. Therefore, it is one of the most important tasks for urban managers to provide urban citizens with a safe and adequate drinking water supply system.

5.3.2 Historical Droughts

Although the Yangtze River basin is rich in rainfall, the temporal distribution of precipitation is very uneven due to the influence of the monsoon climate. In the mountainous and hilly areas where water storage conditions are poor, drought occurs soon after a rainfall, and the frequency of drought occurrences is relatively high. Most of the areas have a probability of drought occurrences above 60%, and droughts may occur in all four seasons, indicating that the probability of drought occurrences in the water resource-rich Yangtze River basin is not lower than that in north China. In recent years, drought events reported by the media in the Yangtze River basin are more than north China. This may be attributed mainly to the fact that the mountainous and hilly areas account for a large proportion of the Yangtze River basin; groundwater makes up a low proportion; and more high water-consumptive crops have been planted. The most serious droughts in the Yangtze River basin generally occur during the spring-summer seasons, or during the summer-autumn seasons.

The differences in occurrence between droughts and floods in the Yangtze River basin are as follows. ① The occurrence of droughts is directly affected by meteorological conditions; many of them are regional; and droughts rarely occur in the entire basin. ② The duration of a drought is generally longer than a flood process, and the loss resulting from a drought increases gradually. ③ A severe drought in the upper and middle reaches will eventually lead to extremely low flowrates and extremely low water levels in the mainstream. ④ Droughts can occur in a same area for several consecutive years, but floods rarely occur in a same river for several continuous years.

The first drought disaster in the Yangtze River basin was recorded in the 4th year of King Xiang of the Zhou Dynasty or 644 BC, which was several hundred years earlier than the time when the first flood was recorded. However, early (before Song Dynasty) disaster records were very brief. The historical severe droughts that had widespread impacts in the Yangtze River basin occurred in 1528–1529, 1640–1641, 1785, 1836, 1876–1877, 1900, 1925, 1928–1929, 1934, 1942, and 1945. After the founding of the PRC, the most serious droughts happened in 1978 and 1959. The 1978 and 1959 droughts have been ranked first and second in severity since the founding of the PRC and their respective frequencies of occurrence are 2.3% and 4.7%. Droughts also occurred in 1961, 1976, 1988, and 2006.

5.3.2.1 1959 Drought

The drought occurred in an area east of Chongqing (Chongqing was a city of Sichuan at the time) and was centered in Yichang of Hubei and Ankang of Shaanxi. The annual rainfall was 46% lower than average. The amounts of monthly rainfalls were 85%, 46%, and 70% below their respective averages in July, August, and September, respectively. It did not rain in the Guiyang area in the upper Yangtze for 31 continuous days, at the Changsha Station in the middle Yangtze for 61 continuous days, and in Nanjing of the lower Yangtze for 43 succeeding days.

5.3.2.2 1963 Drought

The drought was centered in an area east of Chongqing, Sichuan, and south of Yichang, Hubei. The annual rainfall was below average, and Changsha and Nanchang were most severely impacted. The amounts of rainfall were 43% and 46% below their respective averages in Changsha and Nanchang, respectively. The amount of rainfall was 70–80% below average at the Changsha Station during June–August and was 50–70% below average at the Nanchang Station during July–September.

5.3.2.3 1978 Drought

The annual amount of rainfall in all areas but the Chengdu area in the Yangtze River basin was below average, especially in the Nanjing and Anqing areas where the amount of rainfall was both 50% below average and was even 60% below average in Nanjing. More specifically, the amount of rainfall was 50–90% below average in Anqing during June–August and 50% below average in Nanjing during May–September and 70% below average during July–August. It did not rain in Hankou for 47 consecutive days and in Hefei for 54 continuous days. The amount of runoff was the lowest in the middle and lower Yangtze during January–October in 40–50 years, and the water levels were the lowest at the Yichang and Shashi Stations in 70–80 years. The river section downstream of Xuliujing in the Yangtze Estuary was affected by seawater intrusion for 6 months; the Wusong Water Plant could not withdraw water from the river for 142 days because the salinity level of the water exceeded the standards; and seven intake structures along the Huangpu River could not be used as drinking water sources due to high salinity levels caused by the intrusion of salty tidal water. The 1978 drought resulted in the historical low measured flowrate of 4620 m^3/s at the Datong Station on January 31, 1979.

5.3.2.4 2006 Drought

The drought occurred in the summer (or flood season), and the amount of rainfall was 36%, 20.5%, 16.2%, 25.7%, and 14.4% below their respective average in Chongqing, Yichang, Wuhan, Anqing, and Shanghai, respectively. In July, the water levels of the river at the above cities dropped to the lowest in a century during the flood season. The water levels were only 17.07 m and 6.64 m at the Hankou and Datong Stations, respectively. A rare flow process of an "extremely dry flood season" occurred.

In 2006, in the mainstream and tributaries of the upper Yangtze, the measured amount of annual runoff at the Gaoyang Station on the Min River, Fushun Station on the Tuo River, and Wulong Station on the Wu River were all ranked as the lowest in the recorded series of data. The measured annual amount of runoff at the Pingshan Station on the Jinsha River and Beibei Station on the Jialing River were both ranked the third from the bottom of their respective measured series of data. In the tributaries of the middle and lower Yangtze, the relatively drier rivers were the Yuan and Li Rivers in the Dongting Lake system. The measured amount of runoff at the Taoyuan Station on the Yuan River had been the lowest since 1956 and at the Shimen Station on the Li River had been the second from the lowest since 1956. Due to the widespread drought in the Sichuan-Chongqing area, the annual amount of runoff at the Cuntan Station on the mainstream of the Yangtze River, which relies on water replenishment from the upper Yangtze, was only 247.6 billion m^3 that had been the lowest measured value since 1893. The measured annual amount of runoff at the Yichang Station was 284.8 billion m^3, which was 36% below the average and had been the lowest measured value since 1878. The annual amount of runoff at the

Hankou Station was only slightly higher than that in 1900 and 1928 and had been the third driest year since 1865. The annual amount of runoff at the Datong Station was 688.6 billion m^3, which was slightly higher than that in 1978 and had been the second driest year since 1951. The 2006 low flow was mainly caused by the concurrent occurrences of low-flow processes in the mainstream and tributaries of the upper Yangtze.

According to historical records from the fifteenth century to the present, the Yangtze River basin has suffered droughts once every 1.8 years on average and severe droughts once every 7.8 years. Droughts have occurred more frequently in the upper reaches than the middle reaches where droughts have occurred more frequently than the lower reaches, which indicate that droughts have occurred more frequently in the mountainous and hilly areas than the plain areas. However, extreme droughts have occurred more frequently in the middle and lower reaches than the upper reaches, averaging once every 65 years, indicating that severe droughts have had a greater impact on the densely populated middle and lower reaches. Historically recorded droughts were not separate from drought disasters. In fact, the vast majority of drought records were for drought disasters. Because there was no measurement of rainfall or runoff in ancient times, only when a drought disaster had occurred was the drought recorded.

5.3.3 *Drought Characteristics in the Yangtze River Basin and Analysis of Representative Drought Disasters*

The Yangtze River basin has not only suffered from many flood disasters but also experienced many droughts and associated disasters. Droughts in the Yangtze River basin reported by media are more than those in the arid north China. Droughts occur in several areas almost every year. In fact, the characteristics, nature, and severity of droughts in the Yangtze River basin are different from those in north China, and the characteristics of droughts in the Yangtze River basin are summarized as follows.

5.3.3.1 Poor Water-Storing Conditions in Mountainous and Hilly Areas Due to Special Geographical Conditions

More than 70% of the Yangtze River basin is mountainous and hilly lands where precipitation quickly turns into runoff and flows out of the area due to steep topographic slopes. A drought begins to appear when it does not rain for a short time (e.g., more than 10 days). However, in north China's plain areas, most of precipitation infiltrates into the subsurface and becomes recharge to groundwater, while most of the groundwater in the Yangtze River resurfaces at feet of mountains and becomes a portion of the surface water in the lower and middle Yangtze River. If there are hydraulic facilities such as reservoirs and canals, the drought-fighting capability will be significantly enhanced to withstand moderate droughts so that no agricultural droughts will occur. If there are no reservoirs and supporting irrigation canals or water withdrawal facilities, an agricultural drought occurs soon after the crops on hillsides begin to be short of water, and the growth of the crops will be affected. In the event of a severe drought, even if there are some reservoirs in the area, the reservoirs will not be able to provide enough water because it does not rain in the area for a long time, resulting in less incoming water to the reservoirs due to reduced runoff. Moreover, the amount of water use will increase. Consequently, the reservoirs will be withdrawn to the dead water level or even become dry, and it will be difficult to supply drinking water to the residents and their livestock living in the mountainous area, resulting in a serious drought disaster. Traditional drought frequently occurring areas in the Yangtze River basin include the Yunnan-Guizhou Plateau, the Sichuan Basin, and the mountainous and hilly areas in Chongqing, Hengshao of Hunan, northern Hubei, Nanyang of Henan, and southern Jiangxi.

Yunnan and Guizhou are rich in karst landforms, and rainfall water quickly infiltrates into subsurface. However, due to thin topsoil and poor vegetative conditions, the water holding capacity of the soil is low. In mountainous areas where elevation is high and available water is low, there is always a water scarcity problem of physical nature because the cost of pumping water is high and the cost for the construction and operation of a hydraulic project is also high. Therefore, the mountainous areas inevitably have a water scarcity problem of physical nature. In Western countries, mountainous areas are generally sparsely populated, and there is no need to farm on slopes. However, in China's many places, farmlands on slopes extend up to the proximity of mountaintops, where droughts and associated disasters occur frequently. The densely populated China has no way to avoid frequent droughts and associated disasters. Such a situation occurred in the past. Nowadays, due to quick information transmission through news media, through communications, and especially through the Internet, drought events that had drawn little attention in the past are now spread rapidly and receive extensive coverage. As a result, droughts appear to have occurred more in the Yangtze River basin than in north China. In fact, droughts in north China are more severe, but the agricultural production and urban life have become accustomed to the arid climate.

5.3.3.2 Yangtze River Basin Has Less Groundwater than North China

Groundwater that does not contribute to the surface water only makes up about 1% of the total water resources in the Yangtze River basin, and the readily available groundwater is less than that in north China. In the mountainous and hilly areas, groundwater mostly resurfaces at feet of mountains and becomes part of surface water. In the plain areas where plains and lakes were formed by alluvium, with historic warm climate, high rainfall, good forest vegetation, and clayey soil layers that are much more than north China in coverage and thickness, precipitation mainly becomes runoff or is retained in lakes and rivers or shallow subsurface, with very little in deep subsurface. Therefore, the quantity of groundwater available for utilization is very limited. Moreover, reliance on surface water for the vast majority of production and living has been a historic habit. In the middle and lower reaches, once the surface water level is low, it will be difficult to withdraw water for use, resulting in a drought and associated disaster. In Beijing of north China, 60% of current water use is from groundwater, while in the Yangtze River basin, the amount of groundwater use is no more than 5% of the total amount of water use in any large cities, indicating a great difference exists in the use pattern of water resources between south China and north China.

5.3.3.3 Prominent Summer Drought

In the Yangtze River basin, after the summer-rainy season or no occurrence of a rainy season, due to high temperature, the evapotranspiration from ground and plants is high, especially after the summer-rainy season. As the subtropical high pressure controls the middle and lower Yangtze for a long time, a summer drought occurs almost every year in different regions, resulting in difficulties in water supply for crop production and living. Although regions of western Sichuan, Chongqing, the Jianghan Plain, Dongting Lake and Poyang Lake in the Yangtze River basin, and the Yangtze Delta area are rich in total amounts of water resources, they are prone to frequent summer droughts.

5.3.3.4 Large Areas of High Water-Consuming Crops and High Water Demand

The Yangtze River basin is China's most major rice-growing area where the rice-planting area exceeds 18 million ha. Therefore, water demand and consumption are high. Once dry years or dry seasons occur, the normal growth of rice is often affected, and agricultural droughts occur frequently.

5.3.3.5 Aggravated Severity of Droughts due to Water Quality Problems

Water pollution is serious, or eutrophication occurs in natural waters such as lakes near large cities along the Yangtze River. In many large- and medium-sized cities, especially in the middle and lower reaches of Yangtze River, where water pollution of their own municipal sources (local water) is serious, they rely heavily on guest water (upstream water). When guest water is decreased, or an emergent water pollution incident occurs, the resulting water shortage problem is prominent. At the same time, if a local drought occurs, the damage from the drought will be amplified.

5.4 Climate Change and Hydrological Extremes

5.4.1 Climate Change and Climatic Cycle

In the past 20 years, global warming has become the hottest topic and research hotspot in the world, which has not only attracted widespread attention of a considerable number of scholars, environmental protection workers, and the general public, but governments of all countries and the United Nations institutions have also attached great importance to it. The *United Nations Framework Convention on Climate Change* (UNFCCC) was adopted at the United Nations Conference on Environment and Development (Earth Summit) in Rio de Janeiro, Brazil, on June 4, 1992. The convention is the world's first international treaty to stabilize greenhouse gas emissions (carbon dioxide and other anthropogenic greenhouse gases) in response to the adverse effects of global warming on socioeconomic development, as well as a basic framework for international cooperation in tackling global climate change issues by the international community. A new international treaty, Kyoto Protocol to the UNFCCC, was adopted at the Third Conference of the Parties to the UNFCCC in Kyoto, Japan, in 1997, and the treaty has clearly defined the world's emission standards for carbon dioxide, namely, between 2008 and 2012, carbon dioxide emissions in the world's major industrialized countries would be 5.2% lower than the average of the 1990 emissions. These international conventions have not only involved academic implications but have also affected the socioeconomic development of all countries in the world.

The concept of climate change or global warming has been debated. One view argues that climate change is a normal natural phenomenon. From the perspective of the evolution of the earth's climate, climate change has a cyclical characteristic, and there are different timescales of the change cycles. On a short timescale, the magnitude of change is small, and on a long timescale, the amplitude of change is large. Studies on the history of the earth have shown that climate, atmospheric temperature, glaciers, and sea levels have undergone great changes, and it is difficult for human beings to alter the regularity of the large cyclical climatic changes. Another view argues that since the beginning of the human society, due to deforestation and changes in land use, especially since the industrial revolution, the massive use of fossil fuels, such as coal, petroleum, and natural gas, has resulted in emissions of extraordinary amounts of greenhouse gases, such as carbon dioxide, which has led to global warming, polar glaciers melting, and the degradation of the earth's ecosystems. Ultimately, human's sustainable development has been impacted, and a concerted action is required to prevent or mitigate this trend. The latter view is currently dominant.

The term of climate change, when used by the Intergovernmental Panel on Climate Change (IPCC), refers to any change of the climate over time, whether due to natural variability or the result of human activity, as distinct from the definition of the UNFCCC. The UNFCCC defines the climate change as "a change of climate which is attributed directly or indirectly to human activity that alters the composition of the global atmosphere and which is in addition to natural climate variability observed over comparable time periods." Climate change includes the change in the mean and the variability of its properties. The question is how to differentiate the change of natural periodicity from the one caused by human activity and identify the percentage of each contribution. It is difficult to answer the question in a short time, and it will at least take a relatively long time of observations before a regular pattern becomes evident. There is plenty of evidence to show that alpine glaciers such as Himalaya and the Greenland and Arctic ice caps have decreased markedly over the past few decades, indicating that global warming is evident. The relatively large short-term impact from global warming is the increased number of extreme meteorological events. The long-term impact would lead to the rise of sea level, inundation of coastal low-lying lands, extinction of some rare species, or occurrence of large-scale biological succession.

5.4.2 Hydrological Cycle

Climate change and hydrological change are closely related. Therefore, hydrological change is also cyclical, and extreme weather and hydrological events also have periodicity. During the 1050 years from 950 to 2000, the Yangtze River basin underwent 332 years with rainy and flood seasons, accounting for 31.62% or about once every 3 years; 304 years with drought seasons due to scarcity of precipitation, accounting for 28.96%; and 414 years of normal precipitation seasons, accounting for 39.42%. The statistics of the 1050-year data indicates that there were about 115 years of concurrent occurrences of large droughts and floods and about 39 years of concurrent occurrences of small droughts and floods. The data also indicate that during each concurrent occurrence of large drought and flood, three concurrent occurrences of small droughts and floods. During the most recent long drought period of 1871–1930, two droughts and one flood occurred in the 60-year cycle. The most recent long flood period of 1931–1999 included the short cycles of 1931–1957 floods, 1980–1999 floods, and 1958–1979 droughts. Therefore, based on the statistical data, beginning in 2000, the Yangtze River basin should be in a large drought cycle (60 years or so) and a short drought period (about 20 years), and droughts would occur in more regions and more frequently, which was in agreement with the recent trend that more droughts and fewer floods have occurred in the Yangtze River basin in the past 10 years.

5.4.3 *Extreme Weather and Hydrological Events*

5.4.3.1 Extreme Weather Events

Extreme weather events are low-probability events. Presently, the most widely used method to describe extreme weather events in the world is the percentile definition method. Based on actual research needs and various lengths of time series, different percentiles can be used. Generally, 90, 95, 99, etc. percentiles are used. Using the 99 percentile as the standard to statistically analyze the 1960–2005 temperature data from 426 stations and 1960–2006 precipitation data from 437 stations (Yang et al. 2010) indicates that, in the 1960s, extremely high-temperature events occurred in large-scale high-percentile group-occurring areas between the Yellow and Huai Rivers and between Huai and Yangtze Rivers and extended southward, and the area south of the Yangtze River had a slightly weaker intensity of group-occurring than the area between the Yangtze and Huai Rivers, but it was still in the high-percentile group-occurring area. In the 1970s, the high-percentile area of north China and eastern part of central China disappeared but widespread high-percentile areas occurred in northeast China; west China showed a trend of intensifying from the south to the north; and high-percentile group-occurring areas appeared in the middle and upper reaches of the Yellow River. In the 1980s and 1990s, group-occurring was significantly weakened. In the 1980s, west China had obvious high-percentile group-occurring areas; Yunnan of southwest China had a large-scale Level I high-percentile group-occurring area; and south China and northeast China had small-scale high-percentile group-occurring areas. In the 1990s, the region south of the Yangtze River showed the trend of intensifying group-occurring characteristics from south to north; the upper and lower Yangtze had prominent high-percentile group-occurring areas; and north China had a small-scale high-percentile group-occurring area. In the 2000s, the high-percentile group-occurring areas increased in a ladder-type leaping pattern from south to north. Based on the movement pathway of the high-percentile group-occurring area in various time periods, it first moved from east to west in longitude; then it moved from south toward north in latitude; and the intensity of movement was first from strong to weak and then from weak to strong.

Extreme precipitation events also have the group-occurring pattern. In the 1960s and 1970s, major group-occurring zones were concentrated in south and southwest China. In the 1980s, the group-occurring zones moved to the Yangtze and Huai River basins. Since the 1990s, the southern regions have not had any high-percentile group-occurring zone. Based on its characteristics of interdecadal spatial evolution, the major group-occurring zone was located in the southeastern coastal region in the 1960s, moved toward the southwest in the 1970s, progressed northeastward to the Yellow-Huai River basins in the 1980s, relocated to the south in the 1990s, and shifted northwestward with its center basically in the area south of the Yangtze River in the 2000s to complete a round clockwise movement. Overall, the high-percentile group-occurring zone of rainstorm events was mainly concentrated in the region south of the Yangtze River, which may have a certain relationship with its definition. The precipitation events of 100 mm magnitude had a relatively lower extremeness in the middle and lower Yangtze River basin than in northeast and north China. Rainstorm events mainly occurred in the high-percentile group-occurring zone of north and northeast China in the 1960s, in the upper reaches of the Yangtze River and the middle reaches of the Yellow River with the high-percentile group-occurring zone being oriented from the southwest to the northeast in the 1970s, typically more in central China and fewer in north and south China with the high-percentile group-occurring zone mainly in the Yangtze River basin in the 1980s, in the region south of the lower reaches of the Yangtze River with the high-percentile group-occurring zone moving southward in the 1990s, and more in south and north China and fewer in central China in the 2000s. Moreover, in the 2000s, the Level I high-percentile group-occurring zone also happened in northwest and northeast China; large-scale high-percentile group-occurring zones happened in the region south of the Yangtze River; and no high-percentile group-occurring zones happened in the Yellow and Huai River basins in the 2000s.

The "August 1975" extreme rainfall event that occurred in southern Henan in the early August 1975 was the most typical extreme weather event. After the third typhoon of 1975 moved to the Tongbai Mountains and Funiushan Mountains in Henan Province, it weakened into a low-pressure cyclone. Under the influence of the blocking anticyclone east of Lake Baikal, the low-pressure cyclone stabilized in the area of Zhumadian, Henan, caused the torrential rainfall to occur in the Hongru, Shaying, and Tangbai Rivers in the Nanyang Basin. The amounts of 3-h, 6-h, 24-h, and 3-day rainfall measured at the Linzhuang Station near the Banqiao Reservoir reached 494.6 mm, 830.1 mm, 1060.3 mm, and 1605.5 mm, respectively, all the highest recorded in mainland China. Such an extreme rainfall event would have resulted in a huge flood disaster in most parts of the world. The rainstorm and the subsequent flood resulted in dam failures due to overflowing at two large reservoirs, two medium-sized reservoirs and 58 small reservoirs, and 7180 dike breaks. The two large reservoirs were the Banqiao Reservoir in Miyang County of Zhumadian Prefecture and the Shimantan Reservoir in Pingdingshan Prefecture; the two medium-sized reservoirs were the Zhugou Reservoir in Queshan County of Zhumadian Prefecture and the Tiangang Reservoir in Xuchang Prefecture; and the dike breaks occurred in main drainage channels such as the Hongru and Shaying Rivers. The floodwater rushed as if it had had the momentum of an avalanche and inundated more than 10,000 km^2 of land. Nearly 6 billion m^3 of floodwater wantonly ran in all directions, resulting in more than 100,000 deaths. As indicated, extreme hydrological events are rare, but the loss is tremendous once it occurs.

5.4.3.2 Extreme Hydrological Events

The Yangtze River basin has a typical monsoon climate. As the interannual and monthly distributions of precipitation are uneven, coupled with climate change and the impact of human activities, extreme hydrological events have occurred frequently. Table 5.6 lists extreme hydrological events in the mainstream and tributaries of the Yangtze River. Although the statistical data are incomplete or inaccurate, the collected data indicate that the minimum and maximum flowrates for a same river section may differ more than one hundred times and the minimum and maximum water level and total runoff amount are also quite different. The peak flood flowrate in a tributary of the Yangtze River can reach the level of the

Table 5.6 Characteristics of representative extreme hydrological events at major stations of the Yangtze River basin and years of occurrences

Water system	Station name	Year	Water level (m)	Flowrate (m^3/s)	Remarks
Jinsha River	Pingshan	1924	307.30	36,900	Regional large flood
		1995	278	1060	
Yanlong River	Xiaodeshi		1982.71	352	
Min River	Gaoyang	1917	294.85	51,000	Regional large flood
			274.22	364	
Tuo River	Lijiawan	1896	275.35	18,600	
			258	0	
Jialing River	Beibei	1870	214.00	57,300	Regional extreme flood
		1981	208.02	44,700	
		2007	172.01	50	February 26
Wu River	Wulong	1830	214.45	31,000	Regional large flood
			167.11	55.3	
Donting Lake	Chenglingji	1998	35.94		Watershed extreme flood
			17.04	377	
Han River	Huangzhuang	1931		50,000	
			39.96	180	
Poyang Lake	Hukou	1998	22.59	43,430	Maximum inflow, 1954
				24,300	Maximum outflow, 1993
			5.9	−13,700	
Yangtze River	Cuntan	1870	195.15	100,000	Regional extreme flood
		1978	156.42	2270	
	Yichang	1870	59.50	10,5000	Regional extreme flood
		1979	38.07	2770	
	Hankou	1954	29.73	76,100	
		1865, 1963	10.08 (1865)	4830(1963)	
	Datong	1954	16.67	92,600	
		1961, 1979	3.14 (1961)	4620(1979)	

mainstream of the Yangtze River, but the minimum flowrate can be close to 0. Such a huge difference is bound to have an impact on human safety and water use.

5.4.4 Countermeasures for Extreme Weather and Hydrological Events

Most extreme weather and hydrological events are incidental natural phenomena. However, the longer the statistical time series, the greater the degree of extremeness of the events. The impact of an extreme hydrological event on socioeconomic development is not only the magnitude of the numerical value but also depends on its nature and type. For a drought or a low-flow process, if the duration is short and the coverage is not large, the impact will not be large; however, if the duration is long and the coverage is extensive, the loss of the disaster will increase with the increasing time. As for a flash flood or an extreme flood, the impact will depend on the location of occurrence and the conditions of the local socioeconomic development. It is obvious that if it occurs in a sparsely populated mountainous area, the impact will be small, but if it occurs in a densely populated region, the loss will be very large.

For example, during the 1954 flood, the maximum peak flowrate at the Yichang Station was only 66,800 m^3/s, which was equivalent to a recurring event of less than 20 years, but due to the long duration, the flowrate greater than 40,000 m^3/s lasted for 45 days, and the total volume of floodwater was equivalent to a 100- to 200-year event. Moreover, floods occurred concurrently in many tributaries. Consequently, the 1954 flood was severe and resulted in tremendous losses. Extreme droughts are similar. Meteorological droughts are largely synchronized with agricultural droughts; hydrological droughts are the cumulative result of meteorological droughts; and a prolonged widespread drought is bound to lead to a hydrological drought.

Since extreme weather and hydrological droughts are rare and occur only once in decades or centuries, the countermeasures for such events should be an integrated approach, namely, the combination of structural measures and non-structural measures, especially the use of non-structural measures. If the sole structural measures are relied on, because the cost for the construction, operation, and management of engineering facilities is high and the probability that the engineering facilities will function in a long time is small, they are not necessarily cost-effective. It is more effective to improve early warning forecast and implement appropriate measures such as automatic risk avoidance, emergency rescue, and post-disaster assistance and compensation. In the meanwhile, the impact on the ecological environment is minimized. This is the internationally advocated basic idea to change from the control of floods and droughts to the management of floods and droughts.

References

Cheng, Xin, Weilang Huang, Yi Jiang, etc. 2003. Groundwater problems and countermeasures in the Tai Lake Basin, Water Resour Prot, (4): 1–4

Yang P, Hou W, Feng G (2010) Study on group-occurring regularity of extreme weather events in China. Climat Environ Res 15(4):365–370

Yangtze River Water Conservancy Commission (2003) Records of the Yangtze River volume 13 – flood control. China Encyclopedia Publishing House, Beijing

Yangtze River Water Conservancy Commission (2005) Records of the Yangtze River volume 3 – natural disasters. China Encyclopedia Publishing House, Beijing

Yangtze River Water Conservancy Committee (2002) Flood and drought disasters in the Yangtze River Basin. China Water & Power Press, Beijing

Chapter 6
Regulation, Development, and Utilization of the Yangtze River

Abstract The Yangtze and Yellow Rivers are the cradles of the Chinese civilization. Since the Neolithic age, there have been lots of agricultural activities in different areas of the Yangtze River basin, which have begun to impact the Yangtze River, and the intensity of human activities has been closely related to the geographical environment and waters of the basin. Human beings began by living adjacent to waters with flood- and drought-fighting activities; then they started large-scale activities that consisted of logging forests, reclaiming land from lakes, constructing dikes, and developing land from waters; and then they commenced the construction of modern reservoirs and hydropower stations. This chapter overviews the history of human development and utilization of the Yangtze River, especially the changes of the Jianghan Plain and Dongting Lake, development and evolution of Poyang Lake, development and regulation of the Yangtze Estuary, construction of reservoirs and hydropower stations in the Yangtze River basin, and utilization of water resources in the Yangtze River. This chapter also analyzes changes of water use index and structure in the watershed, completed and planned water diversion projects as well as their associated impacts, and the future trend of water resources development and utilization in the Yangtze River.

Keywords The Yangtze River · Changjiang River · Evolution of river system · Basin ecosystem · Water resources utilization · Floods and drought · Ecological and environmental protection · Basin management

6.1 Relationship Between Human Activities and Landform in the Yangtze River Basin

Human beings must develop and utilize water resources in order to live and produce, but too much water (floods), too little water (drought), and dirty water (water contamination) may cause disasters. Therefore, there is a need to regulate rivers and to construct various types of hydraulic projects. However, excessive development and utilization may take away the water from the ecological environment, affecting

J. Chen, *Evolution and Water Resources Utilization of the Yangtze River*,
https://doi.org/10.1007/978-981-13-7872-0_6

sustainable development of the human society. Thus, there is a need to identify the balance point between development and protection so that water resources can be developed and utilized scientifically without further damaging the ecological environment.

6.1.1 Creation of the Human Society

There have always been debates about the human origins and the route for ancient human beings to migrate around the world. According to the general consensus of Western paleoanthropologists and archaeologists, the earliest human ancestors were born in eastern Africa more than 3 Ma and about 70 ka ago, a group of modern people began a migration that led them through the Middle East toward the world gradually. Human births and their global migration were closely related to the natural environment such as climate and geography. As the climate in Africa was warm, clothing was not needed to keep them warm, and there were abundant animals and plants that could serve as food. Therefore, they just needed simple living conditions to survive and appropriate natural conditions to give births, evolve, and develop. In the northern part of the earth, human survival conditions were harsh due to the cold climate; therefore, human beings needed caves, animal skins, and even fire to survive the winter. In about 75 ka ago, the earth entered the last ice age of the Quaternary, and at about 20 ka ago, global glaciers reached the apex. Although human beings living in the northern hemisphere must face the cold, they had been skilled hunters and begun to use fire and animal skins to keep themselves warm in the cold. In the last glacial maximum, sea level dropped sharply, and the neritic zone turned into land, which provided human beings with the conditions necessary to migrate everywhere. North American Indians might cross the Bering "Land Bridge" from Asia to the Americas during that time.

According to China's archaeological findings, Wushan men existed about 2.04 Ma, Yuanmou men existed 1.7 Ma, and Peking men existed in Zhoukoudian about 0.7 Ma. The first two of the mentioned three human species occurred in the Yangtze River basin. Did the human beings originate from multiple sources or a single source? What is the relationship among ape-man, *Homo sapiens*, and modern people? These questions have yet to be answered and need further archaeological discoveries to confirm. Based on the existing archaeological discoveries and research results, China entered the Paleolithic period and human activities began in the Yangtze River basin about 1 Ma. Homo erectus ape-men were dominant during the early Paleolithic age about 0.5–0.4 Ma. Cavemen emerged in the middle to late Paleolithic age about 0.2–0.1 Ma. From 0.1 to 0.01 Ma, the Changyang men of Hubei, Ziyang men of Sichuan, and Lijiang men of Yunnan emerged, indicating that ancient human beings lived in the Yangtze River basin during the Paleolithic age about 10 ka ago. They lived mainly on hunting, fishing, and fruit picking in a matrilineal-commune lifestyle with the family and tribe as the activity center. The Paleolithic culture developed in the context of the Quaternary climate change and the increasingly intensifying tectonic movement. At that time, human beings mainly faced dangers from wildlife, natural disasters and starvation. Due to limited population, they did not interfere with nature, and human beings and nature were in the primitive harmonious state, but the level of human life was very low and in a primeval state.

Since the Neolithic age of more than 10 ka ago, human beings had begun to make and use sharp-edged stones, invented pottery, and started primitive agriculture, animal husbandry, and handicrafts. About 4 ka ago, the matrilineal commune system in the Yellow and Yangtze River basins gradually transitioned to the patriarchal commune system. It is about 5 ka from the legendary Yellow Emperor era and later that Yellow Emperor's grandson Zhuan Xu and great-great-grandson Diku continued to serve as the monarchs of the tribal alliance. Yao, Emperor Diku's son, succeeded Emperor Diku and became a model monarch. Emperor Yao founded the abdication system and passed his crown to Shun. In the Shun period, flooding was widespread. Gun used the blocking method to control floodwater but resulted in more severe flooding. Gun was executed and his son, Yu, adopted the method of diverting and successfully harnessed the flood. As a result, Yu was elected as the ruler. However, his son abandoned the abdication system and self-claimed the king. He established the first hereditary dynasty – Xia. The Xia Dynasty lasted for more than 400 years, and the last Emperor, Jie, was overthrown by the leader, Cheng Tan, of the eastern vassal state, Shang, leading to the establishment of the Shang Dynasty. The Shang Dynasty had continued for more than 500 years until the Western Zhou Dynasty took over. The Eastern Zhou Dynasty followed the Western Zhou. The Qin Dynasty was established in 221 BC. China began its state from the Xia Dynasty. With the national rule and social system of modern significance, the ancestors began to use copper and other metal tools to conduct agricultural activities.

According to the past 10 years' findings of exploratory projects related to the Chinese civilization, the Liangzhu site, which was discovered in 2006, indicates that the Mojiaoshan-centered Ancient Liangzhu City had a total verified area of more than 3 million m^2, which is China's largest relic site of discovery so far. In the investigation of the northern city, a large relic site of a man-made water project was discovered. The project was estimated to be constructed about 4.8 ka ago. This has not only backdated the relics of the early actual hydraulic engineering projects of China for about 2 ka but also proved that at the beginning of the human society, water projects were constructed after a city or town had been established. Therefore, the utilization of water resources has always been with the development of the human civilization.

6.1.2 Relationship Between Human Living Space and Climate, Landform, and Other Factors

As human beings entered the stage of social development, they could engage in organized activities, and, coupled with the use of animal power and metal tools, they could carry out large-scale agricultural cultivation. The change from picking and hunting for food to agricultural farming had significantly increased their ability to transform nature. Based on current archaeological findings, in terms of the locations and geomorphic features of early human activities, there were significant differences in the spatial distributional characteristics of cultural sites in various times and were basically compatible with the level of human productivity and the ability to fight floods. The main features are as follows. ① Neolithic cultural sites are mainly distributed in areas of moderate elevations at margins of plains such as piedmont areas of mountains, hillocks, and high alluvial plains. Cultural sites are rarely located in plain centers of low elevations or mountainous areas of high elevations. Because the low-lying areas were often encroached by floodwater, it was not safe to live there; and in mountainous areas, living and traffic conditions were poor, and food was relatively scarce. ② After the Western Zhou Dynasty, the elevations of cultural sites of the time showed a downward trend, transferring from piedmonts-hillocks and high alluvial plains to low alluvial plains. Human beings began to construct dikes to fight floods or divert floodwater, which was the beginning of dike construction. As increasingly complete dike engineering systems were constructed from the Qin and Han Dynasties to the Ming and Qing Dynasties, cultural sites of the time were identified to be mainly distributed in the low-elevation, gentle-sloping alluvial plains, and lacustrine plains. In summary, the spatial distribution of cultural sites in different periods reflected the response of human activities to changes of the natural environment under various productivity levels.

Neolithic cultural sites in the middle and lower reaches of the Yangtze River basin are mainly distributed in the relatively high piedmonts-hillocks of mountains, terraces and hillocks of plains, and high alluvial plains. The distribution of the Shang Dynasty relic sites is different from that of Neolithic sites, and the Shang Dynasty sites are mainly located in the piedmont areas of the Dabie Mountains north of Wuhan and in the piedmont areas of Yueyang, Changsha, and Changde. Although the spatial distribution of the two periods are different, the landform types of the cultural relic sites of the Shang Dynasty are mainly located in piedmont areas and hillocks of mountains, terraces, and hillocks of plains and high alluvial plains, which were consistent with the landform types of the Neolithic culture.

The Pengtoushan, Chengbeixi, Daxi, Qujialing, and Shijiahe cultures were the five most representative indigenous cultures from the Jianghan Plain to Dongting Lake in the middle reaches of the Yangtze River. Their successions and development constituted the basic contents and characteristics of the Neolithic culture in the middle reaches of the Yangtze River and the Jianghan area. The first one was the late Pengtoushan culture that occurred about 7.5–7 ka ago. The Pengtoushan cultural site is mainly centered in the Li River basin northwest of Dongting Lake. The second one was the third stage of the Guanmiaoshan style of the Daxi culture that occurred about 5.8–5.5 ka ago. The Daxi cultural site is located at the east mouth of the Qutang Gorge on the south bank of the Yangtze River, 15 km west of Fengjie County, and 45 km east of the Wushan County seat. The third one was the third stage of the Qujialing type that occurred about 5–4.8 ka ago. The Qujialing cultural site is located about 30 km southwest of Jingshan County in Hubei Province. The fourth one was the last stage of the late Shijiahe culture to the second stage equivalent to the Erlitou culture in Central China. The Shijiahe culture occurred approximately 4.1–3.8 ka ago and covered a large area. The cultural site is bordered by Macheng, Huanggang, Echeng, and Daye on the east; by the Xiling Gorge on the west; and mainly by the Nanyang Basin north of the Jianghan Plain on the north and northeast. After considering the environmental climate change, the first change occurred at the beginning of the second warm period of the Holocene; the second change occurred at the end of the second warm period; the third change occurred in the late third warm period; and the fourth change occurred in the fourth warm period. The common distinctive characteristics are that all the changes of the four sites occurred during the Holocene warm periods (Cheng 2005).

The Pengtoushan, Zaoshixiaceng, Daxi and Qujialing cultures, and the Longshan culture in the middle Yangtze were the five major stages of the Neolithic period in the Hunan region. With the accumulation of human production experience over time, farming techniques had been improved. Therefore, the primitive agriculture had progressed and been more advanced from one time period to another. The Pengtoushan and Zaoshixiaceng cultures in the Hunan area during the early Neolithic age prevailed in the emergence and development of the slash-and-burn agriculture, respectively. The Daxi and Qujialing cultures in the Hunan area during the middle Neolithic age occurred in the emergence and development of the hoe farming, respectively. The Longshan culture in the middle Yangtze occurred in the Hunan area during the late Neolithic age when the hoe farming reached the advanced stage.

Cultural Sites of Western and Eastern Zhou Dynasties During this period, bronze tools had been extensively used; iron tools had begun to be utilized; farming techniques had been improved a lot; and chariots and animal riding had been widely used for transportation. Cultural sites were mainly concentrated in Jingzhou, Lixian, Changde, and Changsha where the landforms were piedmont areas of mountains, hillocks, high alluvial plains, and low alluvial plains. Compared with the early cultural sites, the landforms of the Western Zhou cultural sites showed the trend of extending from piedmont areas of mountains and hilly areas to alluvial and lacustrine plains.

Cultural Sites of Qin-Han Period to Ming-Qing Period A comparison of the cultural sites' spatial distribution during the Qin-Han period with those of the Western and Eastern Zhou Dynasties indicates more cultural sites of the former were located in low alluvial plains. After the Qin and Han Dynasties, the spatial distribution of cultural sites showed a trend of moving toward the low-lying plains, and the pattern of the Neolithic cultural sites that were concentrated in the piedmont areas of mountains was disrupted. According to an ArcGIS statistical analysis (Deng et al. 2009), cultural sites located in low alluvial and lacustrine plains accounted for 12.96% during the Qin-Han period, 24.23% during the Wei-Jin period and Southern and Northern Dynasties, 35.33% during the Sui-Tang period and Five Dynasties, 21.23% during the Song-Yuan period, and 39.45% during the Ming-Qing period, indicating a general upward trend in the number of cultural sites located in the low alluvial and lacustrine plains.

If the median elevations of the centralized areas of cultural sites in various periods are used for a comparison analysis, the elevation of the cultural sites from the Neolithic period to the Ming-Qing period decreased by about 12 m, of which an approximately 4 m decline occurred in the Western Zhou Dynasty; a roughly 2.5 m decline occurred in the Qin-Han period; and a nearly 5.5 m decrease occurred during the period of the Wei, Jin, Northern, and Southern Dynasties. The decline in the elevation of residential areas indicates that dike construction activities had been intensified to ensure the safety of life and production.

A comparison analysis of the spatial distribution and evolution of the ancient cultural sites in the middle reaches of the Yangtze River and the comprehensive environmental findings of the lacustrine sediments in Gaichengkong Lake of the Jianghan Plain indicates that although the evolutionary process of the spatial distribution of ancient cultural sites was not all correspondent to the regional climate change process, the decrease of the regional temperature, reduced precipitation, and reduced flood risk after the end of the Holocene warm period were consistent with the trend of the decreasing elevation of ancient cultural sites. During the Wei-Jin period and the Western and Eastern Zhou Dynasties, the large elevational decline of cultural sites was correspondent to not only the cold period but also the first massive population migration in China's history. The large elevational decline was also coincidental with the dike construction activities along the Yangtze River. Therefore, all the changes reflected the combined effects of both nature and human activities.

The tectonic evolution of the Han River-Dongting Lake area had a great influence on the local topography and landform, and the most obvious impact was that the Yangtze River changed its course. Since the Holocene, the tectonic subsidence rate of this area has been increasing, which has caused the elevation of the riverbed and the water level to rise and the river landform to change greatly. The tectonic movement-induced course change of the Yangtze River had a great impact on the change in the form and scale of the surrounding river-lake relationship and the living environment of ancient human beings during the Neolithic age. Most of the cultural sites are far away from tectonic fault zones.

The population growth in the middle and lower reaches of the Yangtze River was related to climate change, the level of productivity development, and the national political situation. When the climate became colder, the flood risk was lower and human beings moved to low-lying areas. Similarly, when increased productivity allowed human beings to construct dikes, they could migrate into floodplains and were encouraged to live in low-lying areas. In ancient China, political, economic, and cultural centers were mainly located in north China. Political unrests or invasions by foreign nations forced northern populations to migrate to the Yangtze River basin. For example, the "Yongjia Rebellion" during the Yongjia years of the West Jin Dynasty, the "An-Shi Disturbance" during the Tang Dynasty, and the "Jingkang Incident" during the Northern Song Dynasty were the peak time periods when northern populations migrated to the Yangtze River basin. The population growth naturally intensified competition for land between human beings and water, which sabotaged the harmonious relationship between human and water.

6.2 Regulation and Development of the Yangtze River

6.2.1 Historical Activities of Development and Utilization

Human development of the Yangtze River basin began with land use. When our ancestors began to transform from living on hunting and picking in the Paleolithic age to cultivating agriculture in the Neolithic age, deforestation began, especially when the slash-and-burn farming method was used, and they gradually extended their farmlands from piedmont areas of mountain fronts and hillocks to both hilly areas and alluvial plains. Obviously, cultivating agriculture could not only improve the security of food supply so that human beings could begin to settle but could also extend their range of migration activities, such as reclamation of new farmlands. Cultivating agriculture improved the development of the tribal society, but the variation of the land use pattern was the greatest and the most prolonged change of the human society to the earth environment and also had the greatest effect on the earth's ecology and environment, such as decreased natural forest area, increased soil erosion, and decreased wildlife habitat.

In the past 10 ka since the Holocene, the natural forest cover in the Yangtze River basin has decreased from 80% to less than 5%, and the continuous large forest area has been even smaller. The Yangtze River basin encompasses a total area of 180 million ha, of which 31 million ha are arable land, accounting for 17%. Artificial forest area accounts for 15%, natural forests 5–7%, and other land are mountains, urban areas, highways, railways, water projects, and other infrastructures.

Ancient water conservancy projects were developed primarily for flood control and then for agricultural irrigation and canal transportation. In ancient times, due to the low level of productivity, water conservancy projects began with the construction of dikes and weirs and land reclamation from lakes. In modern times, water conservancy projects started with the construction of flood diversion/retention zones, reservoirs, hydropower stations, and trans-basin water diversion projects. Table 6.1 presents the historical development of water conservancy projects in the Yangtze River basin.

Table 6.1 indicates that before the Qin Dynasty, recorded water conservancy projects had mainly been canals and irrigated areas and their purposes had mainly been for transportation and food production at the time, but there had been few records for the construction of dikes, indicating that people chose to live in high areas to avoid floodplain or flood retention area. In the middle and lower reaches of the Yangtze River basin where the landform was natural alluvial plains, floodwater naturally flowed to low-lying floodplains. The water system was complex, and rivers and lakes were well connected. After the Three Kingdoms and Jin Dynasties, the

Table 6.1 Historic development of water conservancy projects in Yangtze River basin

	Dikes and seawall flood control and drainage	Lake area development	Navigation	Irrigation
Before Qin Dynasty	Emergence of basic water conservancy facilities such as drainage and diversion in settlements; dike construction		Jianghan Canal (535 BC); Xuhe River (11 BC); Xupu River(495 BC); Chaofei Canal (400 BC); Han Canal (486, BC), connecting Yangtze to Huai Rivers	Shang Dynasty: emergence of field ditches and nine-square fields. Dujiangyan Irrigation Project (256–251 BC)
Qin-Han Period	Emergence of dikes and seawalls	Emergence of reclaimed farmland from lakes in middle and lower reaches	Lingqu Canal (219 BC), connecting the Yangtze River to the Pearl River	Hanzhong, Nanyang, and Luan River Irrigation Districts
Three Kingdoms-Jin-Southern Dynasties	Jin Dike (345–365 AD), former Jingjiang Dike	Emergence of berms and agricultural development in lake areas of middle and lower reaches	Navigation as primary transportation means in middle and lower reaches	Occurrence of large irrigation districts in Hubei, Anhui and Jiangsu
Sui-Tang-Song Dynasties	Dikes gradually connected. 90 km Fangongdi (seawall) of Song Dynasty. Accelerated seawall construction	Large-scale land reclamation from lakes, emergence of opposition	Beijing-Hangzhou Grand Canal connected, emergence of ship lock facilities, navigable waters reached current level	Irrigation works all over middle and lower reaches, emergence of aqueduct
Yuan-Ming-Qing Dynasties	Major dikes along Yangtze River were completed during this time period. Outlets on north bank of Jingjing River were blocked during Emperor Jiajing era of Ming Dynasty	Large-scale land reclamation from lakes, constant berm failures and repairs, and occasional returning farmland to lake. Formation of Ouchi River, Songzikou Outlet in 1860 and 1973, respectively. Formation of four outlets from Jingjiang River to Dongting Lake	Marine and inland navigation well developed	Irrigated areas increased

construction of dikes and land reclamation from lakes were gradually increased and people began to move and live in low-lying areas. As a result, the mainstream of the Yangtze River gradually formed, and nearby lakes and wetlands began to shrink.

6.2.2 Development of and Impact on Yangtze River, Jingjiang River, and Dongting Lake

6.2.2.1 Formation of Jingjiang River and Dongting Lake in History

The historical regulation and development of the of Yangtze River that impacted the river-lake relationship had typically occurred in the middle and lower reaches of the Yangtze River, of which the Jianghan Plain, the Jingjiang River, and Dongting Lake were the most representative.

More than 5 ka ago, the Yangtze and Han Rivers emerged in the Jianghan Plain where the Yangtze River did not have a mainstream, and the Jianghan Plain was then a lacustrine wetland (Yunmeng Marsh), while the Dongting Lake area was then a plain studded with river systems and lakes. Due to the long-term deposition of large amounts of sediments in the upper reaches of the Yangtze River and the Han River, the Yunmeng Marsh gradually developed into swampland, which created objective conditions for human beings to use the fertile land. Figure 6.1 illustrates the evolution of the northern Jingjiang River system in the past 2500+ years.

About 2.5 ka ago, there were many waterways across the Jianghan Plain flowing from west to east. For example, the Xia, Yang, and Tong Rivers had once been the channels of the Yangtze River north of the present-day Jingjiang River and had been affected by human land reclamation through berm construction since 2 ka ago. The construction of dikes along the north bank of the Jingjiang River can be dated back to the Jin Dynasty. According to the *Commentary on the Water Classics • Yangtze River* by Li Daoyuan of the Northern Wei Dynasty, when Huanwan (312–373 AD) of the Jin Dynasty was the governor of Jingzhou (345 AD and subsequent years), "southeast of Jiangling City was low, so the Jin Dike was constructed," indicating that the Jin Dike (Wancheng Dike) was the flood wall of Jiangling City. This was the first historical record of the Jingjiang Dike, and thereafter until the Tang Dynasty, no more records about the dike were identified. This might be attributed to the possibility that the sediment content in the water of the Yangtze River was low and the river was stable for a long time. During this time period, the middle reaches of the Yangtze River might still be flowing through two channels (north and south); the north channel was transitioning gradually from flowing at a lower flowrate (through Dongting Lake) to a higher flowrate than the south channel; the north channel became the mainstream of the Yangtze River; and the south channel transitioned gradually to the Hudu River due to sedimentation. Meanwhile, the flowrate through the Jingjiang River significantly increased and the water level rose. During the Five Dynasties, the Cunjin Dike was constructed outside the Longshan Gate of Jingzhou City to block floodwater from the Shu River (Chuan River). From the Southern

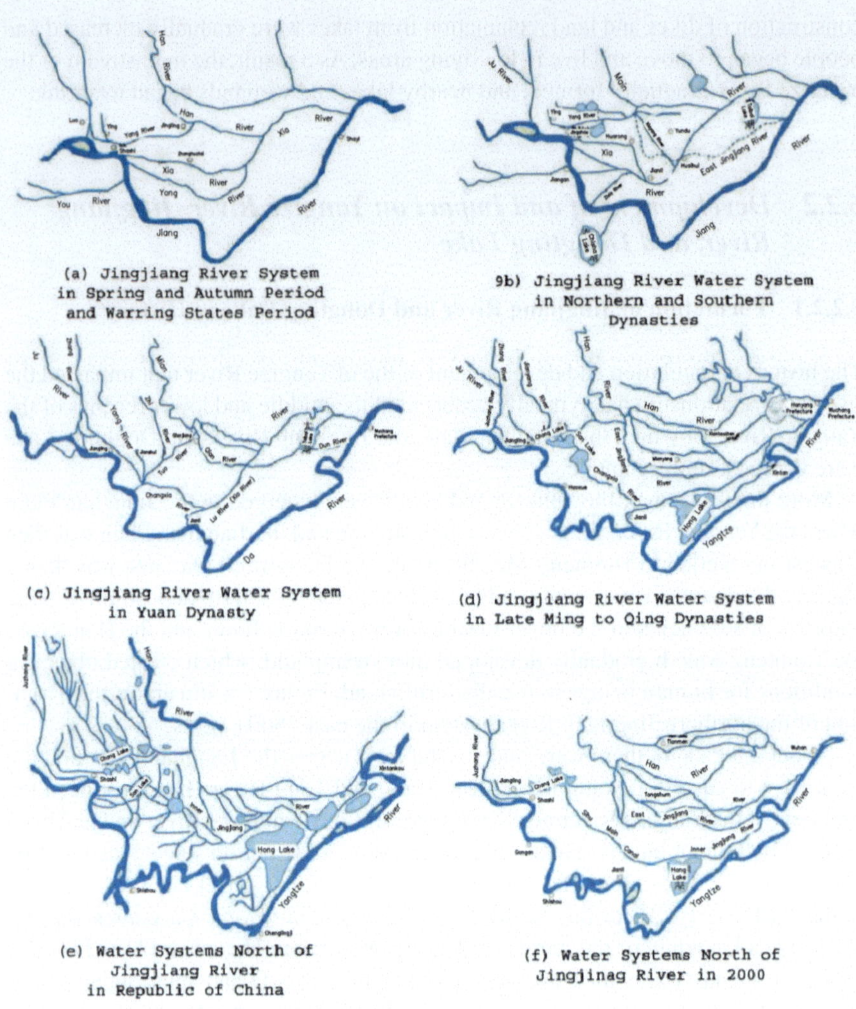

Fig. 6.1 Evolution of the Jingjiang River water system in the past 2500+ years

Song Dynasty to the early Yuan Dynasty, all counties along both banks of the Jingjiang River constructed dikes to facilitate land reclamation. During the Dade years of the Yuan Dynasty (1264–1307 AD), “there used to be nine openings and 13 outlets along the Jinagling River, but now only Haoxue, Chibo, Yanglin, Caixue, Diaoxuan and Xiaoyue have remained open and all the others have been blocked,” indicating that there still used to be some outlets/openings although dikes were constructed along both banks of the Jingjiang River. In the event of a flood, floodwater could still carry sediments through the outlets/openings across the dikes, but this also exacerbated the sedimentation problems at the outlets/openings, resulting in the formation of many natural dikes to various degrees at the mouths of lakes to the river and the conditions for the outlets/openings to be blocked. In the Ming

Dynasty, each opening was gradually blocked and only Haoxue outlet was left open on the north bank. During the Chenghua, Hongzhi, and Zhengde years (1465–1521) of the Yuan Dynasty, due to flooding, the Huangtan Dike was repaired repeatedly, and the new Wencun Dike was constructed on the north bank of the river. In Emperor Hongzhi's 12th year (1499) and Emperor Jiajing's 11th year (1532) of the Ming Dynasty, the river dike west of Shashi broke and floodwater raged for 30 km, resulting in many drowned. The floodwater formed a vast lake around western Shashi and broke the Shang Dike to flow southward. In Jiajing's 21st year (1542) of the Ming Dynasty, the Haoxue outlet was blocked and no opening was left on the dike along the north bank. During the Longqing years (1567–1572) of the Ming Dynasty, only Hudu and Diaoxian outlets were left on the south bank of the river that could divert water to Dongting Lake. Since then, the Jianghan Plain has no longer been used for diversion/retention of floodwater from the Jingjiang River for nearly 500 years.

The Jianghan Plain and the Dongting Lake Plain both resulted from alluvial and lacustrine sedimentation after the Yangtze River channeled through the Three Gorges and were located in a subduction zone, on which the Yunmeng Marsh developed north of the river. However, due to the filling by the sediments transported by the Yangtze River for a long period of time, the Yunmeng Marsh had evolved from the marshland to a plain landform by the Warring States Period before the Qin Dynasty. In the period before the Qin Dynasty, there had been two large plains in the Jianghan area located at the east and west ends of the Yunmeng Marsh. The west plain was the Jingjiang River Delta east of Jiangling and north of Shashi, and the east plain was the floodplain on the west side of the Chenglingji-Wuhan section of the Yangtze River. On these two plains, there had been residential areas and settlements before the Qin Dynasty. Before the Yuan Dynasty or before the Jingjiang Dike was completed, the north and south channels of the Yangtze River had both been navigable for boats, and the two plains had been at the same elevation. In 1542, the Jingjiang Dike was completed. Since then, the area north of the Jingjiang River's north bank has stopped receiving water from the Yangtze River. As a result, the Yangtze River has not transported any sediments to the Jianghan Plain. After hundreds of years of sedimentation on the south channel and the subduction of the north channel, the relative height difference between the two channels reached 5–10 m. If this trend continues for a long time, the elevational difference will become increasingly greater. Then if the Jingjiang Dike broke, the Jianghan Plain would be inundated and the resulting damage would be beyond anticipation, which is why the primary flood control goal of the TGR is to safeguard the safety of the Jingjiang Dike.

As the Jianghan Plain has not undergone any flood diversion/retention since the Jiajing years of the Ming Dynasty, sedimentation has ceased for a long time. Meanwhile, as the Jianghan Plain and Dongting Lake area have been in the tectonic subsidence stage from the viewpoint of tectonic movement, an elevational difference between the Jianghan Plain and the flood level of the Yangtze River has developed gradually, and the flood threat has been increasingly higher. As a result, after a flood event, the Qing authorities did not dare to reopen the blocked openings/outlets in the dike on the north bank to divert floodwater or to block the diversion outlets/ openings in the dike on the south bank. If they had opened the blocked northern openings/outlets to divert floodwater, the floodwater would have rushed out the dike and inundated a large area. Since then, the Dongting Lake has received growingly more sediments, and the lake has become shallower and smaller. In Xianfeng's 2nd year (1852) of the Qing Dynasty, the dike broke at Ouchi; and in Xianfeng's 10th year (1860), a flood resulted in the formation of the Ouchi River that further developed to connect the Yangtze River with the lake. In Tongzhi's 9th year (1870), the dike broke at Songzi during a large flood that occurred in the upper Yangtze. Although the break was blocked, it broke again in 1873, resulting in the formation of the Songzi River with two upper confluent channels. Since then, there have been four rivers to divert water from the Jingjiang River to Dongting Lake. Subsequently, the sediments in the lower Jingjiang River have reduced significantly, and the curvature of the channel has increased, resulting in the formation of the meandering channel. Consequently, during the flood season every year, up to 45% of the Yangtze River's sediments flowed to Dongting Lake, and plus the sediments from the four rivers in the lake system, the annual amount of sediments into the lake was up to 216 million tonnes, most of which was from the four outlet rivers of the Jingjiang River. Since the formation of the Ouchi and Songzi Rivers, a Nan County had developed on the south bank of the Jingjiang River and north of Dongting Lake through sedimentation in only 30+ years. Afterward, due to prolonged land reclamation activities in the mountainous areas of the upper Yangtze, the amounts of sediments in the river were increasing, and Dongting Lake was shrinking in area from 6000 km^2 in 1870 to 5400 km^2 in Guangxu's 20th year (1894), to 4300 km^2 in 1949, and to only 2740 km^2 in 1980. With the lake shrinking, the bottom of the lake has been rising due to sedimentation. It has been pointed out in recent years that the elevational difference between the Jianghan Plain and the bottom of the Jingjiang River and Dongting Lake has reached 5–7 m. Relative to the Jianghan Plain, Dongting Lake may become another hanging lake and will become a great potential threat to the relatively low-lying Jianghan Plain.

Figure 6.2 shows changes in water surface area of Dongting Lake in the past 180 years. Figure 6.2 indicates the following. First, the area of water surface has been apparently reducing; second, the once well-connected large water surface has becoming a complicated water system and the change has mainly occurred since 1871; and third, since 1953, the area of water surface has reduced significantly, and the water system has changed to become relatively simple. These three changes have reflected the natural evolution of the Dongting Lake area and the impact of human activities. In 1870, a 1000-year flood occurred in the Yangtze River, resulting in the formation of nearly ten rivers flowing through the four outlets into Dongting Lake from the south bank of the Jinjiang River. In the 1950s, due to the needs of crop production and flood control, large areas of lacustrine beaches were reclaimed for land, and the complex water system was simplified to facilitate flood control.

During the 100+ years between the nineteenth century when the four outlets formed to the twentieth century, the water flow diverted through the four outlets had been decreasing. For example, on July 21, 1937, when the peak flowrate was 66,700 m^3/s at the Zhicheng Station, the maximum peak flowrate at three outlets was 11,600 m^3/s at the Songzi Outlet, 3140 m^3/s at the Taiping Outlet, 12,100 m^3/s at

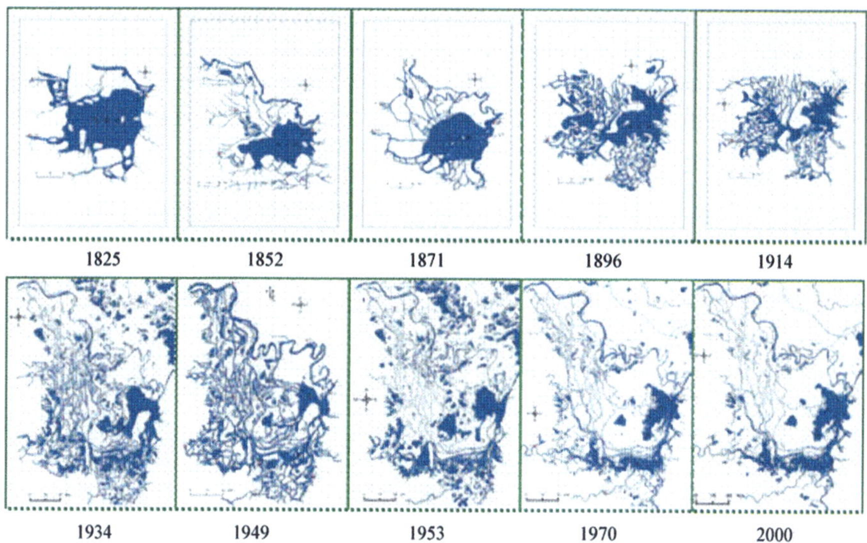

Fig. 6.2 Changes of Dongting Lake in Past 180 Years

Guanjiapu on the east channel of the Ouchi Outlet, and 6870 m^3/s at Kangjiagang on the west channel of the Ouchi Outlet, indicating that the three outlets diverted 50.4% of the flow from the Zhicheng Station. Additionally, the flowrate through the Diaoxian Outlet was 1460 m^3/s. As a result, the four outlets diverted 52.7% of the flow from the Zhicheng Station. During 1981–1994, the average annual runoff amount of the three outlets reduced by 76.3–69.7 billion m^3 from 146 billion m^3 during 1951–1958, which was a 52.65% decrease. The average runoff amount of the Ouchi River during 1981–1994 reduced by 55.9–18.7 billion m^3 from 74.6 billion m^3 during 1951–1958, which was down by 74.9%. The order (from high to low) of percentage of decrease in average annual runoff amount of the three outlet rivers was the Ouchi River, the Hudu River, and the Songzi River, which has revealed the following two aspects. First, the closer to Dongting Lake, the faster the impact of sedimentation on the outlet and the faster the natural degradation; and second, all the changes were also related to the position of the natural meander cutoff of the Jingjiang River, or the closer the meander cutoff position to the downstream side the outlet river's inlet, the greater the impact on the decrease of runoff amount and the faster the degradation. Both the two aspects had the greatest impacts on the Ouchi River, so the Ouchi River suffered the most serious degradation (Han and Zhou 1999).

6.2.2.2 Development of Polders in Lake Area

The major form of development in the Dongting Lake and Poyang Lake areas was polders. A polder is an agricultural zone with an independent irrigation system reclaimed from a piece of shallow water zone near a bank of a lake or a river and protected by berms or dikes from the outside of water. Irrigation and drainage of the agricultural zone is achieved through check gates and culverts in the berms. Large-scale polder development in Hubei and Hunan was recorded for the Shaoding years of the South Song Dynasty. During that time, to resist the Yuan troops from marching southward, it was planned to reclaim land on both sides of the Jingjiang River. Meng Gong, the Ningwu Military Governor, "started large-scale land reclamation, gathered many men to build berms and called in peasants to plant crops. From Zigui to Hankou, 180,280 ha of polders were developed in 20 places and 170 villages were involved."

Large-scale land reclamation from Dongting Lake was conducted from the Ming Dynasty to the early Qing Dynasty. According to records by Chen Siyuan of Huarong in the late Ming Dynasty, during the Zhengtong years (1436–1449), 48 polder zones were developed in Huarong and later expanded to more than 100 zones, of which the largest polder extended to more than 5 km and small ones were about 6.7 ha. Counties along the lakeshore all had a large number of records for the construction of the polders. During the years of Emperors Kangxi (1662–1722) and Yongzheng (1723–1735) of the Qing Dynasty, the Imperial Court promulgated the policy of encouraging reclamation, which further promoted the development of farmland in the lake area. In Kangxi's 55th year (1716) and Yongzheng's 6th year (1728), the Imperial Treasury spent 120,000 liang of gold on the construction of more than 100 polder zones. Polders in Yuanjiang, Yiyang, Anxiang, Xiangyin, etc. had increased a lot during the Kangxi and Yongzheng years. Polder construction had further development during the Qianlong years (1736–1795). In Qianlong's 5th year (1740), the Imperial Court issued an edict which required that isolated land should be reclaimed if possible and the newly-reclaimed land should be exempted from taxes. Since the enactment of this policy, "over the years, the privately-reclaimed land had doubled that by the government… Most of the water zones had become areas for polder development." Since then, privately funded polder construction (polders constructed using the fund from the Imperial Treasury during Yongzheng's 6th year were called official polders, and those constructed thereafter were called private polders) had rapidly increased. According to the report by Yang Xifu, Hunan Governor, in Qianlong's 10th year (1745), "as Hunan is adjacent to Dongting Lake, each county has reclaimed land from the lakeshore and explored land from the water zone, which has caused worries about berm breaks due to overflow." Moreover, to increase the production of crops, even "kilometers of shallow lake zones" or even "hectares of ponds" were filled for croplands. From land reclamation during the Kangxi and Yongzheng Years, more polders had been extended into the water-impoundment zones of the lake in the early Qianlong period and large areas of flood retention zones were occupied.

Farmland reclamation and economic development directly weakened the function of the lake for diversion/retention of floodwater. The trade-off between the pros and cons, and the resulting continuous controversy forced the Imperial Court and local governments to make a choice. In Qianlong's 12th year (1747), Yang Xifu, Hunan Governor, argued that the excessive land reclamation from the lake would endanger the overall situation of flood control. In April of the same year, the Imperial

Court accepted his opinion and ordered "based on inspections, Dongting Lake has been greatly encroached, but it still receives floodwater from upstream provinces. Lakeshore areas that had been used to retain water have been gradually developed into polders by constructing berms, which is called competing land with water... Except the lakeshore areas that have been developed into polders by constructing berms, land reclamation from undeveloped areas on the fluctuation zone shall be strictly prohibited. Future land reclamation that would occupy waterways shall not be allowed." The following year, Peng Shukui, Hubei Governor, also submitted a request to the Imperial Court exhorting a ban of land reclamation from lakes and the reservation of lake areas for flood control and sharply pointed out: "as human beings compete land with water for profit, water is bound to battle for land with human beings for detriment. When rivers are blocked, dikes will break. That's all the reason." His understanding was close to the modern concept of flood management that is "to provide floodwater with space," indicating that the ancient China had already boasted a master of flood control. In Qianlong's 53rd year (1788), the Jingjiang Dike broke, resulting in a catastrophe. The ensuing investigation indicated that the break was related to land reclamation in the mid-channel Jiaojin sandbar of the Jingjiang River. As a result, the ban on land reclamation from lakes was reaffirmed the following year and was reiterated by Emperors Jiaqing and Daoguang.

The objective reason for land reclamation from lakes was the rapid population growth in the lake areas. For example, the total population of Yuezhou, Changsha, and Changde in the Dongting Lake area was 580,000 during the Jiajing era (1522–1566) and the Longqing years (1567–1572) of the Ming Dynasty and increased to 7.22 million in Jiaqing's 25th year (1820) of the Qing Dynasty, indicating the population increased by 12.5 times in less than 300 years. Under the pressure of the population growth, the simple ban had very little effect, and land reclamation from lakes continued. In Jiaqing's 7th year (1802), Ma Huiyu, Hunan Governor, recorded in his *Land Reclaimed from Lakes Encroaching Water Area* that "investigations indicate that the 10 counties of Hunan Province adjacent to Dongting Lake had 155 official polders and 298 private polders, of which 67 private polders destroyed with 91 private polders remaining." The so-called private polders referred to those reclaimed since the late Qianlong era and the height of the protecting berms ranged from only 0.3–0.7 m to 2–2.3 m. These low berms could fall themselves when they were overflowed by floodwater and farming could continue in the protected areas when the water level was low. Ma Huiyu believed that the private polders could be divided into two types. One type that blocked the floodway must be destroyed and could not be reclaimed anymore, and the other that did not block the passage of floodwater could be maintained but the height of protecting berms must be strictly limited to the existing level, and raising of the berms would not be allowed. In fact, that was a compromise between the strict ban on land reclamation from lakes and the need for reclamation and shared the same philosophy with the "farming in flood retention area" strategy advocated in the 1950s. It did not appear that our early land use strategy in lake areas had transcended that of our ancestors and we had just repeated the historical phenomenon. Every time after a flood, some reclaimed lands were "returned to the lake," but not long thereafter, land reclamation from lakes reappeared. In the past, due to the pressure from population growth and food supply adequacy, people were naturally forced to reclaim land from lakes.

6.2.2.3 Land Reclamation Since the Twentieth Century: Major Cause for Degradation of Dongting Lake and Jianghan Plain

In the shrinking process of Dongting Lake, natural sedimentation and artificial land reclamation were two major causes. Natural sedimentation was the main cause before the eighteenth century, and artificial land reclamation was performed on natural sediment deposition. In the early years of Qianlong, artificial land reclamation began through facilitating sediment deposition, which was reported many times by local officials by stating "at first, measures were taken to induce sediment deposition along the shore and in the middle of the lake, and then berms were constructed around to form polders for growth of fish and crops." Even reclaimed land encroached floodway "to block water from passing." Therefore, in Jiaqing's 7th year, Ma Huiyu specifically pointed out in his memorial, "Majestic Emperor, wherever water retention or passing is concerned, local officials should be ordered to personally inspect … land reclamation shall not be allowed." It can be seen that the land reclamation from lakes that encroached the floodway has become a major factor in weakening the flood-control capacity, especially since the twentieth century.

As China experienced a serious food crisis from the 1950s to the early 1960s, land reclamation from lakes caused a similar reduction in lake area to that caused during the previous 120 years. Moreover, the accelerated sedimentation process was accomplished when the total amount of sediments through the four outlets decreased. It is obvious that the lake shrinkage in the 1950s was mainly the result of artificial land reclamation. In the 1950s, the new land reclaimed from the lake was 1432 km^2; the abandoned land was 309 km^2; and the net land increase was 1123 km^2. During the same time period, the area of Dongting Lake shrank by 1209 km^2, almost the same as the net increase of the reclaimed land. Oppositely, as large-scale land reclamation from the lake ceased in the early 1960s, the shrinkage of the lake area declined significantly. The opposing facts of the 1950s and 1960s indicate that land reclamation from the lake played a primary role in the shrinking process of Dongting Lake in the twentieth century.

Decrease in the volume of Dongting Lake also showed the impacts of natural sedimentation and artificial land reclamation. According to the statistical data from 1951 to 1978, the total annual amount of sediments transported through the four outlets and associated rivers into Dongting Lake was about 216 million tonnes, and the annual amount of sediments flowing out of the lake was 63 million tonnes, resulting in an annual sedimentation amount of 153 million tonnes or about 96 million m^3. Based on this estimate, the volume of sediment deposition in 28 years was about 2.7 billion m^3. During the same time period, the volume of Dongting Lake reduced by 11.5 billion m^3. Since natural sedimentation was about 2.7 billion m^3, the remaining volume of 8.8 billion m^3 was undoubtedly caused by artificial land

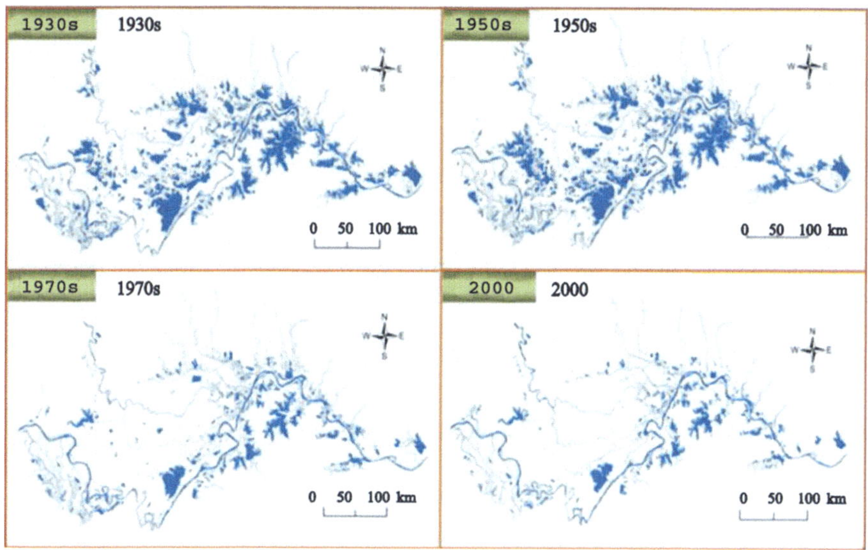

Fig. 6.3 Changes of lakes in the Jianghan Plain during the twentieth century

reclamation. Therefore, the impact of artificial land reclamation was 3.3 times that of natural sedimentation.

The situation of land reclamation from lakes in the Jianghan Plain was similar. Figure 6.3 shows the changes of the lakes in the Jianghan Plain since the 1930s. At the end of the Qing Dynasty and the beginning the Republic of China, there were more than 2000 lakes each with an area of more than 6.7 ha in Hubei, and the total area of the lakes was more than 26,000 km^2. In 1950, there were 1332 lakes each with an area of more than 6.7 ha in the Jianghan Plain, and the total lake area below the median water level was about 8500 km^2, accounting for 88.6% of all the area of 9627 km^2 for rivers, lakes, and ponds. Of the 1332 lakes, 322 lakes each had an area of more than 333 ha, a total area of 7640.6 km^2, a corresponding volume of 13.05 billion m^3, and an effective volume of 11.54 billion m^3. In the 1980s, there were 844 lakes each with an area of more than 6.7 ha, a reduction of 488 from 1950, and the total lake area was 2983.4 km^2, 35% of that in 1950. Of the 844 lakes, 125 lakes each had an area above 333 ha, a total area of 2519.7 km^2 and a corresponding volume of 5.69 billion m^3, 43.6% of that in 1950 (Gu et al. 2009). From the 1950s to the 1970s, land reclamation in the Jianghan Plain was the most intensive because of the shortage of food. The slogans were "growing food as key link," "planting rice seedlings at the centers of lakes and harvesting rice from the centers of lakes." As a result, the lake area reduced significantly.

After the founding of the PRC, in order to control flood and eradicate blood flukes, the "river-lake separation" policy was implemented. As a result, many lakes that had been connected to rivers became separated from the Yangtze River or seasonally connected to the river (via sluices). From the viewpoint of safeguarding food production and improving the health of peasants, the "river-lake separation" policy was somewhat reasonable at the time. Consequently, lakes accelerated their separation from rivers and the lakes shrank. Nobody recognized that the lacustrine wetlands had very important values for the ecology and environment. Moreover, as no large flood had occurred since the 1954 large flood (until 1998), the decrease in lake area in the lower and middle reaches had not shown any adverse impact on flood control or floodwater retention. Furthermore, as water pollution from non-point sources was not so serious, the lake eutrophication problem was not prominent. Therefore, issues associated with lake area reduction and lake-river separation had not been obvious before the 1990s. But now, it appears the impact of the river-lake separation on the aquatic environment and the ecological system has become increasingly serious.

6.2.2.4 Debate on Strategy of Regulating the Yangtze River

The management strategy of the Jingjiang River and Dongting Lake can be best reflected in the debate on the management plan of the middle and lower reaches. Since the dike along the north bank of the Jingjiang River was completed, the pattern of diverting floodwater through both the north and the south banks during a flood event has begun to change, and the long-term strategy to protect the north and divert floodwater to the south has rendered the area north the Jingjiang River becoming increasingly lower and the area south of the river becoming growingly higher, which has not only increased the pressure to the northern Jingjiang Dike for flood control, but also led to more serious sedimentation problems in Dongting Lake and higher losses from flooding in the lake area. As a result, a prolonged debate has continued as for the management strategy of the Jingjiang River and Dongting Lake.

According to historical records, during the middle Ming Dynasty, the construction of the Jingjiang Dike was intensified, blocking all the outlets/openings along the north bank with the political goal of protecting the area north of the river at the price of the area south of the river. Because the residence of Princess Jiajing was in Anlu County on the Jianghan Plain before he became the emperor, the Imperial Court of the Ming Dynasty naturally provided the former residence of the emperor with a special protection. During the Jiajing years of the Ming Dynasty, all the outlets/openings along the north bank were blocked, and an integrated northern Jingjiang Dike was completed, which had profound effects on the management of the Jingjiang River and Dongting Lake in the next 500 years and also caused a prolonged controversy on flood control between Hunan and Hubei.

In the management strategy of Dongting Lake, there have been a variety of ideas in the past 200 years that can be categorized into four major views. ① Return land to the lake to provide space for flood control. ② Block outlets to restore the river. ③ Excavate floodway to divert water from the four outlets into the river. ④ Install sluices at the four outlets to control floodwater into the lake.

During the Daoguang years, Wei Yuan proposed a ban on the construction of collapsible berms in both his *Discussion on Hubei Dikes* and *On Water Conservancy of*

Hubei and Hunan. He advocated "returning reclaimed land to lakes." At that time, the collapsible berms were mainly low-standard dikes for protection of polders. Thereafter, others such as Zhu Kuiji and Huang Juezi advocated a similar opinion, and whenever the lake area suffered from large floods, the opinion reemerged, including the 1998 flood after which, the state also renewed the policy of "returning reclaimed farmland to lakes."

"Blocking outlets to restore the river" was proposed after the formation of the four outlets. In Guangxu's 18th year (1892), Zhang Wenjin, a Hunan native and the Chief of the Criminal Department, Hu Shurong, a Hanshou gentleman, Mei An, and others proposed to block the Ouchi Outlet. As there had been no Ouchi Outlet until its formation in 1860, the proposal was reasonable. However, after Zhang Zhidong, Governor-General of Hubei and Hunan, sent people to inspect the site, they found blocking the outlet was too huge a project to implement. Moreover, as there had been no problem since the formation of the outlet and the associated river for nearly 40 years, blocking it might result in new disputes. Therefore, the proposal was not implemented. During the Daoguang years, Wang Baixin, a Hubei native, advocated in his book *Three Guides to Three Rivers* that "no outlet should be blocked, and it should be let go as its regime guides." He also advocated that floodwater should be diverted to both the north and south areas.

During the Republic of China, the proposal of "returning reclaimed land to lakes" by Hubei and the proposal of "blocking outlets to restore the river" by Hunan caused a prolonged debate which also led to a compromise of the two proposals. In 1932, Wang Huxian of the Hunan Water Conservancy Commission argued in his book *Discussion on Regulating Hunan Waterways* that diversion of water and sediments by the four rivers south of the Jingjiang River had been the major cause of the shrinkage of Dongting Lake and the serious flooding problem in the lake area since the late Qing Dynasty and he proposed to construct a long dike at the border between Hubei and Hunan and excavate a canal to divert the water from the four outlets out of the river to mitigate the damage to Dongting Lake. A similar proposal was broached after the founding of the PRC, but it has never been implemented due to objections.

After the 1935 flood, the YRWCC first proposed to construct sluices at the four outlets. In 1936, Li Yizhi, famous water conservancy expert, presented the following proposal in the management of Dongting Lake. In addition to define the boundaries of the lake and floodway, it would be necessary to maintain the existing water surface and flood storage capacity of Dongting Lake. He also proposed to construct overflow dams at the four outlets and that "as for dam construction, an overflow dam should be constructed at the Ouchi Outlet first; it is not urgent to construct such a structure at the Songzi Outlet; no dam should be constructed at the Taiping Outlet; and the Diaoxian Outlet is suitable to construct a sluice." However, this proposal was opposed by the Hubei water conservancy communities. In 1948, the Yangtze River Hydraulic Engineering Administration (YRHEA) proposed the "Engineering Plan for Regulating Dongting Lake" that was similar to the recommendations by Li Yizhi. Since the proposal of constructing sluices at the four outlets was broached, the idea of blocking the outlets to restore the river has lost its importance. From the viewpoint of flood control, the construction of sluices at the four outlets would provide flexibility in the management of flood control. Therefore, the idea has been supported by Hunan while the YRWCC has been cautious. Nonetheless, the YRWCC has agreed to conduct a preliminary study or to choose the right time to implement it.

In 1947, He Zhitai, Advisor of the YRHEA, proposed three strategies for regulating Dongting Lake. He believed that dredging the lake and constructing dikes would be the last strategy. Efforts to reduce sedimentation and the utilization of the existing lake storage capacity to maintain the function of floodwater retention would be the second strategy. Sedimentation reduction methods would include controlling sediments in the upstream scouring river sections, constructing dams at the four outlets which could also control the amount of sediments flowing into the lake and would be the scope of the second strategy. The newly deposited sedimentary sandbar area within the lake boundaries could be modified for the use of floodwater retention and farmland; and the low-lying lakeside polder area should be warped in rotation in the autumn to improve soil and reduce the amount of sediments to the lake. In the 1950s, Lin Yishan also proposed the strategy. The best strategy would be to comprehensively consider the water and soil resources in the upper and middle reaches and prepare an integrated plan to comprehensively incorporate the role of Dongting Lake. The best strategy proposed by He Zhitai has been of great significance to the modern regulation of the Yangtze River and the comprehensive planning of the Yangtze River basin carried out by the YRWCC since the founding of the PRC.

In 1949, a flood occurred in the middle reaches of the Yangtze River and resulted in a tremendous loss. Therefore, in the first year after the founding of the PRC, the Central Government decided to establish the YRWCC in Wuhan to lead the comprehensive regulation of the Yangtze River basin, such as flood control. At this time, the Jianghan Plain belonged to Hubei and was still a national important base of food production. Wuhan City also became an important industrial metropolitan in Central China. The first river-regulating project since Lin Yishan became the Director of the YRWCC was to establish the Jingjiang Flood Diversion/Retention Zone on the south bank of the Jingjiang River, which later played a crucial role in controlling the 1954 flood. In 1953, Lin Yishan proposed the following "three-stage river regulating" strategy. The first stage would be mainly to construct dikes to the elevation above the actual water levels during the 1931 or 1949 flood (later changed to the 1954 flood). The second stage would be to take advantage of the presence of many lacustrine low-lying areas along the river to develop a floodwater retention-farmland zone for storing excessive floodwater of the 1931 or 1949 flood (the volume of floodwater above the flood control level of the dike). The third stage would be to construct reservoirs in canyons in addition to the projects in the first two stages to completely control the flood of the Yangtze River. Lin Yishan's three-stage strategy was a general concept for flood control in the middle and lower reaches of the Yangtze River, and it was fully integrated into the 1959 *Synopsis of Comprehensive Utilization Plan of the Yangtze River Basin* and the *1990 Summary Report of Comprehensive Utilization Plan of the Yangtze River Basin* of 1990. The three-stage strategy was a development of He Zhitai's best strategy and was gradually implemented from the planning phase to the construction phase in the next 60 years. The dike construction of the first stage was basically completed after the 1954 large flood. After the 1998 large flood, the dike systems on major tributaries of the middle and lower reaches of the Yangtze River were all in full compliance and were further improved. By 2001, the goal of the first stage had been fully achieved. The second stage has also been implemented as planned. In the low-lying areas of the lower and middle reaches, 40 retention zones have been planned to store 59 billion m^3 of floodwater, and the interests of both Hunan and Hubei have been taken into consideration. For example, the planned 10 billion m^3 diversion/retention capacity near Chenglingji was equally divided into Hubei's Hong Lake and Hunan's Dongting Lake, or 5 billion m^3 each. The plan of the second phase was completed with the publication of the *Flood Control Plan of the Yangtze River* in 2008. However, only the Jingjiang and Dujiatai Flood Diversion Zones were constructed as planned but the construction of the other flood diversion/retentions zones was delayed. The major reason for the delay was that, with the completion of the TGP, it has been expected that the TGP should play a greater role in flood control for the middle and lower Yangtze and the planned flood diversion/retention zones should be adjusted accordingly. Some have even hoped that the "planned zones would be removed from their designation." Moreover, no large watershed flood has occurred since 1998, and every local government has become mindfully indifferent to the flood retention projects. As a result, the plan for the second stage has only been partially implemented. The third stage would be to build control reservoirs in the upper reaches with the TGR as the key. Now the TGP has been completed, and control reservoirs on the mainstream and tributaries in the lower reaches of the Jinsha River are under construction. It is estimated that these reservoirs will have been substantially completed by 2020 and fully completed by 2030.

For the safety issue of the Jingjiang Dike, Lin Yishan proposed two strategies: to discharge alluvial sediments to (or warp) the lowlands protected by the north dike and to move the mainstream of the river southward. Warping the protected lowlands would be to mitigate the high elevational difference between the high flood water level and the lowlands protected by the north dike through transporting sediments from the river to reinforce and thicken the north dike. Director Lin's idea has never been implemented for various reasons. The strategy to force the mainstream of the Jingjiang River to move southward through river regulation and control engineering would be to prevent the floodwater from scouring the foundation of the north dike and help develop a floodplain-type dike. This would also improve the safety of the most dangerous north dike. Lin Yishan designated warping the lowlands protected by the north dike to be the best strategy and moving the mainstream of the Jingjiang River southward to be the medium (or second) strategy. All these strategies were centered around the safety of the Jingjiang Dike, especially to ensure the safety of the north dike. Because the opinions of various parties were inconsistent, these two strategies have never been implemented.

6.2.2.5 Construction of Sluices at the Four Outlets

Since the formation of the four outlets of the Jingjiang River, control of the outlets has been proposed many times. In the 1930s, the famous scholar Li Yizhi formally proposed to construct an overflow dam at each outlet to control the outlets, and his proposal gained support from many people at the time. After the founding of the PRC, as the Jingjiang Diversion Project was completed in 1952, the South Sluice (90 km from the Taiping Outlet) was constructed on the Hudu River. In 1958, the YRWCC had several discussions with Hubei and Hunan regarding alternatives for regulating the four rivers south of the Jingjiang River and Dongting Lake. Hubei and Hunan reached an agreement about the control project of the Diaoxian Outlet which has since been implemented. The control projects of the Songzi and Ouchi Outlets have never been implemented due to various reasons, but the research work has never ceased. The evaluation study on the TGP in the 1980s demonstrated that the construction of the TGP would have created the conditions for the construction of control sluices at the four outlets. In the *Summary Report of Comprehensive Utilization Plan of Yangtze River Basin (Revised in 1990)*, an additional study was presented on the construction of sluices at the four outlets. A detailed study was carried out in the preparation of the *Report of Flood Control Plan of the Yangtze River Basin* after the 1998 flood.

6.2.2.5.1 Conditions Without Considering TGR

The *Near Future Plan for Comprehensive Regulation of the Dongting Lake Area* reviewed in 1997 mainly focused on the situation before the TGP started to play a role in flood control. At that time, the flood control situation of the Jingjiang River was very grim. Moreover, the sedimentation problem at Dongting Lake was becoming increasingly serious due to the diversion of water into the lake through the four outlets. To alleviate the sedimentation problem at Dongting Lake while not to increase the burden of flood control on the Jingjiang River, the initial assumptions of constructing control sluices at the four outlets would be that under normal conditions, the water from the upper reaches would flow through the mainstream of the Yangtze River or the Jingjiang River as much as possible to reduce the amount of water and sediments into Dongting Lake to decrease sediment deposition and lower the flood water level in the lake area. In the meantime, after considering that the TGP was not in operation and the amount of floodwater diverted by the three outlets (refer to Songzi, Taiping, and Ouchi Outlets) accounted for a large percentage of the Jingjiang River's floodwater discharge, which was very important to the flood control safety of the Jingjiang River, it would be necessary to open the sluices to divert floodwater when the flood water level is high in the Jingjiang River. Presently, the dike's warning water level in the middle and lower reaches of the Yangtze River is an important characteristic water level for flood control thereof. When the river water rises above the warning level, the risk to the dike will gradually increase. In order to not increase flood control pressure on the mainstream of the Yangtze River, studies were performed to use the warning levels at major control stations on the mainstream of the Yangtze River as the critical water levels to decide whether or not to open the control sluices. When the water level in the river reaches the warning level, the sluices would be open to divert water. When the water level in the river is below the warning level, the sluices at the Songzi and Ouchi Outlets would be partially open to provide necessary water to meet demands for life and production in the downstream areas on both sides of the two rivers. The envisioned operational mode at that time would be: when the water level is expected to reach the warning level of 43 m at the Shashi Station, open the sluice at the Songzi Outlet; when the water level is expected to reach the warning level of 35 m at the Jianli Station, open the sluice at the Ouchi Outlet; when the water levels are lower than the above mentioned water levels, both sluices would be controlled to discharge at 500 m^3/s; and when the diversion capacity is lower than 500 m^3/s, the sluices would be open to discharge the actual flow. According to this operational mode, calculations were performed to estimate changes of the amounts of water and sediments. The following situations would occur when the control of sluices is established:

1. Decrease of Lake Inflow

After the sluices are constructed at the Songzi and Ouchi Outlets the same time, the amount of water flowing into the lake would be reduced by 50% on average and by 66% during the flood season. This option would be conducive to flood control of the Dongting Lake area, but this would also increase the flowrate in the mainstream of the Yangtze River and in turn would back up the water level at Chenglingji and reduce the outflow from the lake through Chenglingji. Moreover, when the water level in the river reaches the warning level, the sluices would need to be open to divert floodwater into the lake. Therefore, the flood water level in East Dongting Lake would not decrease much. The flood water level in West and South Dongting Lakes would decrease some and would be 0.1–0.2 m lower in the cases for 1981 and 1983 floods, indicating this option would be slightly beneficial for flood control in the lake area.

2. Less Amount of Sediment into and Slower Sedimentation in the Lake Area

After the sluices are constructed at the Songzi and Ouchi Outlets the same time, the average annual amount of sediments flowing into the lake would be reduced from 94.35 million tonnes (mean of the calculation series) to 37.5 million tonnes, a 60% decrease. If the current amount of sediment deposition in the lake area is assumed to be 74% of the quantity of the sediments flowing into the lake, the average annual reduction would be 42.1 million tonnes, indicating the effect would be remarkable.

3. Prominent Souring Effect on the Jingjiang River and Sedimentation Effect on the River Section Downstream of Chenglingji

The construction of sluices at the four outlets would reduce the inflow of water and sediments into the lake. Accordingly, the Jingjiang River would discharge more water and sediments. If the sluices are constructed at the Songzi and Ouchi Outlets the same time, the average annual amount of scouring sediments in the Yichang-Chenglingji section would increase by 26 million m^3, and the amount of sediment deposition would increase by more than 15 million m^3 in the Chenglingji-Hankou section and would also increase some in the Hankou-Jiujiang section. After the control sluices are constructed, scouring and sedimentation conditions of each river section would change and the water level in the Yichang-Jiujiang river section would change even when the flowrate is the same. The scouring effect on the Jingjiang River section would result in a lower water level at the Shashi Station, and the water level would be higher at the Luoshan Station due to sediment deposition in the Chenglingji-Hankou section.

The above three conditions indicate that, without the TGR, the construction of the control sluices at the four outlets, on the one hand, would reduce the amount of water and sediments into the lake, reduce sediment deposition in the lake area, and lower the flood water level in South and West Dongting Lakes. However, on the other hand, the amount of water and sediments in the mainstream would increase accordingly. The general trend would potentially lead to increased sedimentation in the Chenglingji-Hankou section and a higher water level at the Luoshan Station when the flowrate is the same. If this happens, the outflow from Dongting Lake would be impacted. The reduced discharge capacity at the same water level at the Luoshan Station would directly result in a significant increase of water volume in the vicinity of Chenglingji, which is unfavorable. Therefore, the construction of sluices at the four outlets should not be considered until the completion of the TGP.

6.2.2.5.2 Conditions with TGP

After 1998, the flood control plan of the Yangtze River basin was re-evaluated and the construction of control sluices at the four outlets after the completion of the TGP was evaluated. With the TGR in operation, great changes would occur to the water and sediment conditions in the mainstream of the Yangtze River and would be mainly manifested by the following. The flood control standard of the Jingjiang River would reach a 100-year flood event, and in the event of a flood similar to the 1870 event, there would be reliable countermeasures. The sediment content of the discharging water from the TGR would be significantly lower than that under natural conditions and would remain at only about 10–20% of that under natural conditions for 20–30 years (if reservoirs continues to be constructed in the upper reaches, the time would be longer). The mainstream channel, especially the Jingjiang River, would undergo increased scouring, which would cause a significant reduction in diversion of water and sediments through the four outlets, especially through the Ouchi Outlet. Therefore, it would not be important to construct a sluice at the Ouchi Outlet.

Thus, the study focused on the construction of a control sluice at the Songzi Outlet, and the operation of the sluice would consider the following conditions. ① West Dongting Lake has the most serious flood hazard problem in the Dongting Lake area. After the completion of the TGP, the construction of a sluice at the Songzi Outlet and the operation of the sluice in conjunction with the TGP can desynchronize the peak flowrates between the Songzi and Li Rivers, thus lowering the water level of West Dongting Lake and reducing flood damage in the lake area. After flowing through the Jiangya and Zaoshi Reservoirs, the flowrate of a 20-year flood at the Shimen Station on the Li River can be controlled to no more than the safety discharge of 12,000 m^3/s. According to the actual flood control capacity in the Songzi River and Li River area, when the floodwater in the Yangtze River is not large, the diversion flow through the sluice at the Songzi Outlet could be controlled so that the combined flowrate from the Songzi Outlet and the Shimen Station would be no more than 14,000 m^3/s, which could guarantee the safety of flood control in the area of the Songzi River and Li River basins. ② In order to not increase flood control pressure on the mainstream, besides considering the desynchronization of the peak flood flowrate in the Songzi and Li Rivers, the warning water level at the Shashi Station should be used as one of the operational control conditions. ③ In order to meet the needs of water use and protection of the ecological environment in the lake area, the ecological flowrate should be maintained in the Songzi River after the construction of a sluice at the Songzi Outlet, such as 500 m^3/s.

Based on the above considerations, the typical operational rules of the proposed sluice at the Songzi Outlet would be that when the water level at the Shashi Station is below 43 m (warning water level), the flowrate at the sluice of the Songzi Outlet would be controlled at 500 m^3/s; when the actual flow capacity of the Songzi River is below 500 m^3/s, discharge the actual flow; when the water level at the Shashi Station is over 43 m, and if the combined flowrate between the Li River (at the Shimen Station) and the Songzi Outlet is not more than 14,000 m^3/s, the sluice at the Songzi Outlet would be open without any control; when the combined flowrate between Li River (at the Shimen Station) and the Songzi Outlet is over 14,000 m^3/s, the flowrate through the sluice at the Songzi Outlet would be controlled to 14,000 m^3/s minus the flowrate at the Shimen Station; when the flowrate from the Li River is above 14,000 m^3/s, the sluice at the Songzi Outlet would be closed; and the reduced flowrate by controlling the sluice at the Songzi Outlet, plus the reduced flowrate to control the water level at the Shashi Station in the mainstream to not exceed 44.0 m, would be discharged through the mainstream or otherwise would be intercepted by the TGR.

These considerations are mainly for the operation of flood control. After the construction of the sluice, the river-lake relationship would be further adjusted, which would need to be monitored. Moreover, after the construction of the sluice, the river-lake separation time would be obviously increased. Therefore, the impact on the ecological environment would need to be carefully considered.

6.2.2.5.3 Construction of a Sluice at Chenglingji

Inspired and influenced by the construction of the eco-economic zone and the installation of a sluice at Hukou of Jiangxi in recent years, Hunan and Hubei jointly prepared a strategic plan for the construction of the Dongting Lake Eco-Economic Zone. The strategic plan considered the construction of a sluice at the Songzi Outlet but also proposed the construction of a sluice at Chenglingji, the exit outlet of Dongting Lake to the Yangtze River. The preliminary study of the plan has begun. It was a new idea in the twenty-first century to construct a sluice at Chenglingji. The main goal of the sluice would be to reduce the impact of the early impoundment stage of the TGR on Dongting Lake during the dry season and in the meantime to solve the problem associated with water supply and water use in the lake area during the dry season.

Presently, as the study on the impact of the sluice construction at Chenglingji has just begun, many technical and managerial problems have been identified. The major problems are as follows. ① There are many technical issues, such as the location and type of the sluice and the selection of the control water level. In the left bank of Chenglingji are tidal flatlands and a flood diversion zone where geologic conditions are complex. The main and secondary dams may be relatively long. If a traditional sluice is used, the flow capacity at Chenglingji would be reduced, and it is possible to elevate the flood level in Dongting Lake and aggravate the sediment deposition problem in the lake area. ② The impact on the ecological environment would be significant. Chenglingji is a major international wetland and the most important biological channel between Dongting Lake and the Yangtze River. Maintaining a large water surface during the dry season is not conducive to the habitat of migratory birds. Moreover, the pressure for eutrophication of the water body and protection of water resources would be increased. ③ The operation of the sluice would be complicated. Not only the relationship between the operation of the control reservoirs in the four rivers (Xiang, Zhi, Yuan, and Li) of the Dongting Lake system and the three sluices south of the Jingjing River would be complex, but also the relationship between Chenglingji and the joint operation of impoundment and discharge at the TGR would be complicated. ④ The cost for the construction, operation, and management of the sluice would be high, and the project itself would be low in efficiency. The beneficiary relationship for the operation of the project would be intricate, and the cost for construction and operation would be high. Therefore, the future operation and management of the sluice would depend mainly on the government.

6.2.2.5.4 Comprehensive Evaluation

The construction of control sluices at the four outlets has long been considered as the key measure to mitigate the flood threat and shrinkage problem of Dongting Lake. The above analyses indicate that it is the construction of the TGP that has created the conditions for the installation of the four sluices, and the joint operation of the TGR and the sluice at the Songzi Outlet can desynchronize the peak flood flow of both the Li and Jingjiang Rivers and has obvious effects on flood control and sedimentation reduction. In the meantime, the regulation of the TGR will not increase the flood control pressure on the Jingjiang River in the event of a flood. However, since the impoundment of the TGP in 2003, great changes have taken place on sediment load conditions from the upstream. Because the upper Jingjiang River has undergone continuous scouring, the amount of water and sediments flowing into the three outlets has been further reduced, especially the amount of sediments flowing into Dongting Lake. As a result, the volume of sediment deposition in the lake area is expected to be only 10% of the average annual value. Therefore, the problem of sedimentation in Dongting Lake is not as serious as previously conceived. One of the main purposes of the installation of the sluices at the three outlets would be to guarantee the water supply to the Dongting Lake area and the four rivers south the Jingjiang River. There are also some problems associated with the construction of the sluices. For example, the construction of the sluices cannot solve the problem of sedimentation in the three rivers south of the Jingjiang River and may create potential problems to the aquatic environment and ecology.

Presently, the exiting major problems of the four rivers south of the Jingjiang River and Dongting Lake are as follows. First, the river system is complex with many dikes, long flood control front, and low safety standards for the protecting berms of polders, and the issue of flood control safety for the urban areas in the lake area has not completely resolved. Second, the time for water flowing from the Yangtze River to the lake is shorter and the no-flow time is long. The aquatic environment of the four rivers has been increasingly deteriorating, and water supply to the areas along the banks of the rivers is not safe or adequate. Third, the river-lake connection is getting worse. Water exchanges between the interior and exterior lakes and between inside and outside polders are minimal. As a result, the water pollution problem is prominent. Fourth, the safety improvements and management of flood detention zones are lagging. Fifth, agricultural development and drainage systems associated with irrigation in the lake area need to be improved. Sixth, problems such as degradation of the wetland ecosystem cannot be solved by the construction of the sluices. Therefore, from the viewpoint of maintaining a robust ecosystem, the construction of the sluices is clearly not beneficial.

6.2.3 Development and Regulation of Poyang Lake

Poyang Lake is fed by the Gan, Fu, Xin, Rao, and Xiu Rivers with a catchment area of 162,000 km^2. The Poyang Lake area includes three parts: a channel to connect to the Yangtze River, a lake basin, and a tail channel of the five rivers. Its boundary is composed of natural low hilly land and man-made dikes. Currently there are about 600 dikes with a total length of about 4000 km, which protects about 8000 km^2 of land, 450,000 ha of farmland, and tens of millions of people in Nanchang, Jiujiang, Shangrao, and other cities. Similar to Dongting Lake, as the Poyang Lake basin is rich in rainfall and its flood control front is long, flood control has always been the highest priority of water conservancy efforts.

Poyang Lake has two prominent characteristics under natural conditions. First, Poyang Lake has the feature of "being a lake when the water level is high and appearing a river when the water level is low." In fact, it is a seasonal shallow water-passing lake, and the variation of its water level is controlled by the water level of the Yangtze River and water received from the five rivers. The water level varies greatly between the wet and dry seasons. The interannual highest and lowest water levels were 22.59 m (above the Wusong Datum and same thereafter) and 5.90 m, respectively, with a difference of up to 16.68 m. During the year, the water level ranges from 8.55 m to 15.36 m with an average of 12.06 m. When the water level was at 22.59 m (July 31, 1998), the lake was 3203 km^2 in area and 27.5 billion m^3 in storage capacity, while when the water level was at 5.90 m (February 6, 1963), the lake was only 26 km^2 in area and only 60 million m^3 in storage capacity. There are as much as 10 km of mudflats and marshes along the lakeshore during the dry season, which are suitable habitats for migratory birds. Second, because of rich water resources and good water quality, the amount of usable water resources is large. While the basin of the lake system and the administrative jurisdiction are highly consistent, the annual and interannual distributions of the water resources in the fiver rivers are uneven. The average total annual amount of water resources (amount of runoff at the Hukou Station) in the province (or lake basin) is about 146 billion m^3 with the amount of the provincial water resource of 3600 m^3 per capita, which is 1.8 times the national average of 2000 m^3.

6.2.3.1 Existing Conditions of Regulation and Development in the Poyang Lake Watershed

Similar to Dongting Lake, during the 1950s to the 1980s, Poyang Lake experienced large-scale land reclamation with a total reclaimed area of 146,700 ha, resulting in a reduction of over 8 billion m^3 in storage capacity (Jin et al. 2012). The construction of water conservancy projects has been continued for decades, and about 3,712,000 water supply facilities and 209,000 water storage facilities have been constructed. By the end of 2009, 25 large, 238 medium-sized, and 8633 small reservoirs had been constructed, leading to a water resources utilization rate of 21%. Important cities, such as Nanchang and Jiujiang, and key dikes in the lake area have met flood control standards as planned.

Control hydraulic projects have been or are being constructed on the five rivers of Poyang Lake, such as Zhelin and Dongjin on the Xiu River; Xiajiang (under construction), Wanan, and Shangyoujiang on the Gan River; Xinzhou, Jiepai, and Da'ao on the Xin River; and Wuxikou (under construction) on the Rao River. The development level of water conservancy is relatively high.

6.2.3.2 Existing Major Problems in the Poyang Lake Watershed

There are many problems in the Poyang Lake watershed. Specific issues are provided below.

6.2.3.2.1 RichWater Resources with Both Flood and Drought Problems

Large numbers of dikes in the lake area that were constructed with relatively low flood control standards may not survive a large flood. Despite years of construction, only 29 key dikes have met the flood control requirement when the water level is 22.50 m at the Hukou Station; some of the dikes have not met the flood control requirement of the 1954 flood; and the remaining dikes have only met the 5- to 10-year flood control standard. The quantity of water resources in a dry year is 1.5 times lower than that in an average water year. During July–August when agricultural water use peaks, agricultural irrigation water use accounts for 66% of the total amount of annual water use, while the incoming water is only 22% during the same time period. Therefore, there is a seasonal water shortage problem.

6.2.3.2.2 Worrying Aquatic Environmental Safety

Pollutants in Poyang Lake mainly come from urban domestic sewage effluents, mining and industrial wastewater, and agricultural non-point pollution sources in the five river basins. In the lake area, farming is the primary industry and aquaculture grows rapidly. Overuse of pesticides and fertilizers on farmland and soil erosion in dryland have resulted in increasingly serious water pollution from non-point agricultural sources. During the dry season and summer drought period, large numbers of fish deaths often occur in the tail channel of the five rivers. It is obvious that water pollution in any locale of the lake area and the upstream 16 million km^2 area would affect Poyang Lake since Poyang Lake is not only a catchment of water but is also an area for the collection of wastewater in the province. Therefore, the situation for pollutant interception and treatment is very grim.

6.2.3.2.3 Schistosomiasis Epidemic Remains to Be Resolved

Schistosomiasis and other diseases still seriously threaten the health of the residents along the lakeshore. There are currently still 12 schistosomiasis-epidemic counties (districts) where 2.5 million residents, 140,000 farming cattle, 20,000 sick cattle, and 73,000 ha of *Oncomelania* snails are threatened by schistosomiasis (Jin et al. 2012). Although the infection rate of human beings and animals has been decreasing, the infection rate of the snails has been increasing.

6.2.3.2.4 Rich Lacustrine Wetland Biodiversity Significantly Impacted by Human Activities

In 1988, Poyang Lake was approved to be a national nature reserve. In July 1992, China acceded to the *Convention on Wetlands of International Importance especially as Waterfowl Habitat* (CWIIWH) and Poyang Lake has since been on the list of Wetlands of International Importance (Ramsar List). In 1997, Poyang Lake was designated to the Northeast Asian Crane Protection Network. In 2000, it was listed by the World Wide Fund for Nature (WWF) as one of the "Global Ecoregions." In 2002, it was included in the important ecological functional reserves of China. In October 2002, Poyang Lake officially joined the "World Life Lake Network" in the World Life and Lake Conference held in South Africa. Poyang Lake is rich in fish and bird species and is the habitat for more than 95% of the world's wintering

Siberian cranes. The lake is also a very important wetland ecological functional area of China and even the world and boasts the functions to divert and retain floodwater, regulate climate, decompose pollutants, and maintain the biological diversity.

6.2.3.3 Water Conservancy Project at Poyang Lake Mouth

6.2.3.3.1 Project Background

Before 1949, it had been proposed to build a dam to mainly control floodwater from the Yangtze River into the lake. In 1972, the Shanghai Institute of Parasitic Diseases submitted a report proposing "to construct a sluice at the mouth of Poyang Lake to impound water for eradication of snails and comprehensively utilize resources." In 1982, the Jiangxi Provincial Academy of Sciences led a comprehensive scientific investigation at the Poyang Lake region, and the "Lake Control Project" was listed as a "strategic subject." In June 1986, the Jiangxi Provincial Water Conservancy Planning and Design Institute (JPWCPDI) evaluated the "Lake Control Project" from four aspects, including flood control, power generation, irrigation, and drainage. In December 1986, the Jiangxi Provincial Government held a conference to evaluate the project and presented 11 special reports, one of which was the *Comprehensive Research Report on Artificial Control of Poyang Lake*. In 1987, during the fifth meeting of the sixth National People's Congress, some delegates presented a proposal "to perform a feasibility study of a water conservancy control project in Poyang Lake." In 1995, the JPWCPDI submitted the *Proposal of Poyang Lake Control Project Planning* based on the results of previous studies and evaluations. After the 1998 flood, the study of the "Lake Control Project" was included in the *Yangtze River Basin Flood Control Planning* as a special appendix report. The following year, the Changjiang Institute of Survey, Planning, Design and Research (CISPDR) submitted a draft special planning study report of the "Lake Control Project" for comments. In 2002, 40 delegates from Jiangxi Province submitted "Proposal #1" to the fifth meeting of the Ninth National People's Congress in Beijing, appealing for the construction of the Water Conservancy Project at Poyang Lake, China's largest freshwater lake, for comprehensive benefits of flood control, navigation, fisheries, and power generation. Experts from other provinces, the MWR and the YRWCC raised questions about "Proposal #1" for the following reasons. The main design concept of the "Lake Control Project" was to block the floodwater from entering Poyang Lake and ease flood control pressure from Jiangxi Province, but it would put more flood control pressure of the mainstream Yangtze on Hubei, Anhui, and Jiangsu. Moreover, for comprehensive benefits such as power generation and navigation, the dam's control water level would be above 18 m, which would lead to a complete separation between the Yangtze River and the Poyang Lake water system, inundation of large areas of lacustrine wetlands, and severe effects on habitats of migratory birds and breeding of aquatic flora and fauna in Poyang Lake. Furthermore, the project would significantly affect the compliance of Poyang Lake with the Ramsar Convention. Since then, the "Lake Control Project" has seldom been mentioned even though studies have still been continuing.

In early 2008, the Jiangxi Provincial Government made a strategic decision to develop the Poyang Lake Eco-Economic Zone. At the end of 2008, the Jiangxi Provincial Bureau of Water Conservancy proposed to construct the "Poyang Lake Ecological Control Project." Compared with the "Lake Control Project," the "Ecological Control Project" had two obvious changes in its design concept. First, the Ecological Control Project would "control the water during the dry season but will not control the floodwater." Specifically, the sluice of the dam would be fully open when a flood occurs so that the "river and lake would be connected," which would not only be conducive to the diversion/retention of floodwater from the Yangtze River and relief of the flood control pressure from other provinces and cities but would also be beneficial to the reproduction of aquatic plants and migratory aquatic animals. Second, the water conservancy project would integrate wetland protection, and the control level would be lowered to 16 m so as to maximize the transitional wetland zone between water and land. The impact of the water conservancy project on the ecological environment of Poyang Lake would be minimized. At the end of May 2008, a consultation meeting on the planning opinion of the Poyang Lake Ecological Water Conservancy Project was held in Beijing, and many academicians and experts in the water conservancy communities considered that the concept of "controlling the water during the dry season but not controlling the floodwater" was a major breakthrough from the original plan and incorporated the concept of scientific development and the harmonious coexistence of human beings and nature. At the same time, the MWR integrated the comprehensive plan of Poyang Lake into the revised comprehensive plan of the Yangtze River basin and assigned the preparation of the project's application report to CISPDR. In 2010, 15 experts from the Chinese Academy of Sciences and other departments sent a petition to the State Council, requesting to stop the project. Afterward, Jiangxi Province communicated with the experts who raised questions and organized a number of special studies and evaluations. As a result, more revisions were made to the original plan, including the elimination of the hydroelectric function and further lowering of the control water level.

During April 12–17, 2010, at the invitation of the Forestry Bureau of Jiangxi Province, the secretariat of the Ramsar Convention Bureau organized a group of experts (referred to as the Expert Group) to investigate the ecological environment of Poyang Lake and the basic conditions of current wintering migratory birds. At the same time, the Export Group met with officials and scientists of governmental departments on the Poyang Lake Water Conservancy Project. During this time, the Export Group visited Poyang Lake and exchanged large amounts of information with the officials and scientists of the government in the form of reports and discussions on the ecological characteristic of Poyang Lake, especially biodiversity, water chemistry, hydrology, etc.

In December 2010, the State Council formally approved the "Plan of Poyang Lake Eco-Economic Zone" (hereinafter referred to as " the Plan"), and the Office of the Poyang Lake Water Conservancy Project was established in January 2010. The organization's leadership consisted of a number of departments, including the leaders from the Jiangxi Provincial Bureau of Water Conservancy that would be responsible for the construction of the Poyang Lake Water Conservancy Project, the

Provincial Bureau of Housing and Urban Development, and the Poyang Lake National Nature Reserve Authority. In February 2012, the MWR reviewed the application report for the "Poyang Lake Water Conservancy Project." In February 2012, the Jiangxi Provincial Bureau of Water Conservancy officially authorized CISPDR to perform the study, survey, investigation, and design of the "Poyang Lake Water Conservancy Project."

6.2.3.3.2 Potential Issues of Project Construction and Operation

The controversial focus of the project was initially about the overall flood control of Jiangxi Province and the middle and lower Yangtze, utilization of water resources, river-lake connection, wetland protection, etc. After the project was changed to only control water in the dry season and not to control water during the flood season, the river-lake connection and wetland habitat protection were left as the major problem of concern. Naturally, there would still be many uncertainties associated with the construction and operation of the project. The following are a summary of the main issues:

1. After the project is completed, the river and the lake would be separated for more than one half of the year, which would impact migratory animals such as the Yangtze finless porpoise and FMCC, and other aquatic animals and plants.
2. Poyang Lake is a water-passing lake with seasonal variations and frequent water exchanges. After the sluice is constructed, because the flow velocity is lowered, the self-purification capability of the lake's water body would decrease. It would be difficult to predict how to avoid similar eutrophication phenomenon that had occurred to Tai Lake. Therefore, the pressure on protecting the aquatic environment would increase.
3. After the water level is raised, the groundwater level would rise, which would cause the low-lying farmland protected by dikes to become gleyic and swampy.
4. The raised water level of the lake would cause the transition zone between water and upland to expand outward, which may cause the spread zone of oncomelania snails to be enlarged or closer to the densely populated areas.
5. Regardless of the height of the dam, as long as the water level is raised during the dry season, the area of mudflats and meadows would shrink, which would inevitably affect the habitat for the wintering migratory birds. During the winter, the water level in the Poyang Lake Nature Reserve and the Nanjishan Nature Reserve would need to be artificially lowered by the construction of dikes and via pumping, which would result in the man-made fragmentation of the wetland ecosystem of Poyang Lake.
6. As the dam would be mainly for water supply and irrigation of the lake area and would not have the benefit of flood control or power generation, the revenue of the project would be low, but the cost for construction, operation, and management would be high. Therefore, the project would burden the fiscal disbursement.

7. After the completion of the project, it would be difficult to determine the impact on the middle and lower reaches of the Yangtze River and the estuary, especially on the ecological system, flood control, and utilization of water resource in Anhui, Jiangsu, and the Yangtze Estuary area of Shanghai, on which no detailed study has been performed.
8. The relationship of the operational technique and management of the Poyang Lake Project, the control reservoirs on the fiver rivers, and the projects on the mainstream of the upper reaches of the Yangtze River, such as the TGR, would be complicated. Moreover, it would be difficult to develop an operational program that would satisfy all watersheds, regions, and stakeholders.

6.2.4 Regulation and Development of the Yangtze Estuary

The Yangtze Delta is one of the fastest economically growing regions in China. The development and utilization of the Yangtze Estuary have been mainly in three aspects: navigation, utilization of water resources, and use of tidal flatlands. Navigation included the construction of deep-water channels and wharfs; development and utilization of water resources were mainly improvements of the water sources on the Yangtze River for Shanghai; and the use of the tidal flatlands was to take necessary measures to rationally utilize the land resources after sedimentation.

6.2.4.1 Construction of Navigable Channels and Wharfs

Since 1998, to establish the Shanghai International Shipping Center, a deep-water channel improvements project at the Yangtze Estuary had attracted the world attention. The project consisted of three phases and was completed in 10 years. The first phase of the project began on January 27, 1998, and was completed in July 2000. The average water depth of the waterway reached 8.11 m. On March 29, 2005, the second phase of the 10-m-deep waterway was fully completed. Since the completion of the first and second phases of the project, the navigability of the 8.5-m- and 10-m-deep waterways have been maintained 100%. The third phase of the project was to deepen the 10 m waterway completed in the second phase to 12.5 m and create a two-way channel with a full length of 92.2 km and a width of 350–400 m. The channel has met the requirements for the third- and fourth-generation container ships and 50,000 tonne ships to sail in two ways during a full tide and allowed the fifth- and sixth-generation large container ships, and 100,000 tonne fully loaded bulk carrier and 200,000 tonne partially loaded bulk carriers to sail during a tide. The total investment of the third phase of the project was 4.767 billion yuan, including Changxing Submerged Dike Project, Waterway Project, Dredging Project, Blow-Mud-to-Beach Project, etc. It took 4 years to complete this phase and it was completed in October 2010.

As for the port construction, through more than 100 years of construction, Shanghai Municipality completed the construction of five ports in the Huangpu River, Zhanghuabing, Jungong Road, Gongqing, Zhujiamen, and Lognwu and three ports on the south bank of the Yangtze River, Baoshan, Luojing, and Waigaoqiao. In addition, Baosteel Group Co., Ltd., Shidongkou Power Plant, Waigaoqiao Power Plant constructed their own special ports. On December 10, 2005, the first phase of the Yangshan Deepwater Port Project was completed and began operation.

6.2.4.2 Water Source Improvement Project for Shanghai

Shanghai is China's largest city by population and an international metropolis. Improvements of water sources are the key factor for the socioeconomic development of the city. Since the construction of the Yangshu Water Treatment Plant in 1883, 12 water plants have been constructed, and the daily water supply capacity has reached over 10 million m^3. The Huangpu River used to be the source of more than 80% of water supply for downtown Shanghai. Because water pollution in the Huangpu River has become increasingly serious, Shanghai Municipality planned to use the Yangtze River where the water quality is better as its main water source. However, as the Yangtze River has been impacted by tides and salt water intrusion, salty water must be avoided during the dry season or high tidal time. Therefore, in the past 30 years, Shanghai has constructed a number of coastal reservoirs on the south bank of the Yangtze River as water sources, including Chenheng Reservoir, Baosteel Reservoir (mainly for Baosteel use), and others. In order to increase the capacity of water sources in the Yangtze River and extend the time to avoid the salty water, the Qingcaozhou Water Source was constructed in the mid-channel sandbar north of the Changxing Island on the North Channel of the South Branch at the Yangtze Estuary. The water source was developed through the construction of a 43-km-long dike whose crest elevation was 8.5 m to encompass a water surface area of 60 km^2. As China's largest mid-river reservoir, the Qingcaozhou Reservoir not only conserves the very valuable south bank of the Yangtze River and land resources along the bank but also lengthened the time to avoid salty water. After the construction of the Qingcaozhou Reservoir, the maximum effective capacity reached 553 million m^3 with a design effective capacity of 435 million m^3. In 2010, with a water supply capacity of 7.19 million m^3/day (the total design water supply capacity from the Huangpu River was 5 million m^3/day) which was above 50% of Shanghai's original water supply demand. The reservoir has supplied water to more than 10 million people in Shanghai's 10 entire administrative districts such as Yangpu and Hongkou and 5 partial districts such as Baoshan and Putuo. Due to natural disasters such as drought or saline-tidal intrusion, the time period during which water could not be continuously withdrawn from the water zone of the Qingcaozhou sandbar has been generally no more than 15 days in a year with the maximum of 38 days. Since the completion of the reservoir, the water level of normal operation has generally been about 3 m with an effective water storage capacity of no less than 387 million m^3. At the estimated water supply rate of 7.19 million m^3/day in 2010, the reservoir would be able to continuously supply water for 50 days without outside water replenishment, which is adequate to withstand the occurrence of a natural disaster or an emergent water pollution accident at the Yangtze Estuary. Using the reservoir's maximum effective capacity of 553 million m^3 at an assumed supply rate of 9.5 million m^3/day, the reservoir would be able to supply freshwater of good quality for at least 68 consecutive days. In the future, the West Dongfengzhou Reservoir will be constructed to the west of Chongming Island to solve the water supply problem thereof.

6.2.4.3 Utilization of Tidal Flatlands

China has a long history of land reclamation from the sea and has reclaimed more than 100,000 km^2 of land from the sea in the past thousands of years, which is about three times the total area of the Netherland. During 1950–2000, the total land reclaimed from the sea in China was 11,000–12,000 km^2. During 1951–2007, the coastal zone of Jiangsu accumulatively developed 203 reclamation areas from tidal flatlands with a total area of 2686.7 km^2. Since 1949, for the purpose of socioeconomic development, land reclaimed from tidal flatlands at the Yangtze Estuary reached more than 1000 km^2 (100,000 ha). As a result, the area of Shanghai has been enlarged by 15.8%.

The utilization model of China's tidal flatlands has undergone a transformation process from "dike construction around an area from tidal flatland-aquiculture-farmland reclamation" to "aquiculture-dike construction around an area from tidal flatland-farmland reclamation" or from the sole agricultural utilization to comprehensive development. Presently, the models commonly used in coastal tidal flatlands are as follows:

① Comprehensive model of "agricultural fish pond" production-oriented agricultural development
② Comprehensive model of protective agriculture
③ Comprehensive model of coastal saline-alkali soil improvements
④ Comprehensive model of seawater intrusion prevention
⑤ Coastal farmland-forest network and forest shelter belt model
⑥ Comprehensive improvements model of coastal grassland
⑦ High-efficiency land-use model of three-dimensional cultivation on tidal flatlands
⑧ Tidal flatland aquiculture model
⑨ Urban park development model
⑩ Nonagricultural use of tidal flatlands (including tourism, salt industry, marine chemical industry, port, etc.) and other development models

Therefore, sediments and tidal flatland resources have played an important role in the socioeconomic development of the coastal areas.

6.2.4.4 Impacts of Water Resources Development on Estuarine Ecosystem

Although the utilization of tidal flatlands is an important contribution to the estuarine socioeconomic development, due to intensive use of the Yangtze River banks, coupled with the waterway regulation and passing ships, the estuarine ecological environment has been adversely affected. The water zone at the Yangtze Estuary (Shanghai section) has long been an excellent traditional fishery of the economic aquatic species such as Japanese grenadier anchovy, anchovy, Salanx prognathus, river crab, and eel fries. In the 1970s, the water zone was divided into nine fishery zones by the Shanghai Municipal Administration. Since 1970, the water zone at the Yangtze Estuary and coastal area has been gradually developed, and many large enterprises and major projects have been constructed along the coast of the Yangtze River such as the Shitousha Anchorage and Yuansha Anchorage, Baoshan Iron and Steel Plant along the coast of the Baoshan District, Pudong Economic Development Zone, Pudong International Airport, Yangtze Estuary Deepwater Waterway Project, Zhenhua Machinery Industry Company, and Jiangnan Shipyard along the southern coast of Changxing Island, Baoshan Chenheng Reservoir, and Qingcaosha Reservoir as well as the tunnel and bridge between Pudong and Chongming Island. All these developments have greatly reduced the area of the original traditional fishery zones. Among the nine fishery zones, zones II, III, and IV (the fishery along the southern bank of the Yangtze River between Bailonggang and Liuhe, and the fisheries along the southern edge of Changxing and Hengsha Islands) have completely disappeared; zones I, VI, and VII have also been affected; and zones VIII and IX along the southern edge of the west side of Chongming Island have not been major fisheries for nearly 10 years due to the changes of topography and water depth. According to preliminary estimates, the current fishery area is less than one-third of that in the 1970s. The fishery villages in Chongming County, especially Haixing Fishery Village in Changxing Township and Hengsha Township Fishery Village, that have long lived on fishing have been greatly impacted, since the two fishery villages used to fish in the original fishery zones II and III in the Yangtze River. Due to the development in the water zone, their fishing area has shrunk to zone I in the area of North and South Passages. Moreover, since 1998, due to the construction of the Yangtze Estuary Deepwater Waterway Project and the plan for the Shanghai Jiuduansha Wetland Nature Reserve, the area for fishing has become increasingly smaller, and the fishermen have become unable to continue their fishing operation.

The traditional fishing sites in the Yangtze Estuary (such as Baoshan, Waigaoqiao, southern edge of Changxing and Jengsha Islands, Xinhe of Chongming Island, North and South Passages) have or will have disappeared. As land reclamation in the north and east beaches of Chongming Island, the east beach of Hengsha Island, Pudong and the sandbanks along the southern edge of the river is being accelerated, the fishery in the water zone of the Yangtze Estuary is bound to decline. The Baoshan District has completed the transformation of the fishermen in Luojing, Yuepu, and Shengqiao who used to live on fishing in the Yangtze Estuary to other work.

Currently, the fishermen in the Yangtze River are mainly from Chongming, Changxing, and Hengsha Islands.

The densely populated Yangtze Estuary region is well-developed in economy and has a long history of transforming nature with high intensities. As the past urban and economic development has somewhat neglected the protection of the wetland ecosystem, the wetland ecology at the Yangtze Estuary has been subjected to a certain degree of destruction; the regional biodiversity index has been trending low; and the quality of the ecosystem has been declining. All this can be attributed to the environmental pollution caused by the regional economic development in the Yangtze Estuary and the irrational exploitation of natural resources such as tidal flatlands. According to Cao and others (2008), the existing major problems in protecting the wetland ecological environment of the Yangtze Estuary include the following:

1. Excessive Reclamation of Tidal Wetland and Overuse of Wetland Biological Resources

Presently, the wetland at the Yangtze Estuary has been seriously threatened by increasingly accelerated land reclamation. Land reclamation first destroyed wetland plants of primary producers, resulting in the reduction or loss of habitats, forage grounds, and spawning sites for waterfowls, birds, amphibians, reptiles, and fish species and leading to the reduction in number or even extinction of biological species. As the tidal and high tidal flatlands are the most concentrated in biodiversity and rich in resources, especially the high tidal flatlands that constitute the dominant area for food chain, they are the areas for the circulation of the biological materials and energy flow. The absence of tidal and high tidal flatlands is bound to affect the biodiversity of the region, resulting in the loss of other functions of wetlands. Over-reclamation of tidal flat wetlands has very destructive impacts on the wetland ecosystem. With the construction of the control cascade reservoirs in the upper reaches of the Yangtze River such as the TGR, and a large amount of sediments deposited in the reservoirs, the amount of sediments transported to the estuary has been greatly reduced, and the middle and lower reaches of the Yangtze River and the estuary have already entered the stage of sediment scarcity, which will not only affect land reclamation at the estuary but may also potentially cause the coastline erosion in the future.

The coastal wetlands in the Yangtze Estuary extend relatively far away from land and some wetland resources have been overused. For example, as local herdsmen have raised many farm cattle, beef cattle, and other livestock, overgrazing has damaged wetland plant resources, and some of the tidal flatlands have been trampled by cattle into barren "mudland." Moreover, local residents have uncontrolledly taken "shellfish" and "snails", resulting in a sharp decrease in birds' food stock. Harvesting of eel fries and crab fries and poaching of birds in the Yangtze Estuary have seriously affected the reserve of wetland biological resources, which has posed a threat to the ecological balance of wetlands in the Yangtze Estuary.

2. Decline of Fishery Resources

The unique saline-freshwater transition zone in the Yangtze Estuary is rich in aquatic resources. According to the ecological attribute of the population, the fishes in the Yangtze Estuary can be divided into four ecological types, namely, freshwater, seawater, brackish, and anadromous types. The fishery production of Shanghai Municipality is mainly from the Yangtze Estuary area and somewhat reflects the change of the aquatic ecosystem in the Yangtze Estuary. According to statistical data, the overall trend of the fishery production for Shanghai Municipality has been generally in a wavy decline. Except in the late 1990s when the fishery production showed a slight upward trend, the output was generally lower than the average production in the other years. The overall fishery resources in the Yangtze Estuary are in a general trend of decline. However, the trend of decline has slowed down in recent years, but the output has been stable only at a low level. This can be mainly attributed to the ongoing intensified fisheries management measures in recent 20 years.

3. Biodiversity Index Decreased and Quality of Ecological Environment in Decline

Presently, the biodiversity index in the Yangtze Estuary has reduced significantly and the number of biological species has also decreased. According to the number of observed species, the plankton species decreased by 69% from 1982 to 1998; the benthic organism reduced by 54% from 1992 to 1998; the benthic biomass dropped by 88.6% from 1982 to 1998; and the forage area has been declining. Some of the animal species under state Level I protection such as Chinese paddlefish, baiji, and roughskin sculpin are close to extinction, and the resources of the salanx prognathus, a historically important economic fish species, have almost been depleted. Overall, the environmental quality of the wetland ecosystem in the Yangtze Estuary area is trending down.

6.3 Reservoirs and Hydropower Stations in Yangtze River Basin

6.3.1 Basic Construction Work of Modern Water Conservancy and Hydropower Projects

The modern water conservancy and hydropower projects were developed based on modern hydrological monitoring, geodetic survey, and geological investigation. It would have been impossible to develop modern water projects such as reservoirs and hydropower stations without river hydrology, geodetic coordinates, and geological investigation.

China has a long history in establishment of hydrological stations, as early as in the third century BC, a "stone man" was set up at Dujiangyan, Sichuan, to monitor the water level. Thereafter, every dynasty installed some apparatuses for water level measurement such as "water ruler," "water recorder," "wooden staff," etc. for the purposes of flood control, navigation, irrigation, and other needs in the Yellow River, canals, Tai Lake, etc. Engraved marks of water levels during floods and droughts can be found throughout many watersheds. Because no continuous monitoring data were recorded from ancient survey points, the recorded data are not so valuable for modern water conservancy works, but they have some values for water culture. China's earliest hydrological monitoring began at the Hankou Station on January 1, 1865. By 1949, 280 water-level stations had been installed. Due to wars, only 104 hydrologic stations and 219 water level stations had been maintained in operation by the founding of the PRC, indicating that the basic hydrological work was weak at the time. For decades since the founding of the PRC, great progress had been made in the construction of hydrologic stations in China. By 2005, a total of 36,012 hydrological monitoring stations had been installed throughout the county, including 3191 hydrological stations (255 basic hydrological stations managed by other departments), 1166 water level stations, 14,373 rainfall stations, 4557 water quality stations, 12,313 groundwater monitoring stations, 349 evaporation stations, and 69 experimentation stations (Wang et al. 2006). In the Yangtze River basin (except the Tai Lake basin), there were 831 basic hydrological stations, which could basically meet the requirements for flood control, drought relief, and the construction of water conservancy and hydropower projects. The only deficiency was that the measured hydrological data series were short.

The geodesy technology was introduced from the West in the late Ming Dynasty, and China's geodetic network began to be installed in the period of the Kangxi era of the Qing Dynasty. During the Republic of China, the precision leveling survey reached 8100 km and the ordinary leveling survey reached 16,000 km. The technology for surveying waterways such as rivers was introduced after the 1840 opium war during which China was invaded by the Western powers. In 1843, the United Kingdom surveyed the sea map from Xuliujing to the East Sea and the waterway map of the Yangtze River from Yichang to Shanghai for the purpose of invasion to and trade with China. During the following decades, the survey of China's river channels was mainly performed by foreign powers. Due to the lack of techniques, professionals, and financial resources, China only performed partial waterway surveys. The survey of the Yangtze River channel was discussed in the Yangtze waterway seminar in 1922. Most of the systematic data have been collected from the existing hydrologic stations since the 1950s. Therefore, the measured hydrological series are all shorter than 100 years.

China's geological research and investigation began after 1840. Western scholars had initiated China's exploration, prospecting, and scientific investigation. After the Republic of China began to set up geological institutions, geologists such as Ding Wenjiang, Xie Jiarong, Ye Liangpu, and Li Siguang began to investigate China's geological characteristics and perform geological mapping and related scientific research. Many of the geological discoveries and basic evolutionary laws were established by the older generation of geologists during the 1920s through the 1960s.

In 1926, the first reinforced concrete sluice was constructed in China. The Minxin Sluice was installed for flood control on the Baputai Dike in Ezhou on the right bank of the middle reaches of the Yangtze River. In 1936, the 30-m-high Baimao Sluice was constructed in Changshu City of Suzhou. During 1910–1937, seven small hydropower stations were constructed, of which six were located in the Yangtze River basin. The first hydropower station is the Shilongba Hydropower Station on the upper reaches of the Tanglangchuan River at the outlet of Dianchi Lake in Yunnan Province. The construction started in July 1910 and power generation began in 1912. The original installed capacity was 480 kW (two 240 kW turbine-generator units, of which the generators were manufactured by German's Siemens and the turbines by Austria). After several expansions, the Shilongba Hydropower Station had become China's largest hydropower plant before the anti-Japanese war. By 1949, the installed capacity had been increased to 2920 kW. After the founding of the PRC, the hydropower station was renovated, including the change of turbine-generator units. By 1958, the installed capacity had been increased to 6000 kW. By 1949, the total installed capacity of all hydropower stations in the Yangtze River basin had reached 13,000 kW, and all of them were small hydropower plants. This total installed capacity was not comparable to that of just one unit in the United States at that time where mega-kW hydropower stations such as Grand Coulee and Hoover Hydropower Stations had been constructed, indicating how backward China was in hydropower development at the time.

6.3.2 *Construction of Reservoirs and Hydropower Stations*

6.3.2.1 China's Reservoirs and Hydropower Stations

To resolve the huge difference in the temporal distribution of incoming water in natural rivers, human beings need to construct reservoirs to ensure the safety of water supply and comprehensively utilize the reservoirs for multiple purposes, such as flood control, power generation, irrigation, and navigation. Globally, the total annual runoff amount of the world's rivers is about 55 trillion m^3, of which about 9 trillion m^3 can be used for hydropower generation. By 2007, the total capacity of global reservoirs had reached nearly 7 trillion m^3, of which about 4 trillion m^3 was the effective capacity, accounting for about 7.3% of the annual runoff amount of the world's rivers. The total area of the reservoirs was 500,000 km^2, equivalent to about one-third of the total area of all the natural lakes on the earth.

After the founding of the PRC, China's economy began to grow relatively fast, especially in the 1950s to 1960s when a large number of small- and medium-sized reservoirs and hydropower stations were constructed nationwide mainly for flood control, urban water supply, agricultural irrigation, and electricity need for economic development. By the end of 2007, a total of 86,353 reservoirs of all types had been completed with a total capacity of 692.4 billion m^3 (not including the reservoirs in Hong Kong, Macao or Taiwan), and China was ranked fourth in the world at the time (Jia et al. 2010).

China is a major energy consumer, but coal consumption accounted for more than 70%. By 2007, the installed capacity of coal-powered plants had reached 554 million kW, accounting for 77.7% of the total installed capacity, hydropower stations 148 million kW, accounting for 20.4%, nuclear power plants 9.068 million kW, accounting for 1.3%, and wind power and other renewable energy plants more than 6 million kW, accounted for only 0.8%. The proportion of the renewable energy was still very low. Worldwide, the total hydroelectric power generation in 2008 was 3.045 trillion kWh, accounting for 34.9% of the economically explorable capacity of 8.728 trillion kWh. Most developed countries had developed more than 60% of their hydropower potential, while most developing countries had less than 30% and Africa had less than 8%. China's economically explorable hydroelectric power generation is 2.474 trillion kWh annually, accounting for 28.3% of the world's, ranked first globally. By the end of 2008, China's hydroelectric power generation had reached 565.5 billion kWh, accounting for 18.6% of that of the world and 22.86% of nationally economically explorable potential, indicating that the potential for more development was still relatively high. In addition, by the end of 2007, more than 40,000 rural small hydropower stations had been constructed in China, equivalent to 36.62% of the potentially explorable installed capacity.

6.3.2.2 Reservoirs and Hydropower Stations in the Yangtze River Basin

By the end of 2009, the Yangtze River basin had completed 47,000 reservoirs of various types with a total storage capacity of more than 250 billion m^3 and a total beneficial storage capacity of more than 120 billion m^3. Of the 47,000 reservoirs, 166 were large reservoirs with a total storage capacity of 190.8 billion m^3 and a total beneficial storage capacity of 96.6 billion m^3. The total storage capacity and total beneficial storage capacity of large reservoirs accounted for 76% and 81% of those of all reservoirs, respectively, indicating that large reservoirs were not many in number, but their storage capacity was high in proportion. The total storage capacity and total beneficial storage capacity of the existing reservoirs on the Yangtze River accounted for 25% and 12%, respectively, of the average annual surface runoff amount of 985.6 billion m^3 and those of the large reservoirs accounted for 19% and 10%, respectively.

Among the administrative regions in the Yangtze River basin, Hunan Province has the largest number of reservoirs with a total of 12,344, followed by Jiangxi Province where there are 9824 reservoirs, and then Sichuan, Hubei, Yunnan, and Chongqing each have more than 3000 reservoirs. Table 6.2 summarizes the regional distribution and regulating parameters of existing large reservoirs in the Yangtze River basin.

Based on the reservoirs' control capacity of their runoff amounts, the Qing, Han, and Wu Rivers are on the top of the rankings, and the regulating capacity of the reservoirs in other water systems and tributaries was still relatively low, less than

Table 6.2 Distribution of large reservoirs in the Yangtze River basin

	Level II tributary	Large 1	Large 2	Total	Beneficial storage capacity (billion m³)	Flood control capacity (billion m³)	Annual runoff (billion m³)	Beneficial capacity/annual runoff (%)
Upper Yangtze	Mainstream	1	1	2	16.58	22.15		
	Yalong River	1	1	2	3.963	0.163	52.4	7.6
	Min/Tuo Rivers	2	4	6	5.26	1.03	106.6	4.9
	Jialing River	3	4	7	4.547	2.42	69.9	6.5
	Wu River	5	8	13	8.498	1.306	55.1	15.4
	Upper reaches	1	7	8	2.28	0.537		
Middle Yangtze	Qing River	2	2	4	4.828	1.244	14.1	34
	Han River	4	22	26	15.666	10.275	55.5	28.2
	Dongting Lake system	5	21	26	19.344	8.573	207.8	9.3
	Poyang Lake system	3	22	25	7.743	4.004	151.3	5.1
	Yichang-Hukou	3	30	33	5.513	3.511		
Lower Yangtze	Downstream of Hukou	3	11	14	4.138	3.555	46.2	10.0
Yangtze River Total		33	133	166	98.36	58.768	985.6	10.0

10%. Large reservoirs in the Yangtze River basin were mostly constructed on tributaries, and quite a lot of them are on secondary or even tertiary tributaries. In these tributaries, if large reservoirs are constructed, their control capacity of the river flow is high. For example, because Geheyan and Shuibuya large reservoirs were constructed on the Qing River, the regulating capacity of the two reservoirs on the Qing River is the highest. From the technical angle of flood control and hydropower utilization, it is generally necessary to build one to two large control reservoirs on a tributary, while on the mainstream of the Yangtze River, because it is long, and many tributaries emerge to it, it is necessary to construct more large reservoirs so as to effectively regulate the flow of the Yangtze River.

Because of the regional distribution and the relationship between the upstream and downstream reservoirs, regulating the upstream reservoir will directly affect the regulation of the downstream river section and reservoir, but the downstream reservoir does not have any direct influence on the control of the upstream water flow. The relatively prominent problems of flood control and drought relief in the middle and lower reaches of the Yangtze River need to be controlled by the regulation of the control reservoirs in the upper reaches, because a reservoir on a tributary can mainly solve the problem in the tributary, cannot solve the problem in the upper reaches, and cannot solve the problem in the middle and lower reaches either. For the drought-relief problem, the total existing beneficial capacity of large reservoirs in the upper reaches is only 41.1 billion m^3, which is 9.5% of the annual runoff amount in the upper reaches, and in the dry years, the proportion is even lower. Therefore, the existing reservoirs in the upper reaches do not have enough capacity to solve the problem associated with low water levels and droughts in the middle and lower reaches of the Yangtze River. For flood control in the middle and lower reaches, as the present flood control capacity of large reservoirs in the upper reaches is only 27.6 billion m^3, in the event of a large flood, the flood control capacity is still limited, and flood diversion zones will still need to be used.

Although there are many reservoirs in the Yangtze River basin, not many large reservoirs have a high regulating capacity, and they are geographically scattered in a large area. As a result, it was natural in the past for each large reservoir to be independently operated in accordance with its own plan, and a joint operation of a reservoir group was mainly limited to power generation scheduling and floodwater regulation during a flood. With the completion of the TGP and a number of large reservoirs in the upper reaches of the Yangtze River, the conflict between the independent operation of individual reservoirs and comprehensive utilization and the protection of the basin has become increasingly prominent, and a conjunctive operation has become growingly more urgent.

6.3.3 Planned Reservoirs and Hydropower Stations

Large hydropower stations planned to be constructed in the Yangtze River basin in the near future are mainly distributed on the mainstream of the Jinsha River, the Yalong, Dadu and Wu Rivers, Yibin-Yichang section on the mainstream in the upper reaches, the Jialing River, and tributaries in western Hunan. Presently, except the mainstreams of the Wu River and the rivers in western Hunan (Zi, Yuan, and Li Rivers) where the developed hydropower capacity over the hydropower potential is relatively high, the other rivers still have a great potential for development.

According to the plan for China's hydropower stations, by 2020, additional 63,930 MW of large hydropower stations will have been installed with an annual electricity generation of 281 billion kWh; additional 17,870 MW of medium-sized hydropower stations will have been installed with an annual electricity production of 83 billion kWh; and additional 11,500 MW of small hydropower stations will have been installed with an annual electricity yield of 55.2 billion kWh. Figure 6.4 is a distribution map of hydropower potential in the Yangtze River basin. Figure 6.4 indicates that the upper reaches are the richest in both theoretical hydropower potential and technically explorable hydropower potential. Large hydropower stations that have been planned to be constructed include Liyuan, Ahai, Jinanqiao, Longkaikou, Ludila, Guanyinyan, Jinsha, Yinjiang, Wudongde, Baihetan, and Xiaonanhai on the Jinsha River; Lianghekou, Yagen, Guandi, and Tongzilin on the Yalong River; Shuangjiangkou, Jinchuan, Houziyan, Changheba, Huangjinping, Yingliangbao, Dagangshan, Zhentouba, and the second stage of Shaping Power

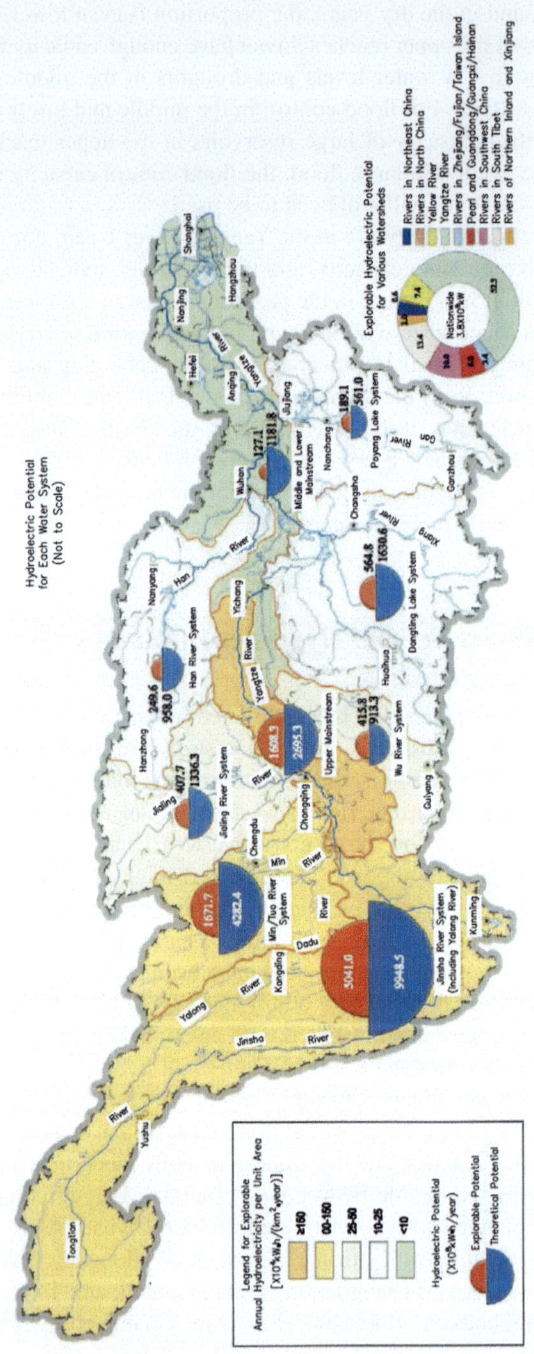

Fig. 6.4 Distribution of hydropower potential of the Yangtze River basin

Station on the Dadu River; Shilipu on the mainstream of the Min River; and Shatuo, Yinpan and Baima on the Wu River.

By 2030, additional 28,310 MW of large hydropower stations will have been constructed with an annual electricity generation of 129.6 billion kWh; additional 12,700 MW of medium-sized hydropower stations will have installed with an annual electricity generation of 63 billion kWh; and additional 13,000 MW of small hydroelectric power stations will have installed with an annual electricity production of 62.4 billion kWh. Large hydropower stations that have been planned to be constructed include Saiwu (Shaila), Enan, Baiqiu (Baili), Jiangquhekou, Batang, Wangdalong (Suwalong), and Rimian on the Jinsha River; Renqingling, Yingda, Xinlong, Gongke, Gongbagou, Lenggu, Mengdigou, Yangfanggou, and Kala on the Yalong River; and Xiaerga, Bala, Dawei, Busigou, Badi, Danba, and Laoyingyan on the Dadu River.

In addition, a large number of medium-sized and small hydropower stations have been or are being constructed on most secondary and tertiary tributaries of the Yangtze River basin. The planning of hydropower development in these tributaries has been basically completed. The impact on the ecological environment of small- and medium-sized rivers is significant.

6.4 Utilization of Water Resources in the Yangtze River Basin

6.4.1 Distributional and Regional Characteristics of Water Resources in the Yangtze River Basin

The Yangtze River basin is rich in total quantity of water resources. Table 6.3 summarizes the quantity and utilization of water resources in secondary regions in the basin in 2007.

In the secondary regions of the Yangtze River basin, the Dongting Lake and Poyang Lake watersheds are the richest in total quantity of water resources. The Jinsha River basin is the richest in quantity of water resources per capita. The Tai Lake watershed is the highest in water use, followed by the Dongting Lake system and the Yangtze River's mainstream section downstream of Hukou. In terms of the source of water use, the Tai Lake watershed and the Yangtze River's mainstream section downstream of Hukou (mainly in Jiangsu) primarily use water from outside their regions, while the Dongting Lake watershed principally use the local water. In terms of the utilization rate of water resources, the Tai Lake region exceeded 100%, indicating that the quantity of water use in the region is well above the availability of local water resources and mainly relies on the water from outside the region to meet the water demand for the regional socioeconomic development. The Han River basin is second in the utilization rate of water resources. If the Central Route of the South-to-North Water Diversion Project and the Yangtze River-Wei River Water

Table 6.3 Utilization of water resources in secondary regions of the Yangtze River basin (2007)

Secondary region	Precipitation (billion m^3)	Total water resources quantity (billion m^3)	Water resources quantity per capita (m^3)	Total water use (billion m^3)	Utilization rate (%)	Proportion of agricultural use (%)
Jinsha River section upstream of Shigu	104.69	41.61	58,494	0.22	0.5	85
Jinsha River section downstream of Shigu	234.18	114.91	5482	6.66	5.8	64.3
Min/Tuo Rivers	175.47	106.61	3015	11.92	11.2	58.3
Jialing River	149.03	69.88	1614	8.28	11.8	53.6
Wu River	101.0	55.11	2488	5.36	9.7	45.5
Yibin-Yichang	114.92	63.48	2015	7.7	12.1	31.8
Dongting Lake system	375.31	208.59	2834	36,85	17.7	62.9
Han River	139.95	57.32	1638	12,74	22.2	55.1
Poyang Lake System	267.01	153.25	3782	21,61	14.1	68.8
Yichang-Hukou section	120.76	59.0	1728	16,74	28.4	49.4
Mainstream Downstream of Hukou	111.27	48.47	1085	26,73	55.1	47.6
Tai Lake system	43.44	17.6	388	37,76	214.5	24.9
Total	1937.03	995.83	2331	192.55	19.3	50.0

Transfer Project are considered, the utilization rate of water resources for the Han River basin would be close to 50%. Therefore, the Tai Lake watershed and the Han River basin are the highest in the utilization rate of water resources.

From the structure of water use, the Tai Lake watershed and the upper mainstream of the Yangtze River have a relatively low proportion of agricultural water use. The Tai Lake watershed has the highest proportion of industrial and domestic water use, indicating that the region is more economically developed. However, the upper mainstream of the Yangtze River runs through mountainous areas where the agricultural water use is not directly withdrawn from the mainstream but mainly from secondary or tertiary tributaries. There is almost no industry in the upper Jinsha River and the domestic water use is limited; and therefore, the proportion of the agricultural water use is relatively high, but the actual total water use is very low. The Dongting Lake and Poyang Lake watersheds still have a relatively high proportion of the agricultural water use.

Table 6.4 Water use data of each administrative region in the Yangtze River basin (2007)

Province (autonomous region or municipality)	Area (km^2)	Population (thousand)	Water use (billion m^3)	Proportion of agricultural use (%)	Mean annual water resources (billion m^3)	Water use/ local water (%)
Tibet	22,933	150	0.073	93.2	8.3	0.9
Qinghai	158,392	250	0.027	70.4	17.94	0.2
Yunnan	109,525	8200	4.606	65.3	42.41	10.9
Sichuan	467,292	81,170	21.372	58.3	256.82	8.3
Chongqing	82,401	28,160	7.744	26.8	56.78	13.6
Guizhou	115,747	27,460	7.207	51.4	67.99	10.6
Gansu	38,484	3400	0.353	70.8	10.04	3.5
Hubei	184,545	56,650	25.790	52.8	103.04	25.0
Shaanxi	72,436	9970	2.452	82.3	30.67	8.0
Henan	27,609	16,210	2.292	46.3	7.13	32.1
Hunan	206,712	63,300	32.006	61.7	164.05	19.5
Guangxi	8399	1320	1.409	87.3	9.76	14.4
Guangdong	339	30	0.017	82.4	0.28	6.1
Jiangxi	163,143	42,880	23.348	65.6	153.21	15.2
Fujian	1050	110	0.073	86.3	1.07	6.8
Anhui	66,097	23,910	12.345	50.2	42.0	29.4
Jiangsu	38,484	36,100	33.345	31.7	13.07	255.1
Zhejiang	12,877	9420	6.070	51.8	8.44	71.9
Shanghai	6341	18,580	12.019	13.7	2.75	437.0
Yangtze River basin total	1,782,807	427,270	192.548	50.0	995.83	19.3

Table 6.4 provides the water use data for each administrative region in the Yangtze River basin.

Table 6.4 indicates that Jiangsu, Hunan, Hubei, Jiangxi, and Sichuan are the largest water users. All the five provinces but Jiangsu have more than 95% of their areas located in the Yangtze River basin, and their total quantity of water use is largely normal. In terms of the structure of water use, where the amount of agricultural water use accounts for more than 70% in the Yangtze River basin, the total area of the region is small, or there are not many important cities or industries. Shanghai, Chongqing, and Jiangsu, where the proportion of agricultural water is less than 40%, are more economically developed regions. From the proportion of water use over the local water resources, the quantity of water use in Shanghai and Jiangsu is much more than the available quantity of local water resources, and they mainly rely on the water from the upstream area to sustain their socioeconomic development. Where the utilization rate of the local water resources is low, the population is small, or the economic development is slow.

6.4.2 Use Efficiency of Water Resources in the Yangtze River Basin

6.4.2.1 Existing Water Use Conditions

Since 1980 when statistical data were available, the quantity of water use per capita in the Yangtze River basin has largely maintained at about 400 m^3 and has trended up from west to east. From the efficiency of water use, the amount of water use for generating every 10 thousand yuan of gross domestic product (GDP) in the Yangtze River basin decreased from 2975 m^3 in 1980 to 240 m^3 in 2009, a reduction of 92%; the quantity of water use for generating every 10 thousand yuan of added industrial value decreased from 946 m^3 to 190 m^3, a reduction of 80%. Table 6.5 presents the major water use values in secondary water resources regions of the Yangtze River

Table 6.5 Water use efficiency values of secondary regions in Yangtze River basin in 2009 (unit: m^3)

Secondary region	Water use per capita	Water use per 10 thousand yuan of GDP	Urban domestic water use PER capita	Water use per 10 thousand yuan of added industrial value	Mean irrigation water use per Ha	Rural domestic water use per capita
Jinsha River section upstream of Shigu	304	569	115	170	6225	58
Jinsha River section downstream of Shigu	315	253	117	129	6465	71
Min/Tuo Rivers	328	186	122	120	6075	66
Jialing River	187	195	119	159	3675	57
Wu River	247	271	125	251	7665	76
Yibin-Yichang	257	223	140	272	3900	63
Dongting Lake system	493	367	164	222	7650	112
Han River	382	265	121	193	5370	63
Poyang Lake System	527	459	146	237	8070	80
Yichang-Hukou	501	307	142	270	6990	79
Mainstream Downstream of Hukou	640	272	149	225	509	79
Tai Lake system	678	118	166	130	445	145
Yangtze River basin total	431	240	142	190	447	81

Note: GDP and added industrial revenue used in calculations of water use are in 2000 price

basin in 2009. In the Tai Lake region, the amount of water use and urban domestic water use per capita was the highest while the quantity of water use for generating every 10 thousand yuan of GDP was the lowest, indicating the efficiency of water use was the highest.

6.4.2.2 Gap from Advanced Regions and Countries

6.4.2.2.1 Comparison with Other Primary Regions in China

Table 6.6 is the water use comparison of China's primary water resources regions from the *China Water Resources Bulletin* (2009). Table 6.6 indicates that the amount of water use for generating every 10 thousand yuan of added industrial value in the Yangtze River basin was clearly higher, but other parameter values of water use were close to the national averages; when compared with rivers in north China such as the Hai, Yellow, and Huai Rivers, all major parameter values of water use in the Yangtze River were behind; when compared with the Pearl River, except the amount of water use for generating every 10 thousand yuan of added industrial value that was clearly higher in the Yangtze River basin, all the other parameter values were close.

Table 6.6 Comparison of water use among China's Primary River basins in 2009 (unit: m^3)

Primary water resources region	Water use per capita	Water use per 10 thousand yuan of GDP	Urban domestic water use	Water use per 10 thousand yuan of added industrial value	Agricultural irrigation water use per Ha	Rural domestic water use
Nationwide	448	178	212	103	6465	73
Songhua River	682	284	160	131	6855	54
Liao River	365	115	201	39	6375	64
Hai River	267	80	166	26	3480	57
Yellow River	344	135	159	41	5865	49
Huai River	320	126	164	41	4200	63
Yangtze River	451	174	230	152	6765	71
Pearl River	497	171	302	95	11,745	122
Southeast Rivers	444	116	206	92	7860	130
Southwest Rivers	499	429	160	169	7875	72
Northwest Rivers	2061	928	180	61	9990	61
Lake Tai	689	79	307	129	7005	104

Note: GDP used in calculations of water use is in 2009 price

6.4.2.2.2 China's Water Use Ranking in the World

In 2009, China's water use quantity of 596.5 billion m^3 was only second to India in the world; the average quantity of water use per capita of 448 m^3 was 74% of the world average; and the proportion of agricultural water use of 62.4% was 5% lower than the world average. The world's average quantity of water use for generating every 10 thousand dollars of GDP was 711 m^3, while China's was 1197 m^3 that was 1.7 times the world average. The world's average amount of water use for generating every 10 thousand dollars of added industrial value was 568 m^3, while China's was 603 m^3 that was slightly higher than the world average. The above-listed parameter values of water use indicate that China's quantity of water use for general industries was large; the efficiency of water use was not high; the contribution to GDP of the tertiary industries was relatively small; and therefore, there was still a relatively large gap in industrial water use between China and the world. In terms of the utilization rate of irrigation water, Israel was 87%; Australia was 80%; France was 73%; Russia was 78%; the United States was 54%; and China was 46%, indicating that China was also at a relatively lower level.

6.4.2.2.3 Comparison with Developed Countries

Table 6.7 presents a comparison in the utilization efficiency of water resources between China and major developed countries. China's average quantity of water resources per capita was only higher than Israel, South Korea, and Germany. Among the countries with water resources scarcity, China's average amount of water use per capita was ranked fourth from the bottom. The proportion of agricultural water use over the total water use was only lower than that of the arid Australia and Spain, indicating that the proportion of agricultural water use in China was still high. China's water use quantity for generating every 10 thousand dollars of GDP was the largest, and the water use for generating every 10 thousand dollars of added industrial value was only lower than Canada, indicating that there was still a relatively large gap in water use efficiency between China and the developed countries. In addition, China's reuse rate of industrial wastewater was 62%, 20% lower than the developed countries. All these parameter values indicate that China would still have relatively large room to adjust the industrial structure and improve the efficiency of water use.

6.4.2.3 Means to Improve Water Use Efficiency

To establish a water-conserving society, China has set forth the following water-conserving goals (Xu et al. 2011). ① By 2030, the national average amount of water use for generating every 10 thousand yuan of GDP will have been reduced to below 70 m^3 or 75% lower than the current level; the efficiency and effectiveness of water use will have been significantly improved; and major water use parameter values

Table 6.7 Comparison in utilization efficiency of water resources between China and developed countries (2008 Data)

Country	Quantity of water resources per capita (m^3)	Mean GDP per capita (dollar)	Average water use per capita(m^3)	Proportion of agricultural use (%)	Water use per 10 =thousand dollars of GDP (m^3)	Water use per 10 thousand dollars of added industrial value (m^3)	Mean urban domestic water use per capita (L)
Japan	3373	36,612	693	62	189	104	374
United States	10,491	33,575	1647	41	491	935	573
Netherland	5,65 1	23,9 15	493	34	206	496	187
Germany	1867	23,037	571	20	248	554	193
United Kingdom	2475	24,286	161	3	66	177	285
France	3409	22,189	668	10	301	980	287
Canada	92,692	23,154	1469	22	635	1313	355
Italy	3290	18,901	765	45	405	523	38 1
Australia	25,097	20,664	1219	75	590	220	492
Israel	281	19,100	289	56	151	27	298
Spain	2682	14,03 1	860	68	613	389	3 17
South Korea	L,474	10,822	393	48	363	146	384
China	2100	3300	446	63	1340	890	131

will have reached or been close to the international advanced levels. To achieve this goal, China will need not only to greatly expand the tertiary industry that uses relatively less water but also improve the efficiency of industrial and agricultural water use. Otherwise, it will be difficult to achieve the goal. ② The agricultural water-conserving goal will be that, by 2030, the water-conserving irrigated area will have accounted for more than 80% of the total effective irrigated area; the water utilization coefficient will have been improved to 0.6; and the average amount of irrigation water use of farmland will have reduced to 5775 m^3/ha. This goal is particularly important to the northern regions, while the southern regions can achieve the goal by mainly adjusting crop planting structures. ③ The industrial water-conserving goal will be that, by 2030, the amount of water use for generating every 10 thousand yuan of added industrial value will have reduced to below 38 m^3 and the amount of water use per unit major product for high water-use industries will have reached or been close to the international advanced level. This will be achieved mainly through improvements of production technology and water reuse rate. ④ The urban domestic water-conserving goal will be that, by 2030, the average leakage rate of the national urban water supply network will have decreased to below 11% and the water use efficiency for the service sector will have reached the international advanced level.

The water-conserving goal in the middle and lower reaches of the Yangtze River basin will be that, by 2030, the amount of water use for generating every 10 thousand yuan of added industry value will have reduced to 50 m^3 (presently at 204 m^3); the reuse rate of industrial wastewater will have reached 87% (currently at 64%); the effective utilization coefficient of irrigation water will have reached 0.62 (currently at 0.5); and the leakage rate of the urban water supply network will have decreased to 11% (currently at 20%).

Although the utilization level and efficiency of water resources in China have been improving continuously, the means to utilize water resources is still extensive and the efficiency is low. When compared with the international advanced level, the gap is still large and the potential to conserve water is high. Conserving water, on the one hand, can reduce waste and ineffective loss and control the uncontrolled growth of water use; on the other hand, part of the conserved water can be used to increase water supply, to expand the agricultural and industrial production scale, or to improve the ecological environment through changing the means of water use.

There are four main means to conserve water. First, water use can be reduced through rational adjustments of the industrial structure and distribution, technological advancement, and improvements of technology and agronomics. In the meantime, through adjustments of the industrial structure, water use can be transferred from low-productivity uses to high-productivity consumptions so as to improve the economic output per unit water use. Second, significant improvements in the efficiency of industrial and domestic water use can be achieved through the improved water demand management, rational adjustment of water price, and market-regulating mechanism. Third, the efficiency of irrigation water use can be improved through structural water-conserving measures, primarily including the implementation of water-conserving renovations on irrigation districts and the development of efficient water-conserving irrigation technologies. The reuse rate of industrial wastewater can be improved through industrial water-conserving renovations. Water conservation can also be achieved through improving urban water supply system to lower the leakage rate and promoting the use of water-conserving equipment. Fourth, a good social manner of water conservation can be established through publicity and public education to promote the public awareness of water conservation and encourage everyone to conserve water.

6.5 Interbasin Water Transfer Projects

If the construction of a reservoir is to resolve the uneven temporal distribution of the incoming water in a river, an interbasin water transfer project is mainly to resolve the uneven spatial distribution of precipitation and the incoming water of rivers. Due to various reasons such as history, arable land distribution, transportation, and culture, when people settled to live in a place in the past, they generally did not consider the restraining role of water resources in socioeconomic development. As a result, the population density and socioeconomic development in many areas were not consistent with the distribution of water resources. Because of the recent advancement of economic development and technology, the construction of interbasin water transfer projects can resolve the issues associated with shortages of water resources. The following is a brief description of several representative interbasin water transfer projects and the existing problems using the Yangtze River basin as the water source.

6.5.1 South-North Water Transfer Project

6.5.1.1 Project Scale and Significance

The general plan for the South-to-North Water Transfer Project (also known as South-North Water Diversion Project) was to connect the Yangtze River to the Yellow, Huai, and Hai Rivers through three water transfer routes: Eastern, Central, and Western Routes to form a basic "national water supply network" consisting of "four east-west lines and three south-north lines." This network would facilitate the redistribution of China's water resources from south to north and between east and west. The planned three routes would have an annual capacity to transfer 44.8 billion m^3 of water by 2050, of which 13 billion m^3 will have been transferred through the Eastern Route (8.9 billion m^3 in Phase I, 10.6 billion m^3 in Phase II, and 14.8 billion m^3 in Phase III), 13 billion m^3 through the Central Route (9.5 billion m^3 in Phase I and 13 billion m^3 in Phase II), and 17 billion m^3 through the Western Route (4 billion m^3 in Phase I, 9 billion m^3 in Phase II and 17 billion m^3 in Phase III). The entire project would be implemented in phases based on the actual situation. Presently, the feasibility study-level cost estimate of Phase I for the Eastern and Central Routes would be about 254.6 billion yuan.

The South-to-North Water Transfer Project is of great significance to China's socioeconomic development. From the allocation strategy of water resources, the project can solve the water scarcity problem in north China, increase the carrying capacity of water resources, and make north China gradually become a water-conserving and antipollution society with a rational allocation of water resources and a healthy aquatic environment. The project would not only alleviate water resources scarcity situation in north China or the constraints of the urbanization process and promote the local urbanization process but also ensure the water source for the year-round navigation through the Jining-Xuzhou section of the Beijing-Hangzhou Grand Canal and strengthen and expand the commodity grain bases in western Shandong and northern Jiangsu. From socioeconomic development, the project would provide water security for the economic development in north China and promote the strategic adjustment of the economic structure. The project would effectively address the naturally occurring groundwater quality problems, such as high fluoride-content water, brackish water, and other problems associated with water sources that contain substances harmful to the human body. In the meantime, the project would increase potential productivity through improved water resources conditions and develop new economic growth. The construction of the project would help increase domestic demands and promote regional economic development. From the ecological environment, the project would help improve the regional ecological environment of the Yellow, Huai, and Hai River basins, increase recharge to groundwater, preserve local wetlands and biodiversity, restore the degraded environment caused by water scarcity, and greatly improve the ecology and environment, especially the conditions of water resources, in north China.

Of course, attention needed to be paid to problems associated with the project. First, the scale of the project would be large and the cost would be huge, but the direct economic return was not high; second, the construction of the project would result in encroachment into a lot of lands, relocation of a large number of residents, and major social impacts; third, the project would need many secondary projects to be constructed along the routes and would result in fragmentation of terrestrial habitat and major impacts on the ecological environment; and fourth, the relationship between the project operation and water price would be complex and there would be many challenges as to how to improve the operation and efficiency of the project.

6.5.1.2 Eastern Route Project

The goal of the Eastern Route Project would be to lessen the water resources shortage conditions in Jiangsu, Anhui, Shandong, and Hebei Provinces and Tianjin Municipality. Water would be withdrawn and lifted at a pump station in Yangzhou from the lower Yangtze River, transmitted northward through the Beijing-Hangzhou Grand Canal and a parallel river channel, lifted by booster stations along the route, and flow through Hongze, Luoma, Nansi, and Dongping Lakes that have regulating and retaining functions. Out of Dongping Lake, the route would be divided into two branches. One would run all the way to the north and cross the Yellow River through a tunnel near Weishan and the other would proceed eastward and divert water to Yantai and Weihai through Jinan using the Jiaodong water transmission pipeline. The construction of the Eastern Route Project started the earliest, and it was relatively small because there had been existing water transmission waterways. The project would include facilities of water conveyance, water storage, and power supply.

6.5.1.2.1 Water Conveyance Project

The water conveyance project would include conveyance channels, pump stations, and the Yellow River crossing facility.

1. Water Conveyance Channels

There would be two diversion inlets: Sanjiangying Inlet from the Huai River to the entrance of the Yangtze River and Liuyu Inlet from the Beijing-Hangshou Grand Canal to the entrance of the Yangtze River. The primary main line of the water conveyance channel would be 1150 km long from the Yangtze River to Tianjin, of which 651 km would be located south of, 9 km across and 490 km north of the

Yellow River. The total length of the secondary main line would be 740 km, of which 665 km would be located south of the Yellow River.

2. Pump Stations

The terrain along the Eastern Route can be characterized by the Yellow River as the backbone from which the terrain slopes toward the north and the south. The water intake site is 40 m lower in elevation than the Yellow River. Thirteen cascade booster stations would be needed to lift water from the Yangtze River to the south bank of the Yellow River with a total lift head of 65 m. After the water crosses the Yellow River, the water could feed Tianjin by gravity. In addition to three booster stations installed among the four lakes south of the Yellow River (one between the upstream and downstream lakes), three cascade booster stations would be installed in each river section. Thirty booster stations would be installed along the main conveyance line south of the Yellow River: 13 on the primary main line and 17 on the secondary main line. The total design pumping capacity of the booster stations would be 10,200 m^3/s, and the required installed motor capacity would be 1.0177 million kW, of which existing seven booster stations would be used with a design pumping capacity of 1100 m^3/s and an installed motor capacity of 110,500 kW. The first phase of the project would still have 13 cascades with 23 booster stations and installed motor capacity of 453,700 kW, of which 5 booster stations would be installed at each shallow water storage inlet lake north of the Yellow River with a design pumping capacity of 326 m^3/s and an installed motor capacity of 14,600 kW.

The booster stations along the Eastern Route of the South-to-North Water Diversion Project can be characterized by low lifts (mostly at 2–6 m), high flowrates (single pump flowrate generally at 15–40 m^3/s), and long running time (about 5000 h/year per booster station south of the Yellow River). Some booster stations would serve as drainage pumps as well, and, therefore, the booster stations would be required to be operated with flexibility and efficiency.

3. Yellow River Crossing Project

Tunnels were selected to be constructed between Dongping and Donga Counties in Shandong Province to underpass the Yellow River. The Yellow River Crossing Project would be 8.67 km long from the outlet sluice of Dongping Lake to the entrance of the Beijing-Hangzhou Grand Canal, of which the inverted siphon tunnel section across the Yellow River would be 634 m and horizontal tunnel sections would be 70 m under the Yellow River bottom, consisting of two tunnels each with a diameter of 9.3 m.

6.5.1.2.2 Water Storage Project

Hongze, Luoma, Nansi, and Dongping Lakes are along the Eastern Route south of the Yellow River. After slight modifications and reinforcement, the lakes would have a total regulating capacity of 7.57 billion m^3. As a result, no additional water storage works would be needed. North of the Yellow River, Tianjin's existing

Baidagang Reservoir would continue to be used; Tianjin's Tuanbo Depression and Hebei's Qianqing Depression would need to be expanded; and five new storage facilities north of the Yellow River, such as Hebei's Dalang Lake and Lang Depression, would be constructed with a total regulating capacity of 1.49 billion m^3.

6.5.1.2.3 Power Supply Project

There would be 30 booster stations along the route south of the Yellow River; new installed motor capacity would be 887,700 kW; and average annual electricity consumption would be 3.82 billion kWh with the highest annual electricity consumption of 5.75 billion kWh. In the first phase of the project, 23 booster stations would be constructed with new installed motor capacities of 343,200 kW and an average annual electricity consumption of 1.9 billion kWh.

The current major problems associated with the Eastern Route Project are that the water quality of the Beijing-Hangzhou Grand Canal and the storage lakes is poor, and the aquatic environment needs to be improved. Moreover, since the project requires to lift water through booster stations, the operational cost would be high.

6.5.1.3 Central Route Project

The Central Route Project would withdraw water from the Danjiangkou Reservoir on the Han River after the dam was raised with an increased storage capacity. The water would flow through the first sluice at the headworks of the Taocha Canal (at Jiuchong Town of Xichuan County of Henan); run through the Fangcheng Pass on the divide between the Yangtze and Huai River basins along the western Tangbai River basin in southwest Henan; then follow the western edge of the Huang-Huai-Hai Plain; cross the Yellow River at Gubozui west of Zhengzhou; and then flow by gravity along the west side of the Beijing-Guangzhou Railroad to the end of the route in Beijing. The Central Route Project would mainly supply water to approximately 20 cities in Henan, Hebei, Tianjin, and Beijing. The construction of the Central Route Project commenced on December 30, 2003, and was scheduled to be substantially completed at the end of 2013 and convey water after the 2014 flood season.

The Phase I of the Central Route of the South-to-North Water Diversion Project would be composed of three major components: water source area expansion, water conveyance facility, and middle and lower Han River improvements. The improvements in the water source area were the continuation of the late-stage construction of the Danjiangkou Water Conservancy Project, and the conveyance facilities would consist of the main trunk canal to divert water from the Han River and the Tianjin Main Canal.

6.5.1.3.1 Facility Expansion in the Water Source Area

The Danjiangkou Reservoir controls 60% of the Han River basin with an average annual natural runoff amount of 40.85 billion m^3. After considering the upstream utilization of water resources and the impact of the water transfer project, it was projected that the inflow to the reservoir would be 38.54 billion m^3 in 2020. This would involve raising the height of the dam from the crest elevation of 162 m to 176.6 m. The addition to the dam's height would allow the design water level in the reservoir to rise from 157 m to 170 m with a total storage capacity of 29.05 billion m^3, resulting in an additional storage capacity of 11.6 billion m^3. Moreover, the raised dam height would increase the effective regulating capacity of the reservoir by 8.8 billion m^3 and the effective flood control capacity by 3.3 billion m^3.

6.5.1.3.2 Compensation Project in the Middle and Lower Han River

In order to lessen the adverse effect of the water transfer project on the middle and lower reaches of the Han River, four compensation projects were planned. They are the Xinglong Water Conservancy Project, Yangtze-Han River Diversion Project, modification of some sluices, and local navigable waterway improvements. The Xinglong Water Conservancy Project was to raise the water level of the reservoir during the dry season and improve the irrigation conditions in the riparian zone and the shipping conditions in the reservoir area. The Yangtze-Han River Diversion Project was to convey water from Longzhouyuan on the Jingjiang River section of the Yangtze River to Gaoshibei on the Qian River section of the Han River, and the full length of the conveyance system would be 67.1 km. The project would help meet the ecological water demand, irrigation, water supply, and navigation for the river section downstream of Xinglong on the Han River, basically solve the "algae bloom" problem in the lower Han River resulting from the first phase of the Central Route water transfer project, solve the irrigation water source problem for the Dongjing River, and restore the water level to a certain extent to improve the navigation safety of the lower Han River. The reconstruction of some sluice stations and modification of the local navigable waterway were mainly aimed to solve the problems of water intake structures associated with aged sluices and river regulation to adapt to the new conditions in the middle and lower reaches of the Han River.

6.5.1.3.3 Water Conveyance Facility

Main Canal The main trunk route south of the Yellow River was restricted by the location of the canal headworks, the Fangcheng Pass of the drainage divide between the Yangtze River and Huai River basins, and the area of the Yellow River crossing. However, the direction of the route was well defined. Two alternatives had been compared for the conveyance route north of the Yellow River: using the existing river channel or constructing a new canal. After considering the water quality and gravity flow conditions, the highline new canal alternative was chosen.

The main canal would withdraw water from the Taocha Canal, extend along the existing 8-km long canal, run northeastward between hillocks and plains along the southern foot of the Funiu Mountains, cross the Tangbai River in Nanyang, and traverse the divide between the Yangtze River and Huai River basins through the Fangcheng Pass into the Huai River basin. The main canal would continue to run through Baofeng, Yuzhou, and Xinzhengxi and cross the Yellow River at Gubaizui northwest of Zhengzhou. Then the main canal would proceed on the plain along the eastern foot of the Taiheng Mountains and the west side of the Beijing-Guangzhou Railroad to the hilly area in Tang County and then cross the Beijuma River into the territory of Beijing. After the canal crosses the Yongding River, it would enter the urban area of Beijing and reach the end at Yuyuantan. The total length of the main canal would be 1241.2 km.

The main canal would connect the Yangtze, Huai, Yellow, and Hai River basins; would cross the mainstream of the Yellow River, 129 other rivers whose catchment area are above 10 km^2 and 44 locations of railroads; and would need to build 571 bridges over highways. Moreover, 936 structures would need to be constructed along the main canal, including control gates, diversion sluices, water release structures, tunnels and culverts, etc., of which the largest structure would be the Yellow River crossing facility. The Tianjin Main Canal would cross 48 large and small rivers and need to construct 119 structures.

Yellow River Crossing Facility The main canal had been planned to cross the Yellow River basin through the Taohuaxia Reservoir. The crossing facility would be large in scale, complex in issues, and high in investment and would be the most critical facility of the main canal. A comparison of multiple alternatives indicated that an aqueduct and inverted siphon tunnels would be both technically feasible. Since the tunnel alternative would avoid nonconformance with the regime and planning of the Yellow River and the tunneling shield construction technology has been successful both at home and abroad, it was recommended that the tunnel alternative be used for the construction of the river crossing at Gubaizui based on the canal alignment on both sides of the river. The Yellow River crossing facility would be about 7.2 km long; its design water conveyance capacity would be 500 m^3/s; and two 8.5 m inner diameter tunnels would be constructed.

Because of the need to raise the Danjiangkou Dam for the Central Route Project, the main canal route would be long, and the number of relocated people would be as many as 345,000.

6.5.1.4 Western Route Project

The planned Western Route Project was aimed to divert water from the Tongtian, Yalong, and Dadu Rivers in the upper Yangtze to the upper Yellow. Dams would be constructed on the Tongtian, Yalong, and Dadu Rivers, and a tunnel would need to be constructed through the Bayan Har Mountains to cross the drainage divide between the Yangtze River and the Yellow River basins. The project was intended mainly to solve water-scarcity problems in the six provinces (autonomous regions) of Qinghai, Gansu, Ningxia, Inner Mongolia, Shaanxi, and Shanxi in the upper and middle Yellow and the Weihe (or Guanzhong) Plain. Combined with the construction of the major water conservancy projects on the mainstream of the Yellow River, the project would also be able to supply water to the Hexi Corridor in Gansu Province near the Yellow River basin and replenish the lower reaches of the Yellow River if necessary. Presently, the project is still in the preliminary planning and research stage and there is no timetable to start construction.

According to the *Planning Outline of the Western Route of the South-to-North Water Diversion Project and Project Planning for Phase I*, the water diversion plan for Phase I would be to pump water from the Yalong River with the exception that no water diversion would occur during December through the following March. The diversion plan from the Dadu River would be to withdraw water at Xieyijia for 9 months annually, and no diversion would occur between December and the following February. Diversion via gravity would occur for 10 months annually and no diversion would occur between December and January. When water is urgently needed during the dry season, no water could be diverted into the Yellow River basin; but during the flood season, more water would be diverted into the river where there is no detention/retention area to regulate the diverted water, which would put additional pressure of flood control to the Yellow River.

Major problems with the Western Route Project included the following. First, the project is located in the Qinghai-Tibet Plateau where the altitude is high; the geological structure is complex; and the intensity of earthquake is high. Moreover, as an approximately 200-m-high dam and an up to 100-km-long tunnel would need to be constructed, the engineering work of the project would be complex, and the cost would be huge. Second, as the project is located in a fragile and sensitive area of China's western ecological environment, the impact on the ecological environment would be significant. Third, after the diverted water is mixed with the water in the Yellow River, water rights status would be complicated, and it would be difficult to determine the beneficiaries and economic benefits. Fourth, the future water use in the source area of the Yangtze River would be in conflict with water transfer for the Western Route Project, and the water transfer would somewhat impact electricity generation of the cascade hydropower stations in the upper Yangtze River.

6.5.2 *Yangtze River-Tai Lake Water Transfer Project*

The Tai Lake watershed crosses Jiangsu, Zhejiang, and Shanghai, has a watershed area of 36,900 km^2, and is one of China's most economically developed areas. The Tai Lake basin has insufficient local water resources where the amounts of water resources per capita and per ha are only one-fifth and one-half, respectively, of the national average. Moreover, as the intensity of water pollution control and water resources protection is far below that of the socioeconomic development in the basin, the conditions of the aquatic environmental quality are not promising. In 2005, 89.3% of the rivers in the basin were assessed to be poorer than Class III in water quality; only 18.6% of the major water function areas were in compliance with water quality requirements; and 34% of the evaluated 47 surface drinking water sources were poorer than Class III in water quality.

After the 1991 large flood, the Tai Lake Key Mitigation Project began to be implemented, which provided the conditions for the construction of the Yangtze River-Tai Lake Water Diversion Project. The diversion project used the completed Wangyu River Project to divert water from the Yangtze River into the river network and Tai Lake and then through the Taipu River, Dongshao Creek, and the dredged outlets on the dike around Tai Lake to divert the clean water from Tai Lake to the upper Huangpu River, Zhejiang's Hangzhou-Huzhou, and areas around Tai Lake, so as to increase the quantity of water resources and improve the carrying capacity of the aquatic environment. The project has been in operation since 2002.

The idea of the diversion project is "to use the moving water to clean the stagnant water, the clean water to dilute the polluted water, and the surplus water to replenish the dry season, so as to improve the water quality." An average annual amount of 2.5 billion m^3 of water has been transferred from the Yangtze River through the Changshu Water Conservancy Facility and 1 billion m^3 of water via the Wangting Water Conservancy Project, which has increased the water supply to the surrounding areas of Tai Lake. In the meantime, the project has supplied 500–700 million m^3 of water annually to the downstream areas such as Shanghai and Zhejiang through the Taipu Sluice to improve the aquatic environment in the Tai Lake and downstream areas. The major findings of the operation are as follows. When the water level at Tai Lake is maintained in an appropriate range of 3.0–3.4 m, the cycle of water exchange is reduced from 300 days to 250 days, which is conducive to improving the aquatic environment of Tai Lake.

The project also has problems. Although the implementation of the project has improved the hydrodynamic conditions of Tai Lake and parts of the river network, and the aquatic environment, the project cannot benefit all the Tai Lake watershed, especially the hydrodynamically "dead end" areas of Tai Lake. In the meantime, the clean water diverted into the lake has mainly played a role in diluting pollutants but cannot really reduce the total amount of sewage into the lake.

6.5.3 Yangtze River-Chao Lake Water Transfer Project

Located in central Anhui Province and close to the Yangtze River, Chao Lake is one of the five major freshwater lakes in China and has a surface area of approximately 780 km^2. The lake bottom is 5–6 m in elevation, and its storage capacity is 1.7 billion m^3 at the normal water level of 8.0 m and 5.2 billion m^3 at the design flood control level of 12.5 m. Major tributaries around Chao Lake include Hangbu-Fengle, Pai, Shiwuli, Nanfei, Shuangqiao, Zhigao, Zhao, Baishitian, and Yuxi Rivers. The runoff amount of the Hangbu, Pai, Nanfei, and Baishitian Rivers accounts for more than 80% of that of the Chao Lake watershed. The average annual precipitation in the Chao Lake basin is 1032 mm; the surface runoff amount of the lake is 3.7 billion m^3; and Chao Lake drains to the Yangtze River mainly through the Yuxi River.

Chao Lake formed about 12 ka ago and is a fluvial shallow water lake. Historically, the lake and the Yangtze River have been naturally connected and water flows between them. The water level at the lake fluctuates as the water level in the river changes. A dry season occurs in the lake for almost 3–5 months every year; the lake is flooded once every few years; flood and drought occur very frequently. In order to alleviate flood and drought disasters, since the 1950s, the Yuxi and Xi Rivers have been modified, the Chao Lake sluice, Yuxi sluice, and Fenghuangjing drainage and irrigation station have been constructed. These facilities have significantly improved the Chao Lake basin's capability of flood control and drainage. However, since the construction of the Chao Lake sluice in 1962, the operation of Chao Lake in accordance with the needs for flood control, irrigation, navigation, and water supply has resulted in the average annual amount of water to the lake from the Yangtze River at only 172 million m^3, just 12.6% of the amount before the construction of the sluice, and the time for one cycle of water exchange of the entire lake has increased from 2.5 years to 25 years, which has greatly reduced the self-purification capability of Chao Lake's water body, extended the retention time of nutrients, and accelerated the eutrophication process in the lake.

Since the 1980s, with the accelerated industrialization and urbanization in the lake basin, intensive use of chemical fertilizers and pesticides on farmlands, and increased discharge of wastewater, the contaminant load to Chao Lake has been increasing. Presently, as more than 66 major wastewater discharge points are distributed upstream of the Chao Lake sluice, the discharge of urban production and domestic wastewater alone has reached more than 300 million m^3 annually. Consequently, problems have arisen. For example, the non-point source pollution is widespread; the aquatic environment is suitable for algae growth, and there is the lack of water exchange between the Yangtze River and Chao Lake.

Water quality monitoring results in recent years indicate that 70% of Chao Lake has become eutrophic; the lake is of Class V in water quality or even poorer than Class V. The concentrations of total phosphorus and total nitrogen have well above their respective standards. The western part of the lake near Hefei has been in the eutrophic to hypereutrophic state. *Cyanobacteria* eruptions occurred in large parts of Chao Lake many times in the 1990s and in 2003, 2004, 2006, and 2007 as well.

Water pollution in Chao Lake is very serious, which has seriously threatened the safe water supply and human health in the urban and rural areas. The state and Anhui Province have attached great importance to the integrated rehabilitation of the aquatic environment in Chao Lake that has been included in the state top priority rehabilitation list of "three rivers and three lakes" and in the Anhui provincial list of "blue sky and green water projects." In recent years, Anhui Province has accelerated the integrated rehabilitation efforts on the aquatic environment of Chao Lake and organized departments of environmental protection and water conservancy to conduct many preliminary investigations. In 2008, the *General Plan for Integrated Rehabilitation of Aquatic Environment in Chao Lake Basin* was completed and submitted to the State Council. Major measures contained in the plan included pollutant discharge reduction, ecological restoration, Yangtze River-Chao Lake water transfer, etc., of which the Yangtze River-Chao Lake Water Transfer Project would lead to the river-lake connection and would be a major measure to improve the aquatic environment of Chao Lake.

Three major alternative routes were analyzed for the Yangtze River-Chao Lake Water Transfer Project. They were Caizi Lake, Baidang Lake, and Fenghuangjing. The Caizi Lake Route was the favorite alternative, under which water would be diverted on gravity flow from the Yangtze River through Zongyang sluice into Caizi Lake; after the diverted water is regulated and stored in the lake, the water would continue northward toward the Luobu River, a tributary of Chao Lake, after crossing the drainage divide through the Kongcheng River; and the water would eventually flow into western Chao Lake through the Baishitian River. The Baidang Lake Route was an alternative used for comparison purposes, which would require to expand the Baidang Lake sluice so as to be used as the inlet for the water from the Yangtze River to flow by gravity into the lake. After the diverted water is regulated and stored in the lake, the water would continue northward toward the Huangni River, a tributary of Chao Lake, after crossing the drainage divide through the Luochang River; and the water would eventually flow into southern Chao Lake at the central point of the southern bank through Huangpi Lake and the Zhao River. The Fenghuangjing Route is the current water diversion route. It starts at the Fenghuangjing drainage and irrigation sluice station; water flows by gravity and is pumped as well into the Zhao River after flowing through the Xi River and eventually flows into Chao Lake. The major problems associated with this route are that the diversion capacity is low and is often affected by floods and droughts.

In retrospect, the sluice was reconstructed to control the water level and storage capacity of the lake, but now structural measures are taken in an effort to enhance the river-lake connection due to the deterioration of the aquatic environment. However, the effectiveness of the measures is unknown. Because the sources of nutrients, sediments, and other pollutants are not removed, whether the problem can be resolved solely through water exchange will need to be monitored and studied.

6.5.4 Han River-Wei River Water Transfer Project

The Guanzhong region in the Wei River basin is an "industrial corridor" and a major food producing area of Shaanxi Province and is also the political, economic, and cultural center of the province. However, the area is scarce of water resources and serious in water pollution. Therefore, Shaanxi Province has hoped to transfer water from the water-rich Han River in south Shaanxi into the Wei River to help solve the water use problem in five cities of Xi'an, Baoji, Xianyang, Weinan, and Yangling, as well as 11 county-level cities such as Xingping, Wugong, and Meixian and 6 industrial parks in the Guanzhong region.

The planned project would be to construct the Golden Gorge Hydroelectric Project on the mainstream of the Han River and the Sanhekou Water Conservancy Project on the Ziwu River, a tributary of the Han River. A portion of the water would flow through a collection pool from the Golden Gorge booster station into the Qinling Tunnel to supply water to the Guanzhong region. The other portion of the water would flow from the Golden Gorge booster station through the collection pool into the Sanhekou Reservoir where the water would be stored. When the amount of pumped water from the Golden Gorge booster station is too small to meet the water supply need in the Guanzhong region, water can be released from the Sanhekou Reservoir to the Guanzhong region through the Qinling Tunnel. The planned annual amount of water transfer for the near future would be 1.5 billion m^3 and the annual withdrawal would be about 1 billion m^3. The flowrate of withdrawal from the Golden Gorge Reservoir would be 52 m^3/s, and the flowrate out of the Qinling Tunnel would be 50 m^3/s. The Qinling Tunnel would be 81.78 km long with a design flowrate of 70 m^3/s. The project was about to start.

Major problems associated with the project are as follows. First, the project would have some impacts on the Central Route of the South-to-North Water Diversion Project and the utilization of water resources in the middle and lower reaches of the Han River. Second, the diverted water would be mainly used for industrial and urban domestic purposes, and more wastewater would be discharged to the Wei River and may accelerate the degradation of the aquatic environment thereof. Third, as the regulation and operation of the project would involve two major basins of the Yangtze and Yellow Rivers, four provinces of Hubei, Shaanxi, Gansu, and Henan and the South-to-North Water Diversion Project, the relationship between the regulation and management of the project would be complex.

6.5.5 Central Yunnan Water Diversion Project

The central Yunnan region is the core of socioeconomic development in Yunnan Province, including Kunming City (except Dongchuan District), Yuxi City (except Eshan, Xinping, and Yuanjiang Counties), Chuxiong Prefecture (except Shuangbai and Yongren Counties), Qujing City (except Xuanwei City and Huize, Luoping,

Shizong, and Fuyuan Counties), Dali Bai Autonomous Prefecture (except Yunlong and Yongping Counties), Honghe Prefecture (except Hekou, Honghe, Pingbian, Yuanyang, Jinping, and Lǔchun Counties), and 49 counties (cities or districts) such as Yongsheng County in the Lijiang area. The existing water scarcity problems in central Yunnan are related to physical conditions, resource scarcity, water quality issues, etc. and a combination thereof. The severity of water scarcity in the Dianchi Lake basin has exceeded that of the Beijing-Tianjin-Tanggu region. In the future 30 years, the major economic areas of central Yunnan will become short of water due to the lack of water sources, and water resources will become a foremost constraint on socioeconomic development. Therefore, it is of great socioeconomic significance to implement the strategic Central Yunnan Water Diversion Project for the sustainable development of the entire Yunnan Province in the twenty-first century.

The idea of "Water Diversion to Central Yunnan" emerged in the early 1950 and was proposed by Zhang Chong, the late Vice Chairman of the Chinese People's Political Consultative Conference (CPPCC) after he inspected the Tiger Leaping Gorge. At the end of 2003, Yunnan Province and the YRWCC signed a contract for the "Central Yunnan Water Diversion Project Planning." It was proposed to construct a hydropower project at the Tiger Leaping Gorge in west Yunnan with the purpose of diverting water into the central Yunnan region as well. In June 2004, the General Institute of Water Conservancy and Hydropower Planning and Design organized relevant leaders and experts to review the plan of the Central Yunnan Water Transfer Project, but the plan was not approved. According to the MWR, after considering the plan was prepared based on a relatively weak basis, it was recommended that some of the topics be further studied in separate individual special plans. The Yunnan Provincial Government had issued a report, claiming that the Longpan (near the Tiger Leaping Gorge) Hydropower Project would have a total storage capacity of 37.1 billion m^3 and an installed capacity of 4200 MW and would be the leading reservoir on the Jinsha River for hydropower generation and the best source of water for the "Central Yunnan Water Diversion Project."

The planned water diversion project in central Yunnan would be to construct a control reservoir near the Tiger Leaping Gorge on the Jinsha River to centralize the water source, which would require the construction of a 479-km-long main canal. The first phase of the project would cost 37.3 billion yuan. By 2020, the project would have had the capacity to divert 2.56 billion m^3 of water annually, and the headworks of the canal would have had the capacity to divert 95 m^3/s. The unit cost of the canal would be 14.6 yuan/m^3. By 2030 when the project is completed, the long-term water receiving area would have been extended to Honghe Prefecture, and the design annual amount of water transfer would have reached 3.42 billion m^3.

The Central Yunnan Water Diversion Project would be a major water conservancy project in Yunnan Province. The project would cost 68 billion yuan and take 10 years to construct. The project would transfer 3.42 billion m^3 of water annually; the headworks of the diversion canal would have a design discharge of 145 m^3/s; and the final main diversion canal would have a total length of 877 km. The project was originally planned to use the Benzilan section of the Jinsha River in Deqin County of Diqing Prefecture as the source of water. Areas to receive water would include 30 counties and districts in six prefectures or cities of Lijiang, Dali, Chuxiong, Kunming, Yuxi, and Honghe. The length of the main canal would be 89.9 km within the Yuxi Prefecture, involving Hongta District, Jiangchuan County, and Tonghai County.

According to the initial plan, a separate water intake structure would be constructed on the Benzilan section in Deqin County of Diqing Prefecture; the diverted water would flow by gravity to Wangchengpo of Shigu Town in Lijiang City; and the water conveyance facility would need to be extended northward for approximately 189 km from Shigu Town. In the meantime, after considering the factors such as serious water scarcity conditions in the whole central Yunnan including Yuxi and Honghe Prefectures and the increased cost for the construction of the project in phases, the project schedule was changed to be constructed in one phase. After the adjustment, the main canal of the water conveyance facility would be 877 km long, and the diversion flowrate and annual diversion water volume at the headworks of the conveyance canal would remain unchanged. On August 6, 2010, the Yunnan Provincial Government convened the leading group for a special conference regarding the preliminary work of the Central Yunnan Water Diversion Project and decided to abandon the previously disputed location of water withdrawal at the Tiger Leaping Gorge and selected a new location for water withdrawal on the upper reaches of the Jinsha River at Benzilan Town in Deqin County of Diqing Prefecture.

The major problem of the water diversion project was the selection of the control reservoir location on the Jinsha River. As the Tiger Leaping Gorge Project is located in a famous scenic area, the scenery is not only unique, but also the Yulong Snow Mountain and Haba Snow Mountain on both sides of the gorge are national nature reserves. Moreover, the reservoir would inundate the minority areas of Longpan Township, Shigu Town, Jinzhuang Township, Judian Town, Tacheng Township, Tiger Leaping Gorge Town of Shangri-La County (formerly Zhongdian County) in the Diqing Tibetan Autonomous Prefecture, Jinjiang Town, Shangjiang Township, Wujing Township, Nixi Township, Tacheng Township of Weixi County in Lisuzu Autonomous County, and Tuoding Lisuzu Township and Benzilan Township of Dexin County, would cause about 100,000 people to be relocated, and would submerge large areas of alluvial plains and nearly 13,000 ha of farmland.

6.5.6 Dujiangyan Project

The Dujiangyan Project is the earliest water diversion project in China and even in the world. The project was originally constructed in 256 BC and its headworks are located in Doujiangyan City. The project diverts water from the Min River to irrigate part of the Chengdu Plain and has constantly expanded its service area. Since the founding of the PRC in 1949, the Dujiangyan Irrigation District has repeatedly undergone remodeling, expansion, and renovation. By the late 1960s, the irrigated area had reached 400,000 ha and all farmland in the Chengdu Plain had been irrigated. Since the late 1960s, the focus of improvements has changed to divert water to the hilly area west of the Longquan Mountains. After the completion of the Zipingpu Reservoir in 2005, the area of the Dujiangyan Irrigation District increased from the original 20,000 km^2 to more than 25,000 km^2, and the total irrigated area of the district reached about 667,000 ha.

The Dujiangyan Irrigation District currently has 55 main and branch canals (used for drainage and irrigation) with a total length of 2437 km. The irrigation district has 11 large- and medium-sized reservoirs, such as Luban, Tuanjie, Jiguang, Sancha, Shipan, and Heilongtan, with a total storage capacity of 1.089 billion m^3, and 364 small reservoirs with a total storage capacity of 151 million m^3.

The Dujiangyan Project diverts water from the upper Min River across three basins to the Chengdu Plain, and the Tuo and Fu Rivers. Through the Dongfeng Canal, water is diverted from the Min River water system to the water-scarce Tuo River system. Through the Renmin Canal, water is diverted from the Min River system to the traditionally water-scarce Fu River system.

According to the General Planning Report of Dujiangyan, Sichuan Province, the irrigated area in the irrigation district will have reached 0.8 million ha and 1.0 million ha, respectively, by 2015 and 2020.

6.5.7 Dadu River-Min River Water Transfer Project

The annual quantity of water diverted from the Min River to the Dujiangyan Irrigation District has reached 11 billion m^3, which is about 70% of the total yearly water for the upper Min River, well above the 40% threshold for water transfer. Moreover, over the past decades, the inflow of the upper Min River has reduced by 26% when compared to the 1930s, but the irrigation district is still short of 700 million m^3 of water annually. Therefore, Sichuan Province has planned to transfer water from the Dadu River to replenish the Min River.

The proposed Dadu River-Min River Water Transfer Project would be to divert water from the Dadu River through the Zagunao River in the Ngawa Autonomous Prefecture to the upper Min River, and then the water would be supplied to the Tuo and Fu River basins through the water network in the irrigation district.

According to the comprehensive plan of water resources for the water supply area and the upper Min River, and the plan for the Dadu River-Min River Water Transfer Project, 3.32 billion m^3 of water per year (in the year of design level) would be diverted from the Dadu River to the Min River to basically meet the projected peak water demand of the Dujiangyan Irrigation District in 2050. However, the Dadu River-Min River Water Transfer Project is bound to reduce the amount of water transfer available to the Western Route of the South-to-North Water Diversion Project. Therefore, the project needs to be further studied.

6.5.8 Great East Lake Connection Project

Wuhan City is known as the Hundred Lake City, but due to continued urbanization and economic development, most of these lakes have been contaminated in the past 30 years. Therefore, Wuhan City has proposed the construction of a water network to achieve the river-lake connection and the lake-lake connection and improve the environmental carrying capacity of the lakes. The Great East Lake Connection Project is one of the representative projects and would divert an average annual amount of 200 million m^3 of water from the Yangtze River to East Lake. The design diversion flowrate would be 40 m^3/s for the sluice at the Qingshan Port and 10 m^3/s for the booster station at the Zengjia Port. The project would divert water through both the sluice and the booster station.

The water diversion alignment would be to seasonally divert water from the Yangtze River at Zengjia and Qingshan Ports into the lakes. The exchanged lake water would be pumped to the Yangtze River through three booster stations at Luojia Road, North Lake, and Xinsheng Road, respectively.

Four water diversion lines were recommended. Line 1 would be the Yangtze River-Zengjia Port-Neisha Lake-Waisha Lake-Sha Lake Port-Luojia Port-Yangtze River, and the main purpose of this line would be to improve the hydrodynamic conditions of Sha Lake. Line 2 would be the Yangtze River-the sluice at Qingshan Port-Qingshan Port-Yangchun Lake-New East Lake Port-Tangling Lake (East Lake)-Guozheng Lake (East Lake)-Shaoqi Lake (East Lake)-New Canal-Sha Lake Port-Luojia Port-Yangtze River, and the principal purpose of this line would be to improve the water quality of Yangchun, Tangling, and Guozheng Lakes. Line 3 would be the Yangtze River-the sluice at Qingshan Port-Qingshan Port-Yangchun Lake-New East Lake Port-East Lake-Jiufeng Canal-Yanxi Lake-North Lake-Yangtze River, and the primary purpose of this line would be to improve the hydrodynamic conditions of East, Yanxi, and North Lakes. Line 4 would be the Yangtze River-the sluice at Qingshan Port-Qingshan Port-Yangchun Lake-New East Lake Port-Tangling Lake (East Lake)-Guozheng Lake (East Lake)-Shuiguo Lake-Shuiguo Lake Canal-Sha Lake-Sha Lake Port-Luojia Port-Yangtze River, and the major purpose of the line would be to improve the hydrodynamic conditions of Yangchun, East, Shuiguo, and Sha Lakes.

The water diversion time would be mainly in the flood season between May and November and the control water level at East Lake would be between 18.15 and 19.15 m (above the Yellow Sea datum). When the water level in the Yangtze River level is 19.65–25.21 m, water would be diverted. The average annual diversion duration would be 68 days.

The project would require 19 canals between ports to be constructed, modified, or expanded with a total length of approximately 48.7 km. In the East Lake water system, eight port connection canals, such as Qingshan Port and Luojia Port, would be modified or expanded (total length of approximately 24.4 km), and the New East Lake Port (1.7 km long) and East-Sha Lake Canal (1.8 km long) would be constructed. In the North Lake water system, seven port connection canals, such as North Lake Port and Zhou Port, would be modified or expanded (total length of 18.7 km). The Jiufeng Canal would be constructed to connect East and Yanxi Lakes (2.1 km long), and it would also connect the East Lake water system and North Lake water system to form the "Great East Lake" ecological water network consisting of the water flow and circulation between the river and lakes and among the water networks. A total of 32 water diversion and drainage structures and a new diversion dike in North Lake would be constructed.

Problems associated with the construction and operation of the project would include the following. ① Presently, except for Yanxi and Yandong Lakes which are Classes III–IV in water quality, all the other lakes are Classes IV–V or even poorer than Class V. Diversion of water may cause poor water to be introduced to good water or nitrogen and phosphorus nutrients in the Yangtze River to be introduced into the lakes. Because concentrations of nitrogen and phosphorus in the water of the Yangtze River are relatively high, they may intensify the lake eutrophication after they are introduced into the lakes. ② As the time for the free flow of water would be short, diversion of water would depend on the water flow and level conditions of the Yangtze River. Therefore, it is not easy to control. ③ Removal of the contaminated lake sediments remains a problem. The bottom sediments of East Lake are up to 4 m in thickness with an average thickness of 1.06 m. The total volume of the sediments is 35.22 million m^3, which is 20% of the total storage capacity of the lake.

Water pollution in East Lake occurred in recent 30–40 years, but the river-lake separation had occurred even earlier. The main cause for the lake water to become black and smelly is that East Lake has been in the eutrophic state and has been becoming hypereutrophic. There is no mature theory that can be used to predict the occurrence of eutrophication in real bodies of water. Several large storms occur in Wuhan every year, and the amount of rainfall is not much less than the amount of water to be diverted but has failed to significantly improve the water quality of East Lake.

6.5.9 Comments on Water Transfer Projects

The fact that large numbers of water diversion projects have been promoted demonstrates that water resources are important requirements for socioeconomic development and regional ecological environmental improvement. However, large water diversion projects often trigger controversy. First, it is easy to ignore the constraints of water resources and the aquatic environment and indiscriminately allow the urban population to grow or highly water-consumptive industries to be introduced. However, when a drought occurs in both the water source and receiving areas, a new water shortage occurs. Second, a large water transfer project has significant impacts on water resources and the aquatic environment downstream of the water source area. It is difficult to assess the long-term impact on the ecological environment of the water receiving area. As the soils, crop structure, and vegetation in an arid area have adapted to the dry environment, the diversion of excessive amounts of foreign water may not be conducive to the growth of the dryland crops and may also result in soil salinization due to the rise of groundwater level. Third, if the diverted water is mainly used for industrial and domestic purposes, the discharge of wastewater into rivers will increase and may aggravate water pollution in the water receiving area. Fourth, although a considerable number of river-lake water diversion projects have helped improve the aquatic environment of lakes, the projects cannot really reduce the amounts of contaminants into the lakes or eliminate the contaminants but can only transfer the contaminants. Therefore, the projects are not conducive to cleaner production or reduction of wastewater discharge. Fifth, the long-distance interbasin water transfer occupies large amounts of land and result in large amounts of workload to handle the relocation of people. Sixth, the long-term impact on the ecological environment along the water diversion project corridor cannot be ignored. Seventh, the cost of the project construction, operation, and management is high, and it is difficult to rely on revenues from sale of the water for operation, maintenance, and management of the project. Therefore, the project will increase the fiscal burdens of the state and local governments.

In fact, to solve the problem of uneven distribution of water resources, in addition to structural measures, very effective non-structural measures are available. For example, first, through trades, water (virtual water) can be embodied in commodities for transfer. The water scarcity areas import "commercial water," and the water resource-rich areas produce water-intensive commodities, such as grain, iron and steel, and chemical products. Second, the constraints of the environmental carrying capacity for water resources and the aquatic environment can be used to force the excessive population and high water-consumptive industries to move to the water resource-rich areas. Third, the development of the water-conserving society and the management of water demands can help improve the efficiency of local water use and reduce the amount of water transfer. Fourth, the use of unconventional water, such as water reuse, seawater desalination, and seawater use, can help reduce the dependence on freshwater and supplement the freshwater resources. For instance, as the operation of seawater desalinization has become a large-scale industry in Israel, the cost has dropped to 3–4 yuan/m^3, which is lower than the combined cost of the water transfer project. When the demand for water resources is still not met after the full adoption of the non-structural measures (i.e., water resources in a stringent sense), it may be the most economical and rational way to implement the water transfer project, so as to minimize the impact on the ecological environment.

6.6 Future Development and Utilization of Water Resources of the Yangtze River

Water is essential to life. It is not only the foundation of the natural ecosystem, but it is also the important resources of the human socioeconomic development. However, freshwater resources are limited, and the high-quality water is especially scarce. Therefore, water resources will have to be protected for their sustainable use. The Yangtze River basin is relatively rich in water resources, but it is not so during the non-flood season, especially in high-quality water. Thus, only when development and utilization of water resources incorporate effective protection can water resources be sustainable.

6.6.1 Existing Conditions and Future of Water Resources Development

6.6.1.1 Establishment of Water Resources Control Targets

Presently, the socioeconomic condition of the Yangtze River basin is in a stage of rapid development. It is projected that the fast development pace will continue in the next 20 year. However, China has a large population but less water. The pressure on water resources and the environment has become increasingly higher. Therefore, China has begun to implement a strict management system for water resources, i.e., establishment of a control system for total water use, a control redline for development and utilization of water resources, a control system and redline for water use efficiency, a control system and redline for the limit of the contaminant load of water function zones, and the responsibilities and an assessment system for water resources management. The four systems have provided the theoretical basis and institutional framework for China to implement strict control of water resources.

Under the *State Council's Opinion on Implementation of the Strictest Water Resources Control System* (No. 3 (2012) of the State Council), a control redline will be established over the development and utilization of water resources so that the total amount of annual water use in China will have been limited to within 700 billion m^3 by 2030. A control redline will be established over the efficiency of water use so that the efficiency of water use in China will have reached or been close to the advanced world level, that the amount of water use for generating every 10 thousand yuan (in the 2000 price, same thereafter) of added industrial value will have reduced to 40 m^3 or lower, and that the coefficient of effective water use in farmland irrigation will have increased to 0.6 or higher by 2030. A redline will be established on the limit of the contaminant load of water function zones so that the total amounts of major contaminants discharged into rivers and lakes will have been limited to within the assimilative capacity of water function zones and that 95% or a higher percentage of water function zones will have been in compliance with the standard of water quality by 2030.

To achieve the aforementioned targets, the total amount of annual water use in China will have been limited to within 635 billion m^3 by all means; the amount of water use for generating every 10 thousand yuan of added industrial value will have decreased by 30% or more from the 2010 level; the coefficient of effective water use in farmland irrigation will have increased to 0.53 or higher; and 60% or a higher percentage of water function zones of major rivers and lakes will have been in compliance with the standard of water quality by 2015. By 2020, the total amount of annual water use in China will have been limited to within 670 billion m^3 by all means; the quantity of water use for generating every 10 thousand yuan of added industrial value will have been reduced to below 65 m^3; the coefficient of effective water use in farmland irrigation will have increased to 0.55 or higher; 80% or a higher percentage of water functional zones of major rivers and lakes will have been in compliance with the standard of water quality; and all water source areas for water supply to urban areas will have been in compliance with water quality standards.

Under the national general requirements, the total amount of annual water use in the Yangtze River basin will have been limited to 206 billion m^3, 225.6 billion m^3, and 231.1 billion m^3, respectively, by 2015, 2020, and 2030. As this goal includes major requirements for strict control over the use of water resources, it is lower than the amount of usable water resources set forth in the *Yangtze River Basin Water Resources Planning*, and it would be great if the goal can be achieved. As a matter of fact, it will not be difficult to achieve the goal for the total amount of annual water use in the Yangtze River basin, but it is difficult to achieve the goal of the 95% compliance level for water function zones.

6.6.1.2 Trend of Supply and Demand of Water Resources in the Yangtze River Basin

The supply and demand of water resources in the future will depend on the change of population, the growth rate and pattern of economic development, water-conserving potential, and human demand for the quality of life. Compared with the northern basins of the Huang, Huai and Hai Rivers, the Yangtze River has a larger proportion of mountainous and hilly areas and a smaller fraction of arable land. Presently, the ratio of arable land over the basin area in the Yangtze River is approximately 11.7%, and the area of arable land will continue to decrease with continued urbanization and industrialization. From the statistics since 2000, agricultural water use in the Yangtze River basin has been decreasing gradually due to the decrease of arable land and the adjustment of crop types. From the structure of water use, the proportion of agricultural water use decreased from 59.1% in 2000 to 50.0% in 2007, while industrial water use increased from 29% to 40.2%, and domestic water use decreased from 11.5% to 9.8%, or around 10%, during the same timeframe. The share of the reduced agricultural water use was just offset by the increased industrial use.

The future quantitative estimation of water demand can be projected using two simple methods. One method is to assume that the population in the Yangtze River basin will reach the maximum in 2030 at an average annual growth rate of 0.8% (average national growth rate in 2000) and the population in the basin is projected to reach 560 million in 2030. Using the average amount of water use per capita in the recent 3 years, the quantity of water use in the basin will reach 236.9 billion m^3 in 2030. After taking into account the 44.8 billion m^3 of water to be transferred through the South-to-North Water Diversion Project, the total amount of water use will reach 281.7 billion m^3 in 2030, resulting in a utilization rate of 27% in the basin. The other algorithm is to assume that the amount of agricultural and environmental water use will remain unchanged and the amount of industrial and domestic water use will increase with an average annual growth rate of 0.8%; the total quantity of water use is projected to reach 201 billion m^3 in 2030. After considering the water diversion by the South-to-North Water Transfer Project, the total amount of water use is projected to reach 245.8 billion m^3, resulting in a utilization rate of 24.7% in the basin. As the former mothed uses the same population growth rate and the current water use parameter value, the projected amount of water use may be too high and may be considered the upper limit, and the latter method only increases the basic domestic water use for the new population and uses the urban standard of domestic water use, and the projected amount of water use may be closer to the real value and may be used as the lower limit. If mandatory water-conserving measures are considered, or the total amount of water use is controlled, the total quantity of water use will not exceed 231.1 billion m^3 in 2030, but it remains to be monitored as to whether it can be controlled. Presently, the water supply capacity of water supply facilities in the Yangtze River basin has reached 247 billion m^3 per year. In terms of the total water supply capacity, it can fully meet the water demand in the basin. However, due to the relatively large difference in available water, water supply, and water use in various areas, new water source projects in different areas in the basin will need to be constructed in the future. Generally speaking, the total amount of water resources will be fully capable of supporting the long-term economic development in the basin.

The above general analysis was only about the total amount of future water use in the Yangtze River basin. As the Yangtze River basin is a typical large river watershed, the natural conditions and socioeconomic development levels of various areas in the upper, middle, and lower reaches of the basin are quite different, droughts and water shortages have often occurred in local areas during dry years and dry seasons, but there has been no overall resource-type water shortage. The traditionally arid and water scarce areas in the basin are the Sichuan Basin, Hengshao of Hunan, Gannan of Jiangxi, and the Sanbei area near the Hubei-Henan border and central Yunnan. These areas are mainly hilly areas where the water storage conditions are poor and there are not enough necessary water conservancy facilities. The water scarcity in these areas is due to physical conditions. Nonetheless, the water scarcity in these areas mainly affects agricultural water use but has a limited impact on the economic development of the whole basin. According to the data for the past 5 years, the total amount of water use in the Yangtze River basin has been increasing very slowly, and the future increase in the amount of water use in the basin has been mainly due to the increased industrial water use.The cooling water use in thermal power plants has increased the most, followed by the general industrial and domestic water use, but the agricultural water use has not increased. If water-conserving and water resources protection measures are implemented, the total amount of water use will not have increased by 2020.

The future water scarcity in the Yangtze River basin will be due to water quality. In the past 30 years, as the amount of industrial and domestic water use has been increasing and the capacity of sewage treatment has been inadequate, the wastewater discharge has been increasing sharply. Moreover, extensive agricultural use of pesticides and fertilizers has resulted in increasingly serious non-point source pollution. Consequently, serious water pollution has occurred in most lakes in the Yangtze River basin, the Yangtze River's mainstream along the banks where urban areas are present and a considerable number of tributaries, and the water quality has been degrading or deteriorating. Furthermore, due to the lack of strict control of toxic and hazardous substances in China, requirements for the seepage control at municipal landfills are not strict. As a result, groundwater contamination has occurred. The drinking water sources for many urban areas have been contaminated or are highly susceptible to contamination, and there is a general lack of backup water sources. As most cities rely on surface water supply, once their primary water sources are seriously contaminated, their water supply will be interrupted. Therefore, the safety of water supply is not highly guaranteed. On the other hand, because the quality of water sources is not good, the cost for withdrawing and treating the water is very high. With the improvements of the human standard of living, people's demand for the quality of drinking water and the aquatic environment will be increasingly higher. As the contamination of lakes is the result of the cumulative effect for several decades, it may take decades to restore the water quality and achieve the control goal for water function zones if the contamination sources are completely intercepted or removed. Therefore, it will be a daunting task and take a long time to achieve the goal of protecting water resources.

6.6.2 Construction Scale of Reservoirs and Hydropower Stations

For the purposes of flood control, water supply, navigation, and the need for clean energy, a large number of reservoirs and hydropower stations will need to be constructed in the Yangtze River basin. Under the *Flood Control Plan of the Yangtze River Basin*, in addition to the completed TGP, the reservoirs that need to enlarge their flood control capacities include Danjiangkou, Wuqiangxi, Zhexi, Wanan, Ertan, Zhelin, etc. and projects that are under construction or will be constructed include the Tingzikou on the Jialing River, Xiluodu and Xiangjiaba on the Jinsha

River, Pengshui on the Wu River, Xiajiang on the Gan River, Liaofang on the Fu River, and Goupitan on the Wu River.

Based on the reservoirs that have been constructed or are under construction, the ratio between hydropower development and hydropower potential on the Wu River basin and the mainstream in the upper reaches of the Yangtze River is relatively high, while the ratio in other river basins is relatively low. However, many more hydropower stations are planned on each river section, especially in the Jinsha River basin where the total storage capacity of planned reservoirs will reach 83% of the annual runoff amount of the river. With the gradual implementation of the planned reservoir projects, the total storage capacity of the reservoirs in the upper reaches of the Yangtze River will reach 61% of the annual runoff amount of the river. Except for a few tributaries such as the Chishui River, cascade reservoirs will be constructed in most sections of the mainstream and tributaries in the upper reaches of the Yangtze River.

The Yangtze River is rich in water resources and hydropower potential. Presently, only 20% of the water resources has been utilized, and 25% of the explorable hydropower potential has been developed in the basin, but only 20% of the explorable hydropower potential has been developed in the upper reaches. Compared with developed countries, the fraction of developed hydropower relative to the explorable potential is low in China. At present, China is growing rapidly in the economy; the demand for water resources (including hydropower) is increasingly large. Therefore, in the future 20 years, the Yangtze River basin will continue to grow rapidly in development of water resources. As a result, hydropower development in the upper reaches, navigation expansion in the middle and lower reaches, and protection of water resources in the entire watershed will be the most important task in the future development and management of water resources. The nodus and focus for the future management of water resources will be how to promote the development while implementing protection measures and how to incorporate protection measures into development. If development results in the conditions of the aquatic ecology and aquatic environment to deteriorate and adversely affect the quality of human life and the improvement of living standards, the development of water resources and hydropower in the future will be affected, especially the construction of hydropower stations. If the relocation of people and the impact on the ecological environment cannot be appropriately coped with, the implementation of planned projects will be affected.

Under the *Plan of Electric Power Development*, by the end of 2015, 29.1%, 57%, and 68.5% of explorable hydropower potential will have been developed in the middle and lower reaches of the Jinsha River, the Yalong River and the Dadu River, respectively. During the "Thirteenth Five-Year" planning period, the installed capacities will be mainly concentrated in the hydropower stations in the middle and lower reaches of the Jinsha River, including the hydropower stations at the Tiger Leaping Gorge, Longkaikou, Ludina, and Xiluodu that will have been in full production. By 2020, the cumulative installed capacity will have reached 32.1 GW, including the Shuangjiangkou and Danba Hydropower Stations on the Dadu River, Lianghekou, and Yagen Hydropower Stations on the Yalong River that will have fully generated electricity. The ratio of developed capacities over the explorable hydropower potential on the Yalong River, the Dadu River, and the middle and lower Jinsha River will have reached 72.9%, 86.9%, and 54.7%, respectively, by the end of 2020. The remaining hydropower stations of Wudongde, Baihetan, Liyuan, and Ahai in the middle and lower reaches of the Jinsha River will start generation after 2020. By 2020, the total installed capacity of hydropower will have reached 328 GW in China; 29 GW in east China where all explorable hydropower potential will have been substantially developed; 72 GW in Central China, accounting for 22% of that of China, where more than 90% of the explorable hydropower potential will have been developed; and 227 GW in west China, accounting for 69% of that of China, where 60.7% of explorable hydropower potential will have been developed. More specifically in west China, the total installed capacities of hydropower will have reached 76 GW, 62.8 GW, and 18.3 GW, respectively, in Sichuan, Yunnan, and Guizhou, and will have reached 63.3%, 61.6%, and 94%, respectively, of their respective explorable hydropower potential.

6.6.3 Shipping Development Trend of the Yangtze River

The navigable waterways in the Yangtze River water system are 66,400 km long, accounting for 49% of that of China's all inland water systems, of which 2837 km is on the mainstream of the Yangtze River. Shipping through waters has the advantage of low cost and large freight volume. The shipping industry in the Yangtze River has developed rapidly in the past 30 years of reform and opening-up. Especially since 2000, just like the economic development, the shipping industry in the Yangtze River has been developing by leaps and bounds. The shipping freight on the mainstream of the Yangtze River was only 400 million tonnes in 2000, went up to 795 million tonnes in 2005, and increased to 990 million tonnes in 2006, of which 550 tonnes were direct shipping between the Yangtze River and the sea. In 2006, the freight on the Yangtze River surpassed that on the Rhine River in Europe and the Mississippi River in the United States. In 2007, the freight through the mainstream of the Yangtze River reached 1.1 billion tonnes, which was twice that through the Mississippi River and thrice that through the Rhine River, and was number one of inland rivers in the world.

In 2004, more than 81,000 ships sailed through the mainstream of the Yangtze River with a total freight of 19.7 million tonnes; the total annual port capacity was 480 million tonnes; there were 143 10,000 tonne ship berths; total warehouse freight yards reached 4 million m^3; there were more than 20 Class I and Class II ports; a number of ports established bonded areas and bonded warehouses; and a shipping system was basically established for shipping of containers, cars with roll-on/roll-off capability, coal, ore, and dangerous chemicals. In 2005, the freight through the mainstream of the Yangtze River reached 795 million tonnes, which was 1.65 times that of 2000; the cargo throughput of major ports reached 655 million tonnes, of which foreign trade cargo throughput was 78 million tonnes, an average annual increase of 22%; and container throughput was 2.6 million TEU, an average annual increase of 30.4%. According to statistics, the shipping industry on the Yangtze River provided large businesses along the river with about 80% of their required iron ore, 72% of crude oil, and 83% of electricity-generating coal.

In 2010, the freight through the Yangtze River reached 1.3 billion tonnes. By 2020, the freight is projected to have reached 1.7 billion tonnes. For some time in the future, great changes will take place in the transportation of liquid, chemical, and crude oil in the bulk goods market, but the shipping industry on the Yangtze River will still control a certain market share in the shipping of crude oil in the Yangtze River basin, and the market for the transportation of refined oil and petroleum products will not change much. The demand for the transportation of coal in the dry bulk market will continue to grow rapidly in the near future, but the long-term growth will slow down due to the influence of "west-east electricity transmission," "west-east gas transmission," and the adjustment of the energy structure along the Yangtze River. With the increasing demand of metal minerals, ore will be one of the driving forces for the throughput growth in ports along the Yangtze River. As the transportation of nonmetallic ore is affected by resource conditions, the increase will be limited. However, with the development of iron and steel industry and cement industry, the demand for mineral and metallurgical supplements for building materials will be strong, which will be the main growth factor for the transportation of nonmetallic ores. The demand for the transportation of building materials such as yellow sand, which are mainly concentrated in the mining regions (Hubei, Jiangxi, and Anhui Provinces) along the river and downward to the Yangtze Delta region, will be rising. The export and import of steel in the market show a rapid growth trend. As the cement industrial zone along the Yangtze River will be built into one of China's foreign export trade bases for building materials industries, the container market will continue to grow rapidly with the expansion of foreign trade.

References

Cheng G (2005) Relationship between distribution of neolithic cultural sites in Jianghan-Dongting Lake Area and Evolution of Rivers and Lakes. J Anhui Norm Univ (Natural Science) 28(2):218–221

Deng H, Chen Y, Jia J et al (2009) Evolution of distribution of ancient cultural sites in plains of the Middle Yangtze since 8,500 Years ago. Acta Geograph Sin 64(9):113–125

Gu Y, Ge J, Huang J et al (2009) Climate change, human activities and evolution of Jianghan Lake Group over 20,000 Years. Geological Publishing House, Beijing

Han Q, Zhou S (1999) Characteristics and evolution of three diversional channels. J Yangtze River Sci Res Inst 16(5):5–8

Jia J, Yuan Y, Zheng C et al (2010) On statistics, technical progress and concerned issues of large reservoirs in China. Hydropower 36(1):6–10

Jin B, Nie M, Li Q et al (2012) basic characteristics, challenges and key scientific issues of Poyang Lake Watershed. Resour Environ Yangtze River Basin 21(3):268–275

Technical Committee of Yangtze River Water Conservancy Commission (2003) Study on comprehensive utilization planning of Yangtze River Basin. China Water & Hydropower Press, Beijing

Wang Z, He H, Wei X (2006) Construction and development of hydrologic monitoring network in China. Hydrology 26(3):42–44
Xu C, Li Y, Sun S (2011) Water-conserving society establishment and water use efficiency. China Water 23:64–72
Yangtze River Basin Planning Office (1979) A brief history of water conservancy of the Yangtze River. Water & Power Press, Beijing

Chapter 7
Protection Objects of the Ecological Environment in the Yangtze River Basin

Abstract In the past 40 years, the rapid economic development in China and the wide array of high-intensity human activities have caused the ecological environment loss of the Yangtze River basin. Although various types of nature reserves have been set up, the area of each individual reserve is small, and many reserves still have many human activities occurring within. As a result, it is difficult to provide the needed protection as the current effectiveness is poor. From the perspective of the overall ecosystem protection in the entire basin, this chapter describes various types of sensitive ecological environments in different areas of the Yangtze River basin, including national nature reserves, aquatic animal and plant reserves, national forest parks, wetlands of international importance, important lakes, scenic sites, world natural and cultural heritage sites, national geological parks, important water sources and conservation areas, etc. Particular emphasis is placed on the need to protect not only flagship species but also the environmental conditions and habitats that the organisms depend on for survival. It is also recommended that, where possible in the future, the extent of the reserves be maximized or linked to enhance their overall ecological function.

Keywords The Yangtze River · Changjiang River · Evolution of river system · Basin ecosystem · Water resources utilization · Floods and drought · Ecological and environmental protection · Basin management

7.1 Impact of Human Activities on the Yangtze Ecological Environment

For the safety of life, production of food, utilization of energy, and expediency of transportation, human beings naturally and justifiably started to regulate, develop, and utilize the Yangtze River. However, from the perspective of eco-systematic health and human sustainable development, as the development and utilization of water resources have been providing human beings with benefits and services, the river's ecosystem and the environment have also been adversely affected. While the

J. Chen, *Evolution and Water Resources Utilization of the Yangtze River*,
https://doi.org/10.1007/978-981-13-7872-0_7

services and benefits provided by the development of water resources may be greater than the negative impacts on the ecological environment, it should be fully understood how and to what extent human activities affect the environment, so as to protect and manage the Yangtze River accordingly.

7.1.1 Effects of Construction of Dikes and Sluices on River's Lateral Connectivity

The continuity of a river is the most basic characteristic of the river ecosystem and is represented on four-dimensional scales, i.e., continuities of the longitudinal upstream-downstream direction, lateral left-right bank direction, vertical depth direction, and time. Moreover, maintaining the continuity of a river can preserve the characteristics of the river's life and the basic environmental elements of the ecosystem. From the viewpoint of the ecosystem, dikes constructed by human beings have restricted water from flowing to floodplains and the surrounding wetlands. Although it is very important to the safety of human settlements and farmlands to prevent floodwater from flowing to the floodplain, it affects the lateral continuity of the river.

Presently, since the mainstream and tributaries in the upper Yangtze River mostly flow through canyons in mountainous areas, only approximately 3000 km of dikes have been constructed generally around urban areas. However, in the middle and lower Yangtze River, more than 30,000 km of various types of dikes have been constructed to protect 140,000 km^2 of low-lying areas, including large areas of lacustrine wetlands and the original floodplains. Before the dikes were constructed, as the floodwater of the Yangtze River had flowed into floodplains and excessive floodwater had spread to low-lying areas and lakes, sediments had deposited in floodplains, and the Yangtze Estuary had expanded slowly eastward. With the gradual completion of the dikes, large areas of fertile land have been provided to human beings, but the difference between the flood water level in the river and the surrounding low-lying areas has become increasingly larger; the threat of flooding has become growingly higher; and large amounts of sediments have either deposited in the river to raise the riverbed and flood water level or flushed to the estuary to accelerate the expansion thereof.

In modern times, for the purposes of flood control, schistosomiasis control, and land reclamation from lakes, sluices were constructed in the majority of the mouths of lakes where they originally connected to the Yangtze River, which have further disconnected the lakes with the river and affected the river-lake continuity. While many sluices have still maintained a seasonal and selective connectivity, they, in fact, have been maintained closed more than open because the control of most sluices is locally managed primarily for the purposes of flood control and drainage. As a result, the biological channels have largely blocked, which have generally put more pressure on the river channel to store and discharge floodwater. Moreover, the cycle of water exchange of lakes along the banks of the Yangtze River has been lengthened, and, coupled with nonpoint source pollution, closed lakes have become generally eutrophic, aquatic environmental conditions deteriorated, and ecosystem degraded. Chao and East Lakes are typical such examples. In the future, if sluices are constructed at the mouths of Poyang and Dongting Lakes that are connected to the Yangtze River, the aquatic environment of the lakes would be deteriorated.

7.1.2 Significant Impact of Construction of Reservoirs and Hydropower Stations on Longitudinal Connectivity of Rivers

While reservoirs and hydropower stations have played an important role in flood control, water supply, power generation, and navigation, they have significantly altered the hydrological and hydrodynamic processes of the rivers and affected migratory organisms, especially the life passages of migratory fish species, such as the Chinese sturgeon that used to spawn in the 1000 km river section between the lower Jinsha River and Yichang. The construction of the Gezhouba Project has left their spawning river section so short as less than 10 km that it is difficult for the species to survive. Reeve's shad, which once had been the Yangtze River's very important economic fish, mainly lived in the Gan River of the Poyang Lake system. However, due to the construction of the Wanan Reservoir and Xiajiang Water Project on the very important spawning river sections, plus overfishing, the fish species has now been basically extinct.

The completion of the TGD has an obvious effect on the hydrological process in the middle and lower reaches of the Yangtze River. The processes for small- and medium-sized floods are evened, and the discharge of lower-temperature water during the spring has affected the spawning and fries of the FMCC, which are very important economic fishes. According to the monitoring data collected between the end of June and early July during 2004–2006, the peak flood process downstream of the TGD was weakened. As a result, the correlation between the peak flood process and the maximum spawning quantities of the FMCC was markedly decreased. During May–July of 2003–2006, the conditions of spawning sites of the fishes that lay floating eggs in the middle reaches of the Yangtze River were investigated, and the results indicated that there were 13 fish species producing floating eggs in the middle reaches of the Yangtze River, including the following eight major economic fish species: black carp (*Mylopharyngodon piceus*), grass carp (*Ctenopharyngodon idellus*), silver carp (*Hypophthalmichthys molitrix*), bighead carp (*Aristichthys nobilis*), mandarin fish (*Siniperca chuatsi*), Barbel chub (*Squaliobarbus curriculus*), bream (*Parabramis pekinensis*), and bronze gudgeon (*Coreius heterodon*). Compared with the 1970s, the number of fish species decreased by more than ten, but the yellowcheek (*Elopichthys bambusa*) that had not been seen for many years were observed. More than ten spawning sites for the FMCC were distributed in the approximately 300 km Yichang-Chenglingji section of the middle Yangtze River, indicating little changes occurred in the topographic distribution of spawning sites when compared with the conditions before the impoundment of the TGR. During 2003–2006, 143,720 x 10^4 floating eggs were spawned in the middle reaches of the Yangtze River, of which 108,069 x 10^4 were laid by the FMCC. Compared with the number of eggs before the impoundment of the TGR, the spawning quantities of the FCCC in the middle reaches of the Yangtze River decreased markedly. The average annual total quantity of the FMCC eggs in 4 years during 2003–2006 decreased to 42.84% of that during 1997–2002 and 56.88% of that in 2002 before the impoundment of the TGR.

In the TGR area, as the valley-dammed reservoir is about 600 km long upstream of Sandouping, the depth of water has increased; the velocity lowered; sediment deposited; the organism composition of food altered; and the ecological environment of the water zone changed markedly. As a result, some of the original endemic fish species could not adapt to the ecological environment in the reservoir area and have disappeared in the TGR or become extremely rare. According to preliminary statistics, since the completion of the TGR, the habitat area of the fish species has been reduced by one-fifth to one-fourth, and the population of fish species has been reduced accordingly.

7.1.3 Increased Soil Erosion and Nonpoint Source Pollution Due to Changed Land Use

The earliest human activities were deforestation and the development of cultivated agriculture, which increased soil erosion and sediment yield in mountainous and hilly areas. Since the Neolithic age, human beings have started to spread from low hillocks in hilly areas to plains and mountains, harvested trees from primeval forests, and performed slash-and-burn farming in mountainous and hilly areas, resulting in increased soil erosion and, naturally, increased sediment transport in rivers. Consequently, alluvial plains have been enlarged; lacustrine sedimentation has been increased in the middle and lower reaches of the Yangtze River; and storage capacities of river channels and lakes have decreased. The quantities of water and sediments through the four outlets from the Yangtze River to Dongting Lake have decreased markedly in the past 100 years, which is due to sedimentation that have occurred in the four rivers south of the Jingjiang River and Dongting Lake. Moreover, Dongting Lake that used to be the largest freshwater lake in China became the second in 50 years, which was mainly caused by lacustrine deposition of the sediments transported by floodwater from the upper Yangtze River and land reclamation from the lake.

Increased soil erosion has changed the physicochemical properties of water bodies. Especially mining, smelting, and the extensive use of fertilizers and pesticides in farming have added nutrients, minerals, and toxic substances to the eroded soils and significantly altered the water quality of the Yangtze River. For example, the phosphorus concentration in the water of the Xiangxi River, a tributary to the TGR area, has increased significantly due to a large phosphorus mine present in the upper reaches of the river. Before the impoundment of the TGR, eutrophication had never occurred in the Xiangxi River. However, eutrophication happened and intensified after the impoundment of the TGR because the higher water level in the lower reaches of the Xiangxi River slowed down the water flow; the flowing water changed to static water; and the aquatic environment has transformed from a typical water body of a flowing river into a slow-flowing lake. Every spring, the diatom and dinoflagellate algal blooms occur in the entire bay and last a long time, and the water surface appears dark or light soy sauce color (Yang et al. 2011).

7.1.4 Increased Water Pollution Due to Urbanization and Industrialization

Industrialization and urbanization have boosted rapid socioeconomic development but resulted in increasingly serious water pollution at the same time. The seriousness of water pollution in the Yangtze River basin is directly proportional to the economic development level. Based on the evaluation of the water quality in the Yangtze River basin in 2010, 67.4% of the evaluated major sections on the mainstream and tributaries of the Yangtze River were better in water quality than Class III, while only 49.7% of the assessed mainstream sections of the Yangtze River downstream of Hukou was better than Class III, and even only 13.67% of river sections in the Tai Lake watershed was better than Class III, indicating that the more developed the economy, the more serious the pollution of the water bodies in the area. In the sections on the mainstream and tributaries of the Yangtze River that flow through urban areas, the water quality was also poor with a 700-km-long riparian water pollution plume, which not only affected the ecosystem but also often impacted the safety of urban drinking water sources.

7.1.5 Lowered Riparian Ecological Barrier Function Due to Utilization of Riparian Zone and Sandbank

The development and utilization of riparian zones are mainly port terminals, followed by factories, warehouses, bridges, and others such as river-crossing pipelines, intake and drainage structures, ash fields of power plants, docks, and bank alteration projects.

Both banks along the mainstream of the Yangtze River downstream of Yichang is 3322 km long, of which 661 km has been utilized, resulting in a utilization rate of 19.9%. The lower reaches of Yangtze River have a higher bank utilization rate than the middle reaches, and the utilization rate of urban riparian zones is more than 50%. For example, by the end of 2003, Jiangsu Province had developed and utilized approximately 272 km of the 815 km riparian zone on the mainstream of the Yangtze River, or a utilization rate of 33% (Zhou 2007), of which the utilization rate had reached 51% for Level I riparian zones (refer to the deepwater section where the river regime and river banks are basically stable; the contours of iso-depths are continuous; there is an adequate water and land area; and the bank conditions are suitable for the construction of 10,000 tonnes or larger berths) and 41% of Level II riparian zones (refer to the moderately deepwater section where the river regime and river banks are relatively stable; the contours of iso-depths are continuous; there are appropriate water and land areas; and the bank conditions are suitable for the construction of 5000–10,000 tonnes berths).

The vegetation on the bank slopes is not only the important habitat for amphibians, insects, and birds but also the purifying filtration belt for rainwater and wastewater. Various types of bank use and protective hard slope surface of the banks have affected the functions of the green ecological barriers and basic biological habitats. As a result, the ecological and environmental functions of the river have declined.

Sandbars and sandbanks are the most important biological habitat or ecological speckles of a river, which are not only important temporary retaining and passing zones of floodwater but also suitable habitats for wildlife. However, if exploited, the most important habitat environment of the river will be lost. On the mainstream between Yichang and the estuary in the middle and lower reaches of the Yangtze River, there are 120 mid-channel sandbars with a total area of about 2700 km^2 and many sandbanks with a total area of about 1770 km^2. There is a wide disparity in size of sandbars and sandbanks. The largest mid-channel sandbar is Chongming Island that has now been protected from flood hazards, and the smallest is the Liyu sandbar. A considerable number of sandbars and sandbanks have been transformed into farmland or exploited, which has not only occupied important habitats of wildlife but also affected the floodway.

7.2 Ecologically Sensitive Zones to Be Protected in the Yangtze River Basin

7.2.1 Objects to Be Protected

As a result of the rapid development of productivity, human beings have become increasingly more capable of transforming nature. If necessary reserves are not established or protected objects are not clearly defined, not only important wildlife will lose their last home, but human beings will also lose the natural environment and space for tourism and leisure. Consequently, human beings may make irreparable mistakes. Therefore, any country should establish a certain range of protected areas. Of course, the protected objects will need to be determined based on closely related factors such as the national socioeconomic development level, human standards of living, and the quality of life required by the majority of the citizens. In fact, there are many options, and different people may have various options. From the attitude toward the quality of life and the ecological environment, human beings can be divided into three categories. The first category refers to those who are willing to live a poor life to protect the ecological environment. There are many such people in developed Western countries, but there are not many in China where most people used to be so poor that they want to live a rich life by growing wealth first and then considering the protection of the ecological environment. The second category includes those who are willing to protect the ecological environment as long as economic development and growing wealth are not affected. The third category refers to those who consider the well-being of human beings above all, do not care about the ecological environment, and consider living in an artificial environment acceptable. From the doctrine of the mean, most people are certainly in the second category, but it is the most difficult to implement the choice of the category in reality because development and protection are a pair of contradictions. Some consider that development occurs first and protection can follow later. However, this approach is very risky because once the ecological environment is destroyed, it will be very difficult to restore it, or even if it can be restored, the cost will be very high, and some wildlife cannot be restored after extinction. Therefore, the protected objects should be selected first; certain protected areas are established; and these areas will be avoided from the economic development and utilization of natural resources.

Theoretically, if conditions permit, all relatively natural regions, rivers, and living things are worth protecting, and whatever is protectable should be protected even if the environment and wildlife which do not appear to be of a great value at the present time should also be appropriately protected. For example, even in the most economically developed United States, wetlands were not recognized to be valuable more than 100 years ago. As a result, more than 50% of the wetland areas in the United States were lost, and only parts of the lost wetlands have been restored with a lot of manpower and money spent. It is worth it to be conservative in the protection of the ecosystem and avoid making historical mistakes. Natural landscapes, the environment, and living things are worth protecting, such as curvy rivers, biological habitats, wildlife, natural forests, endemic species, etc., which should all be considered as protected objects.

7.2.2 Major Protected Areas in the Yangtze River Basin

7.2.2.1 Types of Reserves

Based on the difference of the protected objects and responsible departments, China has various types of reserves, administered by various ministries and provinces and at the international and state levels. They are national nature reserves, aquatic animal and plant reserves, scenic sites, world's natural and cultural heritages, national geologic parks, important cultural protection units, national forest parks, major concentrated drinking water sources, international wetlands directories, important aquatic landscapes, important humanities landscapes and water landscape, etc. There are 561 various types of reserves in the Yangtze River basin with a total area of 32.8801 million ha, including 59 national nature reserves with a total area of 22.8906 million ha and 202 provincial nature reserves with a total area of 6.435 million ha. The nature reserves in the Yangtze River basin are mainly distributed in the provinces located in the upper Yangtze River, such as Qinghai, Sichuan, and Gansu. Qinghai boasts 60% and Sichuan 20.6% of the area of the nature reserves.

There are four types of aquatic nature reserves, including national nature reserves, such as the Rare and Endemic Fish Nature Reserve in the upper Yangtze River established to preserve aquatic organism resources, Hubei Yangtze Baiji Dolphin Nature Reserve in the Yinluo section, Hubei Yangtze Tian'ezhou Baiji Dolphin National Nature Reserve, and Anhui Tongling Freshwater Dolphins Nature Reserve, and provincial and municipal nature reserves such as the Yangtze Yichang Chinese Sturgeon Nature Reserve, Zhenjiang Yangtze Baiji Dolphin Nature Reserve, Shanghai Yangtze Estuary Chinese Sturgeon Nature Reserve, Poyang Lake Freshwater Mussel Nature Reserve, and Sichuan Tianquan River Rare Fish Provincial Nature Reserve.

To preserve wetland resources, the following nature reserves have been established: Anhui Alligator National Nature Reserve, Anhui Shengjinhu National Nature Reserve, Shanghai Jiuduansha Wetland National Nature Reserve, Chongming Dongtan Bird National Nature Reserve, Hunan East Dongting Lake National Nature Reserve, Hunan Zhangjiajie Giant Salamander National Nature Reserve, and Jiangxi Poyang Lake National Nature Reserve. Provincial nature reserves include Anhui Anqing River Waterfowl Protection Area, Anhui Guichi Eighteen Cable Reserve, Anhui Dangtu Shijiu Lake Reserve, Hubei Wanjiang River Giant Salamander Reserve, Hubei Hong Lake Wetland Nature Reserve, Hubei Zhongjian River Nature Reserve, Hubei Longgan Lake Wetland Nature Reserve, Hubei Liangzi Lake Wetland Nature Reserve, Shanghai Jinshan Sandao Marine Ecological Nature Reserve, and Poyang Lake Nature Reserve.

Some of the nature reserves involve rivers, including the national and provincial nature reserves such as Sanjiangyuan Nature Reserve, Baima Snow Mountain National Nature Reserve, Yulong Snow Mountain National Nature Reserve, and Haba Snow Mountain Scenic Area along the mainstream of the Yangtze River; Baodinggou Nature Reserve along the mainstream of the Min River; Huanglong National Nature Reserve and Guanwushan (Fog View Mountain) Scenic Area along the mainstream of the Fu River; and Hanzhong Crested Ibis National Nature Reserve along the mainstream of the Han River. Although these reserves do not directly involve the mainstreams of the rivers, they were bordered along the rivers. Therefore, projects involving development and utilization of water resources should protect the nature reserves in accordance with relevant state laws and regulations and minimize any impact on them.

7.2.2.2 Included in the UNESCO "Man and Biosphere Reserve"

(1) Wolong Giant Panda Reserve in Wenchuan County of Western Sichuan Province

The Wolong National Nature Reserve has been on the UNESCO's World Network of Biosphere Reserves since January 1980. It covers an area of approximately 200,000 ha and primarily protects the natural ecosystem and rare animals such as giant pandas in the alpine forest region of southwest China. It is located in a subtropical transition zone toward the Qinghai-Tibet Plateau. Because the high mountains block the easterly airflow from the Pacific and the circulation of the westerly wind, a year-round warm and humid subtropical mountain climate has developed, forming complete vertical distributional bands from feet to mountaintops. The major rare animals and plants include the giant panda, snub-nosed monkey, antelope, red panda, Thorold's deer, Sichuan redwood, Min River cypress, kingdonia, birthroot, camphor trees, etc.

(2) Fanjingshan National Nature Reserve

The Fanjingshan National Nature Reserve is located at the junction of Jiangkou, Yinjiang, and Songtao Counties of Guizhou Province and has been on the UNESCO's World Network of Biosphere Reserves since 1986. The reserve is one of a few relatively complete ecosystems in a subtropical region and covers an area of 41,000 ha to primarily protect the precious animals and plants such as the Guizhou snub-nosed monkey and the dove tree. Similar to the grand panda, the Guizhou snub-nosed monkey is a species associated with the quaternary time and a bona fide "living fossil." Currently only an estimated 100–500 Guizhou snub-nosed monkeys live in the Fanjingshan forest area, and they are highly valuable to the study of ancient creatures and their living environment. The natural dove tree only survives in China and its largest area is in Fanjingshan. It is known as the "dove tree of China" because its flowers appear as if they were the dove's extended wings. In the late nineteenth century, it traveled across the sea to foreign countries, and it has now become the world's famous ornamental tree.

(3) Shennongjia National Nature Reserve

The Shennongjia National Nature Reserve is located at the junction of Fang, Xingshan, and Badong Counties of Hubei and has been on the UNESCO's World Network of Biosphere Reserves since 1989. It covers an area of about 74,000 ha to mainly protect rare species such as the snub-nosed monkey and the dove tree. The Shennongjia National Nature Reserve is known as "Five Fantasies of Shennong" because of the presence of peaks, passes, clouds, caves, and trees in the reserve. There are more than 2200 precious tree species of the local origin, of which some are endangered ancient species, such as the dove tree, iron cedar tree, tetracentron, caramel tree (ever incense tree), and ranunculales. There are more than 570 animal species, of which more than 20 species are experiencing peculiar whitening phenomena, such as the white bear, white snake, white water deer, white snub-nosed monkey, and white stork, and the cause of the whitening is still an unresolved mystery.

(4) Yancheng National Nature Reserve

The Yangcheng National Nature Reserve is located in Yancheng City, Jiangsu, and has been on the UNESCO's World Network of Biosphere Reserves since 1992. The reserve mainly protects the rare birds such as red-crowned cranes. The 70,000 ha reserve consists of *artemia halodenderon* and grassy and reedy marshes, provides suitable habitat for birds, and is the world's largest wintering habitat for the red-crowned crane. In addition to the red-crowned crane, other rare species include white stork, Siberian crane, white-shouldered vulture, hooded crane, white-naped crane, black crane, gray crane, swan, etc.

(5) Maolan National Nature Reserve

The Maolan National Nature Reserve is located in Libo County, southern Guizhou Province and has been on the UNESCO's World Network of Biosphere Reserves since 1996. The reserve is the only one highly concentrated, highly primitive, and relatively stable karst forest ecosystem that has survived in the subtropical zone of China and even in the same latitudinal zone of the world.

(6) Jiuzhaigou National Nature Reserve

The Jiuzhaigou National Nature Reserve is located in Nanping County, Sichuan, and has been on the UNESCO's World Network of Biosphere Reserves since 1997. The reserve covers an area of approximately 60,000 ha and is the first nature reserve in China to protect the natural scenery as its main purpose. Jiuzhaigou is a valley with a depth of more than 60 km. According to legend, there used to be nine Tibetan villages. Jiuzhaigou (literally Nine Village Valley) took its name from the nine villages. The most unique here are the scenic lakes and mountains. The 108 lakes are known as 108 sons of the sea by local residents and are connected by dozens of waterfalls. The blue lakes, red willows, flying waterfalls, and green cypress are harmonious in color, balanced in movement and quietude, and natural in combination.

(7) Huanglong National Nature Reserve

The Huanglong National Nature Reserve is located in Songpan County, Sichuan Province, and has been a UNESCO Biosphere Reserve since 2000. Adjacent to the Jiuzhaigou National Nature Reserve, this reserve is a gallery of beautiful and colorful pools. Within the reserve, flowers and trees cover everywhere; Chinese wisteria and vines spread all over the ground; and the blue water from green springs form rolling waterfalls. At elevations ranging from 2123 to 3576 m on the long yellow slope, there are more than 4300 cascade pools decorated with pearl and jade-like colors that form various shapes due to the abundance of twists and turns. For example, some are shaped like a horse palm; some are a crescent moon; and some are a diamond lotus. The pools also have different depths: some are more than 3 m deep, and some are as shallow as a few cm deep, but they are all clear. The most fascinating is how the pool water changes color: at various instances it can go from red to green, green to yellow, emerald green to gold, blue to dark green, and blue to white. The color is always ideal, never too dark or too light. While all water comes from the same source, the water colors of the pools differ and change endlessly.

(8) Yading National Nature Reserve

The Yading National Nature Reserve is located in the southern Ganzi Tibetan Autonomous Prefecture of Sichuan and has been on the UNESCO's World Network of Biosphere Reserves since 2003. The reserve is located in the eastern part of the famous Qinghai-Tibet Plateau in the middle of the Hengduan Mountains and is bordered by Muli County of the Liangshan Prefecture on the southeast, Xiangcheng County and the adjacent Zhongdian County of Yunnan Province on the west, and Litang County on the north. The reserve covers an area of 7323 km^2 and has an average elevation of 3750 m, with the highest elevation at 6032 m. It has an alpine monsoon climate. Most of the time, the weather is good, the sun shines, and the natural scenery is beautiful. The "Ancient Daocheng Ice Cap" is especially famous in the world. A long and wonderful mountain pass appears to traverse through the mundane world and extend into the heaven. Countless devout pilgrims have come to this holy land through hard work and an arduous trek to dedicate their physical kowtows. Daocheng, the ancient name "Rice Dam," means an open basin at the mouth of a valley in Tibetan. According to the *Xikang maps*, "in order to wish a successful pilot cultivation for rice, the area was renamed Daocheng in Guangxu's 33rd year."

Yading in Tibetan means "the land of the sun." In Daocheng, Yading appears long and wide; the world seems boundless; rocks are overwhelmingly scattered everywhere; and everything is dazzling. The blue water surface of 1145 alpine lakes scattered among rugged rocks appears to be jade. The scenery is extremely spectacular and is known as "the Last Shambhala." The Haizi Mountain, also known as the "Ancient Daocheng Ice Cap," is a natural stone carving park and is the remains of the ancient ice bodies left by the Himalayan orogeny. It ranges in altitude from 3600 to 5020 m and spans 3287 km^2. Standing on the Haizi Mountain, the world appears to be boundless when one looks into the distance, and the scenery is ethereal.

(9) Foping National Nature Reserve

The Foping National Nature Reserve is located in Foping County of Hanzhong City in southwest Shaanxi Province and has been on the UNESCO's World Network of Biosphere Reserves since 2004. The Foping National Nature Reserve is located in the southern slope of the middle section of the Qinling Mountains. The major protected objects are the giant panda and its habitat, and the reserve covers an area of 29,240 ha with a vegetation cover of more than 90%. There are many flora and fauna species in the reserve. There are 22 state-protected rare plants in the reserve such as Taibai redwood, Qinling fir, and kingdonia. There are also state-protected animals in the reserve such as the giant panda, snub-nosed monkey, golden cat, and golden eagle. The reserve is beautiful and fraught with countless green trees and crystal clear water, fantastic stone peaks and forest, and flowing springs and powerful waterfalls. The Foping National Nature Reserve is rich in species and biological resources. It is the central region for the distribution of the giant panda in the Qinling

Mountains. In the core area, there is a giant panda in an area of 2.5 km^2 on average, and this density of the wild giant panda ranks first in China. The brown and white giant panda was once observed in the reserve. Therefore, the reserve has since been considered "the most promising land for the wild giant panda to survive and prosper."

7.2.2.3 List of Wetlands of International Importance

The *Convention on Wetlands of International Importance Especially as Waterfowl Habitat, or Ramsar Convention on Wetlands*, or Convention on Wetlands, was signed by representatives from 18 countries in Ramsar, a southern coastal small town of Iran, on February 2, 1971. The purpose of the convention is to protect and rationally use wetlands. The Convention on Wetlands came into effect on December 21, 1975. Currently there are 158 contracting parties, and China joined the Convention on Wetlands in 1992.

By the end of 2008, China had had a total area of more than 38.4855 million ha of various wetlands that are each greater than 100 ha in area (excluding paddy fields), making up more than 10% of the total area of the world wetlands, of which marshes were 13.7003 million ha in area, lacustrine wetlands 8.3516 million ha, riverine wetlands 8.207 million ha, marine wetlands 5.9417 million ha, and reservoirs and ponds 2.285 million ha. By 2009, a total of 37 wetlands had been on the List of Wetlands of International Importance (Ramsar List) with a total area of more than 3.8 million ha.

The List of Wetlands of National Importance in China includes 173 wetlands, 29 of which are located in the Yangtze River basin. They are Longbaotan, Yirancuo Lake, and Duoergaicuo Lake of Qinghai; Jiuzhaigou of Sichuan; Dianchi, Chenghai, Lugu, Bita, Napa, and Lashi Lakes and Huize Black-Necked Crane Habitat and Zhaotong Dashanbao of Yunnan; Yang County Crested Ibis Habitat of Shaanxi; Hongfeng Lake and Weining Caihai of Guizhou; Danjiangkou Reservoir Area; Dongting Lake; Wang Lake, Hong Lake, Liangzi Lake Group, and Shishou Swan Island of Hubei; Shijiu Lake, Yangzi Alligator Nature Reserve, Chao Lake, Taiping Lake, and Shengjin Lake of Anhui; Poyang Lake of Jiangxi; and Chongming, Changxing, and Hengsha Islands of Shanghai. The following nature reserves in the Yangtze River basin are on the Ramsar List:

(1) East Dongting Lake National Nature Reserve

The East Dongting Lake National Nature Reserve is located in Yueyang City, northeastern Hunan Province, with a total area of approximately 190,000 ha. There are 1186 species of 159 families of vascular plants, 114 species of 23 families of fishes, and 158 species of 41 families of birds, among which there are 32 bird species under the state priority protection. The nature reserve is an important wintering habitat for migratory birds with about ten million migratory birds wintering here each year.

(2) Poyang Lake Nature Reserve

The Poyang Lake Nature Reserve is located in Yongxiu County, northern Jiangxi Province, with a total area of approximately 22,400 ha. As the lake area is affected by the Xiu River system and the Gan River system, sandbars and sandbanks emerge when the water level in the protected area falls during the dry season, resulting in the formation of nine independent lakes. The nine lakes merge to form Poyang Lake during the wet season. This reserve is a wintering habitat for migratory waterfowls and important migratory birds. There are nearly 250 bird species in the reserve, including 108 waterfowl species. The major waterfowl species include Siberian cranes, white storks, swans, and wild ducks. There are 122 fish species in the lake, many of which are economic fish species. According to the observation conducted in the winter of 1998, there were nearly 100,000 wintering migratory birds, including more than 1500 Siberian cranes, over 1000 white-naped cranes, above 2000 little swans, in excess of more than 2000 white spoonbills, and greater than 3000 wild geese and ducks.

(3) Chongming Dongtan Nature Reserve

The Chongming Dongtan Nature Reserve is a lowly-lying alluvial island at the easternmost tip of Chongming Island with an area of 32,600 ha. Due to the deposition of sediments from the Yangtze River, large areas of marshes of freshwater to brackish water, tidal ditches, and intertidal flatlands formed. There are lots of farmlands, fish ponds, crab ponds, and reed ponds in the reserve. The marshes are rich in vegetation and benthic animals. It is an excellent resting place and stopover for migratory birds in the Asia-Pacific region in the spring and autumn and an important wintering habitat for migratory birds.

(4) South Dongting Lake Wetland and Waterfowl Nature Reserve

The South Dongting Lake Wetland and Waterfowl Nature Reserve are located in Yuanjiang City, Hunan Province, with an area of about 168,000 ha. The wetland is in the southern part of Dongting Lake, the largest water-passing freshwater lake in the middle reaches of the Yangtze River. The reserve is extremely rich in biodiversity and is an important habitat for many waterfowls, such as white stork and Siberian crane. The productions of economic animals and plants are high with great economic values. The wetland is also extremely important for the regulation and retention of floodwater from the Yangtze River.

(5) West Dongting Lake (Muping Lake) Nature Reserve

The West Dongting Lake (Muping Lake) Nature Reserve covers an area of 35,000 ha. The wetland is an inseparable and important part of the entire Dongting Lake wetland and is one of the typical subtropical inland wetlands. The wetland contains abundant biological resources and has important values for protection and scientific research.

(6) Bita Lake Wetland

The Bita Lake Wetland is located in Diqing Tibetan Autonomous Prefecture of Yunnan Province. It covers an area of 1985 ha with an average elevation of 3568 m. It is a well-preserved wetland of a closed alpine freshwater lake. It is the habitat for rare fish species such as ptychbarbus chungiensis and *Diptychus chungtiensis Tsao* and is also an important resting place and wintering habitat for many rare birds, including the black-necked cranes. As an important catchment area of the Qinghai-Tibet Plateau, it plays an important role in the retention of floodwater and the regulation of water balance in the middle and lower reaches of the Yangtze River.

(7) Napa Lake Wetland

The Napa Lake Wetland is located in the Diqing Tibetan Autonomous Prefecture of Yunnan Province and covers an area of approximately 3434 ha with an average elevation of 3568 m. It is a karst-type seasonal reed wetland composed of meadows, swamps, water surface, and forest around the lake. Due to obvious differences in altitude, the wetland is rich in vegetation diversity. It is an important wintering habitat for rare and endangered birds such as black-necked cranes. It has a natural floodplain of wet grassland with seasonal inundation, which has important effects on the regional hydrology and climate regulation.

(8) Lashi Lake Wetland

The Lashi Lake Wetland is located in Lijiang Naxi Autonomous Prefecture of Yunnan Province and covers an area of 3560 ha with an average elevation of 3100 m. It is an alpine freshwater lacustrine wetland composed of swamps, water surface, and forests around the lake. It is the habitat for the state-protected important wild animals such as black-necked cranes, bar-headed geese, and whooper swans. As it is a catchment area of the Jinsha River basin, it plays an important role in soil conservation and flood control. It also plays an important role in water balance and water level regulation in the middle and lower reaches of the Yangtze River.

(9) Shanghai Yangtze Estuarine Wetland Nature Reserve for Chinese Sturgeon

The Shanghai Yangtze Estuarine Wetland Nature Reserve for Chinese Sturgeon is located in the Yangtze Estuary in Chongming County of Shanghai with a total area of 27,600 ha. The major protected objects of the reserve are Chinese sturgeons and their habitat of the natural ecological environment where Chinese sturgeons spawn and their juveniles grow. The reserve is also an important channel for other fish species to migrate and an important place for them to spawn and browse for food. It is the first wetland of international importance in China's aquatic wildlife reserve.

(10) Hubei Hong Lake Wetland Nature Reserve

The Hubei Hong Lake Wetland Nature Reserve is located in southeastern Hubei Province and in the Jianghan Lake Group in the middle reaches of the Yangtze River. The lake is the seventh largest freshwater lake in China. The reserve covers an area of 37,088 ha. The purpose of the reserve is to protect the lacustrine wetland ecosystem composed of aquatic and terrestrial organisms and their habitat, as well as unpolluted freshwater resources, wetland ecosystems, and biodiversity, especially the state-protected important birds, fishes, and plants.

(11) Dabaoshan Wetland Reserve

The Dabaoshan Wetland Reserve is the source of many rivers in the Jinsha River system of the upper Yangtze and is located in Dabaoshan Township in the northwestern part of Zhaoyang District, Zhaotong City, Yunnan. There are many rare species in the reserve. The reserve has been on the "List of Wetlands of International Importance" since 2005. Major rivers through the reserve include the Tiaodun River, the Dahaizi River, and Dayadong Gulley. The Tiaodun River flows westward into the Niulan River, the Dahaizi River flows northward and is one of the sources of the Daguan River, and Dayadong Gulley flows westward after it emerges into Cha Gulley and joins the Longshu River in Ludian County. Many wetlands are scattered throughout the reserve. However, wetlands with large areas are mainly concentrated in the Tiaodun and Dahaizi Rivers and Lelezhai, Qinjiahaizi, and Yanmaidi Reservoirs. Among them, the Tiaodun and the Dahaizi Rivers are the largest in area and boast a considerable area of water surface. The area and extent of the wetlands vary with the seasonal change in water level. As the water level falls in the winter, the area of shallow water zones increases, which is a suitable habitat for black-necked cranes to stay during the nighttime. Wetlands of a small area associated with springs are scattered throughout the hillside, range in area from more than 10 m^2 to dozens of m^2, and are important habitat suitable for cranes to stay during the daytime.

The Tiaodun River Reservoir has a catchment area of 17.7 km^2, a water surface area of 3.375 km^2, a storage capacity of 12.36 million m^3, a discharge of 0.2 m^3/s in December and 15.1 m^3/s in July, and a water depth of approximately 6.5 m. Moreover, the shallow water zone along the edge of the reservoir is large with surrounding large areas of marshes, which are suitable wintering habitats for black-necked cranes to stay during the nighttime. The Dahaizi Reservoir has a catchment area of approximately 3.5 km^2, an average water depth of 2.5 m, and a surface water area of 0.8 km^2. Furthermore, because the water level in the reservoir is low, the area of the shallow water zone and surrounding wetlands are large, and they have become the most concentrated area for black-necked cranes to stay in the winter. Lelizhai and Yanmaili Reservoirs are not large in area or water surface area, but the meadows surrounding the reservoirs are also one of the habitats for black-necked cranes to forage.

In the reserve, the water quality is excellent, the water temperature is 6.0 °C, and the pH value is 8.2. Moreover, there is no industrial or agricultural production nor domestic sewage. Therefore, the water can be used as a drinking water source for human and livestock as the water quality meets the state's Class I standard. Owing to the regulation of groundwater, the depth of water in the Dabaoshan Wetland is in the range of 0.8–3 m year round. As the water level is relatively stable year round, droughts have never occurred.

7.2.2.4 Cultural and Natural Properties Inscribed on UNESCO "World Heritage"

On December 12, 1985, China ratified the *Convention Concerning the Protection of the World Cultural and Natural Heritage* and became a state party. Since 1986, China has been submitting to the World Heritage Committee applications for properties to be listed on the cultural and natural heritage of the world. From 1987 to June 2007, 35 properties from China were included in the "World Heritage List", 13 of which are located in the Yangtze River basin (see Table 7.1). The following ten World Heritage Properties are entirely in the Yangtze River basin: Ancient Building Complex in the Wudang Mountains, Dazu Rock Carvings, Huanglong Scenic and Historic Interest Area, Jiuzhaigou Scenic and Historic Area, Lushan National Park, Wulinyuan Scenic and Historic Interest Area, Old Town of Lijiang, Mount Qingcheng and the Dujiangyan Irrigation System, and Mount Emei Scenic Area, including Leshan Giant Buddha Area and Sichuan Giant Panda Sanctuaries. The following three world heritage properties are partially in the Yangtze River basin: Imperial Tombs of the Ming and Qing Dynasties, Three Parallel Rivers of Yunnan Protected Areas, and South China Karst.

The famous scenic and historic interest areas in the Yangtze River basin include Zhangjiajie and Wulinyuan in Hunan; Jiuzhaigou, Huanglong, Dazu Rock Carvings, Mount Qingcheng including Dujiangyan, and Mount Emei Area, including Leshan

Table 7.1 List of World Heritage Properties in Yangtze River basin

Name	Year of listing	Location	Remarks
Ancient Building Complex in the Wudang Mountains	1994	Hubei	
Old Town of Lijiang	1997	Yunnan	
Dazu Rock Carvings	1999	Chongqing	
Imperial Tomes of the Ming and Qing Dynasties	2000	Hubei, Hebei, Nanjing, and Beijing	Partially in Yangtze River basin
Mount Qingcheng and the Dujiangyan Irrigation System	2000	Sichuan	
Jiuzhaigou Scenic and Historic Area	1992	Sichuan	
Huanglong Scenic and Historic Interest Area	1992	Sichuan	
Wulinyuan Scenic and Historic Interest Area	1992	Hunan	
Three Parallel Rivers of Yunnan Protected Areas	2003	Yunnan	Partially in Yangtze River basin
Sichuan Giant Panda Sanctuaries	2006	Sichyan	
South China Karst	2007	Yunnan and Guizhou	Partially in Yangtze River basin
Mount Emei Scenic Area, including Leshan Giant Buddha Area	1996	Sichuan	
Lushan National Park	1995	Jiangxi	

Giant Buddha Area in Sichuan; Wudangshan in Hubei, Lushan in Jiangxi; Classical Gardens of Suzhou in Jiangsu; and Old Town of Lijiang in Yunnan.

World geologic parks in the Yangtze River basin include Zhangjiajie in Hunan, Mount Lu in Jiangxi, Wenxing in Sichuan, and Funiu Mountains in Henan.

7.2.2.5 River Sections to Be Protected

Theoretically, the river sections that intersect the terrestrial ecologically sensitive areas should be protected or preserved, thus forming a continuous and relatively complete protected area of land and water. Based on international experiences, it is recommended that the following types of river sections be protected. ① The sections of water conservation and ecological fragile zones on the mainstream and major tributaries of the Yangtze River generally should be protected where any development activities should be prohibited. ② The river sections near ecologically sensitive areas and important tourist attractions. ③ The river sections in sparsely populated areas where there are no highways. ④ The following are various established protected areas:

1. The 1174-km-long section from the source to Zhimenda on the mainstream of the Yangtze River as it is the source of the Yangtze River and three parallel rivers (National Nature Reserve)
2. The 295-km-long Xiangjiaba-Songgai section (Xiangjiaba-Mashangxi of Chongqing section is 353.16 km) The 628.23-km-long mainstream of the Chishui River and the Rare and Endemic Fish National Nature Reserve (Zhejiang-Leibojiang section) in the upper reaches of the Yangtze River
3. Yichang Chinese Sturgeon National Nature Reserve (80 km) in Hubei
4. Tianezhou Chinese Paddlefish National Nature Reserve (89 km and 21 km of the old channel)
5. Xinluo Section Chinese Paddlefish National Nature Reserve (135.5 km) in Hubei, Tongling Freshwater Dolphin National Nature Reserve in Anhui, and Jiuduansha National Wetland Nature Reserve.

7.2.3 Important Habitats for Fish Species

Fish habitats of the Yangtze River include three sites and one pathway.

7.2.3.1 Important Spawning Sites

The spawning sites for several important fish species include the following: the spawning sites for the Chinese paddlefish are concentrated in the section of the Yangtze River near Jiangan County and the section of the Jinsha River near Baishuxi of Yibin County. The spawning sites for the Chinese sturgeon are the natural breeding sites in the 5-km-long section downstream of the Gezhouba Dam. The Dabry's sturgeon is an anadromous fish species in the river, and the sexually mature individuals migrate upriver to lay sinking and sticky eggs in the upper reaches of mainstreams of rivers during the breeding season. The spawning sites of the Dabry's sturgeon are mainly distributed in the Chuan River and the lower reaches of the Jinsha River. The spawning sites of the Chinese sucker are mainly distributed in the upper reaches of the Jinsha River, the Min River, and the Jialing River. Before the construction of the Gezhouba Dam, no spawning sites of the Chinese sucker had been identified in the middle and lower reaches of the Yangtze River. Since the construction of the dam which blocked the upriver migration path of the fish, the population of the Chinese sucker in the upper reaches of the Yangtze River has declined sharply. The gonads of the Chinese suckers that live in the river section downstream of the Gezhouba Dam can reach maturity and have established spawning sites in that river section. The spawning sites of the Chinese sucker downstream of the Gezhouba Dam are mainly distributed in the following river sections: Gezhouba Dam-Xiaoziyan, Yanzhiba-Huyatan, Honghuatao-Houtuojiang, Baiyang-Louzihe, and Zhicheng section. Figure 7.1 is an illustrative map for the distribution and scale for the spawning sites of the FMCC. The black, grass, silver, and bighead carps usually feed and grow in lakes and seasonal floodplains, but mature individuals migrate to the mainstream of the Yangtze River for breeding.

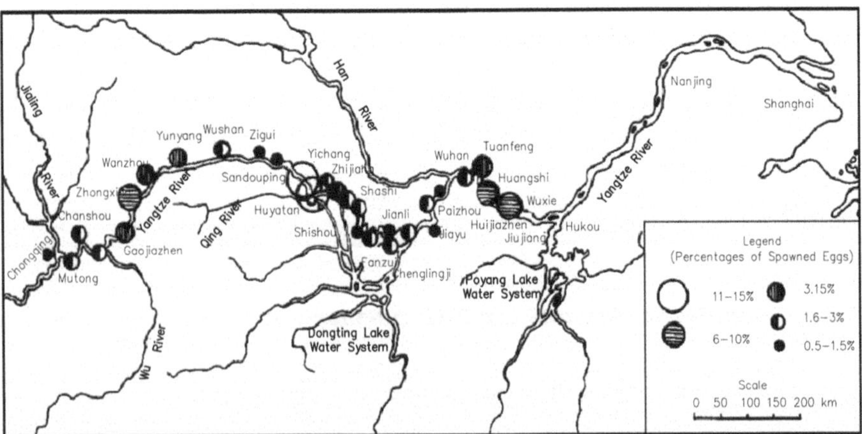

Fig. 7.1 Illustrative map for distribution and scale of spawning sites of FMCC in the Yangtze River

7.2.3.2 Fish Forage and Wintering Sites

The complex water system in the floodplains and lakes connected to the Yangtze River in the middle and lower reaches are important forage sites for the juveniles of migratory fish species. The reservoirs at cascade hydropower stations on the mainstream and tributaries in the upper reaches of the Yangtze River may possess a higher biological productivity and provide more food organisms than natural rivers. The slow-flowing areas, such as forks and tails of reservoirs, can increase food and provide suitable wintering habitats for the fish that live in the static or slow-flowing areas. However, for many fish species that favor to live in the flowing water in the upper reaches of the Yangtze River, their forage sites have often moved to the tails of the reservoirs or the tributary rivers.

7.2.3.3 Migratory Pathways

Migration is a life habit developed to adapt to the characteristics of the ecological system during the long evolution process of the fish. It is an active, directional, and collective cyclical movement and is repeated every year as the life cycle of the fish shifts. Based on the purposes, fish migration can be divided into breeding, food searching, and wintering. Maintaining the four-dimensional connectivity of rivers is the fundamental measure to protect the migratory channel of fish species, and the river-lake connection channel, the main channels of rivers, and the slow-flowing areas are basic migratory pathways.

Figure 7.2 shows the areas of national nature reserves for rare and endemic fish species in the upper Yangtze River.

Figure 7.3 are photographs of typical endemic fish species in the upper Yangtze River, most of which favor to live in the fast-flowing water and prefer to spawn in the substrate of gravel.

7.2.4 Expanded List of Objects to Be Protected in the Future

As human activities have become increasingly intensified, the relatively natural sections and their environments of the water systems in the Yangtze River basin are worth protecting. Moreover, some semiartificial or artificial ecosystems are also worth protecting. As more than 160 large reservoirs and thousands of medium-sized reservoirs have been constructed in the basin, cascade reservoir groups have developed on rivers. As a result, most of the rivers have become artificial and semiartificial wetland ecosystems and formed certain amounts of biomasses of new ecological zones. In addition, relatively more reservoirs have functioned as water supply (drinking) sources, and therefore, they should be protected.

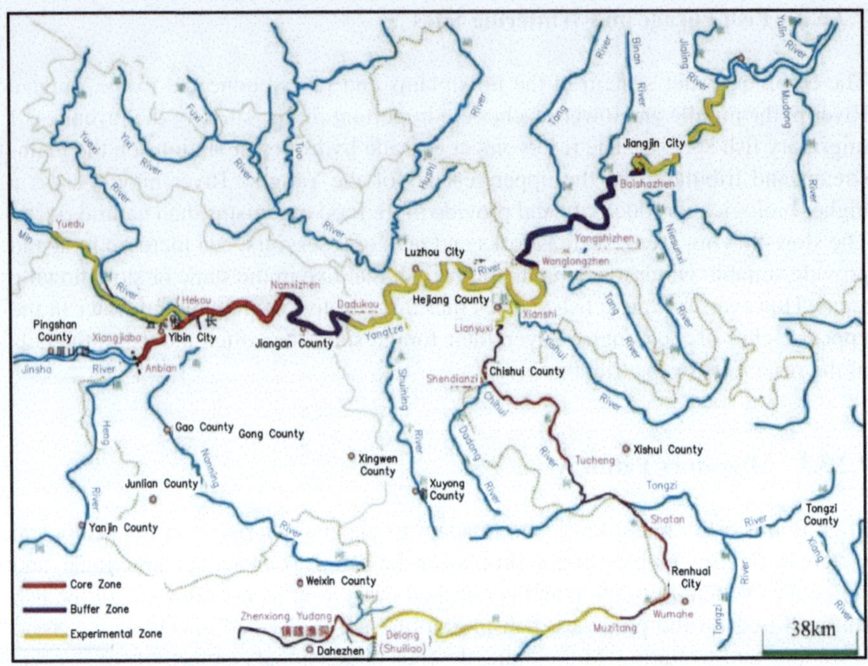

Fig. 7.2 National natural reserve for fish species in the upper Yangtze

Fig. 7.3 Photographs for typical endemic fish species in the upper Yangtze

7.3 Environmental Conditions and Species that Need to Be Protected

7.3.1 Hydrological Conditions

The rapid-flowing environment in the upper reaches and the lacustrine wetlands in the middle and lower reaches of the Yangtze River are the results of the alternate processes of floods and droughts and the effects of the water flow and sediment transport under the influence of tectonic movement and climate change. The rapid flow, complex river-lake relationship, and water system connectivity have created a rich variety of biological habitats and biodiversity. Historically, the lacustrine wetlands in the middle and lower reaches of the Yangtze River have remained connected to the Yangtze River system to varying degrees. It is the regulation of the lakes along the Yangtze River that has made it possible for the low-lying areas to retain and detain floodwater of the Yangtze River during the flood season and replenish water to the river during the dry season so as to ensure the safety of human production and life. Moreover, when the water levels of the wetlands on both sides of the river are maintained, the sandbanks and floodplains along the Yangtze River are vibrant with life. Compared with the rapid-flowing waters of the Yangtze River, the lakes connected to the river maintain an environment of slow-flowing or relatively static waters. Some important fish species, such as the FMCC, need to spawn in the rapid-flowing waters of the Yangtze River and then migrate to the static waters in lakes to live and grow. Thus, the river-lake connection provides many migratory and semi-migratory fish species with "three sites and one pathway" (forage site, breeding site, wintering site, and migratory pathway), and the static and flowing aquatic environments constitute a very rich wetland environment and a biological system in the middle and lower reaches of the Yangtze River. Poyang and Dongting Lakes are both water-based lacustrine wetlands and can be characterized by "an appearance of a vast sea during the flood season and a linear feature during the dry season." During the dry season, large areas of flatland and shallow water zones become an internationally important wintering habitat for migratory birds and one of the most important aspects of the global ecological system. Maintaining the connectivity of a water system is to maintain the smooth flow of "blood" arteries of wetlands, make the water level of lakes change with that of the Yangtze River, and preserve the periodical change of wetlands' water surface. The varying water surface, water level, flowrate, flow velocity, and water temperature are the hydrological and aquatic ecological elements needed by wetlands and are also the vital environment for the aquatic ecosystem. Moreover, the river-lake connection can shorten the cycle of water exchange, improve the water quality, prevent the lakes from becoming eutrophic and swampy, maintain the vitality of the lakes, and extend the life of the lakes.

The Yangtze River is rich in fish resources, and the fish life process is very dependent on the hydrological and hydrodynamic processes of the river. For example, the emergence of fish fries needs seven ecological and hydrological aspects: the process of a flood peak, the initial water level of the flood peak, the daily rising rate of water level, the initial flowrate of a cross section, the daily increase rate of flowrate, the duration of water level rise, and the interval between two consecutive flood peaks. Drifting eggs swell after the membranes of the eggs absorb water, and under external conditions such as water flow, the fertilized eggs are incubated during the drifting process with the water flow. Based on the type of spawning, there are the following two kinds of drifting eggs in the Yangtze River: ① pure drifting eggs whose membranes are not sticky in which the eggs are separated from their membranes in the water and are not adhered to other objects, such as the FMCC, bream, and barbel chub, and ② slightly sticky drifting eggs whose stickiness is relatively weak. They will soon detach from the objects they adhered to during drifting and develop while drifting, such as turtle and topmouth culter (*Culter alburnus*). However, the Chinese sturgeon spawns only when the water level is in a declining process. Therefore, whether it is a flood process or a low-flow process, as long as the hydrological processes are natural, they need to be properly preserved and protected since they are the hydrological processes needed by the aquatic organisms.

7.3.2 Preservation of River Continuity

The four-dimensional continuities of a river refer to the longitudinal, lateral, depth, and temporal continuousness. The longitudinal and temporal continuities are essential elements, and they are the most basic characteristic of a river ecosystem. Presently, there are no dams only on the Tongtian River, the upper reaches of the Jinsha River, the upper reaches of the Yalong River, the mainstream of the Chishui River, and the mainstream of the middle and lower reaches of the Yangtze River whose longitudinal continuities have been maintained. In the upper reaches of the Jinsha and Chuan Rivers and their tributaries, there are about 180 fish species, a considerable number of which have been adapted to the rapid-flowing environment and fed on raw algae or benthic invertebrates. The fishes that live on the bottom of the channel or crawl on rocks require a continuous rapid-flowing environment. Before the construction of the Gezhouba Project, the breeding group of the Chinese sturgeon used to migrate upriver to the spawning sites in the mainstream of the upper Yangtze River to breed, and some Japanese eel (*Anguilla japonica*) individuals migrate upriver to the upper reaches to grow.

The lateral continuity includes not only the different types of floodland but also the river-lake connection. Preservation of the shallows and uplands will need to keep a little extra distance between dikes across the river, and the riparian zone, river forks, and mid-channel sandbars should not be encroached for land resource reasons. A small portion of fish resources in Dongting and Poyang Lakes are from tributaries but most of them are from the Yangtze River. The fish exchange between

Fig. 7.4 River section rich in ecological speckles

the mainstream of the Yangtze River and the lakes has its inherent pattern, i.e., the fries, juveniles, and adults of the Yangtze River migrate into the lakes through Chenglingji and Hukou. The fries and juveniles of the FMCC and other economic fish species drift to the lakes with the flowing water where they live and grow. During 1996–2005, the amount of water diverted through the three outlets and the harvest of the FMCC in the river section downstream of the Gezhouba Dam indicated that the general trend was that the larger the amount of water diverted through the three outlets, the higher the FMCC yield. The correlation coefficient between the amount of water diverted through the three outlets and the FMCC yield was 0.92. Therefore, the river-lake connectivity has direct impacts on the growth and yield of the FMCC. When the river-lake connectivity is low, the juveniles of the FMCC and other fishes cannot enter the lakes to feed or grow, while the flow velocity of the river is too high with lower water temperature and less food for the juvenile to be adequately fed and thus their survival rate is low. Figure 7.4 shows a river section that is rich in ecological speckles. The diversity of habitats is the basis of biodiversity.

The temporal continuity requires not only the flow of a river throughout the year but also requires the flood and low water processes as the season changes. Presently, due to the construction of many diversion-type hydropower stations in some secondary and tertiary tributaries in the Yangtze River basin, large sections of the rivers have had reduced flowrates, but even dry sections have occurred. The reservoirs with high regulation capabilities also have impacts on the temporal continuity of the rivers because they can lower the flood flowrate and the peak flowrate, weaken the flood process, increase the flowrate during the dry season, and decrease the range of hydrological change throughout the year. For example, the FMCC spawn only when the water level is in the rising process. Therefore, the natural hydrological rhythm is the basic environmental conditions of the river ecosystem.

7.3.3 Protection of Riparian Zone and Lacustrine Fluctuation Zone

The water fluctuation zone along the banks varies with the seasonal hydrological change, which is the zone where the gradient of change in a river and a lake is the highest. The riparian zone or the lacustrine fluctuation zone is not only the transition zone between water and land and the filtering zone but is also the richest zone of biodiversity. Although human beings are now paying attention to wetlands, but there are many misunderstandings about the relationship between wetland ecosystem and hydrology. Some argue that wetlands always need to maintain a large area of water surface, or the lack of water surface during the dry season is unfavorable to the ecosystem and so on. These understandings are one sided and sometimes harmful. Because wetlands not only have to have water but also need a large transition zone (or water fluctuation zone), the area around wetlands needs mudflats, grassland, and dry upland vegetation zone. Therefore, the water surface and the surrounding changing fluctuation zone comprise a complete wetland environment.

Wetlands must have water, but not necessarily a constant large water surface, rather a water surface that varies as season changes. During the flood season, wetlands should have a larger water surface where not only a large number of fish and other aquatic organisms can be raised, but also some dominant species can be inhibited, and the wetland biodiversity is preserved. During the dry period, the water surface should be greatly reduced, even to only disconnected small ponds or puddles, and large areas of mudflats, grassland, and bushes emerge in the wetland transition zones, which are the habitats of migratory birds, amphibians, reptiles, and insects. Cyclic exposure of the lake bottom to the atmosphere is beneficial to the decomposition of contaminants on the bottom and improvements of the lake water quality. For human beings, the weed-laden transition zone seems to be a "useless" land, but in fact it is a very important biological habitat and a filtering zone of the wastewater flowing into the lake. The fluctuation zone between land and water is a seasonal inundation zone. Due to the change of the hydrological process and rich habitat environment, the fluctuation zone is the area with the highest environmental gradient and is the richest in organisms and even higher in biomass and diversity than the water zone. Therefore, not only the water zone should be protected as the core of a wetland, but the fluctuation zone and riparian vegetation zone should also be protected as part of the core of the wetland. The riparian vegetation zone plays an indisputably important role in protecting the structure and integrity of the wetland ecosystem. Figure 7.5 shows an ideal transition zone between land and water. Different plants between the slope of the bank and the area beneath the water can provide various animals with different plants and habitats. Moreover, the vegetation

Fig. 7.5 Ideal transition zone between water and land

on the slope of the bank can filter contaminants, and the aquatic plants can absorb nutrients in the water. As a result, they can purify the water and ensure a good water quality suitable for the drinking water source of mankind and aquatic life.

7.3.4 Protection of Indicator Species

In the Yangtze River, many species need to be protected. There are as many as 300 species of fishes alone. In the Tuotuo River and its tributaries in the source area of the Yangtze River, there are only a few fish species which consist mainly of small alpine loach genera. In the upper reaches of the Tongtian and Jinsha Rivers, there are not many fish species which are composed mainly of those that have adapted to the cold and rapid-flowing aquatic environment, such as *Diptychus kaznakovi*, gymnodiptychus pachycheilus, and *Schizopygopsis malacanthus*. The middle and lower reaches of the Jinsha River and the Chuan River system are the richest in endemic fish species of the Yangtze River basin, and the majority of the fish species prefer the rapid-flowing conditions, such as *Schizothorax dolichonema*, *Schizothorax kozlovi*, Schizothorax wangchiachii, labeoninae and barbinae subfamilies of the Cyprinidae, hillstream loach, grass carp, etc. The economic fish species in the upper reaches mainly include common carps, *Coreius zeni*, and bronze gudgeon. In addition, the common economic fish species include Chinese longsnout catfish, *Spinibarbus sinensis*, *Onychostoma simum*, *Sinilabeo rendahli*, *Garra lamta*, rock carp, Yangtze River sturgeon (also known as Dabry's sturgeon), *Discogobio* genus, and *Takifugu* genus. Anadromous fish species in the upper reaches used to be mainly the Chinese sturgeon. Some permanent-resident Salmonidae, such as *Hucho bleekeri* (Sichuan taimen), are sporadically distributed in the Min River and its tributaries Dadu and Qingyi Rivers in the upper Yangtze River.

The middle and lower reaches of the Yangtze River are quite rich in fish resources, and there are many economic fish species, as well as anadromous species such as Reeve's shads, bigmouth grenadier anchovies, Japanese eels, *Fugu obscurus*, roughskin sculpins, and noodlefishes. There are also migratory fish species between lakes and the Yangtze River such as black, grass, silver and bighead carps, yellow-cheeks, *Ochetobius elongatus*, breams, *Xenocypris argentea*, mandarin fish, and St. Peter's fish and permanent resident fish species such as crucian carps, Spanish mackerels, erythroculter, *Hemiculter leucisculus*, northern snakeheads, and yellowhead catfish. Seven rare fish species are in the Yangtze River basin. They are the Chinese sturgeon, Chinese paddlefish, Yangtze River sturgeon, Chinese sucker, *Hucho bleekeri*, *Brachymystax lenok tsinlingensis*, and roughskin sculpin, of which the Chinese paddlefish and roughskin sculpin are China's endemic rare fish species and are mainly produced in the Yangtze River. While they are present in the upper, middle, and lower reaches, the two species mostly live in the mainstream and major tributaries in the upper reaches, and their known spawning sites are also in the upper reaches. Additionally, the Yangtze finless porpoise is currently also an important protected species.

7.4 Protection of Water Resources

7.4.1 Major Pollution Sources in the Yangtze River Basin

The sources of water pollution in the Yangtze River basin mainly include point source, nonpoint source, and endogenous source. The point sources of pollution mainly include industrial pollution sources and urban domestic pollution sources. Contaminant loads of nonpoint sources mainly include urban surface runoff pollution, farmland runoff pollution, rural domestic sewage and domestic waste pollution, soil erosion and associated pollution, and decentralized livestock feeding lot pollution. Due to the poor air quality in China, nonpoint source pollution can also be caused by deposition or precipitation. Endogenous pollution is mainly the internal contamination caused by the long-term retention of dissolved and deposited pollutants in the water body or accumulation on the bottom of the water body. Endogenous pollutions often occur in lakes near urban areas because of the long-term reception of wastewater.

In 2007, the total amounts of chemical oxygen demand (COD), ammonia nitrogen, nitrogen, and phosphorus from point and nonpoint sources in the Yangtze River basin were 6.122 million tonnes, 615,000 tonnes, 2.334 million tonnes, and 320,000 tonnes, respectively. Point sources contributed 54.3% of COD and 58.2% of ammonia nitrogen to the river, indicating that COD and ammonia nitrogen were mainly from point pollution sources. However, nonpoint sources contributed 78.6% of total nitrogen and 74.7% of total phosphorus to the river, indicating that total nitrogen and total phosphorus were mainly from nonpoint pollution sources.

Due to the harm of lindane to human body and organism, China has stopped producing and using the pesticide since 1983, but its impact had continued until the 1990s. At the beginning of this century, 308 organic contaminants were detected in the mainstream of the Yangtze River (Kong et al. 2007). From the 1950s to the early 1980s, the contamination of organochlorine pesticides in the Yangtze Estuary showed an increasing trend and then began to decline. However, it still has an impact on ecosystems. At the same time, the contamination of polychlorinated biphenyls (PCBs) in the Yangtze Estuary during the 1990s was more serious than that of the 1980s. In the 1950s and 1960s, oil pollution occurred in China's coastal areas, and the contamination became the most serious in the 1990s. The contamination of polycyclic aromatic hydrocarbons (PAHs) in the Yangtze Estuary was very minimal in the 1970s, but with industrial development in the Yangtze River basin, the contamination began to worsen in the 1980s and began to lessen in the mid-1990s.

In the past, water pollution control in China focused on the hygienic standard of COD, ammonia nitrogen, and coliform bacteria that would impact human drinking water. However, China did not pay enough attention to contamination of organic compounds, heavy metals, hormones, or drugs such as contraceptive drugs. Because these toxic and harmful contaminants have a long-term effect on human, it is difficult to determine the extent of the effect. On the other hand, the detection of these contaminants requires special equipment and testing methods. In fact, due to the rapid industrial development of China in the past several decades, these "mutagenic, teratogenic, and carcinogenic" toxic and harmful substances have posed a threat to human health. For example, China has been a country with high occurrences of cancers, which is related to the environmental pollution, especially the water pollution. Moreover, the extensive disorderly uses of hormones and other drugs have caused the increase of hormone concentrations in the environment and led to increased cases of precocious puberties and infertilities.

7.4.2 Major Protection Objects of Water Resources

According to the monitoring and evaluation results of water resources conducted in 2011, only 46% of China's water function zones were in compliance with the water quality standard. Based on the planned goals of water pollution control, by 2020, no less than 80% of the urban sewage effluents in the Yangtze River basin will have been treated, and by 2030, treatment plants for urban sewage effluents will have had the capacity up to the actual amounts of all urban sewage effluents. During the construction of sewage treatment facilities, attention should also be paid to the improvements of the associated pipeline network, so that the treated and raw effluents can be conveyed separately and the utilization efficiency of the facilities can be improved.

Drinking water sources are the key objects of water resources protection. Drinking water sources in the Yangtze River basin are mainly from natural surface water bodies, such as rivers and lakes, reservoirs, and with a small portion from groundwater. However, the Yangtze River basin boasts many natural and artificial water surfaces; vast areas and many sites need to be protected; and the river receives large amounts of wastewater. Therefore, it is a daunting task to protect the water resources in the basin.

7.4.2.1 Protection of Water Sources

The major water sources in the Yangtze River basin are surface waters, mainly rivers, followed by lakes. These waters are also the receptors of wastewater and areas for shipping. As the transition zone set forth for the water function zone is not long, especially in the mainstream and tributaries of the middle and lower reaches of the Yangtze River where the urban density and the utilization of riparian zones are high, almost all water sources are not far from upstream contamination sources or in the water zones where the potential polluting vessels are passing. Therefore, it is difficult to protect the water sources from contamination. It is recommended that longer buffers and source protection zones be established by installing physical or biological isolation facilities along the boundaries of the protection zones to prevent interference, such as human activities, with the protection and management of water sources, and to intercept contaminants from directly entering the protection zones of water sources. Protective isolation works include both physical isolation and biological isolation. The physical isolation refers to the use of a fence or a net as a mechanical enclosure around the protection zone of a water source. As for the biological isolation works, a suitable protective belt consisting of woods and grassy vegetation can be selected based on site-specific conditions. Figure 7.6 shows a water source in a well-protected environment.

Fig. 7.6 Green water source

7.4.2.2 Eutrophication Control in Lakes and Reservoirs

As a lake is a centripetal water system with a surrounding catchment area, contaminants can flow into the lake with the precipitation and runoff. As China's agriculture relies heavily on the use of fertilizers and pesticides, all the lakes are facing the threat of eutrophication. Most of the lakes in the Yangtze River basin are mesotrophic and have the trend to become eutrophic. The lakes that have been hypereutrophic include Dianchi, Chao, and Tai Lakes and those within urban areas. Remediation of these lakes will not only need a large sum of capital but will also require decades of time. The first step is to cut off the sources of pollution, and the second step is to resolve the endogenous pollution issue by concurrently dredging the lakes and restoring their aquatic ecology for a prolonged period of time. Therefore, the ecological restoration of the lakes will be planned as if it were a "prolonged war," and a successful restoration of the lakes will require to address the funding source for restoration and protection and to innovate the institutional management mechanisms.

7.4.2.3 Protection of Groundwater

In 2005, the amount of groundwater withdrawal in the Yangtze River basin was 8.279 billion m^3, of which 7.848 billion m^3 was from the shallow aquifer, accounting for 94.8% of the total groundwater withdrawal, and 432 million m^3 was from deep confined aquifers, accounting for 5.2% of total groundwater withdrawal. The total quantity of groundwater use was 8.279 billion m^3, of which 1.754 billion m^3 was agricultural use, 2.775 billion m^3 industrial use, and 3.75 billion m^3 domestic use, accounting for 21.2%, 33.5%, and 45.3%, respectively, of the total amount of groundwater use. According to the statistical data of water resources in secondary regions, the total amount of groundwater withdrawal in the Dongting Lake system was the highest at 2.082 billion m^3, accounting for 25.15% of the total amount of groundwater use in the Yangtze River basin, followed by the Han River basin where the amount of groundwater withdrawal was 1.648 billion m^3, accounting for 19.9% of the total amount of groundwater use in the Yangtze River basin. The secondary region in the Jinsha River section upstream of Shigu only had a groundwater withdrawal of three million m^3.

The shallow aquifer in the Yangtze River basin is rich in groundwater and receives abundant recharge from precipitation and surface runoff. Compared with other basins in China, especially north China, the level of groundwater utilization in the Yangtze River basin is relatively low. The amount of groundwater withdrawal was only about 3% of the groundwater resources in 2005, and the amount of groundwater supply makes up less than 5% of the total quantity of water supply.

In most of the areas with groundwater depression cones in the Yangtze River basin, land subsidence occurred to various degrees. The Suzhou-Wuxi-Changzhou region of Jiangsu, the Hangjiahu Plain of Zhejiang, and Shanghai within the Yangtze Delta are the most seriously affected by land subsidence. In addition, land subsidence also occurred in Nanchang, Jiujiang, Ganzhou, and Pingxiang of Jiangxi. The area of cumulative subsidence depressions demarcated by the 4 m contour was more than 1000 km^2 in Shanghai, and the maximum cumulative subsidence was 2.6 m. In the Suzhou-Wuxi-Changzhou region, more than 5000 km^2 of land cumulatively subsided by more than 200 mm; approximately 100 km^2 of land cumulatively dropped by 1 m; and the maximum cumulative subsidence was about 2 m. The land subsidence in the Hangjiahu Plain has spread over the entire region; more than 2500 km^2 of land cumulatively sunk by more than 100 mm; and the maximum cumulative subsidence was 1.11 m. In some areas, ground fissure disasters occurred, which were the results of differential settlements caused by the overdraft of groundwater. The total area of 19 ground fissures in the Suzhou-Wuxi-Changzhou region was approximately 1.1 km^2. In recent 10 years, the overdraft of groundwater has been alleviated in Shanghai and the Yangtze Delta region through strict control and management measures.

The areas with ground collapses in the Yangtze River basin are mainly located in Wuhan and Xianning Prefecture of Hubei, Ganzhou of Jiangxi, and the Kunming area of Yunnan. In Wuhan of Hubei, the area of sinkholes (karst terrain) was about 80 km^2; 30 ground collapses occurred, mainly located in Fenghuo Village-Maojiagang of Qingling Township and South China Steel Rolling Mill-Lujia Street where two deep confined aquifers were seriously over-extracted; and the maximum collapse depth was 10 m. The Beihongqiao-Guanbuqiao area of Xianan District of Xianning City is located in a karst-rich zone where the daily withdrawal of groundwater was 21,900 m^3 with a mining coefficient of above 1.3. The overdraft of groundwater caused more than 20 ground collapses that involved an area of 10 km^2, and the maximum collapse depth was 15 m. In Ganzhou of Jiangxi, due to the serious overdraft of shallow groundwater, more than 70 ground collapses occurred with the maximum collapse depth of 15 m.

Due to the lack of systematic monitoring data, the groundwater contamination conditions in the Yangtze River basin are unknown. Therefore, it is necessary to monitor groundwater where groundwater use is high, and groundwater is used as a backup urban drinking source, followed by taking appropriate protection measures.

7.4.2.4 Management of Toxic and Hazardous Substances

According to the 2005 statistics, as many as approximately 10,000 of China's 20,000+ chemical enterprises were located in the riparian zone of the Yangtze River basin which posed a potential contamination risk to the aquatic environment of the Yangtze River and increased the probability of emergent water pollution accidents. In the meantime, the probability of contamination accidents, such as water and ground transportation accidents, dangerous solid waste stockpiling around reservoirs, artificial poisoning, etc., was not low. Therefore, there is a need to improve the emergency response capability to handle water contamination incidents, strengthen the protection system to promote safe drinking water, improve the emergency response capability, enforce the regulation of the transportation of dangerous substances, etc. It is necessary to learn the experience from developed countries and implement the "womb-tomb" control of toxic and harmful substances. The entire process of production, transportation, storage, consumption, destruction, and disposal at a landfill should be registered, tracked, monitored, and controlled so as to effectively prevent water contamination.

7.4.3 Protective Countermeasures for Water Resources

Water pollution is not a problem of natural sciences but mainly a problem of management. Based on the existing conditions of water pollution in the Yangtze River basin, the following comprehensive protective measures for water resources must be taken.

7.4.3.1 Step-Up Urban Water Pollution Control Efforts

More efforts should be made on the construction of urban sewage treatment facilities to increase the capacity of sewage treatment. During the construction of sewage treatment facilities, attention should be paid to the installation of the associated pipeline network so as to convey the treated and raw effluents through different pipelines and improve the utilization efficiency of sewage collection and treatment facilities. In the remediation of the lakes and river sections within urban areas, the relationship with the mainstream and tributaries of the Yangtze River should be incorporated. Major efforts or engineering works should not always be on water diversion projects from the Yangtze River to lakes or using the simple water exchange practices so as to avoid transferring local contaminants to the mainstream of the Yangtze River and the downstream areas.

7.4.3.2 Increase Control of Industrial Pollution, Increase Production While Reduce Pollution, and Improve Water Use Efficiency

Each city should speed up the adjustment of the industrial structure and develop industries with less pollution. The existing enterprises should gradually adopt clean production techniques and technology to increase the reuse of industrial water and reduce the pollutant discharge by means of technological renovation, production process improvements, comprehensive resource utilization, and water conservation. Each city should continue to increase the treatment capacity of industrial wastewater to fully meet requirements of industrial wastewater discharge. As conditions permit, no wastewater will be discharged. Management on permits for water withdrawal and outlets for wastewater discharge should be strengthened, and the pollutant discharge from new or expanded projects should be strictly controlled.

7.4.3.3 Implement Modification Projects at Sewage Discharge Outlets

Presently, some of the cities along the Yangtze River have unreasonable layouts of intake structures for water withdrawal and discharge outlets for wastewater, and the quality of the withdrawn water is impacted as a result. During the construction of sewage treatment facilities and dike embankments, the existing intake structures and discharge outlets should undergo optimized adjustment and modification in accordance with the requirements for the water function zones. The wastewater discharge outlets, ports, and garbage yards that are not in compliance with the requirements for drinking water source protection should be completely removed.

7.4.3.4 Proactively Promote Water-Conserving Urban Development

Each city should reform the water price structure, gradually establish the water price mechanism to encourage the conservation of water, step up the management of water-conserving construction projects, promote the adjustment of industrial structures, step up the implementation of water-conserving technology policies and technical standards, and promote the use of water-conserving technology.

7.4.3.5 Improve Ecological Environment and Control Agricultural Pollution Sources

The Yangtze River basin is one of the areas seriously affected by soil erosion in China. Soil erosion destroys the ecological balance and causes serious natural disasters. In the meantime, the lost sediments carry large amounts of pollutants into the water bodies of rivers and contaminate the water. Efforts should be made to expand natural forests and soil conservation forests and to implement comprehensive remediation measures in the areas where soil erosion is serious, such as the modification of slopes to terraces, installation of drainage ditches, construction of siltation basins and retention ponds, etc., which cannot only protect water sources, control soil erosion, improve the ecological environment, and reduce floods, mudslides, and other natural disasters but can also reduce the amount of pollutants from the watershed migrating into rivers and improve water quality. Moreover, well-maintained vegetation can also play an important role in absorbing and decomposing pollutants.

Great efforts should be made to develop an ecological agriculture with the emphasis on reducing applied quantities of fertilizer and pesticides, increasing the utilization efficiency, and promoting the production and use of organic fertilizers. The implementation of organic fertilizer industrial projects can reduce the amount of fertilizer application and promote the use of high-efficiency and low-residue pesticides.

7.4.3.6 Control Contamination from Ships

Almost every year, the Yangtze River experiences pollution accidents from shipping, which poses a serious threat to the water quality of the Yangtze River. In the meantime, large amounts of waste and wastewater are discharged from ships into the Yangtze River, which also have a great impact on the water quality of the river. The Department of Transportation Management should carry out the inspection and registration of ships that transport toxic and hazardous substances and prohibit those vessels that do not have the conditions for transportation of poisonous or hazardous substances or do not have adequate preventive measures for accidents from transporting poisonous and harmful substances. Great efforts should be made to collect and treat wastes from ships and to prohibit all types of vessels from discharging wastewater to rivers.

7.4.3.7 Strengthen Publicity for Protection of Water Resources

Every means should be used to educate the public about the importance of protecting water resources, to promote the public awareness of environmental protection, and to advocate the public participation in and supervision of the protection efforts of water resources.

7.4.3.8 Improve Policy and Regulatory System to Protect Water Resources of the Yangtze River Basin

In order to ensure an orderly implementation of the protection work of water resources in the Yangtze River basin, it is necessary to prepare regulations for the watershed in accordance with the new *Water Law* and *Water Pollution Prevention and Control Law* and based on the actual conditions of the Yangtze River basin. It is recommended that the State Council organizes and formulates the *Regulation on Water Resources Protection in the Yangtze River Basin* as soon as possible. The Regulation should be applicable to the characteristics of the Yangtze River and incorporate the policy of "Prevention First, Followed by Integrated Prevention and Control," strictly implement the *Measures of Monitoring and Management of Water Function Zones* and *Measures for Monitoring and Control of Sewage Outlets on Rivers*, establish and improve the regulatory system of water resources protection in the Yangtze River basin, and safeguard the protection work for the water resources within the watershed by a legal system.

7.4.3.9 Strengthen Supervision and Control of Water Resources Protection

Organizations for water resources protection in a watershed should work closely with the local administrative departments of water conservancy and the environment in the watershed; strengthen the monitoring and control of water withdrawal, sewage discharge, and water function zones for construction projects; strictly administer the review and approval process; and prevent new contamination events from occurring. At the same time, efforts should be made to strengthen the law enforcement team for protecting water resources, establish the procedures of emergency response to water pollution accidents, and enhance the rapid response capability of protecting water resources. More attention should be paid to the supervision by media and publicity so as to allow the society and media to play the role of supervision.

7.4.3.10 Establish Input Mechanism Adaptive to Market Economic System

Protection of water resources is a public welfare undertaking, which should be organized and implemented by the government. Based on the status and responsibility for protecting water resources by the Central Government, local governments, enterprises, institutions, residents, and other market subjects, the relevant costs should be appropriately shared. Efforts should be made to improve the policy of taxes and fees for water resources protection, reform the water price structure, and especially ensure the improvements of soil conservation and the ecological environment, integrated remediation of water bodies, and management, monitoring, and scientific research of water resources protection. In the meantime, the implementation of the projects such as sewage treatment, waste disposal, sewage reuse projects, and a market mechanism should be introduced so as to achieve the marketization, commercialization, and intensification of investment for construction, operation, and management. It is necessary to attract foreign investment and social capital to the construction of water resources protection projects to form a multichannel, multilevel investment, financing, and operation mechanism.

7.5 Protection of Lakes and Wetlands

7.5.1 Wetlands in the Source Area of the Yangtze River

In the source area of the Yangtze River, there are large areas of plateau wetlands, especially the south source Dangqu River that is also known as the Akedashu River, meaning “marshy river” in Tibetan, because it flows through large areas of marshes.

Due to the alpine cold and oxygen deficiency in the area and poor traffic conditions, it is difficult for scholars to conduct scientific research in this area. In fact, the wetland is still a "blind spot" in terms of scientific research, and the information about wetland evolution, biology, and environment is scarce. As mentioned above, in the area, there are not only glaciers, large areas of permafrost and marshes, and a complex water system where large amounts of water are stored but also many rare and endemic animals and plants, all of which have protective values.

7.5.2 Wetlands in the Yunnan-Guizhou Plateau

The wetlands in the Yunnan-Guizhou Plateau are different from the lacustrine wetlands in the middle and lower reaches of the Yangtze River, those in the plains of north China and those on the Ruoergai Plateau. The wetlands in the Yunnan-Guizhou Plateau have developed in a unique landform and are located at low latitudes but high elevations. Various large and small basins of lakes have resulted from different origins at different altitudes in the Yunnan-Guizhou Plateau, and the lacustrine wetlands are mainly located in the areas of flat depressions, of which individuals are small in area and are scattered everywhere. Moreover, the lacustrine wetlands are not connected with any waterways. Therefore, they are closed or semi-closed wetlands and constitute the complex varieties of habitats when rivers, lakes, meadows, marshes, mountains, and forests are considered. For example, the wetlands in Yunnan are all in the mountainous environment with elevations above 1500 m which is not conducive to the overall development of aquatic vegetation, but the diversity of wetland plants is very rich, especially in northwest Yunnan in the upper reaches of the Yangtze River where the altitude is high, and there are 35 different communities of plants, including 10 communities of emergent plants, 7 communities of floating plants, and 18 communities of submergent plants. On the other hand, the aquatic vegetation communities of the wetlands in Yunnan are relatively complex and encompass six different geographic components, including world widely distributed, tropical, northern temperate, East Asian, alpine, and freshwater lake aquatic plants, and are very obvious in vertical differentiation. Moreover, the number of rare, endangered, and endemic species accounts for a high percentage of the total number of species, including not only most of the aquatic plant communities of wetlands in China but also the alpine-arctic aquatic plant communities that do not occur in the plain wetlands in the middle and lower reaches of the Yangtze River, such as the common mare's tail (*Hippuris vulgaris*) community, the *Ottelia acuminata* community that is endemic to the Yunnan-Guizhou Plateau, and the *Isoetes hypsophila* community that is endemic to the alpine wetlands.

The animals that live in the wetland habitats of Yunnan are very rich in diversity. There are 201 species of birds that are important components of wildlife, of which 136 species are key species under state protection, indicating a very high proportion of rare and endangered species and accounting for 67.8% of those protected by Yunnan Province (Peng 1996). The fish species in the wetlands of Yunnan account for 43.2% of those of freshwater fish species in China, and there are up to 260 endemic species, accounting for 60% of the number of fish species in Yunnan. Such a high percentage of occurrences of endemic species is rare at home and abroad. Moreover, both animals and plants have formed a number of unique species or communities, such as water chestnut flower that is a relic plant from the tertiary age and is endemic only to Cibi Lake, the *Ottelia acuminata* that is endemic to the plateau of Lugu Lake, the ptychbarbus chungiensis that is endemic to a small area in Shangrila, and various waterfowls that have been designated as important wild animals under state Levels I and II protections.

Yunnan has established six provincial wetland nature reserves and several other wetland reserves of national and international importance such as Dabaoshan, Napa Lake, and Lashi Lake. As mountainous land accounts for 94% of Yunnan's total area, there is limited land available for utilization. As a result, the topographically flat low-lying and scattered wetlands have become the first choice for development and utilization. However, as the wetlands in the Yunnan Plateau are small in area and not connected one another by any waterway, they are in closed wetland environments. Moreover, they are located not only in ecotones between farmlands and livestock grazing zones but also in ecologically fragile zones. Due to the impact of human activities, wetlands have been increasingly disturbed, and many lakeside marshes have been drained for land reclamation. Consequently, the area of wetlands has greatly reduced, the area of the wetland environment suitable for survival and reproduction of a variety of rare animals and plants has increasingly been shrinking, and biodiversity has been seriously threatened.

Special mentions should be given to Dianchi Lake, also known as the "Sparkling Pearl Embedded in a Highland." The lake is 1886 m in altitude, 330 km^2 in area, 5 m in average water depth with the maximum depth of 8 m, and a storage capacity of 1.57 billion m^3. It is also known as "250-km Dianchi Lake." Major rivers that drain to the lake include the Panlong, Jinzhen, Baoxiang, Haiyuan, Maliao, Luolong, and Laoyu Rivers, and the outlet of the lake is the Tanglang River. Dianchi Lake drains to the Pudu River at Haikou, then into the Jinsha River, and is part of the Yangtze River system. The lake is the sixth largest freshwater lake in China. There are ten peaks of various sizes around Dianchi Lake. The lake is surrounded by mountains and moving clouds in the sunny sky with casting shadows to the water surface, which constitutes a beautiful natural picture. This is a place suitable for human settlement. Therefore, Dianchi Lake was one of the early developments in Yunnan Province and was originally inhabited by a tribe called "Dian" or "Dianji" and invaded by troops led by Zhuang Qiao, a general of the Chu Kingdom during the Warring States Period. However, in the past, the area around the lake often used to have flooding and waterlogging problems. Thus, as early as 1262, the Songhua Dam was constructed on the Panlong River; and in 1268, the Haikou River was excavated to increase the outflow from Dianchi Lake and reduce waterlogging. Since 1955, more than ten large- and medium-sized reservoirs have been constructed on various river upstreams of the lake, and tens of pumping stations have been installed along the shore of the lake for the purposes of drainage and irrigation. While these projects have helped resolve the flooding and waterlogging issues and ensure the water supply for farmland irrigation, urban industries, and domestic water needs, the hydraulic continuity between the lake and the surrounding water systems has gradually been disrupted. In particular, the original flooding-prone fluctuation zone along the shore of the lake was transformed into farmland where flowers and grasses were planted, and the long-term use of chemical fertilizers caused serious nonpoint source pollution. Moreover, Dianchi Lake is located downstream of Kunming and has been the outlet of sewerage of Kunming. As large quantities of untreated waste sewage have been discharged into the lake over the past decades, the originally beautiful and biodiversity-rich lake has become the most seriously eutrophic lake in China. Although the central and local governments have invested a lot for the ecological restoration over recent 10 years, the effect is not significant, indicating that once the lake is polluted, it may probably take decades or even up to 100 years to restore the lake to its "original" conditions.

Over the recent 50 years, the large-scale development of wetlands by human beings has caused great changes in the regional environment, and the development process of wetlands under natural conditions has been severely impacted, which has changed the normal development process of wetlands. As mesophytes and xerophytes invaded, the wetlands have shrunk and degraded; the wetland environment has changed; water pollution has occurred in the rivers and the lake; eutrophication of the lake has intensified; freshwater resources have been depleted; regulating function of the wetlands has declined or disappeared; and human's living environment has been deteriorating gradually.

7.5.3 Lacustrine Wetlands in the Middle and Lower Reaches

7.5.3.1 Dongting Lacustrine Wetlands

The wetlands in Dongting Lake are a series of large shallow lakes, countless small lakes, a large freshwater marsh, wet meadows, and river-connected waterways. The north lake is fed by several channels northwest of the lake from the Yangtze River and the four large rivers, Xiang, Zi, Yuan, and Li Rivers, that originate in the mountains south and west of the lake. The lake discharges through a wide waterway in Yueyang to the Yangtze River from the northeast. Dongting Lake is 5–20 m in depth and 6.8–8.6 in pH value. The water level of the lake varies with the water level of the Yangtze River and the contributing water from the four rivers and fluctuates in a very large range within the year. As the season changes, it shows the natural landscape of "a lake when the water level is high and a dryland when the water level is low."

The wetlands in Dongting Lake are a very important wintering habitat for migratory waterfowls, especially herons, storks, ducks, cranes, and gulls, and 120 waterfowl species have been recorded to occur in the lacustrine wetlands for wintering, including Dalmatian pelicans, black storks, white storks, mandarin ducks, Baer's pochards, and four species of cranes: common cranes, hooded cranes, white-naped cranes, and Siberian cranes. The endangered Chinese mergansers are also wintering birds in the nature reserve, and great bustards are wintering birds on the adjacent plains.

The endangered baiji dolphins and Yangtze finless porpoises occur in East Dongting Lake and the adjacent Yangtze River. Other mammals include ground squirrels and water deer. About 200 fish species have been recorded to occur in the reserve, including Chinese sturgeons, Chinese paddlefish, bigmouth grenadier anchovies, common carps, crucian carps, flathead gray mullets, St. Peter's fish, and Reeve's shads, of which more than 20 fish species have economic values. The protected fish species include the Chinese sturgeon and the Chinese paddlefish. Reptiles include freshwater turtles. More than 400 species of mollusks have been recorded to occur in the reserve.

For the protection of Dongting Lake, in addition to scientific research and monitoring, it is most important to improve the public understanding of the importance of the lacustrine wetlands and standardize development and utilization activities. Moreover, efforts should be made to the following aspects:

(1) First of all, the natural characteristics and evolution of wetlands should be understood; development and utilization activities in the lake area should be standardized; and the management system for the sustainable utilization of the wetland resources in the Dongting Lake area should be further improved.

As it is not long since the concept of "wetland" was introduced into China, wetland protection has begun only in the 1990s. Before implementing measures to protect wetlands, the lacustrine wetlands in Dongting Lake had administratively involved more than ten counties and four reserves in two provinces. The lacustrine wetlands in Dongting Lake should be considered as an entire watershed so as to determine the boundaries of the wetlands, including the blue line (water surface line) and the green line (lakeside vegetation zone and transition zone). The continued protection and management of the wetlands should be carried out in accordance with the management ideas for natural resources and ecosystems, and the human needs for wetland resources and the ecosystems' needs for the wetland environment should be clearly and closely linked. Specifically, attention should not only be paid to small areas of the ecosystem process, but also problem-solving considerations and protection efforts should be applied on a large spatial and temporal scale so as to maintain the productive potential of the ecosystem in the long run.

(2) A model for sustainable utilization of wetland resources in Dongting Lake should be established.

It is necessary to use the ecological viewpoint to study and utilize the wetland resources in Dongting Lake so as to make the full use of the potential value of the resources, proactively protect the resources, and maintain the ecological balance of the wetland area in Dongting Lake. First, the system and mechanism of an integrated watershed management should be established, and an adaptive management model should also be developed. Second, water pollution should be controlled, and the wetland environment should be improved. The serious pollution in a local area of the wetlands in Dongting Lake threatens the biodiversity of the local area and affects the sustainable utilization of the wetland resources. Therefore, integrated prevention and control efforts should be carried out in the Dongting Lake watershed to prevent contaminants in the sources in the entire watershed from migrating into the wetland areas and to protect the transition zone between land and water. Third, through multiple measures such as scientific research and ecotourism, the sustainable utilization of resources can be achieved, and the cultural value can be exploited. Fourth, the structure of the agricultural industry should be adjusted, and the ecological agriculture should be vigorously promoted. Hazard-free green food production should be developed and promoted; the comprehensive utilization and control of the livestock and poultry manure should be strengthened; large-scale and intensive animal breeding industries should be advocated; and the use of pesticides should be greatly reduced. Fifth, nonnative species should be regularly monitored to prevent and mitigate the harmful effect of invasive nonnative species and to minimize or avoid the significant loss of the ecological environment, human health, and socioeconomic condition caused by invasive nonnative species.

(3) The legal system for the management of Dongting Lake and the sustainable utilization of the wetland resources should be improved.

The laws and regulations on wetland protection should be fully implemented, and wetland management should be legalized, standardized, and institutionalized. Efforts should be made to carry out publicity and public education so as to enhance the public understanding of wetlands and legal knowledge, promote public awareness of the protection and rational use of wetland resources, and ensure the successful implementation of wetland protection regulations. The law enforcement agencies and teams should be established based on the government with support from departments to strengthen the function of law enforcement and to protect the wetland resources in Dongting Lake. The following laws and regulations should be fully implemented: *Environmental Law*, *Wildlife Protection Law*, *Regulation on Wildlife Resources Protection of Hunan Province*, and *Regulation on Wetland Protection of Hunan Province*.

7.5.3.2 Poyang Lacustrine Wetlands

Poyang Lake consists of a large freshwater lake and surrounding marshes and wet meadows and is fed by the Gan, Fu, Xin, Rao, and Xiu Rivers from the south. At the highest water level, the lake is about 170 km in length and stretches 74 km at its broadest width. As the water level in the rivers fluctuates as the season changes, the water level in the lake also varies. During the peak rainy season (April–July), the water level of the lake can reach up to 21 m and covers an area of about 4000 km^2. However, during the dry season, the water level drops to 13 m, and the lake turns into many small lakes and ponds, large area of mudflats, sandy dikes, and wet meadows. The nature reserve is in the northwestern corner of the lake area, covering 22,400 ha. During the winter when the water level in the lake is low, the water zones in the nature reserve are divided into nine lakes of various sizes, including Dacha Lake (8500 ha), Dahuchi Lake (3000 ha), Bang Lake (7300 ha), Sha Lake (1400 ha), Zhonghuichi Lake (600 ha), Changhuchi Lake (700 ha), Meixi Lake (350 ha), Xiang Lake (400 ha), and Zhushi Lake (200 ha). At the end of the winter, as the water level drops to the lowest with lake depths only 20–80 cm, the entire lake only covers 50,000 ha, and two shallow lake-turned ponds and Sha Lake are artificially drained for fish harvest.

Fifty-four families and 154 genera of phytoplankton and 38 families and 102 species of aquatic vascular plants have been recorded to occur in Poyang Lake. The dominant aquatic plants are bitter grass, *Potamogeton distinctus*, green algae, cyanobacteria, and colonies of reeds. Rich grasses are supported on the lakebed flooded in the rainy season. Although most of the surrounding areas are now cultivated or covered with grasslands, artificial pine forests, and secondary shrubs, the lake is located in the subtropical deciduous broad-leaved forest and evergreen forest with 28 genera and 79 species of wet plants.

More than 150 species of birds, including 20 endangered species of China, have been recorded to occur in the nature reserve. The lake is famous for large numbers of wintering cranes, especially Siberian cranes. About 1350–1784 hooded cranes wintered there during each of the winter seasons of 1984–1985, 1985–1986, and 1986–1987. According to the calculation conducted in January 1989, no less than 2600 hooded cranes and 110 common cranes wintered there during the 1985–1986 winter season. Moreover, the lake is also a very important wintering habitat for gray geese and ducks. It is estimated that about 500 swans, over 100,000 gray geese, and more than 600,000 ducks stayed here in the coldest winter season. Most of the swans were young ones (e.g., about 3450 in January 1988). The gray geese included more than 50,000 swan geese, at least 15,000–20,000 greater white-fronted geese, and a small number of taiga bean geese, graylag geese, and young greater white-fronted geese. In January 1988, the report of 14,000 young greater white-fronted geese occurring in the nature reserve was particularly interesting. Northern pintails were often the dominant ducks. However, there were also many Eurasian teals; Indian spotted-billed ducks, mallards, and garganeys; many ruddy shelducks, common shelducks, Eurasian wigeons, falcated ducks, gadwalls, northern shovelers, and common mergansers; and a small number of Baikal teals, tufted ducks, and smews. Other wintering waterfowls included cockatoos, Dalmatian pelicans (up to 25 in recent years), Eurasian bitterns, great egrets, gray herons (in hundreds), black storks (usually 5–10), white storks (usually 100–200 but 758 in January 1988), and common spoonbills (more than 1000 during the 1987–1988 winter season). Furthermore, there were also many wintering snipe and gull species, including many spotted redshanks, pied avocets, pin-tailed snipes, common snipes, dunlins (in thousands), European herring gulls, and black-headed gulls (over 10,000). Sixteen Chinese black-headed gulls were also observed in January 1986. Grasslands, especially the floodland at Linggongzhou, have provided wintering habitats for more than 100 great bustards (e.g., 125 in January 1988) and a small number of local various marsh grassbirds (at least seven observed in one area in January 1986). Figure 7.7 shows the Siberian cranes in Poyang Lake.

Fig. 7.7 95% of world's Siberian cranes wintering in Poyang Lake

The wetlands in Poyang Lake are rich in zooplanktons, including 112 species of 22 families. The recorded occurrences of zooplanktons in the reserve included 65 species of freshwater mollusks and 126 species of fish (21 families), including black, grass, silver, and bighead carps. Forty species of mammals have been recorded to occur in the reserve, including Yangtze finless porpoises (occasionally), water deer, and ferrets. Other vertebrates include rhacophoruses and freshwater turtles. At the end of the winter, outside the reserve to the south and east of the lake, there are many great bustards, juvenile swans, and a small number of Siberian cranes, hooded cranes, common cranes, white storks, and black storks.

The preservation focus of Poyang Lake is to maintain the water-based and transitional lacustrine wetland characteristics of "a vast lake during the wet season and a linear feature during the dry season," standardize the development and utilization of the lake, reduce disturbance to the wetland ecosystem, strictly control waste and sewage discharge to the five rivers, continuously improve the compliance rate of the water function zones, maintain the river-lake connection and the transition zone between land and water to the maximum extent practically possible, and preserve the gradient for a large environment and the diversity of the habitat environment.

7.5.4 Wetlands in the Yangtze Estuary

The wetlands in the Yangtze Estuary are mainly natural wetlands, including coastal, neritic, and estuarine wetlands. The coastal and neritic wetlands are divided into the neritic zone, intertidal wetlands, and sandbar wetlands. Estuarine wetlands mainly include perennial rivers, floodplains, islands, and beaches.

The wetlands in the Yangtze Estuary can also be divided into two categories: riparian and coastal wetlands and estuarine sandbar and island wetlands.

7.5.4.1 Riparian and Coastal Wetlands

The riparian and coastal wetlands in the Yangtze Estuary have a total area of 85,691.72 ha, making up 24.26% of the total area of the natural wetlands in the estuary, and are mainly distributed in the riparian and coastal areas along the south bank of the Yangtze Estuary (from Baimaokou on the west to Luchaogang on the east) and northern Hangzhou Bay (from Jinsiliang Bridge on the south to Luchaogang on the north). The riparian and coastal wetlands in the estuary consist primarily of riparian wetlands that make up about 81.3% of the total area of the riparian and coastal wetlands. The riparian wetlands are composed principally of the Nanhui sandbank that accounts for more than 80% of the riparian wetlands.

7.5.4.2 Estuarine Sandbar and Island Wetlands

The estuarine sandbar and island wetlands in the Yangtze Estuary have a total area of 267,463.67 ha, accounting for 76% of the total area of the natural wetlands in the estuary, and consist mainly of sandbars that are separated from islands. Some of the exposed sandbars have been developed into islands by human beings, such as Chongming, Changxing, and Hengsha Islands; some of the exposed ones have remained undeveloped, such as Jiuduansha and Qingcaosha sandbars; and the others are unexposed sandbars, such as Biandansha, Zhongyangsha, Tongsha sandbars, etc. The majority of Chongming, Changxing, and Hengsha Islands have been developed into villages and farms, and some natural wetlands have remained only along the outer edge of the tidal dikes of the islands. These wetlands and other sandbar wetlands constitute the estuarine sandbar and island wetlands in the estuary.

In the Yangtze Estuary and the adjacent waters, there are 52 families and 86 species of ichthyoplanktons, and, except for January–February, their eggs and larvae appear and reach the maximum during May–August. According to the ecological and distributional characteristics, ichthyoplanktons can be divided into four categories: freshwater species, such as stone moroko, pseudolaubusca sinensis, and clearhead icefish, whose larvae generally occur in the channel of the lower reaches; brackish water species, such as anchovy, mullet, bass, Japanese puffer, barracuda, icefish, etc., which are mainly distributed in low-salt waters; coastal species, such as long-jawed anchovy, silver pomfret, white pomfret, common hairfin anchovy, bummalo, big-head croaker, Belanger's croaker, silver white croaker, large yellow croaker, redlip croaker, etc., which are mainly distributed in the muddy waters west of longitude 122.30°E; and offshore species, such as eel, chub mackerel, skinnycheek lanternfish, etc., which are mainly distributed in the mixing zone between the freshwater of the Yangtze River and the seawater.

There are 208 fish species from 2 classes, 22 orders, and 67 families in the Yangtze Estuary area. The pachycormus and the Indian anchovy are the newly recorded species of the region. The fish species are mainly composed of Cypriniformes and Perciformes, and the ecotype is dominated by freshwater fish species. There are some differences in the composition of fish species among the South Branch, North Branch, and the estuarine mouth of the Yangtze River. The Cypriniformes is the major species in the South Branch, while the Perciformes is the dominant species in the North Branch and the estuarine mouth. Compared with the survey data before 1990, the number of fish species in the wetlands of the estuary decreased. The cartilaginous fish and abalone fish decreased relatively more than others. Some important economic species have gradually become rare species, even become endangered, such as the Reeve's shad, *Salanx prognathus*, and *Takifugu obscurus*. The decline in diversity of fish species in the estuarine wetlands may be mainly attributed to overfishing, water pollution, and habitat loss (Zhang et al. 2009).

The wetlands in the Yangtze Estuary are mainly influenced by the natural factors in the upper and middle reaches of the Yangtze River such as runoff, sediment transport, tide, wave, and salinity and are also affected by human activities such as overharvesting of fishery resources, shipping, wharf, water diversion, and waste sewage discharge. The impact from human activities has been increasingly intensified. In the long run, any ecological information in the Yangtze River basin will eventually be transmitted to the estuary and offshore where the health of the Yangtze River will be generally reflected. Therefore, the success for the protection of the estuarine wetlands relies on the integrated efforts in the entire basin of the Yangtze River.

References

Kong D, Li D, Wu Y (2007) Records of major organic contamination events in the Yangtze Estuary in past 50 years. Trans Oceanol Limnol (2):94–101

Yang M, Bi Y, Hu J et al (2011) Daytime and nighttime vertical distribution and migration of phytoplankton during Spring Algal Bloom at Xiangxi River Bay within three Gorges Reservoir. J Lake Sci 23(3):375–382

Zhang H, Zhu G, Lu J (2009) Analysis of composition and diversity of fish species in wetlands of the Yangtze Estuary. Biodivers Sci 17(1):76–81

Zhou D (2007) Bankline utilization and channel alteration of mainstream of the Yangtze River within Jiangsu Province. Yangtze River 38(6):47–50

Chapter 8
Integrated Management of the Yangtze River Basin

Abstract Currently, the problems of water pollution and declining ecosystem function caused by overexploitation and utilization of the Yangtze River are due to the ineffective system of watershed management and the lack of important mechanisms, such as the participation of various regions, departments, and stakeholders in management and ecological compensation. The watershed's natural attributes have determined that the watershed management belongs to the category of natural resources and ecosystem management. Thus, management needs to learn from the internationally popular concept of integrated watershed management. This chapter systematically introduces the basic concept and major experiences of modern integrated watershed management and, based on the existing conditions of China, promotes the establishment of the system in the Yangtze River basin by the provision of the important mechanisms mentioned earlier and building capacity for integrated watershed management, development of a water right system and the principle and methods for water allocation, management of three redlines for water resources, joint regulation of cascade reservoirs for flood control and drought relief in the Yangtze River, prevention and control of flash flood geological hazards, construction and management of hydraulic projects, soil conservation and sediment management, etc.

Keywords The Yangtze River · Changjiang River · Evolution of river system · Basin ecosystem · Water resources utilization · Floods and drought · Ecological and environmental protection · Basin management

8.1 System and Mechanism of Integrated Watershed Management

As the Yangtze River basin is a typical large river watershed, the development, utilization, and protection of water resources involve many regions, departments, and stakeholders and need to coordinate well with all interested parties. The scientific utilization and effective protection of the Yangtze River will need to establish an

J. Chen, *Evolution and Water Resources Utilization of the Yangtze River*,
https://doi.org/10.1007/978-981-13-7872-0_8

integrated system and mechanism adaptive to the national conditions of China and socioeconomic development needs in the basin. Because each country of the world has a different system, the connotation, contents, means, and implementation procedures of integrated watershed management vary. The connotation and means of integrated watershed management which are widely advocated and used in the world are mainly influenced by developed Western countries which have basically completed their large-scale development of water resources and infrastructures and established their system and mechanism of integrated watershed management which involve the participation of relevant stakeholders because of their long-term tradition of democratic consultation. However, if the experiences of developed countries are simply applied, it may not necessarily be conducive to developing countries. For example, in 1993, the United States Bureau of Reclamation issued a public statement that "the future construction of irrigation water projects will no longer be funded" based on the fact that the United States had not only addressed food security issues but also become the world's leading food exporter. However, some African and Asian countries that have still been struggling to produce enough food cannot solve their problem of food supply without irrigated agriculture. Therefore, integrated watershed management should be scientifically developed in consideration of the country's and the watershed's characteristics, and the scientific utilization and effective protection of the watershed should be carried out without compromising the ecosystem of the river.

8.1.1 System of Integrated Management in the Yangtze River Basin

Based on the experience of the world and the national conditions of China, the core of integrated watershed management is the management of natural resources with the core of water circulation, which differs significantly from the governmental administration of a jurisdictional region. As a watershed is the natural catchment area of surface water and groundwater, the main objects of the management are the water resources from surface runoff and groundwater. Because water is the resource needed by human production and life, and the natural ecosystem, watershed management involves many stakeholders and needs to establish a system involving participation, coordination, and management of all stakeholders.

The concept of integrated watershed management was introduced from the West. The management of China's natural resources is administered by different departments, and more emphasis is given to the administrative management. As for watershed management, there are some misunderstandings. Some argued that as long as the management of all affairs related to water is unified, integrated management can be achieved. However, it is neither feasible nor possible to do so in reality. Even in the United States where a small government with a big society has been advocated, water affairs are also managed by multiple departments, but not managed by a unified agency (the management of Tennessee Valley is an exception). Water affairs in the United States are managed by more than ten independent divisions from several departments or agencies such as the Department of the Interior, the Environmental Protection Agency, the Department of Energy, and the Federal Emergency Management Agency. Nonetheless, a system has been established so that the management of a specific type of water affairs is led by an appropriate division and participated by others through consultation. For large water affairs and specific water projects in a watershed, a decision can be made after consultation with each interested party. For example, America's Everglades Wetland Restoration Program in Florida was jointly administered by the Department of the Interior, the Environmental Protection Agency, and the Florida state government and scholars in various fields and was implemented by the US Army Corps of Engineers. It is impossible for any country to manage all affairs involving natural resources by one department, because it is very difficult and impossible for a single department to manage natural resources such as water whose circulation and utilization involve every sector of the government and everyone in the society. Therefore, integrated water management should be administered mainly by a single water department but rely more on a mechanism that provides involvement and cooperation from various departments and major stakeholders and depend on the entire society's ethic to water resources and the aquatic environment to regulate the behavior of each department or community.

Watershed management is not administration but is rather the management of public resources. The main function of a watershed management agency is to manage the public resources with water resources as the core, and the management method is to mainly regulate the development, utilization, and protection of water resources by laws, technologies, consultations, and economy, as well as a small amount of necessary administrative management of water. For example, direct and organize various regions and departments to carry out the comprehensive utilization of water resources and conservation planning, direct or consult all regions and departments to perform the professional planning of water resources, participate in other departments' planning, and provide technical guidance and develop contingency plans for all levels of governments for major water matters. The main function of a future watershed organization is to provide the government with planning and technical programs; supervise and manage the development and utilization of water resources in the basin according to laws such as the *Water Law*, the *Flood Control Law*, and the *Water Pollution Prevention and Control Law*; assist the governments at all levels and the flood and drought control headquarters to make flood and drought forecasting, prepare contingency plans, and provide technical advices; and perform the long-term monitoring and analysis of the watershed hydrology, the aquatic environment, and the aquatic ecological conditions so as to provide technical guidance and services to the state, the local governments, and the society.

The core of watershed management is to provide a platform for consultation and exchange among regions, departments, and stakeholders. There should be a platform for consultation and information exchange between the upstream and downstream regions, between the left and right banks, among the various regions in the watershed, and among the departments for important matters regarding the development, utilization, and protection of the watershed and for the supervision of the entire process of major water projects' planning, construction, operation, and management so as to ensure the participation, information exchange, and benefit compensation of various stakeholders.

Watershed institutions can obtain the basic financial support through the state transfer payments and water resources fees and can also maintain their own operations by providing technical services to the society and the public. The source of funds for the operation of watershed institutions is mainly through the state fiscal appropriations and the state investment on major projects. The funds should be mainly obtained from the society by providing quality services and technical advices to all regions, departments, and the public in the basin, such as the provision of basic information, planning, design consultation, and special studies for major projects.

8.1.2 Mechanism for Integrated Watershed Management of the Yangtze River

Since the essence of integrated watershed management is the management of natural resources, it is necessary to establish various mechanisms for stakeholders to participate in the management, including the following major aspects:

8.1.2.1 Planning Coordination Mechanism

Planning is the main technical basis for the development, utilization, protection, and management of a watershed, and the watershed planning is divided into comprehensive planning and professional planning. Comprehensive planning refers to the preparation of a general plan and an implementation program for the development, utilization, conservation, and protection of water resources and flood and drought prevention and control based on the needs of the socioeconomic development and the existing conditions of water resources development and utilization in a basin or a region. Professional planning refers to the planning for flood control, waterlogging control, river channel modification and control, bank utilization, irrigation, shipping, water supply, hydroelectric power generation, transport of felled bamboos and trees via water, fisheries, water resources protection, soil conservation, sand and desertification control, water conservation, etc. The professional plan should be subject to the general plan, and the regional economic development plan should reference the comprehensive watershed plan and the water resources plan.

Both the comprehensive planning and professional planning involve various departments, regions, and stakeholders. If major stakeholders do not participate in the planning process or agree on the content of the plan, the implementation of the completed plan will confront difficulties and problems. In the past, some of the plans were basically prepared by individual departments. After the completion, the plans were sent to other departments to solicit comments, and then all the departments would sign before they were submitted to the State Council for approval. This working process rendered the process of review and approval a prolonged period of time. Moreover, if other departments had relatively significant comments after reviewing the plan, the plan was returned for revision before it was signed by all departments. This process was time-consuming and laborious. As a result, it might take 2–3 years to prepare an important plan, but it could often take up to 3–5 years to have the plan reviewed and approved. During the review and approval process, great changes may have taken place in socioeconomic conditions, and some contents of the approved plan became outdated. Therefore, the approved plan would need to be re-revised and the guiding role of the plan was reduced. It has been proved that a plan is feasible only if all parties reach a consensus. Therefore, it is necessary to establish a mechanism to allow various departments and major stakeholders to participate in the planning process. If a plan involves the interests of the general public, the plan should be discussed, reviewed, and approved by the local people's congress at all levels, and a public notification protocol for planning results should be established to allow major stakeholders and the society to understand the planning contents so as to facilitate collection of different opinions and improve the planning contents. It is not only quick to have such a plan approved but it is also easy to have the plan implemented.

8.1.2.2 Coordination Mechanism for Development and Utilization Projects

The development and utilization of water resources generally need to be achieved through a number of engineering projects. However, the new projects will occupy or inundate land resources, and the relocation of residents may affect the interests of a third party and will have some impacts on the ecological environment. For example, the construction of a reservoir will need to submerge land, relocate residents, and need to consider the upstream and downstream impacts. Therefore, a coordination mechanism will need to be established for a hydraulic and hydroelectric engineering project. Even the construction of a bridge across a river would inevitably narrow the cross section of the river channel because piers would need to be installed; thus, the river regime would be changed and even the safety of flood control would be affected. As a result, any project of water resources development or any project that involves water or a river would affect the ecosystem of the river and may cause the interests of other stakeholders to be damaged. Therefore, it is necessary to establish the mechanism of justification, public notification, and coordination for development and utilization projects.

8.1.2.3 Coordination Mechanism for Watershed Protection and Management

In the past, a watershed plan was mainly implemented through structural measures to achieve development, management, and utilization. However, the regulation of the activities related to development and utilization of a river basin and the protection of the ecological environment need more non-structural measures. Non-structural measures are generally aimed at all aspects of the society, involve social and public interests, and can only be implemented by joint efforts by all departments and the entire society. For example, the prevention and control of water pollution need to build wastewater treatment plants and monitor and manage the outlets of wastewater discharge; require each enterprise to carry out clean production, conserve water, improve the efficiency of water use, and reduce the discharge of wastewater and sewage; and can only be achieved by the concerted efforts of the entire society. The implementation of the effective protection and the regulation of enterprises and individuals' behaviors need to establish a mechanism for the entire society to participate in protection and management. For example, the United States *Clean Water Act* provides citizens with the power to sue and prosecute businesses and organizations who have illegally discharged polluting matters and to participate in and monitor water pollution. The *Clean Water Act* also provides the procedures for citizens to use the power so that illegal acts to discharge polluting matters in violation of the act can be exposed and punished in time.

8.1.2.4 Mechanism for Investment, Financing, and Compensation

The development, utilization, and protection of water resources can be accomplished through both structural and non-structural measures, which generally require investment and financing. Profitable hydropower projects can be carried out through the market and competition, while the public welfare-based projects or the projects whose beneficiaries are difficult to determine such as flood control, water supply, irrigation, water resources protection, and ecological rehabilitation generally need governmental investment, supplemented by other investments and financing. The government can set up public welfare funds for the construction of water conservancy and ecological rehabilitation projects through collection of taxes, resource fees, and funds from beneficiaries and can also entrust enterprises to operate and manage so as to provide compensation and capital to the parties to whom damages would occur as a result of the implementation of the projects. In the future, there will be a need to establish a scientific, reasonable, and mutually beneficial investment, financing, and benefit compensation mechanism of projects.

8.1.2.5 Mechanism for Information Sharing and Public Participation

In order to obtain support from stakeholders and the whole society for the development, regulation, protection, and management of a watershed, it is required to establish a mechanism for information sharing and public participation regarding important water-related matters in the watershed. To promote the right for the public to participate in and to know and to supervise the regulatory institutions and law enforcement personnel as to whether they follow laws and regulations in administration and management, a mechanism should be established for information sharing and communication regarding the development, utilization, protection, and management of public resources so as to notify the public of important information regarding the development and utilization of resources and to prevent major engineering matters that need to be disclosed from being kept as internal or confidential matters that are only known by the level of experts and officials. Meanwhile, all other relevant departments, regions, and stakeholders should have the right to participate in and to know any important event involving water resources. As public resources should belong to the public and should serve the society and the public, they, especially the news media, should play a supervisory role in the development, utilization, protection, and management of natural resources. Of course, efforts should be made to provide public education on the scientific knowledge of public resources. Publicity and public education should also be provided in primary and secondary schools and universities regarding natural resources and ecological protection so as to improve the public capability of participating in the protection and management of resources.

8.1.3 Capability Building for Integrated Management of the Yangtze River Basin

8.1.3.1 Network of Monitoring and Observation

Without measurement and monitoring, there would be no real management. The conditions of natural precipitation, inflow and water quality, and the supply, use, consumption, and drainage of human water resources need to be monitored through the construction of monitoring and metering facilities, including monitoring station networks for climate, hydrology, water quality, and aquatic ecology and monitoring and metering systems for water users' diversion, use, consumption, and discharge of water, which are the basis of water resources management. A monitoring system of water resources includes two levels. One is the macro-level for the monitoring of the flowrates, water quality, and water levels at the control cross sections of a river, and the other is the micro-level for the monitoring of diversion, use, and drainage of water by water users. The focuses of monitoring watershed resources include the following:

1. Monitoring of River Control Sections

The monitoring of each control section can measure the impact of the inflow and water use from the upstream of a river or an important water zone from a macroscopic perspective, identify the responsibility of the upstream area's discharge of polluting matters and emergent water-pollution accidents, as well as provide dynamic guidance for water diversion, use, and drainage in the downstream area.

2. Beneficial Storage Capacity and Impoundment and Discharge Conditions of Large Reservoirs

The beneficial storage capacity of large reservoirs is an important means to solve the water-resource shortage problem in the Yangtze River basin. Due to the restriction of the administrative authorities, the current operation and management of large reservoirs are administered by various departments and regions. In the future, an integrated monitoring and operational plan will need to be prepared for the beneficial storage capacity, impoundment, and discharge of large reservoirs in the watershed, and in case of a special drought, integrated operation and management are required.

3. Monitoring of Interbasin Water Diversion

Presently, the total amount of interbasin water diversion for completed and planned projects does not account for much of the total amount water resources in the Yangtze River basin, but during an extremely dry year or season, the amount of the water to be transferred out of the basin needs to be controlled for a short time to ensure that the basic water demand in the watershed is met. The monitoring data can also be made available to the society with explanations to the public regarding the impact of the amount of the outflow.

4. Monitoring of Sluices Along the River

A large number of streams and lakes connect to the Yangtze River in the middle and lower reaches. The amount of inflow and outflow through most of the streams and lakes is controlled by sluices and culverts. The water diversion capacity of completed projects has reached up to 20,000 m^3/s. In case of a dry or extremely dry season, it will be necessary to initiate the emergency water resources management plan and reappropriate the water resources inside and outside the Yangtze River basin. At this time, the outflow may have a substantial impact on the middle and lower reaches and the estuary and must be monitored and controlled. Otherwise, the sole joint operation of the large reservoirs in the upper reaches of the Yangtze River cannot solve the widespread drought and water shortage problems.

5. Monitoring of Major Water Users

In the past, as applications for water diversion permits did not specifically emphasize the installation and records of gauging facilities, many water diversion permits were issued but not many gauging facilities were actually installed. Moreover, only a few were in effective operation, and even fewer were in the daily monitoring or management. To carry out the monitoring of the total water-use quantity control, it is necessary to require each water user to install gauging facilities for water diversion and drainage and gradually achieve real-time online monitoring so as to carry out the management of water resources.

8.1.3.2 Management Information System

While monitoring and measuring network systems are established, an information system for live data collection, analysis, monitoring, and management should be established using the modern 3S technology, Internet, and a computer system, which is the most important technical support system for hydraulic engineering modernization and integrated watershed management.

8.1.3.3 Professional Team Building

The management of water resources involves multiple disciplines such as geology, meteorology, hydrology, hydraulic engineering, economics, law, ecology, and environmental science, and watershed management needs a large number of professional and technical personnel in watershed planning, flood control and drought relief, ecology and environmental protection, legislature and law enforcement, etc. Therefore, there is a need to adjust the curricula of universities regarding the education of water resources, re-configurate the existing professional structure, and tweak disciplines so as to develop professional personnel who will not only have the technical knowledge of hydraulic engineering but will also master the knowledge of environmental, ecological, and social management. The current educational system of hydraulic engineering institutes in China needs to be adjusted. Major problems include outdated textbooks and single discipline. In the past, the educational purpose was mainly to cultivate engineers for the development and utilization of water resources from the engineering perspective. However, efforts should be made to teach multidisciplinary knowledge, such as water resources, ecology, environment, management, and law so that hydraulic engineering professionals can learn more knowledge about the sustainable use and protection of water resources.

8.1.3.4 Evaluation and Assessment System

An integrated watershed management is mainly for public welfare services such as flood control, water supply, irrigation, and ecology, and the effect of management implementation needs to be reported to the state, all levels of governments, and the society. At the same time, the public need to understand the construction of each important project, reservoir operation, or the effect of a management system implementation, including the economic input, loss, beneficiaries, and the benefit to affected parties. A system to evaluate or assess the effect of implementing an integrated watershed management will need to be established preferably by a technical evaluation team and a technical standard system composed of third-party experts and professional consultants.

8.1.3.5 Financing Platform and Asset Management

After the completion of a project for watershed management, development, and protection, there is a need to establish an engineering management organization to report to the state or investors on the investment, operation, maintenance, and management of the project and the revenue of the project assets and, at the same time, to effectively manage the raising and distribution of compensation funds for the parties affected by the construction and operation of the project, as well as the revenue of the project. As all water conservancy projects and eco-environmental projects are mainly invested for public welfare, it is sometimes difficult to directly estimate their economic value. With the increase of social wealth, increasingly more private and charitable funds will be invested into the protection of water resources and the restoration of the ecological environment. In the future, these project funds will mainly from the state fiscal fund. In addition, a management mechanism will also need to be established for private funding and investment so that the whole society can participate in ecological environment protection and river regulation.

8.2 Management of Water Resources

8.2.1 Several Basic Concepts

8.2.1.1 Water Rights

Water is a natural resource and is also a valuable special commodity. In order to effectively appropriate water resources and improve the use efficiency of water resources, a well-defined property-right-type water right system should be established. Presently, people are more familiar with the land right. Recently, as China began to reform the forest right system, people began to understand the forest right. In fact, the water right is also a very important concept of property rights involving natural resources. However, because water flows through forests and land, water rights are often subordinate to land or forest rights, but water rights are quite different from land and forest rights. Based on the development of foreign water right systems, there are several types of water rights.

1. Riparian Rights

The riparian right has its origin in English common law. The riparian right gives all owners of land contiguous to bodies of water the right to withdraw and use the water. As long as the water that flows back to lakes or rivers has no substantial qualitative and quantitative changes, the occurrence of return water often does not need prior approval. Riparian rights, whether they are exercised or not, are normally not lost if unused, and they do not establish priority by the use of time. Riparian rights are attached to the natural runoff of rivers and are also attached to the amount of stored water. Riparian rights themselves do not require the effective use of water resources, do not consider the relationships between the upstream and downstream or between the left- and right-bank water flow or water zones, and are practically derived from land rights. It is clear that riparian rights are not suitable for arid areas or densely populated areas. They were adopted in eastern United States but have never been adopted by the arid western United States. In recent years, many countries have abolished riparian rights or made restrictive provisions. For example, some states of Australia have abolished riparian rights, and in the other states, riparian rights only apply to small amounts of water use, such as domestic water use, but cannot be transferred or traded.

2. Prior Appropriation Rights

In the early days of westward expansion in the United States, most of the land was government-owned or no-man area. In order to encourage western migration and expansion, the US government repealed restriction provisions associated with riparian rights for the withdrawal and use of water resources in land reclamation and utilization in the western region. Whoever occupies the land first owns the land and the associated water rights. Therefore, this simple "take-and-use" form for the use of water resources developed into the doctrine of prior appropriation, or the "first in time, first in right," or whoever comes first has the priority to own water rights. Subsequently, this water right was often used to notify the intent of water withdrawal, was recorded in the local judicial department, and was standardized in the form of notification. Recently, while the court did not recognize the possession of water bodies, it recognized the beneficial use of the right. In effect, the possession and use of the water right were separated. Each of the arid western states of the United States has established a management system and legal procedures, respectively, for the prior appropriation rights of water resources.

3. Prescriptive Rights

Prescriptive water rights are another type of water resources use right. Riparian rights or prior appropriation rights may be lost as a result of prescriptive water rights, while legitimate prescriptive water rights are gained through judicial acts rather than administrative procedures. In the United States, the generation of a prescriptive water right must meet five conditions. The condition of water withdrawal shall be sustained for a longer time period (usually several years); acquisition of the water shall be carried out peacefully; relevant laws and regulations shall be followed; the use of water shall be open to the public; and this right must relapse another right.

Since water is a basic natural resource and flows, water rights differ significantly from land rights or forest rights, and most countries now separate the ownership, use, and management of water. Ownership, especially a relatively large water source, is generally owned by the state, while the right to use and management can be assigned to water users and can be transferred or traded in the water market.

China's water right system has not been established because the revised *Water Law* of 2002 did not incorporate the concept of water rights. The concept of water rights will need to be introduced when the state revises the water resources law, but the process may take a long time. However, the concept of water rights was mentioned in *Document No. 1 of the Central Committee in 2011: Decision of the Central Committee of the Communist Party of China on Accelerating the Reform and Development of Water Conservancy* and has since been in the practical use. There is a need to obtain experiences from actual uses and cases, and then the concept can gradually be promulgated.

The current major characteristics of water rights in China are as follows: ① Water rights are non-exclusive. From the legal point of view, a legally binding water right has unlimited exclusivity, but there are still many problems in China's current management system of water rights. Theoretically, water rights are owned by the state or collectively owned but they are essentially owned by departments or the local government. In practice, water rights are non-exclusive, or namely, when a water user has obtained a water right, other water users can no longer use part of the water right. ② Water rights are separative. Based on the existing situation in China, the ownership, right to the use and management of water resources are separated, which is decided by the unique management system of water resources in China. ③ Water right has external nature or external effect. A water right has certain externality. It has both the positive external economy (benefit), and the negative impact on the ecology and environment. When human beings use too much of it, less is left to the ecological environment, and the quality of the water degraded significantly after the human use. ④ Trade of water rights is imbalanced. The two sides of a transaction are two different representatives of the interested parties and are different in personal status. Also, a successful transaction of a water right depends on necessary engineering conditions.

Because of the particularity of water flow, the appropriation of the initial water rights is very complicated. In Western countries, because the land is privately owned, the landowner obtained the initial water right of the water zone inherent to the land when he occupied it. According to the experience of the countries that have implemented the water right system, water right is entirely privately owned, which is also problematic. In the event of water shortages resulting from a widespread drought, there is no way for the countries or governments to unify the regulation or distribution of the scarce water resources. For example, the Chile government often has to purchase water rights from private owners, and the government's right to allocate water is subject to great restrictions. In Australia, water rights are administered by each state, and thus the adjustment of water rights needs to be not only approved by each state but also requires the water rights to be purchased by the government from the water users who own the rights. Since most water rights (such as the Murray Basin) have been appropriated to water users, the government does not have many water rights. On the other hand, as the establishment of the water market needs necessary conditions such as water conservancy projects, it is not easy, even in the United States, to trade a water right, because water is not a general commodity. Generally, only in a drought and water shortage is there a need for a trade of water rights. Most of the time, there are not many "customers." Moreover, it takes time to store and transmit water. Even if a water right system has been established, there are many legal, technical, and commercial issues that will need to be addressed regarding the establishment of the water market and trade of water rights.

In China, since a contractual system has already been established for land, this system will continue to exist for a long time. The ownership and the right to use the land have been separated, and water rights should also be separated similarly. In the future, China's system of water rights should be managed by different levels. The ownership and the right to use large rivers, major lakes, and groundwater should be managed by the state, watershed management institutions, or local governments, while small-scale waters and the associated groundwater within the contractual land of farmers can be appropriated as initial water rights that are tradable. Transactions of large-scale water rights, even if they are available from the market, need to be reviewed, evaluated, and approved by the state and watershed management institutions because of the exclusivity and external effects of water rights.

The appropriation of the initial water rights must follow the principle of "balanced fairness and benefits." First of all, the basic right of water use for everyone should be considered. Second, existing water rights and customary water rights should be considered. Third, water should be appropriated to the users of high efficiencies and high benefits. Fourth, the natural conditions of the water flow and the conditions of hydraulic projects' regulation and operation should also be considered. Without canals, sluices, or gauging facilities, there is no way to make a trade. Therefore, a trade of water rights requires certain conditions.

The purpose of water right appropriation is mainly to improve the utilization efficiency of water resources, to protect water resources, to establish a water market when the conditions permit, and to further improve the utilization efficiency of water resources through the market mechanism. However, as water is a very special commodity, its transaction is not only restricted by local water resources and hydraulic engineering conditions but is also closely related to the method, the price, and the time of the transaction. A small market, such as a village-level peasant water exchange market, is easy to establish, while a transaction of large amounts of water or a large scale of water rights is constrained by many conditions. Therefore, the allocation of water resources should still be jointly carried out by the state, local governments, watershed management agencies, and the water market so as to take the advantage of different methods.

8.2.1.2 Water Quantity Allocation

Water quantity allocation is the basis for establishing the system of water rights and implementing the combined system of total water quantity control and quota management, as well as technically defines the amount of water that can be used or consumed in each watershed or administrative region. Article 45 of the *Water Law of the People's Republic of China* stipulates that water allocation "shall be formulated on a watershed basis in accordance with the watershed plan and the medium- and long-term plan for the supply and demand of water." Article 2 of the *Interim Measures for Water Quantity Allocation* promulgated by the MWR stipulates that "the water quantity allocation is to allocate the total usable amount of water resources or the total allocable water quantity to administrative regions level by level and to determine the shares of consumptive water quantity for life and production or the shares of water withdrawal and use by administrative regions." Two different allocation recipients (watersheds and administrative regions) were mentioned. In fact, water quantities should be allocated separately in watersheds and administrative regions.

The difficulty in water quantity allocation or total water quantity control is not the determination of water quantity allocation share but whether the method, process, and result of allocation can be accepted by all parties, how the water quantity allocation is determined fairly and scientifically, and especially how the allocation principle is determined, or the determination of priority is the most important. According to domestic and foreign experiences and China's existing conditions, the principle of water quantity allocation should include the following three aspects:

1. Principle of Fairness

The principle of fairness includes three different connotations. First, each person or region has the right to use water, and such a right should be based on equality for all and basic equality on all regions, in which drinking water for human and livestock must be met. Second, water right holders (i.e., permittee for water withdrawal) have advanced priority to use water. Third, water-generating regions or water basins themselves have advanced priority to use water. The three connotations warrant the principle of fairness in water allocation.

2. Principle of High Water-Use Efficiency

On the basis of fairness, the water users, departments, and regions with high water-use efficiency should be given priority.

3. Principle of Sustainable Development

The principle of sustainable development contains two connotations. First, it is necessary to reserve a basic water quantity and a natural flow process to the ecological environment and to coordinate the relationship between the human water use and the water need for the ecological environment. Second, it is necessary to reserve a certain quantity of water for economic development for future generations and to balance the water-use relationship between contemporary generation and future generations of human beings.

There exist contradictions among the three principles of water quantity allocation, and there is a need to balance and determine priorities among the three conditional principles. In the principle of fairness, drinking water for human and livestock has the highest priority. For production water use, the quota of water quantity should guarantee the principle of fairness, and for additional water use, priority should be given to the highly efficient water use. In the event of a severe drought or a severe water shortage, water use for life has the highest priority, and the water use for production and the environment can be properly reduced.

The main contents of water quantity allocation include the following: ① Determine the total amount of usable water resources or allocable water in a watershed or an administrative region. ② Determine the shares of water quantity for each administrative region and its corresponding control index of the river sections, reservoirs, lakes and the zones of groundwater withdrawal. ③ Determine the principle for adjusting and regulating the annual quantity of water use for each administrative region corresponding to different frequencies or satisfaction rates of incoming water quantity. ④ Determine the reserved shares of water quantity and the requirements of their corresponding river sections, reservoirs, lakes and groundwater withdrawal zones. ⑤ Determine the discharge, runoff, ecological flowrate, minimum flowrate, early warning flowrate, minimum control water level, water quality parameters, etc. of the cross sections at the boundaries of the rivers and lakes that traverse administrative regions, and the control parameters, such as the water level and quality of the groundwater within administrative regions. As shown above, water quantity allocation can be carried out in very details, but large amounts of technical work need to be performed to satisfy all parties. However, the most difficult task is to coordinate the opinions of all parties for a consensus.

Water quantity allocation in China is mainly performed by the government. Presently, water quantity is mainly allocated to the administrative regions above the prefectural level. If water quantity is allocated to specific water users, water rights of the water users are involved. If water quantity is allocated to the water users for the first time, this is known as "initial allocation of water rights," which is the starting point and basis for the future trade of water rights.

Although water quantity allocation defines the technical boundary of water quantity, it does not define the legal boundary of water quantity (such as water right). Water quantity allocation is refined by different guaranteed rates of water quantity allocation schemes and water quantity regulation schemes, which is the same as the comprehensive planning of water resources and relatively macroscopic. Water quantity allocation is also to allocate the amount of water resources, which is the basis of the initial water right allocation in the future.

8.2.1.3 Total Water-Use Quantity Control

Article 47 of the 2002 *Water Law* stipulates that the state shall implement a system that combines the total water-use quantity control and quota management. It was the first time that the system of total water-use quantity control was explicitly prescribed in a law. Due to many reasons, however, the system was not truly implemented only until in the 2009 National Water Conservancy Work Conference when Vice Premier Hui Liangyu and Minister Chen Lei, respectively, proposed to "implement the most stringent system of water resources management." The main content included the management of three redlines such as total water-use quantity control.

As a result, the total water-use quantity control has become the basic content of strict water resources management and has been implemented everywhere.

The total water-use quantity control incorporates the constraints on water resources, the aquatic environmental carrying capacity, and the water-conserving requirements such as water-use efficiency in a watershed or an administrative region. Its purposes are not only to conserve water and provide services for daily water resources management such as water withdrawal permitting but also to ultimately promote the improvement of the aquatic environment and the sustainable utilization of water resources. The total water-use quantity control is generally based on the water quantity allocation scheme or comprehensive water resources planning to determine the parameters of total water quantity control for each region in a water planning period. In the meantime, an annual water-use plan is reasonably determined within the control parameters based on the current conditions and water-use quantity in a region, and a management system that combines the total water-use quantity control and annual water-use management is implemented.

The total water-use quantity index is a relatively stable hard cap, but the annual water-use index is relatively dynamic and adjustable. The main purpose of the indexes is to examine a region's level and capability to manage water resources. In the absence of special circumstances (such as a prolonged severe drought or an adjustment of national major industrial layout, etc.), if the total water-use quantity of a watershed or region reaches or exceeds the total water-use quantity index, new water diversion projects for the watershed or region can be approved with stipulations; and if the water-use quantity of a watershed or region exceeds its annual water-use index, the watershed or region can be penalized in accordance with strictly implemented water resources management.

8.2.1.4 Relationship Among Several Concepts

A comprehensive water resources plan, water allocation scheme, total water-use quantity control index, and allowable water diversion amount are closely related. From the ideal sequence of work, a watershed or region first has a socioeconomic development plan approved by the state, second formulates a comprehensive water resources plan, third allocates water quantity, and last estimates the total water-use quantity control index, but the allowable water diversion amount depends mainly on the total water-use quantity control index. However, due to rapid socioeconomic development in the past, various conditions have changed so fast that often planning could not catch up; and the upper level planning could not catch up with the lower-level planning. As the system for the plan preparation and revision was not perfect and the lower level often prepared its plan without prior guidance from the upper level, it was difficult to conduct the planning work with great changes occurring.

A comprehensive water resources plan is formulated based mainly on the water resources conditions, the status of water resources development and utilization, and the current situation of water use in a watershed or region. Moreover, in considering the water-conserving potential, the plan projects additional needs of water resources for various planning years of socioeconomic development and proposes the general construction layout for new water source projects to meet the projected demands of water resources and adjustment of management methods. The cycle for the development, approval, and revision of a comprehensive water resources plan generally takes a relatively long time, possibly up to 10–20 years.

Water quantity allocation is to allocate the total usable quantity of water resources or the total allocable water quantity to watersheds or administrative regions level by level and to determine the shares of consumptive water quantity for life and production or the shares of water withdrawal and use by watersheds or administrative regions. Water quantity allocation in China is mainly carried out through the administrative method. The usable quantity of water resources is allocated to administrative regions level by level and approved by the State Council or all levels of governments for implementation. In fact, it is through the administrative method to stipulate the shares of water use across the administrative regions, and the shares are generally allocated based on the average annual amount of water use. The main difference between water quantity allocation and water resources planning is that water quantity allocation, if allocated to specific water users or legal parties, has the bearing of water rights in a legal sense and is the basis of the appropriation of initial water rights. Therefore, water quantity allocation is more rigid than water resources planning. Water quantity allocation in Western countries generally starts from the low level, and a specific water user can directly obtain water rights through a legal process, but in China, water quantity allocation is mainly implemented progressively from the top to the bottom and has no connection to water rights appropriation or water diversion permits. Between water quantity allocation and total quantity control, the former is more macroscopic, more rigid, and not easy to change, but the allocation process is long. For example, the 1987 Yellow River water allocation program was basically recognized by all provinces involved after a long period of operation and due to constraints on the total amount of water resources, and it is still in use today although all parties involved wanted to make adjustment.

The greatest difference between total water-use quantity control and water quantity allocation scheme is that the former emphasizes water-conserving requirements, is the index of water-use allocation from the top to the bottom, and mainly plays the role of monitoring and management. The content and method for the allocation of total control index are not much different technically from the water quantity allocation but are more concrete. The index contains more contents, such as the quota of water, ecological flowrate, minimum flowrate, minimum water level, and water quality for cross sections of a river channel.

8.2.2 Allocation of Total Water-Use Control Index in the Yangtze River Basin

The major problems associated with water quantity allocation and total quantity control in the Yangtze River basin are as follows: ① The Yangtze River basin is relatively rich in water resources when compared with the river basins in north China. According to data from the *National Water Resources Comprehensive Planning* and from the *Water Resources Bulletin*, the current and future water-use quantities in the majority of the rivers are lower than their usable water quantities. ② Surplus water quantity is saved as a reserve of water and is not allocated, which is opposed by all parties and makes coordination difficult. ③ Even if all water quantity is allocated, it is impossible to monitor or manage because there is no hydrological station at the cross section between two regions. ④ Without water storage facilities and ancillary works, even if it is possible to monitor and manage, problems associated with physical water scarcity can still not be solved. ⑤ If the local water is seriously contaminated and water supply is mainly relied on the guest or the external water, monitoring and management are mainly directed at the upstream and lacks the gripper for the management of the local aquatic environment.

In order to facilitate management, the water quantity allocation in the Yangtze River and other rivers in southern China can be carried out based on the watershed and regional characteristics, the conditions and key conflicts in the development and utilization of water resources, and the needs of water resources management. Moreover, the contents of water quantity allocation or total quantity control index can be selective and be expanded with the core emphasized. The main features of water quantity allocation in the Yangtze River basin are as follows: ① Vast majority of rivers are not scarce of total water quantity. Therefore, there is a need to not only allocate total water quantity, but also to allocate the process (space and time) of total water quantity. ② Key periods of water-use quantity management are the dry years and dry seasons. ③ The water storage and discharge plan (start time, process, end, and risk sharing) for large and medium-sized reservoirs should be considered. ④ The conditions of water conservancy projects and ancillary works should be considered. ⑤ The benefits and costs of water diversion and use should be considered.

In order to increase the momentum of total control in the Yangtze River basin, index allocation must consider the following: ① Total control management should be combined with the management of water function zones. ② Total quantity control should be combined with water quota and efficiency index. ③ Total quantity control should be combined with the improvement of hydraulic facilities' regulating capabilities. ④ Total quantity control should be combined with the management of inter-watershed water diversion. These four combinations can promote the implementation of total water-use quantity control. Table 8.1 shows total water-use quantity control indexes determined for planning purposes of provinces or municipalities within the Yangtze River basin.

Table 8.1 Total water-use quantity control indexes of provincial-level administrative regions in the Yangtze River basin

Provincial administrative region	Total water-use quantity (billion m³)			
	2010	2015	2020	2030
Qinghai	0.03	0.034	0.04	0.044
Gansu	0.31	0.477	0.648	0.749
Shaanxi	2.54	2.664	2.791	2.926
Tibet	0.07	0.092	0.111	0.116
Yunnan	4.66	6.076	7.498	7.974
Guizhou	7.26	8.465	9.676	10.246
Sichuan	22.45	27.274	32.121	33.90
Chongqing	8.46	9.083	9.713	10.559
Guangdong	0.01	0.021	0.031	0.028
Guangxi	1.19	1.139	1.089	1.038
Hunan	31.72	33.648	35.605	35.496
Hubei	26.4	31.411	36.451	36.741
Henan	2.41	2.774	3.143	3.277
Jiangxi	23.04	24.356	25.691	26.144
Fujian	0.09	0.069	0.046	0.041
Anhui	13.70 (13.71)	13.435 (13.458)	13.184 (13.218)	13.263 (13.296)
Zhejiang	0.04 (5.33)	0.039 (5.567)	0.04 (5.811)	0.037 (5.861)
Jiangsu	10.26 (24.95)	11.547 (26.965)	13.465 (29.006)	13.488 (29.282)
Shanghai	0.61(11.94)	0.397 (12.428)	0.675 (12.928)	0.662 (13.354)
Total	155.26 (186.59)	173 (206)	192.019 (225.62)	196.73 (231.07)

Note: Data in parentheses include the water-use quantity from the Tai Lake basin

8.2.3 *Three Redline Management in the Yangtze River Basin*

8.2.3.1 Connotation of Three Redlines

Total water-use quantity control is to manage the total water-use quantity in a watershed or region, is the macro control index of water resources management, and is to control the water withdrawal and water-use scale outside a river channel through calculating the development and utilization rate of the river's water resources. This redline mainly considers the water quantity but does not consider water quality.

The water-use efficiency redline is a comprehensive index and can include the water-use quantity for generating every 10 thousand yuan of GDP and every 10 thousand yuan of added industrial value, industrial wastewater reuse, water-use quantity per ha of farmland, and average water-use quantity per capita. Moreover, it includes the industrial water-use quota and can be used as a macro control index to examine a watershed or regional water-use efficiency. Furthermore, it can be used as a micro-index to assess the water-use efficiency of an industry, an enterprise, or even a water user. The redline cannot only directly control water-use quantity but can also indirectly take quality into account, because a high water-use efficiency means a high reuse rate and less wastewater discharge, which is conducive to improving water quality.

The redline for limiting the assimilative capacity of water function zones is also a relatively comprehensive index, can be used as a macro index to evaluate the effect of interregional protection of water resources through the management of Level I water function zones, and can also be used as a micro-index to evaluate the water quality conditions of the same water zones and assess the wastewater discharge reduction conditions of the same water-use department through the management of Level II water function zones. Although this redline is mainly aimed at the assimilative capacity of a water zone to protect water quality, it can also serve the purpose of protecting the aquatic ecosystem. Restriction of wastewater discharge can indirectly control the total amount of industrial and domestic water use, because when the water consumptive rate is fixed, the more water is used, the more water is withdrawn, and the more wastewater is discharged. The management of the water compliance rate for water function zones can indirectly control the total amount of water use in the region.

8.2.3.2 Relationship Among Three Redlines

The three redlines are to manage the use and protection of water resources from different angles. There are differences among the three, but they are closely related. For example, when the management of water function zones is in compliance, the amount of water supply can be increased, and water shortage problems associated with water quality can be solved. Increased water-use efficiency can effectively help total quantity control and reduce the amount of wastewater discharge into the river. On the other hand, good total quantity control can also help to increase water-use efficiency and reduce wastewater discharge.

Managerially, total water-use quantity control can be used for the macro-management of water resources in a watershed or large region; water-use efficiency can be used for the micromanagement of water use in a specific industry or water user; and limiting assimilative capacity can be used for the management of the quality of the water body. The three redlines together constitute a complete water resources management system.

For problems associated with water resources in a specific region, the three redlines can be applied together or can be applied successively or individually. For a water user who has obtained a water-use permit, total water-use quantity and the installation of the wastewater discharge outlet must be resolved before obtaining the water diversion permit; the remainder is mainly to have the water diversion and wastewater discharge monitored and inspected; and the key of management is to improve the water-use efficiency of the water user. For a resource-deficient region, total water-use control is the most basic control redline. For the river network and delta areas where the waters are severely polluted, and the main problem is water-quality-related water scarcity, the control of pollutant discharge into the rivers is the major control factor. The water-use efficiency in China is well below that of developed countries. Reducing the amount of wastewater discharge and improving water-use efficiency will be long-term efforts for all regions, departments, and the whole society.

8.2.3.3 Regional and Industrial Differences of Thee Redline Controls

Although the three redlines are applicable to all China's regional water resources management, due to various conditions and development and utilization levels of water resources in different watersheds and regions, the control focus of the three redlines also varies, and even for a single redline, the way and content of the control also vary.

In the resource-scarce regions in northern China, water demand is much larger than available water supply, and the total quantity control index is not only necessary but also easy to achieve results. For example, under the Yellow River water quantity allocation plan, 37 billion m^3 of the 58 billion m^3 average annual amount of surface water in the river has been allocated to nine provinces (regions) and the adjacent Hebei Province and Tianjin Municipality that have source-scarce water shortage problems and reserved 21 billion m^3 of surface water for the ecological water need such as sediment transport within the river channel. Strict total quantity control management has prevented the Yellow River from stopping flowing for more than 10 years and gradually improved the ecological environment in the Yellow Estuary and Delta area. However, for the Yangtze River and other rivers in southern China, the existing water-use quantity and future demand are less than the usable water quantity. Therefore, the focus of total quantity control is not the total quantity but rather the process and water quality, especially in physical water-scarce regions during the dry years and dry seasons. In some rivers and water zones, total quantity control is not necessarily aimed at water quantity problems, but it may be to ensure that the water zones have adequate assimilative capacity so as to ensure the compliance rate of the water function zones.

There is little regional difference in industrial and domestic water-use efficiency, but the northern regions use more groundwater. Thus, control of groundwater overdraft is an important part of the management of total water-use quantity. However, as the proportion of groundwater utilization is relatively small in the southern regions, the focus of control is on surface water. There is a large regional difference in agricultural water use that is a large water user. As the northern climate is dry, and the evaporation is high, more than one half of the precipitation evaporates or infiltrates into ground and non-crop water consumption is high. Therefore, water consumption management is the focus, and the reduction of water conveyance loss through irrigation canals and the improvement of crop water-use efficiency are the important contents of water resources management. In the south, water shortage is mainly of a seasonal, regional, and physical type. The mountainous and hilly regions are high in elevation and the water is low. Either storage conditions (such as karst area) are poor or the cost for water conveyance projects is too high. In the plain areas, as water systems are well developed, the area of waters is large, but water pollution is severe. Although there are seasonal water shortages, they are mainly of a water-quality type. Hence, the control of contaminant discharge into rivers is the focus of water resources management. The total water-use quantity control of the southern rivers must combine the water-use efficiency index with the management target of the water function zones so as to achieve results, because considerable constraints of regional water-resources management are related to low water-use efficiency and serious water pollution. Therefore, the strict management of water resources needs to consider regional differences.

Presently, water-use quota management has been used in various industries, but it is difficult to compare water-use efficiencies among the industries. However, the relatively low water-use efficiencies in some industries are still noticeable, because the water-use quota of each industry in China is well behind that of advanced countries and that of some industries is even worse. Therefore, not only China's overall water quota is well behind that of advanced countries, but also, the water-use efficiency in various industries is also behind. There is a need for a department to comprehensively evaluate and manage the water-use efficiency index of various industries.

8.2.4 Means to Implement Management of Three Redlines

The implementation of strict water resources management should not only consider the characteristics of watershed and regional water resources but also need the cooperation of all sectors of the country and the participation of the whole society. The means of implementation are recommended as follows.

8.2.4.1 Total Water Quantity Control

Total water-use quantity control is to determine the demand based on supply, control the total human water-use quantity, and ensure to leave the necessary water-use quantity for the ecological environment in a river's water zone under the premise of water-conserving and high-efficiency water use. The implementation of management needs to prepare the total annual water-use quantity plan, to install gauging facilities at the segments of water diversion and water use for the monitoring of actual water-use quantity, and to perform post-use inspection and evaluation in accordance with the framework of the comprehensive water-resources use plan, the water quantity allocation scheme, and the water diversion permits in a watershed or region. The comprehensive water resources use plan and the water quantity allocation scheme are used to control the medium- and long-term total water-use quantity in a watershed or region and to mainly control the average annual quantity of water resources, which are the technical basis for the total allowable quantity control of water diversion. The difficulty of total quantity control management is how to conduct the interregional and interdepartmental allocation of water-use quantity and make adjustment based on the change of demand. Therefore, first of all, there is a need to establish a network to monitor the flowrate, water diversion, and water use of a water zone to provide the necessary basic information and technical solutions to the problems associated with total quantity control. In the meantime, there is a need to establish the administrative and technical committees for the allocation and adjustment of total quantity control indexes to address administrative and technical issues associated with water quantity allocation and adjustment. In order to address the conflict between total quantity control and economic development, there is a need to establish a water right system and a water right transaction system to transfer water to industries and regions with higher efficiencies and better benefits under the precondition of guaranteeing basic human and livestock water use.

8.2.4.2 Water-Use Efficiency Control

Water-use efficiency is mainly based on the quota and index of water use. If the index is not scientific or advanced, the water-conserving standard or water-use efficiency cannot be determined. A large gap still exists in water-use index between China and advanced countries. Therefore, there is a need to re-evaluate water-use indexes for various industries and regions and support the advanced level of regional water-use quota in various industries based on the water-use quota and index of advanced water-conserving countries and China's actual situation. For the agricultural water-use quota, field experiments need to be conducted based on regional water and soil conditions and crop types, especially in southern regions where there are few experiments of the water-use quota on paddy fields. As a result, there has been the lack of scientific experimental data for a long time. For the management of water-use efficiency, a management system should be established so that various departments and regions will perform itemized management and the water conservancy department will perform unified review and evaluation.

8.2.4.3 Water Function Zone's Pollutant Load Control

The goal of a water function zone's pollutant load control is to ensure the water function zone in compliance, to implement management, and to determine the assimilative capacity of a river and major water zones based on the management objectives of the water function zone. Because the assimilative capacity is related to not only the aquatic environmental carrying capacity but also the hydrological and flow processes, it is technically difficult to accurately determine the assimilative capacity of a water zone. Therefore, there is a need to perform the phase-by-phase and step-by-step calculation and determination of the assimilative capacity for the major water function zones of water bodies and then establish a monitoring system, install gauging devices at the main outlets of wastewater discharge, live monitor and evaluate the water quality conditions of major water function zones, establish a rapid monitoring and assessment method for emergent water pollution accidents, clearly define responsibilities, formulate an emergency remediation plan, and develop a professional remediation team. In the region where water function zones are not in compliance, a strict pollutant discharge reduction system should be developed, including laws, market access, pollutant discharge right trading, and public participation, to encourage pollution reduction enterprises and units.

8.3 Joint Regulation of Cascade Reservoirs

8.3.1 Present Conditions of Reservoir Group Construction and Necessity of Their Joint Regulation

By the end of 2009, 47,000 reservoirs of all kinds had been constructed in the Yangtze River basin with a total storage capacity of more than 250 billion m^3 and total beneficial storage capacity of more than 120 billion m^3. Of the 47,000 reservoirs, 166 were large ones with a total capacity of 190.8 billion m^3 and total beneficial capacity of 98.3 billion m^3, accounting for 76% and 81%, respectively, of the total reservoir capacity and beneficial capacity of the reservoirs in the basin. Large reservoirs played an important role in the utilization of water resources in the basin and were the key objects of the joint regulation of reservoir groups.

The Yangtze River basin is relatively rich in water resources, the average amount of annual surface runoff reaches 985.6 billion m^3, and the total storage capacity and total beneficial capacity of the existing large reservoirs in the basin accounted for 19% and 10%, respectively, of the basin's total amount of surface runoff. Therefore, the actual available storage capacity that can be used for regulation is limited. With the gradual completion of large reservoirs in the upper reaches of the Yangtze River, the beneficial capacity will increase, but the total storage capacity will still not be large, while the expectation of reservoir regulation in all the regions of the basin is very high with an increasingly higher demand.

With the completion of large control reservoirs, such as the TGP, in the upper reaches of the Yangtze River, the conflicts between the impoundment and beneficial regulation of the reservoir and the public welfare regulation in the basin and the water use in the middle and lower Yangtze have become increasingly more prominent, which has made the joint regulation of the control reservoirs very necessary, but there are still many technical and managerial problems. In flood control, the main control reservoirs in the upper reaches of the Yangtze River have a total flood control capacity of more than 59.1 billion m^3 distributed in more than ten reservoirs on the Yangtze River. A unified regulation is required for various areas and different flood events. From the impoundment and discharge of the reservoirs, the regulating capacity of the reservoirs in the upper Yangtze will be more than 100 billion m^3. If all reservoirs start to store water during October and November, the flowrate will need to be 20,000 m^3/s. Such a concentrated water impoundment operation will have a great impact on the river channel and the regional water use in the middle and lower reaches of the Yangtze River.

8.3.2 Distributional Characteristics of Large Reservoirs in the Yangtze River

The regulation of cascade reservoirs can be divided into the river regulation and watershed regulation of control reservoirs. Obviously, the latter one is more complex. In the Yangtze River basin, cascade reservoirs are mainly constructed in the mountainous sections and mainly concentrated on the mainstream and tributaries in the upper Yangtze, with a very few in the plain areas of the middle and lower reaches of the river. Presently, there are only 1–2 control reservoirs with good regulating functions in most of the tributaries. Figure 8.1 illustrates the distribution of major control reservoirs in the upper Yangtze, and Table 8.2 summarizes the basic information of the major control reservoirs in the upper Yangtze.

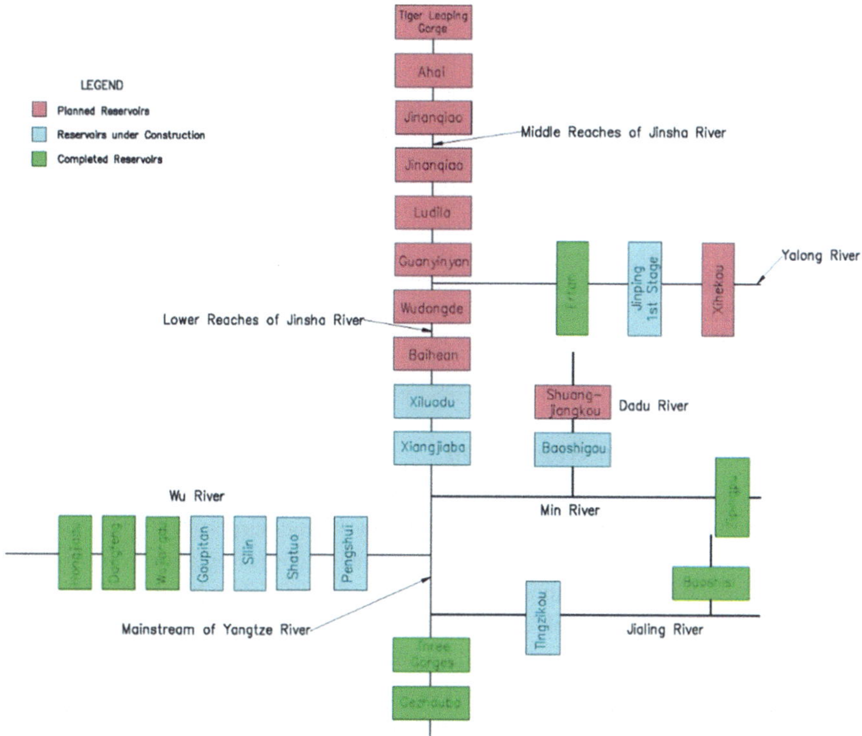

Fig. 8.1 Illustrative chart for distribution of major reservoirs in upper reaches of Yangtze

Table 8.2 Basic information of major control reservoirs in the upper Yangtze

River	Cascade name	Normal level (m)	Dead level (m)	Regulating capacity (billion m³)	Installed capacity (MW)	Flood control capacity (billion m³)
Mainstream	Tiger Leaping Gorge	1950	1880	14.734	6000	5.859
	Ahai	1620	1556	4.209	4750	3.152
	Jinanqiao	1410	1400	0.158	2800	0.158
	Longkaikou	1297	1289	0.130	1900	0.130
	Ludila	1221	1212	0.574	2200	0.574
	Guanyinyan	1132	1120	0.542	3000	0.542
	Wudongde	950	925	1.662	7000	1.611
	Baihetan	820	760	10.032	12,000	7.324
	Xiluodu	600	540	6.460	12,600	4.65
	Xiangjiaba	380	370	0.903	6000	0.903
	Three Gorges	175	155	16.500	18,200	22.15
	Gezhouba	66	63		2715	
Yalong River	Lianghekou	2860	2780	6.275	2700	2.0
	Jinping Stage I	1880	1800	4.910	3600	1.1
	Ertan	1200	1155	3.370	3300	0.9
Min River	Zipingpu	877	817	0.774	760	0.167
	Shuangjiangkou	2510	2430	2.152	2000	0.51
	Babugou	850	790	3.882	3300	0.66
Jialing River	Baozhushi	588	558	1.340	700	0.28
	Shengzhong			0.672		0.27
	TIngzikou	458	438	1.750	1100	1.46
	Caojia	205	195	0.658	600	0.658
Wu River	Hongjiadu	1140	1076	3.357	540	
	Goupitan	630	585	3.154	3000	0.4
	Silin	440	431	0.317	1000	0.184
	Shatuo	360	350	0.413	1000	0.204
	Pengshui	293	278	0.518	1750	0.232
Total				89.446	104,515	56.078

8.3.3 Need for Joint Regulation

The joint regulation of a reservoir group is mainly for the purposes of flood control, power generation, navigation, water resources, and ecology. Regulations for hydroelectric power generation and navigation involve a few sectors, while the other regulations of public welfare involve many stakeholders and the effectiveness of the regulation has a great impact on the society and the ecological environment, but it is hard to estimate the economic value. From the viewpoint of regulated objects and mutual influences, hydroelectric regulation can be carried out not only by regulating the upstream and downstream water flow in the river but also by jointly regulating the electric grid in the air. However, the other regulations must be carried out from upstream to downstream and from tributaries to mainstream, level by level. The major need for the joint regulation of reservoirs in the Yangtze River basin is reflected in the following aspects.

8.3.3.1 Flood Control Regulation

Under the *Flood Control Law of the People's Republic of China*, flood control regulation shall be jointly carried out by the state, river basin, provinces, and cities of all levels. With the construction of the TGR and other large reservoirs in the upper Yangtze, the flood protection areas in the middle and lower reaches, especially the planned flood diversion/retention zones, hope the TGR to play a greater role in flood control, and expect the reservoir to intercept not only large floods, but also small- and medium-sized floods, which involve major issues in the basin such as adjustment of flood control strategy, flood risk control, and benefit adjustment. From the angle of the flood control system, flood control regulation involves issues such as the locations and the using order of dikes, reservoirs, and flood diversion/retention zones and technical issues such as the optimal regulation of the flood control capacities of large reservoirs in the upper Yangtze.

8.3.3.2 Regulation of Reservoir Storage and Preflood Discharge

As the runoff of the mainstream and major tributaries of the Yangtze River are large, it is difficult to construct multi-year regulating reservoirs. Therefore, most of the reservoirs have only an incomplete annual (seasonal, monthly, daily, and runoff) regulating capacity, which means that for a large reservoir with the flood control function, the water level of the reservoir must be lowered to the flood control level before the flood season starts so as to reserve the flood control capacity to regulate floodwater. Even for the reservoir that does not have the flood control function, the operating water level also needs to be lowered before the flood season starts for the purposes of dam safety and sediment discharge, and it can start storing water at the end of the flood season, generally in September–November, until it reaches the design level. Therefore, the reservoir group in the basin has a problem associated with concentrated and competitive water storing. As individual reservoirs are managed by different entities and regions, if a unified impoundment regulation is not adopted, it is not only difficult to ensure each reservoir will be fully filled with water after the flood season, but also dry season may occur earlier in the downstream river channel, affecting the water-use safety and ecological safety in these regions. However, there are also problems associated with lowering the water level at each individual reservoir through discharging water before the flood season; if this is not considered in a unified way, high water levels may occur earlier in the river channel and increase the flood control pressure on some regions in the middle and lower reaches.

8.3.3.3 Regulation for Drought Relief and Water Replenishment

In recent years, the drought problem in the Yangtze River basin has become increasingly prominent, such as the Sichuan-Chongqing drought of 2006 and the southwest regional drought during 2009–2011. In the past 10 years, low water levels and droughts have occurred in Dongting and Poyang Lakes for many consecutive years, and a severe drought occurred in the middle and lower reaches in 2011. As a result, new requirements have been brought up for the TGR and other large reservoirs to be jointly regulated for drought relief and water replenishment.

8.3.3.4 Ecological and Environmental Regulation

Due to dam construction, the river's hydrological and physiochemical process have been changed. There is an increased need of the ecological regulations for the processes of the water level, flowrate, and water temperature to meet the requirements for the rare fish and the FMCC to spawn and for aquatic animals and wetland waterfowls. At the same time, environmental regulation is urgently needed for each reservoir to prevent the occurrence of algal blooms and emergent pollution accidents.

8.3.4 Technical Conditions and Constraints of Joint Regulation

8.3.4.1 Joint Regulation of Flood Control

The flood control system of the Yangtze River consists mainly of three parts including dikes, reservoirs, and flood diversion/retention zones. Based on the flood size and the region of occurrence, three aspects of measures are generally utilized comprehensively. Dikes are essential, reservoirs are the major regulating means, and the flood diversion/retention zones are the last means to ensure safety. When a flood occurs, generally dikes are first used to block the floodwater, and the storage capacity of river channels can be used to withstand small- and medium-sized floods. Next, reservoirs are used to store and regulate medium-sized to large floods. Finally, in case of a severe flood when the flood control capacity of the reservoirs is insufficient and the discharge in the downstream river channel has reached the safe flowrate, the flood diversion/retention zones are needed to ensure the safety of the dams and the dikes in the key regions. Therefore, no matter how many reservoirs are constructed, a considerable number of flood diversion/retention zones will need to be maintained to protect against rare flood events.

Approximately 30,000 km of dikes have been constructed in the flood control system of the Yangtze River, and the flood storage capacity of the river channels at the design flood level of the dikes is nearly 100 billion m^3. However, since the dikes were constructed gradually in different historical periods, their quality varies widely.

When the water level reaches the warning level, a large number of personnel are needed onsite to check the dikes. Nevertheless, the dikes are still the most basic flood works that should be used first. The total flood control capacity of large reservoirs constructed in the Yangtze River is 58.8 billion m^3, of which only 27.6 billion m^3 is in the upper reaches. During the 1954 large flood, more than 100 billion m^3 of floodwater above the storage capacity of the river channels needed to be regulated. Additional flood capacity of 34–36 billion m^3 will be provided by the reservoirs that are under construction or will be constructed in the mainstream and tributaries in the upper Yangtze in the future, but their flood control capacities are combined with their beneficial capacities, which is different from that of the TGR, and are mainly used for the flood process before September with the primary goal to defend local floods and limited their flood control function in the late flood season. Therefore, these reservoirs can only be used to supplement the TGR in flood control regulation, and the total flood control capacity is not as important as the numbers indicate, depending on where and when the capacities can be used. Although the flood diversion/retention zones constructed and planned to be constructed in the middle and lower reaches of the Yangtze River will be able to store 62.7 billion m^3 of floodwater, the use of them will be very costly because the construction is delayed and nearly 7 million people live in the diversion/retention zones. Not much flood diversion/retention capacity can now be used. Thus, joint flood control regulation will need to individually consider the spatial and temporal distribution of reservoir groups to intercept floodwater. Namely, the optimal regulation of flood control capacities should be used differently between the regional and watershed floods and between the early and late floods.

Joint flood control regulation includes two technical aspects: one is the regulation that considers the dikes, reservoirs, and flood diversion/retention zones, and the other is the one that involves the joint regulation of reservoir groups. Obviously, the former is more macroscopic and has strategic significance, while the latter emphasizes the tactical aspect. With the construction of the TGR and other large reservoirs in the upper Yangtze, the regions and communities in the middle and lower of the Yangtze River expect the reservoirs to play a greater role in flood control and hope that the warning water level of the dikes will not be exceeded while the probability to use the flood diversion/retention zones will be reduced. These expectations may be reasonable from socioeconomic benefits but may be problematic from the viewpoint of technology and risk control. Overreliance on reservoirs for flood control may not be conducive to the construction, maintenance, and management of the dikes and flood diversion/retention zones or the maintenance of floodways or the health of the river ecosystem. When the floodwater does not reach the floodplain, the river channel is prone to shrinking and being used, affecting the connectivity of the river channel in the lateral direction and depth.

The main objective of the original flood control design for the TGR was to intercept medium-sized and large floods with the regulating policy of both storing and discharging floodwater and mainly discharging. If the policy for regulating floodwater is changed to mainly storing, increased sedimentation in the reservoir and higher flood control risk to the dam and the reservoir will occur, which is not conducive to the flood control function of the river channel or the management of the sandbars or floodplains in the middle and lower reaches of the Yangtze in the long run. In order for the TGR and other large reservoirs in the upper Yangtze to play a maximum role in flood control, it is necessary to improve the accuracy of flood forecast, control the flood risk of the reservoirs when continuous flood peaks occur, and solve the problem of sediment discharge and silt reduction. For flash floods, due to their instantaneity and occurrences in remote gullies, non-structural measures are the main control means. For large floods in the Chuan River, the Chongqing section, and the upper Yangtze, flood control is mainly carried out by the joint regulation of the large reservoirs in the Jinsha, Yalong, Min, and Jialing Rivers, and the TGR can play a role in regulating and storing floodwater to reduce the impact on the middle and lower reaches. If a 1000-year or larger flood (such as the 1870 flood) occurs, the flood diversion/retention zones in the middle and lower reaches will need to be used. For a regional flood in Dongting Lake, Poyang Lake, and the Han River, flood control mainly relies on the regulation of local reservoirs, supplemented by the large reservoirs such as the TGR in the upper reaches by controlling discharge to ease the flood control pressure in middle and lower reaches, and local flood diversion/retention zones will be used when necessary. For a watershed flood, it is necessary not only to apply the joint flood control regulation of all the reservoirs in the entire watershed but also to use the flood diversion/retention zones to ensure the safety of the dams and important flood control areas. Table 8.3 presents the joint regulation and other flood control measures during typical floods in the Yangtze River basin.

8.3.4.2 Joint Regulation of Water Resources

Traditional arid and water-scarce areas in the Yangtze River basin include the central Yunnan Plateau, the Sichuan Basin, the central-southern Hunan, the Jitai Basin of Jiangxi, the Northern Hubei-Southern Henan Basin, etc. They are all located in relatively arid regions, or hilly regions and watershed divide areas where it is difficult to construct large reservoirs or irrigation canals. In the recent 30 years, the use quantity of water resources in each region has increased due to rapid socioeconomic development, and serious droughts have occurred frequently in Chongqing, Yunnan, Guizhou, Sichuan, Dongting Lake, Poyang Lake, and the Jianghan Plain because of the impact of climate change. The large reservoirs that were constructed more than 30 years ago have water supply and irrigation functions with associated water diversion irrigation canal systems, but most of the large reservoirs that have been completed in the recent 30 years or are under construction are dominantly for hydroelectric power generation and are not aimed at irrigation or drought relief. When droughts occur in the downstream areas, these reservoirs cannot supply irrigation water to the drought-occurring areas due to the lack of supporting irrigation canals. As a result, the only thing that can be done is to increase the discharge to the downstream river channel to indirectly participate in drought relief. For example, the TGR and other reservoirs in the upper Yangtze cannot directly supply water to the drought-occurring regions in the middle and lower reaches, but they are

Table 8.3 Joint regulation and other flood control measures for typical Yangtze floods

Flood region	Dike	Upper reaches and local reservoirs	TGR and other reservoirs on mainstream	Diversion and retention zones	Non-structural measures	Constraints
Flash flood	√	√			√, early warning, forecast, and evacuation	Large reservoirs are hardly useful
Flood in Chuan River and Chongqing	√	Reservoirs in Jinsha, Yalong, Min, and Jialing Rivers	√, regulation, and storage		√, early warning, forecast, and evacuation	Conflict between flood control and reservoir impoundment in late flood season
Flood in upper reaches	√	√, reservoirs in upper reaches	√, regulation, and storage	√, flood diversion in Jingjiang, Dongting Lake, and Hong Lake for >1000-year flood	√	Conflict between flood control and reservoir impoundment in late flood season
Flood in Han River	√	√, Danjiangkou Reservoir	√, control discharge	√, Dujiatai, and others	√	
Flood in Dongting Lake	√	Dongjiang, Wuqiangxi, Jiangya, Zaoshi, and Zhexi Reservoirs	√, control discharge	√, Dongting Lake and Hong Lake diversion/ retention zones	√	
Flood in Poyang Lake	√	Zhelin and Wanan Reservoirs	√, control discharge	Kangshan and other diversion/ retention zones	√	
Large watershed flood	√	All reservoirs	√, regulation and impoundment	√, subregional diversion, level by level	√	High loss in flood diversion and retention zones

Note: √ indicates "to be used"

generally used to raise the water level in the mainstream so as to indirectly alleviate the local drought conditions.

Drought relief generally needs to take comprehensive measures to be effective, and reservoir regulation needs to meet certain conditions to be effective. These basic conditions are: ① A downstream reservoir cannot help an upstream drought-occurring area. ② A low-altitude reservoir cannot help irrigate the higher-altitude farmland unless economically favorable pumping stations and irrigation canals have been constructed. ③ The reservoir in the upper reaches cannot directly help solve the problem of saltwater intrusion in the estuary by increasing discharge because a salty tide is mainly affected by the marine tide and is a natural phenomenon. Using the reservoir water to suppress the seawater intrusion is not only ineffective but also not cost-effective. ④ Drought relief should first use local reservoirs and thereafter use other regional reservoirs.

As the TGR is 1750 km from the Yangtze Estuary, it takes 10–15 days for the water discharged from the reservoir to flow to the estuary. Moreover, the relationship of the TGR with several water systems along the river between the TGR and the estuary, such as Dongting, Poyang, and Tai Lakes and Huai River, is complex and water exchange occurs a lot. Furthermore, there are many rivers that connect to the Yangtze River and water diversion facilities in the middle and lower reaches. The diversion capacity of these hydraulic works is more than 20,000 m^3/s, while the TGR has a beneficial capacity of only 16.5 billion m^3 and can only have additional discharge of 1000–3000 m^3/s during January–April which is well short to maintain the flowrate of more than 15,000 m^3/s at the Datong Station (below this flowrate, a tidal seawater intrusion may occur at the Yangtze Estuary). Therefore, although the TGR is the most important control reservoir on the mainstream of the Yangtze River, its capability of drought relief is limited. For droughts in Dongting and Poyang Lakes and low water levels in the middle and lower reaches of the Yangtze River, the TGR can increase its discharge to alleviate the conditions to a certain extent, but the joint regulation should use local reservoirs first to provide water for drought relief, followed by the participation of the TGR. More importantly, there is a need to manage the demand for water diversion and water use in the drought-occurring areas at the same time. When a prolonged severe drought occurs in the middle and lower reaches of Yangtze River, the demand management of water diversion and water use in the riparian zone is more effective than the joint regulation of reservoir groups. Therefore, for a drought in the middle and lower reaches of the Yangtze River, local reservoirs should be regulated first, supplemented by the TGR. As Dongting and Poyang Lakes are connected to the Yangtze River, except when floods occur, most of the time, the water levels in the two lakes are higher than that of the Yangtze River. The main role for discharging more water from the TGR is to raise the water level of the Yangtze River so as to delay the time for the water in the two lakes to flow into the Yangtze River. Table 8.4 presents comprehensive drought-relief measures to be taken in a typical drought-occurring region in the Yangtze River basin.

Table 8.4 indicates that no matter where a drought occurs, local reservoir regulation and water resources demand management measures should be adopted first, and the TGR and other reservoirs on the mainstream of the Yangtze River may play an important role in water replenishment mainly during the period from November through next May.

Table 8.4 Reservoir regulation and other countermeasures for typical droughts in the Yangtze basin

Drought-occurring region	Upstream and local reservoirs	Mainstream reservoirs including TGR	Water resources management and other countermeasures	Constraints
Local drought	√		√	Limited role of mainstream reservoirs
Upper reaches	√		√	
Dongting Lake	√	√, increase discharge	√	TGR during November–May
Han River	Danjiangkou		√, limit water diversion through the Central Route Diversion Project	Dry season
Poyang Lake	√	√, increase discharge	√	TGR during November–May
Middle and lower mainstream of Yangtze	√	√, increase discharge	√, control diversion quantity from Yangtze River	TGR during November–May
Yangtze Estuary	√	√, increase discharge	√,control diversion quantity from Yangtze River	TGR during November–May

Note: √ indicates "to be used"

8.3.4.3 Regulation for Hydroelectric Power Generation and Navigation

The regulation for hydroelectric power generation is mainly carried out through the power grid, but the cascade hydroelectric stations in a river still have the potential to optimize the regulation. When a control reservoir is constructed in a river, it will significantly improve the downstream hydropower station's guaranteed output and hydroelectric generation capacity, and a joint regulation should be carried out in water flow connection and comprehensive hydropower-utilization efficiency. It should be noted that the regulation of hydroelectric generation needs to be coordinated with water resources regulation and shipping regulation and generally should follow the rule that the regulation of hydroelectric generation should be behind the regulations of water resources and beneficial regulation should be subordinate to welfare regulation, which naturally involves the coordination of economic interests and integrated management.

The navigable waterways in the mainstream and tributaries of the Yangtze River amount to more than 67,000 km long, accounting for 52.8% of China's inland waterways. With the completion of the cascade reservoirs, the satisfaction rate of navigable water levels will be significantly improved. The water-flow connection between cascade reservoirs and the prevention of the impact from unsteady flow in the downstream channel caused by the peak regulation of hydropower stations are major issues that need to be considered in the joint regulation.

Shipping has specific requirements for the water depth, flowrate, and evolution of the river channel. For example, since the construction of the TGR, shipping in the reservoir area and across the dam has developed rapidly and has exceeded the original design level (the shipping freight across the dam toward the downstream will have reached 50 million tonnes by 2030). However, the freight load through the ship locks reached 78.8 million tonnes in 2010. If the load across the dam toward the upstream is added, the total load reached 87.94 million tonnes. Moreover, ships across the dam are apparently trending up in size. The original design was mainly for fleets of less than 1000 tonnes, but now the maximum load of a single ship has reached 5000 tonnes. In 2011, 21% of the vessels across the dam were less than 1000 tonnes and 54% were above 2000 tonnes. Therefore, the requirements for the water depth and water level fluctuation range are higher. At the mouth of the navigation channel, the longitudinal flow velocity shall be no greater than 2.0 m/s, the lateral flow velocity no larger than 0.03 m/s, the return flow velocity no greater than 0.4 m/s, the fluctuation range of daily water levels at the deeply cut gage on the Gezhouba Dam no more than 3.0 m/day, and the maximum hourly fluctuation of water levels no greater than 1.0 m/h. For the downstream approach channel at Sanjiang of Gezhouba, the TGR needs to discharge 6000 m^3/s to maintain the water level of 39.0 m in the navigation channel, and the navigable channels of the mainstream, tributaries, and other river sections of the Yangtze River also have regulation requirements.

8.3.4.4 Ecological and Environmental Regulation

Ecological regulation especially needs the joint regulation mechanism of reservoir groups. For example, the discharge of the ecological flowrate, the generation of a flood process, etc., need to coordinate between the upstream and downstream reservoirs. So far, due to the complexity of the aquatic ecosystem, the ecological regulation of large reservoirs in the Yangtze River basin has not been really incorporated into the regulation procedures, except to ensure the minimum discharge (often called the ecological base flowrate). With socioeconomic development, the understanding of the river ecosystem's importance has been gradually improved. As a result, increasingly more attention will be paid to the ecological regulation to rehabilitate and improve the river ecosystem and to maintain the environmental flowrate.

The minimum ecological flowrate discharged from a reservoir is the primary stage of the ecological regulation and includes the water-use demand of downstream human life and production and the environmental water-use quantity of water function zones, but it is still well short of ensuring the aquatic system's structural integrity and desired objective.

For example, fish species in the Yangtze River need breeding, spawning, wintering sites, and migratory pathways. Fish species not only need a certain flood process but also need specific water quality, water temperature, and other physiochemical processes. A spawned drifting fish egg usually expands rapidly after absorbing water, forming the perivitelline space. The unit weight of the drifting egg is slightly higher than water, but it does not sink in the flowing water. The eggs and fries separate while they are impacted by the water flow. Fries from sinking sticky eggs tend to be benthic, and most of them collectively stay in the vicinity of their spawning sites. Only when the larvae have gained strong swimming skills do they swim to the food-rich riparian zone and downstream channels to feed. Even a seemingly simple fish spawning process requires so much, and the entire aquatic ecosystem is more complex. Therefore, it is difficult to carry out ecological regulation. Table 8.5 shows the hydrological characteristics and time required for spawning of several typical fish species in the Yangtze River.

Table 8.5 Basic water flow requirements for important fish species in the Yangtze River

Protected fish species	Water temperature (spawning time)	Water level	Flow velocity	Egg features
Chinese sturgeon	17–20 °C (October–November	Flood receding and water level decreasing to a certain level	Surface flow velocity 1.1–1.7 m/s	Sticky sinking eggs
Chinese paddlefish	>17 °C (March–April)			Sinking eggs
Dabry's sturgeon	(October–November)			Sinking and sticky eggs
Chinese sucker	>13 °C (March–April)	Spawn in rocky rapids when water level and temperature are relatively stable		Sticky eggs
Rock carp	>10–12 °C (February–March)		1 m/s	Sticky eggs
Coreius zeni	>17 °C (April–July)		Turbulent rapid flow	Drifting eggs
Prenant's schizothoracin	9–19 °C (April–June)		Rapid Flow	Slightly sticky sinking eggs
Longnose catfish	(April–June)		Turbulent rapid flow	Sticky eggs
Varicorhinus angustistomatus			Turbulent rapid flow location	Sticky eggs
Elongate loach	>22 °C (April–June)		Turbulent rapid flow	Sticky eggs
FMCC	>18 °C (April–June)	Water level rising for a while	≮1–1.5 m/s	Drifting eggs

Table 8.5 indicates that fishes spawn from March to November each year. If breading and wintering sites and migration pathways are considered, the requirements are even higher. Presently, the joint regulation of reservoirs cannot fully meet these requirements, and continuous efforts should be made toward these requirements in the future ecological regulation.

Besides ecological regulation, the emergency regulation for responses to reservoir bays' algal blooms and emergent water pollution accidents are important scopes of the environmental regulation of reservoir groups. The response to algal blooms is of a great technical difficulty and needs to consider the reservoir's multipurpose requirements and bank problems and use appropriate timing through regulation to speed up the flow velocity of the water body in the reservoir area. However, the response to emergent water pollution accidents must be considered in the emergency regulation plan of all reservoirs.

8.3.5 Regulation Issues of TGR

The TGR is the most important control reservoir on the mainstream of the Yangtze River. Under the original design specifications, the TGR would provide the regulating functions of flood control, hydroelectric generation, and navigation, but now it has other additional functions, such as drought relief (release water to replenish the downstream area), the aquatic environmental protection of the reservoir area, and the ecological environmental protection in the middle and lower reaches of the river. The beneficial use of the reservoir is practically in significant conflict with watershed flood control, water replenishment, and ecological and other public welfare regulations, as mainly reflected in the following aspects.

8.3.5.1 Role of Flood Control

Under the original design regulation plan, the major goal of flood control would be to defend against large flood, such as the 1954 flood, and intercept excessive floodwater so as to maintain a safe discharge of 56,700 m^3/s through the Jingjiang River. After the flood peak has passed, the water level in the reservoir would be lowered to the flood control level of 145 m in preparation for the next possible flood process until the end of September. Now the situation has changed. On the one hand, the downstream regions (mainly Hubei Province) hope that the water level of the Jingjiang River will not exceed the warning water level in addition to the interception of large floods so as to save the manpower on dike patrol. In the meantime, the Three Gorges Corporation also needs to maintain a relatively higher water level in the reservoir and discharge less water so as to generate more hydroelectric energy and ensure a full reservoir after the flood season, which raised a question as to whether the TGP should defend against both large- and medium-sized floods or mainly large floods. On the other hand, Hunan Province hopes that the TGR would mainly provide compensatory regulation to the Jingjiang River as well as Chenglingji. If the sediment load to the TGR is reduced, sedimentation issues are lessened, and flood forecasting is more accurate, the regulation of the TGR may be modified to make all parties satisfied. As a result, the flood control level and flood control regulation method will need to be adjusted, but the current relatively accurate hydrological forecast is good for only 7 days and at most up to 10 days. Since a higher flood water level means a smaller flood control capacity, in the event of a large flood, the downstream flood diversion volume would increase. Moreover, raising the impoundment water level may not only increase sedimentation but may also increase the flood risk in the late flood season.

8.3.5.2 Hydroelectric Generation

Due to the installation of six additional units beneath the right bank of the TGD, the water utilization rate of the Three Gorges Hydroelectric Station has been further increased. Moreover, with the improvements of the hydrologic stations in the upper reaches, the forecast period for floodwater and incoming water has been improved. Furthermore, as the sediment load to the reservoir has been reduced markedly, the flood control level can be appropriately raised, and impoundment can start earlier to change the relatively rigid regulation plan under which only large floods are regulated. The greatest benefit of this change is to significantly increase hydroelectric generation at the price of an increased flood control risk and the loss of more small- and medium-sized flood processes in the river channel of the middle and lower reaches.

8.3.5.3 Shipping Regulation

The shipping industry in the TGR area and the middle and lower reaches of the Yangtze River has developed so rapidly that the shipping has exceeded the original design in capacity and scale. It was originally projected that the annual shipping freight over the dam toward the downstream would have reached about 50 million tonnes by 2030, the ships through the dam would be mainly composed of fleets below 1000 tonnes, and the satisfaction rate for 10,000-tonne fleets to reach the Chongqing Port would be 50%. In 2010 when the water level in the TGR reached 175 m for the first time, the freight through the ship locks reached 78.8 million tonnes and cargoes over the dam reached 86 million tonnes (Yangtze River Yearbook Editorial Board 2011). Moreover, ships over the dam changed from the traditional small-ship fleets to large-vessel fleets, such as 3000-tonne ships. As a result, higher requirements for the water depth and increased minimum discharge are needed during the impoundment period. However, the maximum navigable flowrate of the ship locks of the TGP is 56,700 m^3/s, and when this flowrate is exceeded, shipping has to be stopped.

Since the TGP started operation, the society has paid great attention to the operation of the TGP, and the regions in the middle and lower reaches have been particularly sensitive to the post flood-season impoundment. In order to reduce the impact of the impoundment of the TGR on the middle and lower reaches and increase the probability of a full reservoir, impoundment at the TGR has started earlier and the starting water level has also been raised. Table 8.6 shows the impoundment conditions of the TGR in recent years.

Table 8.6 indicates that, according to incoming water forecast, the starting time was much earlier, and the starting water level was much higher for impoundment. The benefits of the adjustments were to ease the pressure for the October impoundment, further reduce the impact of the TGR impoundment on the middle and lower reaches, and increase the hydroelectric generation. However, the resulting problems include the increased risk of flood control in the late flood season and slightly higher sediment deposition in the reservoir.

8.3.6 Summary

The joint regulation of large reservoirs is needed to maximize multi-objective comprehensive benefits, such as flood control, water supply, irrigation, drought relief, hydroelectric generation, shipping, and ecology. However, to achieve the desired goal, many scientific, technological, and managerial issues will need to be studied and addressed. From the scientific and technological perspective, first, there is a need to establish and improve the automatic monitoring network of weather and hydrological stations in the whole watershed, establish a watershed precipitation-runoff model, and improve the accuracy of the medium-term hydrological forecast so as to provide the joint regulation of reservoirs with the timely and accurate basic data of forecast regulation. Second, there is a need to study and develop the initial conditions (domain value) for the joint regulation of reservoirs for various river sections (regions), levels, and objects and establish the monitoring and evaluation method for the effectiveness of joint reservoir regulations so as to provide the basis for the coordination and compensation of interests. Third, there is a need to study and come up with regulation methods and implementation procedures to effectively combine public welfare regulation, such as flood control, ecology, and drought relief, with beneficial regulation. From the managerial perspective, the comprehensive benefit of the joint regulation of large reservoirs is certainly better than the

Table 8.6 Impoundment conditions at TGR in recent years

	2008	2009	2010	2011
Starting water level (m)	145.32	145.87	160.2	150.15
Starting date	September 28	September 15	September 10	September 1
Ending water level (m)	172.80	170.93	175	174.49
End date	November 10	November 2	October 26	October 27

regulation of individual reservoirs, but because each reservoir belongs to a different region and a different owner, the better overall comprehensive benefit does not necessarily mean a better benefit for each individual reservoir. A better social benefit does not necessarily mean a better economic benefit to everybody since the joint regulation involves many stakeholders and the relationship of interests is complex. Therefore, it is difficult to comprehensively manage and coordinate interests for joint regulation. As flood control, drought relief and ecological regulation are basically public welfare, and the social and indirect economic benefits are large with a wide range of beneficiaries, but it is difficult to determine the major beneficiaries and evaluate the economic benefits. In addition, the technical requirements for the joint regulation of a reservoir group are high, the incoming water condition and the benefit and impact of the regulation have relatively high randomness, and regulation itself also bears a certain risk. Therefore, the greatest difficulty of a reservoir group's joint regulation is the establishment of a management system mechanism and an interest coordination and compensation system.

8.4 Flood Control and Drought Relief in the Yangtze River Basin

8.4.1 Strategy and System of Flood Control for the Yangtze River

8.4.1.1 Flood Control Situation of the Yangtze River Basin

The Yangtze River basin is rich in rainfall with many mountainous and hilly areas and is home to more than 400 million. The protection area below the flood control level is 153,800 km^2, of which 13,400 km^2 is in the upper reaches and 140,400 km^2 in the middle and lower reaches, indicating that the key flood control zones are in the middle and lower reaches of the river. These flood control regions do not include those in the mountainous and hilly areas that are prone to flash floods. Therefore, the Yangtze River basin has had and is having and will continue to have flood control risk, but the risk has only been rationally transferred and dispersed through man-made structural measures. As a result, the overall risk of flood control does still exist, and the situation is grim.

The overall strategy and principle of flood control in the Yangtze River are "to combine retention and discharge and mainly to discharge." At the same time, flood control also considers the principle to "benefit both rivers and lakes," balance the right and left banks, and systematize the upper, middle, and lower reaches. This principle has undoubtedly resulted from many years of implementing the flood control strategy. However, as to how to apply this principle during a specific regulation in various flood situations, there are many operating variations. Based on the level of national socioeconomic development, steps to establish the flood control system of the Yangtze River are as follows. First, rationally raise and fortify dikes, and regulate river channels so as to fully utilize their storage capacities. Second, arrange and construct flood diversion/retention zones in the low-lying areas where flooding has occurred frequently historically. Third and last, construct large control reservoirs on the mainstream and tributaries of the Yangtze River to combine flood control with other beneficial purposes. The current flood control strategy in the middle and lower reaches is as follows. Dikes are the basis, and the TGR is the key and incorporates other reservoirs on the mainstream and tributaries, flood diversion/retention facilities, river channel regulation, and non-structural measures. This strategy has resulted from long-term flood control planning and practices since the founding of the PRC, and it has largely remained unchanged in the 1990 *Comprehensive Plan of the Yangtze River Basin*, the 2000 *Yangtze River Basin Flood Control Plan*, and the 2011 *Comprehensive Plan of the Yangtze River Basin*, but some variations have occurred as to how to apply the strategy. For example, the basic principle "to combine retention and discharge and mainly to discharge" has been continuously modified after the completion of the TGP because the communities in the middle and lower reaches and the public hope that the TGR should first intercept and retain the floodwater regardless of the size of the flood, coupled with the power generation company's desire to intercept floodwater and raise the flood water level to increase hydroelectric generation. Another problem is that due to the need of economic development, all the regions along the Yangtze River hope to raise the flood control standard for dikes so as to utilize the floodland outside the dikes, lower the probability of flood diversion in flood diversion zones, and use reservoirs, especially the control reservoirs, to intercept floodwater as the major flood control means. This would actually transfer excessive flood risk to the reservoirs and increase flood control risk and pressure to reservoirs.

Presently, major problems associated with the flood control strategy of the Yangtze River basin are as follows: ① Although the flood control storage capacity of reservoirs has increased, the available capacity of flood diversion/retention zones has been decreasing because improvements on flood diversion/retention zones have fallen behind or the zones have not been used for a long time, and it is very difficult or very costly to resume the use of the planned or even most of the completed flood diversion/retention zones, or nobody dares to use them. Although the flood control storage capacity of reservoirs has increased, the total increased capacity of flood diversion/retention is limited. As a result, in the event of a large flood, the flood control situation is still not optimistic. ② The flood control system consists of structural and non-structural measures. In the past, more emphasis has been placed on the structural measures, but non-structural measures have played an inadequate role. Non-structural measures include soil conservation in the upper reaches, scientific land use in flood control areas and flood diversion/retention zones, the improved accuracy of rainfall forecast and early warning and information release, the scientific regulation of flood control projects, etc. It is necessary to attach great importance as to how to fully utilize non-structure measures in the future. In fact, there is a need to transfer from flood control and prevention to flood management. ③ In the structural system of flood control, how to comprehensively operate the dikes, reservoirs, flood diversion/retention zones, and the river channel regulation is not only a scientific and technological issue but is also a managerial issue. How to comprehensively utilize all these means in flood control is quite an art.

8.4.1.2 Flood Control Strategy of the Yangtze River

8.4.1.2.1 Flood Control Strategy

Actually, human beings experienced several stages in flood control strategies. The early method was mainly to improve the flood control capability of structural measures. Later it was found that as more flood control projects had been constructed, flood damage had become increasingly higher, and it was gradually realized that flood control should be transitioned to flood management. The Netherlands began to perform cost-benefit analysis for flood control projects in 1953, and the United States began to introduce flood insurance in 1968. Western countries began to adopt the methods of flood risk management in the 1990s, such as restricting and standardizing land development and utilization in floodplains so as to reduce flood losses. The Netherlands proposed to give space to rivers (floods) by widening floodways and restoring wetlands and floodplains in 1995. After the large flood in the Mississippi River in 1993, the Americans realized that structural measures alone could not reduce the loss resulting from floods and non-structural measures must be stepped up. Presently, flood management has become the mainstream strategy in the field of flood control throughout the world.

Located in the subtropical monsoon climate zone, the Yangtze River basin has more and concentrated rainfalls during the flood season. Moreover, there are many mountainous and hilly areas and more floods in the basin. Therefore, flood control has always been an extremely important matter. Even with the completion of the control reservoirs, such as TGR, flood control still cannot be ignored. The flood control of the Yangtze River needs a completely new strategy to incorporate the concept of flood risk management that has commonly been used in the world, in which both structural and non-structural measures are equally important. Future flood control should be carried out in the following three levels:

The first level is about the long-term strategy. Floodwater should be given space and sufficient retention/detention locations and capacities. Enough space should be reserved in rivers, reservoirs, and diversion/retention zones to divert and retain extremely high volumes of floodwater from extremely large floods, especially in the low-lying flood diversion/retention zones such as lakes in the middle and lower reaches of the Yangtze River where enough space should be reserved to increase the retention/detention capacities of lakes and wetlands. In the meantime, more time should be given to floodwater by intensifying the protection of natural forests and soil conservation in the mainstream and major tributaries in the upper Yangtze River basin so as to increase the water holding capacity of land and reduce the flood peak or delay the flood process.

The second level is tactical. To improve the regulation of dikes, reservoirs, and flood diversion/retention and detention zones, the storage capacity of the floodway between the dikes will be used as the basis, reservoirs will be the key, the flood diversion/retention zones will be used when necessary, and the flood diversion/retention zones should be ensured to be available with low losses in case of a large flood. The river channel and floodway should be strictly managed; the agricultural development and utilization of the land in the flood diversion/retention zones should be restricted; and the land use in flood protection zones should be standardized and regulated. The advantages of the hydrological forecast and early warning capability, and the joint regulation of reservoirs should be utilized, and various flood processes should be scientifically defended.

The third level is the emergency action. Due to the objectivity of flood and the continuity of socioeconomic development, a flood control emergency event may occur at any time, and the loss resulting from the flood is not only inevitable, but the loss will also increase with economic development, and land development and utilization in low-lying areas. Therefore, there is a need to establish a professional team of flood control and rescue, utilize every level of governments and organizations to play a role in mobilization, transfer and risk control, and establish a flood insurance system and a rapid disaster compensation mechanism, so as to start restoration and reconstruction efforts as soon as possible after the disaster.

8.4.1.2.2 Flood Control Standard of the Yangtze River

The determination of flood control standards involves not only technical and economic issues but also other factors such as the strategy of flood control, the construction of flood control systems, the level of socioeconomic development in the watershed, and the social requirements for flood control. Therefore, a comprehensive determination can be made after the flood control requirements, and capabilities of the whole watershed are incorporated. According to the experience of developed countries, the urban flood control standard is generally a 50- to 200-year flood event. Some countries stipulate that the standard should be a 500- to 1000-year event for especially important cities, while the agricultural area (or rural area) is generally a 10- to 50-year event. Presently, China's flood control standard ranges from a minimum of a 20-year event up to a maximum of a 200-year event for urban areas and from a 10- to 100-year event for rural areas. As for the flood control standard itself, there is not much difference between China and other countries with only one difference in the fact that the flood control standard in some rural areas of China is higher than foreign standards. The main reason for this is that China's rural areas are more populated, and therefore, the standard of flood control should be higher.

Flood control of the Yangtze River basin is mainly in the middle and lower reaches. The current standard of flood control was determined mainly according to the political and economic status of the plain areas in the middle and lower reaches and the historical floods and associated damages that occurred in the nineteenth and twentieth centuries under the *Flood Control Standards of the People's Republic of China* and *Document No. 12 (1999) of the State Council*. The overall flood control standard set forth for the middle and lower reaches of the Yangtze River was that "the flood control standard for the Jingjiang River section on the mainstream of the Yangtze River is at least a 100-year event and efforts should be made to create conditions to ensure that the Jingjiang River section can safely pass the floodwater, no natural failure occurs to either north or south major dikes, or no devastating disaster occurs in the event of a flood similar to the 1870 flood (equivalent to about a 1,000-year event). The flood control function and regulation plan of the TGR has been set up based mainly on this standard. This standard is relatively higher for the Jianghan Plain because the Jianghan plain has 733,000 ha of cultivated land and a population of more than 8 million. For the river sections downstream of Chenglingji, the flood control standard was to defend an event same as the 1954 flood. Wuhan, Nanjing, Shanghai are very important urban areas where the flood control standard should be even higher when conditions permit. Other regions should have various standards according to their status and the potential damage resulting from flood hazards."

After dikes were constructed according to the standard, the flood control capacities in major sections of the Yangtze River are summarized in Table 8.7.

After the 1998 flood, the state appropriated huge funds in the reinforcement of the dikes of the Yangtze River. By the end of 2002, the cross sections of the dikes along the mainstream of the Yangtze River in Hubei, Hunan, Jiangxi, Anhui, and Jiangsu Provinces had basically met the standards set forth for the height, width, and slope. Figure 8.2 shows the reinforced dike of the Yangtze River.

Since the TGP was put into use, the flood control capacity in the middle reaches has been improved greatly, especially in the Jingjiang River area where the flood control situation has changed fundamentally. For the Jingjiang River area, when a less-than-100-year flood event (such as the 1931, 1935, 1954 floods, in which the 1954 flood was a below 100-year event in the Jingjiang River area) occurs, the water level at the Shashi Station will not exceed 44.5–45 m, and the Jingjiang flood diversion zone will not need to be used. When a 1000-year flood event or the 1870 flood

Table 8.7 Flood control capacities in major sections without TGP

Section	Flood control capacity		Remarks
	Dikes only	With diversion/ detention zones	
Jingjiang River	10-year flood	40-year flood	Higher capacity requires that the south dike will be properly raised and weakness fixed
Chenglingji	10- to 15-year flood	1954 flood	
Wuhan	20- to 30-year flood	1954 flood	200-year flood for 30-day floodwater volume when diversion/detention zones are properly used
Hukou	20-year flood	100-year flood	1954 flood if all upstream flood diversion/ retention zones are properly used

Fig. 8.2 Reinforced dike of the Yangtze River

occurs, the flowrate at the Zhicheng Station will not exceed 80,000 m^3/s, and with the use of the Jingjiang flood diversion zone, the water level at the Shashi Station will not exceed 45.0 m to ensure the flood control safety of the dikes along both sides of the Jingjiang River. In the vicinity Chenglingji, in an average year, flood diversion is not required (except for the tail of each tributary). When an extremely large flood occurs, such as the 1931, 1935, 1998, or 1954 large flood, the regional volume of flood diversion/retention and the area of inundated land will be reduced. In the Wuhan area, because the flood in the upper Yangtze has been effectively controlled, the threat that the floodwater resulting from a dike break of the Jingjiang River would directly charge into Wuhan can be avoided; the regulation of the TGR has improved the ability of flood control in the Chenglingji area; and with flood diversion/retention at the Danjiangkou Reservoir and the area near Wuhan, an uncontrolled water level in Wuhan can be avoided. In addition to the original flood diversion/retention capacity, the completion of the TGP has added the flood control volume of 22.15 billion m^3 in the area upstream of Wuhan, which has undoubtedly greatly improved the flexibility of the flood control regulation for Wuhan and has played an important role in flood control security for Wuhan. The dikes along the mainstream in the middle and lower reaches of the Yangtze River are divided into different levels based on the importance of the protected. Level I dikes include the Jingjiang River Dike (see Fig. 8.3), the Wuwei Dike, the Nanxian Dike, and the dikes for the key flood control cities, such as Yueyang, Wuhan, Huangshi, Jiujiang, Anqing, Wuhu, Nanjing, and Shanghai, along the Yangtze River. Level II dikes include the Songzi River Dike, Jingnan Yangtze River Mainstream Dike, Honghu-Jianli River Dike, Hunan Provincial Yangtze River Dike, Siyigong Dike, Bapu Dike, Huangguang Dike, Jiangxi Provincial Yangtze River Dike, Tongma Dike, Guangjiwei River Dike, Zongyang River Dike, Hexian River Dike, Wudang River Dike, Jiangsu

Fig. 8.3 Jingjiang River dike in the middle reaches of the Yangtze River

Yangtze River Dike, and the dikes along the river in Yangxin, Chizhou, Tongling, and other urban areas. The other dikes along the mainstream of the Yangtze River are of Levels III and IV. The design standards and design water levels also vary for different levels of dikes. The freeboard is 2.0 m for Level I dikes along the middle and lower reaches of the Yangtze River, 1.5 m for Levels II and III dikes, and 1.0 m for the other levels of dikes.

Based on many years of investigations and research, the *Integrated Plan of the Yangtze River Basin* stipulates the design flood level at four major control stations in the middle reaches of the Yangtze River, namely, 45.00 m at Shashi, 34.40 m at Chenglingji, 29.73 m at Hankou, and 22.50 m at Hukou. The flood control standard for the river section upstream of the Jingjiang River before the completion of the TGP was to defend against the flood flowrate of 80,000 m^3/s (about a 40-year flood event) at the Zhicheng Station with a prescribed design water level of 51.75 m at the station. The river section downstream of Hukou was controlled by the water level of 22.5 m at the Hukou Station, and the corresponding water level at the Datong Station should be 17.10 m according to a recent hydrological analysis. The river section downstream of Datong is a tidal section where the flood level is influenced by the runoff, tidal flow, typhoon, local runoff, and other factors, while the section upstream of Jiangyin is primarily controlled by the runoff, and the section downstream of Jiangyin is mainly controlled by the tidal flow and the effect of typhoon.

8.4.1.2.3 Dikes and Storage Capacity of Channels

Various dikes of the Yangtze River system are more than 30,000 km in length, and the dikes along the mainstream and major tributaries of the Yangtze River are more than 3000 km long. As a result, the channels in the mainstream and major tributaries of the Yangtze River have a storage capacity of more than 100 billion m^3, and the dikes play a basic flood control role in the flood control system and can withstand small- and medium-sized floods. The design water levels of the dikes include the flood control action water level, warning water level, and maximum safe water level.

1. Flood Control Action Water Level

The flood control action water level is equivalent to the level when the floodwater has just covered the floodplain, and the corresponding flowrate is the channel-forming discharge. When the river flow covers the floodplain, the foot of the dike is in contact with water, and the local dike may be in danger, indicating the protection of the dike into a state of emergency. Flood control personnel begin to patrol the dike and need to prepare for emergency manpower and material. However, it should be noted that some rivers in China, such as the Yellow River in north China, have wide and shallow channels and it is hard to differentiate their channels from their floodplains. For such rivers, it is not appropriate to use the flood control action water level as the control water level to start flood control work. In fact, due to decades of reinforcement and maintenance, when the water level reaches the flood control action water level, there is no major safety issue, and the water level should be allowed to rise to the flood control action water level so that, on the one hand, the channel's storage capacity can be utilized to remodel the river channel and riverbed, reduce flood control pressure on reservoirs and avoid the development and utilization of the floodplain; and, on the other hand, it is conducive to the lateral connection of the river, the river-lake connection, and the health of the riparian wetland ecosystem.

2. Warning Water Level

When the warning water level is reached, dike protection efforts should begin, and the dike embankment has already blocked water. Old traditional dikes may be in a dangerous situation or even in a serious danger; close attention needs to be paid to the change of hydrologic conditions, work conditions, and dangerous situation; more dike inspections need to be performed to include day and night dike patrols; and emergency manpower and materials need to be more readily available. Since dike reinforcement efforts after the 1998 flood, most dikes along the mainstream and major tributaries of the Yangtze River have been up to standard. The dikes that are still below standard are mainly secondary dikes or polder-building berms in lake areas. Therefore, even if the warning water level is reached, the situation does not necessarily constitute a danger to the dikes. In fact, when the warning water level is reached, the situation allows the river channel to play its role to store floodwater, and the floodwater with sediments can advance rapidly toward the downstream, which can test the safety level and quality of the dikes and is conducive to the lateral connection, the reforming of the river channel and riverbed, and the emergence of new habitats. Thus, don't be afraid of the occurrence of the warning water level. The regular occurrence of the warning water level can avoid land use on the floodplain and maintain a certain extent of the floodway. Therefore, flood control regulation should not be aimed to avoid the occurrence of the warning water level but rather should allow moderate and timely occurrences of the warning water level to test the dikes and allow the river channel to play its role to store floodwater.

3. Maximum Safe Water Level

The maximum safe water level is a standard flood water level that the dike was designed to withstand, and the corresponding flowrate is the safe discharge of the river channel. When the maximum flood water level is reached, indicating that the dike is in its safety limit, flood control enters a state of emergency, and serious danger may occur to the dike at any time. The flood control department should take all measures to ensure the safety of the dike, if necessary, declare the state of flood control emergency, and take flood protection measures and evacuate citizens from dangerous areas. When the maximum safe water level, or the dike design water level, is reached, the dike's maximum capacity for flood protection is reached, which should be avoided generally. Of course, if it occurs, actions should be taken to safeguard the dike. In fact, dikes generally have a freeboard with the ability to withstand the check flood level. For example, a Level I dike has a 2 m freeboard; Levels II and III dikes have a 1.5 m freeboard; and other dikes have a 1 m freeboard. Moreover, important dikes generally have a higher design safety level, and many have reinforced concrete cutoff walls for seepage control. As a result, even if the maximum safety water level is reached, the situation does not mean that a dike break would be imminent.

8.4.1.2.4 Flood Control Using Reservoirs

After several decades of construction, flood control reservoirs in the Yangtze River began to take shape. On all the rivers that are historically prone to large floods, flood control reservoirs have been constructed to a certain scale, such as the Danjiangkou Reservoir on the Han River, the Tingzikou Reservoir on the Jialing River, and the TGR on the mainstream of the Yangtze River. As a result, the flood control regulating capacity in the middle and lower reaches of the Yangtze River has significantly improved. By 2020, the upper reaches of the Yangtze River alone will have had more than 50 billion m^3 of the flood control storage capacity, and the major tributaries will also have had flood control reservoirs. They will be able to regulate the floodwater of tributaries and help the TGR, and other mainstream reservoirs alleviate the flood process in the middle and lower reaches. The advantage of building reservoirs for flood control lies in the fact that reservoirs are generally built in the mountainous areas where the population is small and the inundation loss is low, but the disadvantage is that the floodwater is stored at a high altitude and the risk of flooding does not decrease but is borne by the reservoirs, which increases flood risk and pressure to the reservoirs.

8.4.1.2.5 Construction and Use of Flood Diversion/Retention Zones

Presently, there are 40 planned flood retention/detention zones in the Yangtze River basin (since the Hong Lake flood, retention/detention zone was divided into three zones, and the total number of the zones has become 42). If so many zones are all constructed, the investment is huge. Owing to limited state and local fiscal resources, the construction of flood retention/detention zones has progressed at a slow pace, and completed facilities are far from meeting the safety need for flood diversion. A preliminary estimate indicates that an additional investment of 60–70 billion yuan will be required for the construction of flood retention/detention zones, including safety improvements (safety zones and platforms), project construction (dikes, flood inlet, and outlet facilities), and population relocation (building towns for the relocated).

Based on the probability and importance of using flood retention/detention zones, the zones can be divided into three categories: key zone, common zone, and planned reserve zone. A key flood retention/detention zone is that which is more likely to be used under present conditions (usually within a 20-year event). A common flood retention/detention zone is that which will need to be used to defend a flood event equivalent to the 1954 flood in addition to the use of key flood retention/detention zones with the completion of the TGP. The planned reserve zone refers to that which will be needed to defend against a flood above the design standard or an extremely large flood with the completion of the TGP.

The flood retention/detention zones in the Yangtze River basin can be divided into four large regions in Jingjiang, Chenglingji and adjacent area, Wuhan and adjacent area, and Hukou and adjacent area, and they divert and retain local excessive floodwater separately.

The 1931, 1954, and 1998 floods are of a watershed type; the 1935 flood is a severe regional event that occurred in the Li, Qing, and Han Rivers in the middle reaches of the Yangtze River; and the 1981 flood is a regional event in the upper reaches. With the normal operation of the TGP, when an event equivalent to the 1945 flood occurs, 39.8 billion m^3 of floodwater will need to be diverted and retained if the TGR is regulated to supplement the Jingjiang River Dike, or 33.6 billion m^3 of floodwater will need to be diverted and retained if the TGR is regulated to compensate the Chenglingji Dike. When an event equivalent to the 1954 flood occurs, flood diversion will not be required in the Jingjiang region; flood retention will be needed in the middle subzone at Hong Lake and flood diversion/retention zones near Dongting Lake such as Xicheng, Mingzhu, Weidi Lake, Linan, Xiguan, Jianxin, Jianshe, Quyuan Farm, Jiuyuan, and Luchengyuan of Jiangnan after 10 billion m^3 of floodwater is detained and retained near Chenglingji; and flood retention/detention may be arranged by planning in Wuhan and Hukou and their adjacent areas based on the real-time flood situation.

With the completion of the TGP and according to the compensatory regulation of the TGR for the Chenglingji region, when an event equivalent to the 1954 flood occurs, there will be no excessive floodwater near the Jingjiang region and 21.8 billion m^3 of excessive floodwater near the Chenglingji region, a reduction of 10.2 billion m^3. The excessive volume of floodwater near Wuhan and Hukou has not changed before and after the TGP was built. Therefore, the expanded diversion zone at Liushi, the diversion zone at the Renmin polder area, and the Huxi reserve in the Jingjiang region have been designated as flood retention/detention reserves. The Jingjiang flood diversion zone is special and has been listed as an important flood retention/detention zone. The volume of floodwater diversion near Chenglingji has been reduced according to the principle of equivalence in floodwater diversion between Hubei and Hunan. It has been planned that the western flood retention/detention subzone of Hong Lake (effective volume 5.225 billion m^3) and 11 flood retention polder fields in the Dongting Lake area, including Junshan, Jichenganhe, Nanhan, Hekang, Anhua, Anli, Anchang, Beihu, Yihe, Nanding, and Liujiaoshan (effective volume 5.121 billion m^3) will be flood retention/detention reserves. There are 15 flood retention/detention reserves. In addition to the abovementioned flood retention/detention zones, the other flood retention/detention zones are the ones that will need to be used to protect against the occurrence of an event equivalent to the 1954 flood after the TGP was completed.

The planned diversion/retention of 10 billion m^3 of excessive floodwater in the flood retention/detention zones (three flood retention polder fields at the Qianliang Lake, Gongshuangcha and east of Datong Lake in the Dongting Lake area, and the east subzone at Hong Lake) can play a very important role in the defense against an event similar to the 1998 flood, and the probability of using the facilities is about 5% or once every 20 years on average. The newly constructed Jianxin and Jianshe polder fields at the outlet of Chenglingji are good in controlling the water level at Chenglingji, and the probability of using the facilities is also about 5% or once every 20 years. Flood retention/detention polder fields (Chengxi, Minzhu, Weidihu, Linan and Xiguan) in the tail zone of the four rivers out of Dongting Lake can reduce the amount of flood diversion near Chenglingji and will also have major flood control effects on the tail zone of the four rivers. Moreover, the polder fields at Weidihu, Linan, and Xiguan have already completed and are empty polder fields waiting to be used. Therefore, the abovementioned 11 facilities are planned key flood retention/detention zones.

The Dujiatai flood retention/detention zone near Wuhan has frequently used with a use probability of about 5% or once every 20 years, has played an important role in flood control for the Yangtze and the Han Rivers, and has been proposed to be a key flood retention/detention zone. The Kangshan flood retention/detention zone near Hukou has the largest capacity of the four flood retention/detention zones in the Poyang Lake area with good operation results. Moreover, more efforts have been made on safety improvements in the Kangshan zone than the other flood retention/detention zones in the region. To ensure the flexibility of flood regulation in the Hukou area, the Kangshan flood retention/detention zone has been proposed to be a key flood retention/detention zone. The use probability of the Jingjiang flood diversion zone was about 2–5% or once every 20–50 years (when the first stage impoundment occurred at the TGR) and has been 1% or once every 100 years since the TGP was completed. When the Xiluodu and Xiangjiaba Reservoirs are completed and jointly regulated with the TGR, the use probability of the Jingjiang flood retention/detention zone will be even lower. The Jingjiang flood retention/detention zone should not have been classified as a key zone based on the classification criteria. In case of a large flood like the 1860 and 1870 large floods, the Jingjiang flood retention/detention zone will still be a major flood control structural measure; therefore, it is not appropriate to lower its classification right now. To sum up, the abovementioned 14 flood retention/detention zones in the Yangtze River basin should be key flood retention/detention zones, and the other 13 flood retention/detention zones are common ones, including the central subzone of Hong Lake near Chenglingji, Quyuan, Jiuyuan, and Lucheng flood retention/detention zones of the Dongting Lake area; Xiliang Lake, Wu Lake, Zhangdu Lake, Baitan Lake, and Dongxi Lake near Wuhan; and Zhu Lake, Huang Lake, and Xietong of Fangzhou and Huayang River flood retention/detention zones near Hukou.

8.4.1.3 Flash Flood Control Strategies

8.4.1.3.1 Characteristics of Flash Floods

1. Widespread Distribution and Large Quantity

Flash floods are particularly prominent and prevalent in creek flood disasters. The topography and geology of China are complex and diverse, rainfall is uneven in time and space, and flash flood disasters can be characterized by diverse in type, large in quantity, and widespread in distribution. According to the survey statistics of flash floods, China has 18,901 gullies where creek floods have occurred and 81,360 disasters have occurred; China has 11,109 gullies where debris flows have occurred; 13,409 debris flow disasters have occurred; and 16,556 landslide disasters have occurred (Zhao and others 2009).

2. Highly Instantaneous and Difficult to Predict and Prevent

China's rainstorm is high in intensity and highly instantaneous; mountainous and hilly areas have steep slopes and deep valleys with highly undulating ground, resulting in fast runoff development in mountainous and hilly rivers with high flow velocities and the instantaneity of induced flash flood disasters. It generally takes several hours, even less than 1 h and no more than 24 h, from rainfall to occurrence of a flash flood disaster. Moreover, as the network monitoring coverage of mountainous and hilly areas is currently low, it is very difficult to predict and prevent the occurrence of flash flood disasters.

3. Disaster Develops Fast and Is Strongly Destructive

Because slopes in mountainous and hilly areas are steep, creeks are densely distributed, floodwater develops rapidly, and the population and properties are located in limited low flatlands in valleys, flash floods can be characterized by rapid developing. Debris flows have a strong ability to scour and destroy. When a large- or medium-sized sticky debris flow occurs, all facilities, roads, and farmland on its running path will be wiped out, forming the landscape of a stone sea and causing a serious disaster. Although it may take a certain amount of time for a landslide body to slide, for example, a few centimeters a day or even a few centimeters a month, the movement of the landslide may speed up suddenly due to external factors, and it slides rapidly, resulting in huge losses.

4. Strong Seasonality and High Frequency

Because the flash flood disaster is caused by a rainstorm, the occurrence of a flash flood is highly correlated in time with the occurrence of a rainstorm. China's rainstorms and extreme rainstorms are mainly concentrated during May–September, and flash flood disasters also frequently occur during this time period. It is during June–August that more flash flood disasters occur.

5. Obvious Regionality and High Susceptibility

The occurrence of flash floods is closely related to rainfall, topography, geology, and human activities, but rainfall, topography, geology, and human activities have various characteristics in different regions of China. Therefore, flash flood disasters are highly regional. Flash flood disasters are concentrated in mountainous and hilly areas in southwest China, the Qinling Mountains area, the hilly areas south of the Yangtze River, and the coastal areas in southeast China; their frequency of occurrences is high; and the susceptibility is also high. Flash flood disasters in northwest China and Qinghai-Tibet Plateau regions are relatively dispersed, and the frequency of occurrences is low.

8.4.1.3.2 Flash Flood Control Measures

Based on the characteristics of flash flood hazards and the economic comparison of technical control and prevention measures, flash flood control should be primarily aimed to minimize casualties and should adhere to the principle of prevention first and combined prevention with control and non-structural measures first and combined non-structural measures with structural measures.

For specific preventative measures, structural measures include ecological regulation measures and structural preventative measures. The ecological regulation measures mainly include the restoration of natural forests, soil erosion prevention, reforestation of cultivated land, maintenance of creek channels to restore flow conditions, and forestation efforts. Structural preventative measures primarily include repairs and reinforcement of problematic reservoirs, regulation of medium-sized and small creek channels, construction of riparian dikes, dredging of flash floodways, and reinforcement and stabilization of potential sliding slopes. However, because the cost of structural measures is generally high, but their hazard prevention efficiency is not high, it is very difficult to reach the design benefit requirement in the relatively sparsely populated remote areas. Moreover, as for the current structural measures for flash floods, more attention is paid to construction than maintenance, and more attention is paid to preparation and research of measures than inspection, management, and implementation. As a result, the proportion of problematic reservoirs is high, and phenomena such as land reclamation from creek channels and steep slopes, garbage dumping into creek channels, construction of houses on creek channels, or land-sliding bodies are still prominent.

Non-structural measures include the following:

1. Planning and Management of Land Use in Mountainous and Hilly Areas

With socioeconomic development, the intensity and scope of human development and utilization activities have been increasing with growingly more irrational activities associated with land development and utilization, resulting in more occurrences of flash flood hazards with losses trending up year by year. Therefore, of the non-structural measures for flash flood disasters, the most important is to well plan and manage the land use in mountainous and hilly areas by the means of relevant laws and regulations to strictly control the irrational human activities of land use, such as prohibiting land reclamation on steep slopes, prohibiting the construction of houses on creek channels or near potential sliding slopes, and prohibiting the activities that would affect the flow of creek channels such as reclamation of land on creek channels and dumping garbage to creek channels. In the meantime, a plan of land development and utilization should be prepared to provide zonings that show various land use zones such as permitted development, restricted development, and prohibited development. Those who have lived in the prohibited development zone should be relocated with an appropriate compensation. For the restricted land development zone, economic means should be used to prevent land development and utilization from occurring. Rational land development and utilization should be encouraged in the risk-free or low-risk permitted land development zone. Moreover, supervision and management for the implementation of the land use plan and policies should be strengthened to ensure the implementation and avoid or minimize the occurrence of flash flood hazards to the extent possible.

2. Risk Zoning for Flash Flood Hazard

The occurrence of flash flood hazards is affected by many factors, such as rainfall, topography, landform, vegetation cover, rock structure, and residential distribution. Comprehensively considering the above factors and combining the frequency and losses of the historically investigated occurrences of flash flood hazards, a regional risk zoning map for flash flood hazards should be created to help the public have a more intuitive understanding of the risk distribution of flash flood hazards and provide a theoretical reference for the government to prepare a land use plan. The past risk zoning of flash flood hazards mainly included sub-watersheds or subregions of one large watershed or region and delineated the key protection zones and common protection zones of flash flood hazards. However, the risk of flash flood hazards of various locales in a small watershed or region may be different. Therefore, it is necessary to do more detailed risk zoning for the areas with frequent occurrences of flash flood hazards. Moreover, timely investigations should be performed on the change of hazard risk caused by altered geological conditions and human activities, and the risk zoning map should be updated accordingly.

The rainfall intensity and geological conditions are the most important causes for the formation of flash flood geological hazards. As the topography, vegetation cover, and subsurface conditions such as the geological structure are different for various areas, the intensity of flash flood and debris flow hazards will also vary. Therefore, in order to better protect against flash flood hazards, in addition to doing well in the risk zoning for flash flood hazards, a critical rainfall zoning map for flash flood hazards should be prepared based on statistics of historical rainfall data and hazards so that people in various zones can take corresponding measures according to the actual rainfall conditions observed.

3. Improvements of Hazard Monitoring, Early Warning, and Communication System

Monitoring and early warning of flash flood hazards are important non-structural measures to control the hazards. If a flash flood can be accurately forecast and the early warning information is sent out in time, the loss of the hazard can be greatly reduced. Presently, the overall monitoring and early warning system in China is still relatively backward and is short of more advanced early warning devices and equipment. Moreover, flash flood hazards in China are often accompanied with geological hazards such as landslides or debris flows. Therefore, there is a need to monitor the regional rainfall and geological conditions. Previous monitoring and early warning systems were designed to only monitor hydrologic conditions such as rainfall and did not monitor geological conditions. The hazard monitoring and early warning system needs to be further improved to better investigate and monitor the geological conditions of mountainous and hilly areas and reinforce and control the potential hazard-occurring sites to improve the early warning capability for the geological hazards of flash floods and debris flows.

After a flash flood hazard has occurred, certain damage to the local communication system may also happen, resulting in difficulties to hazard warning and rescue efforts. Therefore, the communication system in the hazard-prone areas should be improved, and an emergency or standby communication device should be provided. In the meantime, for the areas where the monitoring and early warning system needs improvements, the local governments should educate the local residents in the mountainous and hilly areas about the preventative knowledge of flash flood hazards by various means, intensify hazard prevention drills, and improve the awareness and capability of hazard prevention.

In the areas where hazards occur frequently, each village should be equipped with a large trumpet for hazard warning, and each household should be provided with a rain gauge and a contingency plan booklet. Based on the intensity of the rainfall, appropriate measures should be taken, such as to report emergency, evacuate, etc.

4. Establishment of Risk-Sharing Mechanism for Flash Flood Hazards

By establishing the risk-sharing mechanism for flash flood hazards and encouraging people to participate in flood insurance, the loss caused by the hazards can be shared by the state, society, and individuals, and the loss to the residents in the area of hazards can be reduced to a certain extent. In Europe and the United States, the flood risk-sharing mechanisms have been relatively mature, and their programs have been relatively perfect. Their mechanisms and systems can be introduced and referenced to establish an insurance program for flash flood hazards suitable to China based on the characteristics of flash flood hazards in China and implement it on a pragmatic basis to minimize the loss resulting from the hazards.

8.4.1.4 Urban Flood and Waterlogging Hazard

With rapid socioeconomic development and urbanization, the loss and social impact of urban flood and waterlogging hazards in China have been increasing in the past 10 years. For example, the rare rainstorm in Beijing on July 21, 2012, caused 77 deaths and tens of billions of yuan of direct economic loss. Urban flood control and drainage have become an important part of flood control efforts. However, because the construction of urban drainage systems in China used to be managed by urban improvements departments in the past and the standards adopted were mainly from the former Soviet Union in the past, the drainage standards were too low and well below the flood control standards of large rivers for key cities. For instance, the current urban drainage standards are generally to use a 1- to 3-year storm to design the drainage system that drains the stormwater in 1–3 days. Now a large 1- to 2-h waterlogging event can paralyze the traffic system (i.e., several stalled cars, blocked bridge underpasses), and serious losses can be result within less than 1 day.

The means to reduce the loss of urban flood and waterlogging are still the combination of structural measures and non-structural measures. Structural measures include the following: ① Raise drainage system standards. The standard for the drainage system should be raised to a 50-year event for key districts of mega cities, a 20-year event for the majority of the main urban area, and a 5- to 10-year event for the suburbs. ② Dredge or clean up urban channels, sewer pipelines and open drainage channels to improve the urban drainage network system. ③ Expand urban lacustrine wetlands and other low-lying water retention areas and increase the volume of rainwater collection. ④ Increase the permeable land area.

Non-structural measures include the following: ① Establish a feasible and reliable emergency management system, and practice regularly. ② Improve rainfall forecast and warning, and extensively rely on the network, mobile phones, television and other means of communications so that the government, units and individuals get prepared in advance. ③ When an extremely large rainstorm is imminent, suspend or postpone going to school, work or meeting and other business work. ④ Cease all conducts that may obstruct the traffic, such as suspend collecting road tolls. ⑤ Prepare urban flood risk maps, inform the public of the location and frequency of stagnant water, flooding and waterlogging, and instruct the public how to scientifically avoid hazards. ⑥ Step up public education, training and practices to improve the public awareness of hazard control and mitigation, and self-protection capability. ⑦ Establish a flood and waterlogging insurance system. After a hazard event has occurred, rapidly process claims and compensation so that normal work and life can be restored from the hazard event.

8.4.2 Drought Management

8.4.2.1 Drought Relief Strategy and Measures of the Yangtze River

There are two types of measures for the drought problem. The first type is the structural measure, such as the construction of reliable water sources (i.e., reservoirs, pumping stations, wells, etc.), irrigation canals, and field irrigation systems or the construction of interbasin water diversion and pumping station projects, water transfer and pumping from a faraway or low-altitude place, etc. In fact, the mainstream and major tributaries of the Yangtze River are not without water, but rather the cost of lifting and pumping water and the construction of irrigation districts are high, or the water is low and the field that needs water is high, or there is no area equipped with irrigation facilities. Structural measures generally take effects quickly, but the cost is high, and there are some negative ecological and environmental effects. The other is the non-structural measure, such as water conservation and demand management, change of the crop structure, establishment of a drought insurance and compensation system, etc. Non-structural measures have very positive effects and fewer negative effects. The combined use of the two types of measures can result in the best effect. Relying on one type of measure is not scientific or rational. For example, if a city is seriously scarce of water, demand management can be implemented under the condition of ensuring the domestic water use for residents, such as an appropriate increase of water price, restricting water-use quantity by high-water-consuming industries, etc. As a result, the urban water-use quantity will be reduced immediately. We cannot always ask for from nature but rather should first look at how much we have wasted in our production and life. After the drought is over, the normal water price can be restored, and the impact will not be too large.

Different measures can be adopted for agricultural drought and urban water scarcity. The short-term urban water scarcity can be resolved through water conservation and demand management, but for a long-term shortage of water, the loss would be very large. Long-term water shortages need to consider the construction of alternative water sources, such as new reservoirs, the restoration of water-source functions of nearby lacustrine wetlands, etc. For rural and agricultural droughts, the construction of water storage and irrigation facilities can resolve the common droughts. As severe droughts are large in scale and long in time, small- and medium-sized reservoirs, lakes, and ponds in the drought-occurring areas are not full of water, or the water use is too fast that they are very low in water level, or even dry; then pumps or pumping stations can be used to draw lower water or even pumping groundwater through installed wells; and if the drought still persists, crops will have to be lost for a season, and the government and insurance companies will compensate for losses to minimize the losses of the peasants.

Emergency measures for water supply should first ensure the drinking water for human beings and livestock with a satisfaction rate being generally above 99%. The satisfaction rate should be generally above 90% for domestic water demand, above 80% for industrial water use, and about 75% for agricultural water use. With the measures, the overall economic loss can be minimized in case of a severe drought. To be economically rational, the cost cannot be ignored for drought relief. For example, for an agricultural drought relief effort, no matter what the cost is, the agricultural production in one place could be insured to be not affected by the drought, but if the cost is far greater than the value of the crops, it would be economically irrational. Drought is a natural phenomenon, and it is impossible to eliminate drought. Drought relief efforts cannot be aimed at eliminating droughts but rather having a long-term plan, an effective contingency plan, and a rapid reconstruction and other comprehensive measures after the drought, just like the idea of transfer from flood control and prevention to flood management. This transfer should gradually start from the simple drought relief and prevention to drought hazard management.

The average annual rainfall in the Yangtze River basin is over 1000 mm, and the vast majority of regions in the basin are in wet zones where crops and aquatic organisms are accustomed to the wet environment. If the rainfall is greatly reduced, a drought will occur. As large- and medium-sized cities generally have relatively reliable water sources, drinking water for human beings and livestock should not be a big problem. The areas impacted most by the drought are small- and medium-sized urban and rural areas where agriculture is the most damage-prone. For the drought-prone areas, in addition to the construction of necessary water conservancy projects, a comprehensive compensation program consisting of drought insurance and governmental assistance should be established to minimize peasants' loss resulting from droughts.

8.4.2.2 Future Drought and Water Shortage Conditions in the Yangtze River Basin

8.4.2.2.1 Agricultural Drought Relief

Due to the uneven spatial and temporal distribution of rainfall and runoff in the Yangtze River basin, drought and water shortage occur everywhere. Drought first affects agriculture and the solution to the agricultural drought problem is multifaceted. Structural measures can be taken, such as the construction of more reservoirs and other water storage facilities, improvements of canal networks and ditches, reduction of seepage loss from canals and paddy fields, improvements of rural and field rain collection facilities and water ponds, etc. Many non-structural measures can also be taken, such as the accurate measurement of water demand for field crops, the optimization of irrigation timing, and water conservation. If droughts occur frequently, there is a need to change the type of crops and increase drought-tolerant crops. Efforts can be made to develop controlled-environment agriculture or intensive agriculture and increase the grain production per unit area or reduce the quantity of water use and water consumption per unit area. An agricultural insurance program should be developed so as to begin reconstruction work as soon as the drought is over and reduce the loss resulting from the drought.

8.4.2.2.2 Urban Water Supply

China has a large population. There are more than 50 large cities each with a population of more than one million, ranking first in the world. Presently, China is experiencing rapid economic development, the urban population and extent are expanding, and the urbanization process is accelerating. As a result, the urban water demand is increasing, and the demand for a considerable number of cities has reached the limit of their available water resources and even cannot continue to add any new water source. With increasingly severe water pollution, the water sources near most cities have been contaminated, or the satisfaction rate of their water supply is trending down. Of China's 669 cities, 400 are short of water supply and 110 are seriously short of water. In the 46 key cities, 45.6% of them have poor water quality. Nine of the 14 coastal cities that are open to the world are seriously short of water. Beijing, Tianjin, Qingdao, and Dalian are most seriously short of water. Although some cities are not short of water currently, due to their single water source or uneven distribution of their water sources, some urban areas would be cut off if an emergent water pollution accident or other natural disaster occurs. Thus, their satisfaction rate of water supply is not high. It is a priority for the megacity governments to improve the satisfaction rate of water supply and to ensure the sustainable socioeconomic development and urbanization process. Therefore, they need to plan, construct, and protect backup water sources to solve the water supply safety problem.

There are three types of urban water scarcity in China: resource, water quality, and physical conditions. For the resource-scarce cities, the major countermeasure is to adopt the strict management of water resources, vigorously promote the development of a water-conserving society, optimize the industrial structure, restrict the development of water consumptive industry, control the population growth when necessary, and need to identify and plan on new water source projects. For the cities with insufficient surface water sources, it is essential to explore possible groundwater sources, build pumping stations, and install pipelines to connect to water treatment plants. For the cities with insufficient surface water and groundwater, they not only need to improve the wastewater reuse rate but also need to consider the construction of trans-regional water diversion projects. The coastal cities may consider the construction of seawater desalination projects. For the water-scarce cities due to water quality, the key to solving the problem is to optimize the industrial structure, improve the reuse rate of industrial wastewater, and restrict the amount of wastewater discharge into rivers and lakes. For the contaminated waters, the quality of the surface waters can be gradually restored by pollutant interception, aquatic environmental protection, and ecological restoration, and the restorable waters can be used as the major backup water sources. For the large cities without new water sources, they can add new water sources through the construction of wastewater treatment plants. Countries such as Singapore and Australia already have good hands-on experience in treating wastewater to high quality and reusing it for residential purposes, even as drinking water. Physical water-scarce cities can be divided into two categories. One category refers to the cities where there are water sources nearby, and there is a need to change the original purpose of water use, such as a water-use change from agricultural irrigation to urban production and domestic use which needs to construct water conveyance canals or pipelines. The other type is due to the irrational distribution of a city's water sources or a single water source for the city; therefore, there is a need to construct new water source projects or to construct a water pipeline network to connect the entire city so that the city's main water transmission pipeline has the interconnection function to enhance the capability to cope with emergent water pollution accidents.

When conducting the planning, construction, and protection of backup water sources, attention should be paid to the following issues: ① Large cities, especially megacities each with a population of more than five million, should be considered first and then medium-sized and small cities. As the population of a large city is generally distributed in different urban districts, the main urban area should not only have its own independent water sources but also have a water supply network that connects with the other urban districts. Moreover, there is a need to improve the joint operation of the urban water supply network and distribution capacity and improve the overall satisfaction rate of urban water supply. ② The size of a backup water source should be moderate and should not be the same as the primary water source. The purpose of the backup water source is mainly to guarantee the basic water demand of the city in emergency situations, especially the demand of drinking water for human beings and livestock, but it is not for all the water demand of the entire city to avoid unnecessary waste. ③ The backup water source is not a secondary water source that can be waters with relatively poor quality. However, as a backup drinking water source, the water quality should be better, and the protective requirements of the water source and the surrounding waters should be higher. ④ A backup water source is mainly used for emergency and should not be used as a common water source. When there are problems with the main water source of an urban area, it is impossible to solve the water supply problem of the entire city completely by using the backup water source, but instead strict demand management and emergency management measures should be jointly used, and the focus should be on ensuring the living water uses of domestic, hospital, and food production needs in emergency situations.

With the increase of the urban size and population, the increasingly serous water pollution, and the threat of natural disasters, vulnerability of the urban water supply is increasing, and the safety and satisfaction rate of water supply are decreasing. In case of an emergent accident, the water supply for life and production of urban residents may be interrupted. When the interruption time exceeds a certain time (such as 12 h), the quality of life and health of urban citizens would be seriously affected, which could result in a potential social instability. Therefore, a large city needs not only to formulate a strict management program of water resources and emergency but also needs to plan and build a backup water source and associated water supply facilities, so as to continuously improve the satisfaction rate of urban water use and to sustain socioeconomic development.

8.5 Construction and Management of Hydraulic Projects

8.5.1 Management of Hydraulic Projects

In order to ensure the safety of flood control, water supply, irrigation, and electric power delivery, it is necessary to construct hydraulic and hydroelectric projects. There are many types of hydraulic projects, such as flood control projects, water supply projects, irrigation projects, hydroelectric projects, navigation projects, pumping station projects, well drilling and water pumping projects, etc. Large hydraulic and hydroelectric projects often integrate multiple functions. For example, the TGP includes flood control, hydroelectric generation, shipping, water supply, and other functions. Small projects may have a single function, such as a well, an irrigation canal, etc. The greatest characteristic of hydraulic projects is that they are basic and for livelihood. In addition to the function of hydroelectric generation, hydraulic projects cannot be constructed and run by the way of market competition, but they need to be funded by the state and local governments for construction, maintenance, and management and have the characteristic of public welfare. Because of these characteristics, hydraulic projects often have more social benefits than economic benefits and more indirect economic benefits than direct economic benefits, or the beneficiaries are primarily the society and the environment. Therefore, it is not easy to identify the direct beneficiaries, or it is difficult for the projects to rely on fees collected from the beneficiaries.

Presently, major problems in the construction and management of hydraulic projects are as follows. On the one hand, the majority of hydraulic projects are funded by the government through taxation; most of the projects are not good in operational benefits; or their operating costs are relatively high. As a result, great efforts have been made to apply for and construct projects everywhere, but not many efforts have been devoted to the scientific management of the projects. A considerable number of completed projects have become a fiscal burden of governments of all levels due to the exaggerated benefits claimed during the design phase that cannot be achieved during the operation or due to too many management personnel or overstaffing or poor performance. Moreover, an operation and management mechanism of government funding and private operation has not been established. On the other hand, it is not always the cast that the more the number of hydraulic projects, the better, but the scientific planning and appropriate placement of the projects are needed to minimize their impacts on the social and ecological environment and maximize their benefits.

8.5.2 *Management of Water Transfer Projects*

China is one of the countries in the world with the earliest completed water transfer projects. Hangou Ditch is the earliest recorded man-made canal that connects the Yangtze and Huai Rivers; it was excavated to connect many lakes for navigation purposes in 486 BC and is presently named Jiangsu Provincial Jiangbei Canal. The Dujiangyan Project is China's famous water transfer project. Since the founding of the PRC, especially since the reform and opening up to the world, in order to solve the water shortage problem in water-scarce cities and regions, China has built more than 20 large interbasin water transfer projects, such as Jiangsu's Yangtze River South-North Water Diversion Project, Tianjin's Luan River-Tianjin Water Diversion Project, Guangdong's Dong River-Shenzhen Water Supply Project, Hebei's Yellow River-Weiyun River Water Transfer Project, Shandong's Yellow River-Qingdao Water Diversion Project, Gansu's Datong River-Qingwangchuan Basin Water Transfer Project, Shanxi's Yellow River-Shanxi Water Transfer Project, Liaoning's Biliu River-Dalian Water Transfer Project, Jilin's Songhua River-Changchun Water Transfer Project, Jiangsu's Jingdian Lift Irrigation Project, and Yangtze River-Tai Lake Diversion Project. Interbasin water transfer projects that are now under construction include the Central and Eastern Routes of the South-to-North Water Diversion Project, and the planned projects include the Central Yunnan Water Diversion Project, Han River-Wei River Water Transfer Project, Yangtze River-Chao Lake Water Diversion Project, Wuhan's Great East Lake Connection Project, Dadu River-Min River Water Transfer Project, etc.

There are many famous water transfer projects in the world, and 39 countries have already completed 345 large water transfer projects (annual water transfer amount for each project is more than 10 million m^3). The five countries with most major water transfer projects (excluding China) are Canada with water transfer amount of 141 billion m^3/year, India with 138.6 billion m^3/year, Pakistan with 126 billion m^3/year, former Soviet Union with 86.2 billion m^3/year, and the United States with 36.2 billion m^3/year. Various countries have different objectives for water transfer. Canada's objective of water transfer is mainly for hydroelectric generation to deliver electricity to the United States; water transfer in India and Pakistan is mainly for agricultural irrigation; the former Soviet Union's objective of water transfer was mainly for irrigation and also for water supply, hydroelectric generation, and shipping as well; water transfer works of the United States are mainly for urban and industrial water supply. It is noteworthy that most of the foreign water transfer projects were built in the 1980s, and now developing countries, such as

India and Pakistan, are still constructing new water transfer projects, while developed Western countries have seldom constructed because of the increased awareness of ecological environmental protection and improved water conservation.

An overview of China's water transfer projects indicates that most of the projects have played a major role for the development of the economy and the security of people's livelihood; a few projects have not reached their design expectations; and the existing water transfer works are mainly for urban water supply and agricultural irrigation. However, many of the planned future water transfer projects are to improve the aquatic environment of lakes, and such projects will be beneficial to lakes, but they will not have the function to reduce the amounts of contaminants, since the projects are aimed to mainly dilute and transfer contaminants, which can be used as one of the short-term pollution control measures. The most important thing is to reduce the amounts of contaminants discharged into the lakes, dredge the endogenous pollution source and rehabilitate the aquatic ecosystem. Therefore, water transfer projects should adhere to the "three first's and three then's" principle. Namely, first conserve water and then transfer water; first remediate pollution and then divert water; and first protect the environment and then use the water. In operation and management, efforts should be made to consider the comprehensive utilization of the local water and transferred water and balance the relationship between the water use in the source area and the water use in the water-receiving area. Moreover, the water-receiving area should help the socioeconomic development of the water source area by means of water source protection and ecological compensation and improve the living standard and quality of the people in the water source area.

8.5.3 Management of Dams and Reservoirs

A dam and a reservoir are two basic components of a hydraulic project. Where there is a reservoir, there is a dam. Therefore, the two are inseparable. In fact, a large hydraulic project may also include a hydroelectric station, ship locks, spillways, diversion pipelines, etc. Thus, the dam herein is a generalized term and it includes these hydraulic structures. The extent of a reservoir is even larger. For example, the TGR is up to 600 km long and its water surface area is several hundred km^2.

The management of dams and reservoirs involves many aspects but can be divided mainly into safety management, operation and regulation management, economic management, and social administration.

The management of benefit generation and safety is the central function of managing a dam and the associated reservoir to fully utilize the reservoir's functions. The basic function of a reservoir is to serve for socioeconomic development. However, because artificially raising the water level, storing floodwater, and regulating floodwater and the runoff process transfer the risk associated with the potential energy of the floodwater and the water body to the dam and the reservoir, many safety issues need special attention. Large issues include the flood discharge capacity and structural and foundation stability of the dam, the side slope stability of the reservoir, downstream safe discharge, navigation safety in the reservoir area, etc., and small issues include the safety of metallic structures such as generator units, sluices, ship locks, etc.

Large hydraulic projects are generally multi-objective, and how to fully utilize the large projects to generate socioeconomic benefits is an eternal subject. Flood control, drought relief, and ecological regulation are all for public welfare and are mainly reflected in social benefits. However, hydroelectric generation, water supply, and shipping are reflected in economic benefits. How to balance the interests of all aspects is a complex subject and is also closely related to the socioeconomic benefits of the reservoirs. Except for the hydropower stations each with a relatively large power generation capacity, most of large- and medium-sized reservoirs do not make profits, and their management organizations are institutional and are funded from fiscal appropriations from all levels of governments. The economic benefits for dams and reservoirs are generally not good. Raising funds for construction, costs for operation and maintenance, and the stability of management personnel need institutional and mechanism innovations of reservoirs' economic management. If conditions permit, private enterprises or community funds can be used.

The social management of dams and reservoirs has just started in China and in fact has not done much. In the United States and other developed Western countries, since stakeholders and the public started participation in dam and reservoir management, hydraulic projects have not only obtained increased understanding, recognition, and support from the public, but the operation and management of the hydraulic projects have also gained improved understanding and have even obtained certain financial assistance.

8.5.4 Environmental Protection of Reservoirs

The environmental protection of reservoirs is an important aspect of reservoir management, and it will become increasingly more important in the future, because the public are most concerned about the real impact of the completed reservoirs on the ecological environment of the rivers. The management of the reservoir environment mainly includes two parts. One is the environment of the reservoir area, and the other is the impact on the downstream ecological environment.

Environmental protection in a reservoir area is focused on the prevention of geological hazards (landslides, reservoir-induced earthquakes), the protection of the aquatic environment and aquatic ecology of the reservoir area, etc. As the water flow velocity slows down in the reservoir area, an algal bloom is likely to occur if the water zone has too much nutrients. Meanwhile, potential contamination enterprises and shipping in the reservoir area pose a risk of emergent water contamination incidents. Therefore, the environmental protection in the reservoir area is a daunting task.

The downstream effects include the long-term river channel scouring caused by the discharge of clear water and adjustment of the river-lake relationship, the evenness of the water flow process, and the changes of water bodies' physiochemical properties. These factors will inevitably affect the change of the life history of aquatic organisms, and it is especially necessary to consider discharging the ecological flow to meet the needs of aquatic organisms. Large controlled reservoirs, such as the TGR, involve the problems of the entire watershed such as the river-lake relations in the middle and lower reaches and the adjustment of the estuary.

The primary scope of the environmental management of a reservoir includes the establishment of a monitoring system and an information system for hydrology, river channel, water quality, aquatic organism, and natural hazards (landslide, bank collapse, induced earthquake, etc.) in the upstream and downstream impact areas of the reservoir; analysis of environmental information data, preparation of environmental bulletins, and establishment of a notification mechanism; protection and restoration of the ecological environment; establishment of a fund for environmental protection and emergency environmental disposal and establishment of a mechanism for the public participation in reservoir environmental protection; etc.

8.6 Soil Conservation and Sediment Management in the Yangtze River Basin

The Yangtze River basin is an area of serious soil erosion in China. Most of the mitigated soil erosion quantity is offset by new soil erosion each year, but the sediment contents and quantity of sediment transport in the majority of the river channels in the Yangtze River system have continued to reduce. The two statements appear to be contradictory. In reality, due to human activities (regulation, dam installation, road construction, etc.), most of the eroded soils have deposited in the mountainous areas, ravines, reservoirs, and ponds, and sediment transport channels have been blocked. Moreover, sedimentation in reservoirs and sand mining in river channels have reduced the amount of sediment transport through river channels. It appears that the short-term sediment conditions have been improved, but in a long run, many potential problems have left behind. Therefore, it is necessary to scientifically manage sediments so that the sediment distribution in rivers is more reasonable, or otherwise in a century (when a balance between sediment scouring and sedimentation is reached), due to the uneven distribution of sediments, the ecological environment of rivers will not only be affected, but also human production and life safety in mountainous areas and flood control safety in the middle reaches (discharge of clear water, evolution of the river-lake relationship) will also be impacted.

8.6.1 Soil Conservation

Land is the most important natural resource besides water, and the value of land resources is mainly embodied in the soil quality and quantity. Because forests, crops, and grasslands all need soil to grow, and it often takes a few decades or even hundreds of years to generate 1 cm of soil, soil conservation is not only an important ecological and livelihood project, but it is also the precondition for managing the sediment of river channels.

The Yangtze River basin is one of the areas with serious soil erosion in China, and soil erosion is dominated by hydraulic erosion, coupled with wind erosion, freeze-thaw weathering, landslide, gravitational erosion, debris flow, hillock collapse, and mixed erosion. The areas of mild, moderate, heavy, very heavy, and intense hydraulic erosion are 215,000 km^2, 212,000 km^2, 76,000 km^2, 17,000 km^2, and 4000 km^2, respectively, accounting for 41%, 40.5%, 16.5%, 3.2%, and 0.8% of the total hydraulic soil erosion area, respectively.

Soil erosion, especially surface erosion, is the most common type of soil erosion and occurs everywhere. Gully erosion generally occurs in river terraces, alluvial fans, deep residual slope deposit layer, and exposed weak rock formations that are vulnerable to weathering such as shale, gneiss, and granite. Freeze-thaw weathering, glacial erosion, and wind erosion mainly occur in the Qinghai-Tibet Plateau and alpine regions. Presently, new soil erosions are mainly those caused by development and construction projects or the associated induced landslides and debris flows. The man-made new soil erosion area in the Yangtze River basin is about 1200 km^2, the new soil loss is about 150 million tonnes, and the erosion modulus is often more than 10,000 tonnes/km^2/year. The new soil erosion has caused great impacts and harms and is the focus of soil erosion control and prevention. Figure 8.4 shows a typical area of stony desertification in the upper reaches of the Yangtze River.

Fig. 8.4 Stony desertification in upper reaches of the Yangtze River

Landslide and debris flow hazards mainly occur in areas where the ground surface is severely cut, the topographic slope is steep, and the topographic relief is relatively high and are concentrated in the faulted and seismically active zones. These hazards often occur in the mountainous areas of eastern Yunnan, western Sichuan, southern Gansu, and the Three Gorges area. According to the statistics of survey data, 150,000 landslide sites and 10,000 debris flow gullies have been identified in a one-million-km^2 area in the upper Yangtze. According to the survey of 202 countries (cities and prefectures) in Yunnan, Guizhou, Sichuan, Gansu, Shaanxi, Chongqing, and Hubei in the upper Yangtze conducted in 2004, 13,641 landslide sites (including unstable rock mass sites), each of which had a mass volume of more than 10,000 m^3 and posed risk to more than one residence, were identified with a total landslide volume of 1.6689 million m^3 that directly threatened the safety of properties worth 17.322 billion yuan. A total of 6801 of the 13,641 landslide sites, each with a mass volume of more than 100,000 m^3 and posing a threat to more than 30 people, were found to have obvious recent activities. A total of 3186 debris flows, each with a basin area greater than 1 km^2 and posing risk to more than one residence, were discovered with a total area of 50,400 km^2 that directly threatened the safety of properties worth 12.06 billion yuan, and 1491 of the 3186 debris flows were those each had a basin area greater than 5 km^2 and posed a direct threat to more than 30 people.

The collapse of hillocks is a special type of soil erosion in southern China and develops in exposed granite, sandy gravel, sandy shale, muddy shale, and especially granite, in tropic and subtropical regions. While the area of soil erosion associated with hillock collapses accounts for a small proportion of the total soil erosion area, the erosion modulus is huge, and the development is rapid. Hillock collapses of the Yangtze River basin mainly occur in Jiangxi, Hunan, Hubei, and Anhui. According to the results of the latest landslide survey, 73,800 hillock collapses, each of which has an area of more than 60 m^2, have occurred in the Yangtze River basin with a total area of 225.92 km^2.

Programs to control soil erosion generally begin in a small watershed to implement integrated treatment measures. The management of sloping farmland is the key, and runoff control measures are the main means, combined with treatment measures such as control of nonpoint pollution sources, comprehensive treatment of sloping farmland, water and soil, creation of soil-conserving woods and grasses, and gully control. In the gentle-sloping farmland with a thick soil cover, the slope can be changed to terraces with a supporting water system on the slope surface and improved field roads. The terraces may consist of earth steps and stone steps based on local conditions, and the earth-stepped terraces must be covered with protective vegetation. The water system works on the slope surface may include irrigation and drainage ditches, water storage cellars, sediment basins, etc. and should incorporate the local topographic conditions to facilitate flood control, sediment interception, and water storage. Field roads can be laid out in combination with ditches and can be constructed in conjunction with roads for farming tillage machinery where conditions permit. However, for relatively steep-sloping (e.g., greater than 15 degrees) arable land, the change of sloping land to terraces may be successful in erosion control in the short term, but the long-term effect is not ideal, because steep-sloping terraces will be washed away in the event of a rainstorm. As a result, the soil loss will "accumulate," and the ultimate erosion will not be reduced. Therefore, while the short-term effect of erosion treatment in a small watershed appears successful, the erosion reduction is limited from the perspective of a larger area for a longer term. The best effect of erosion control for steep-sloping farmland will result from turning the farmland into woods and grassland with less human disturbance, or otherwise soil conservation work will never end.

Forest-and-grass measures for soil conservation mainly include soil conservation forests, economic orchards, and planted grasses. Soil conservation forests are mainly developed in lands of sparsely distributed young trees, barren hills and slopes, and abandoned steep-sloping farmlands. Grass planting should be performed mainly using the local native grass species; in some places with relatively better conditions, livestock (forage) grassland should be developed. In the relatively gentle-sloping abandoned farmlands where water source conditions are good, economic orchards should be developed as part of the local economic development plan, and fish-scale pits, contour ploughing, and terraced orchards should be used to prepare land for soil erosion control. The development of forest and grassland for soil conservation needs to achieve multi-varieties of species with shrub and grass.

The construction of sediment barriers and check dams on an eroded gully, if conditions permit, can be combined with gully protection to transform the inefficient or ineffective broad gully channel of poor quality into highly efficient and high-quality land. The development of highly efficient agriculture can become an effective approach to the increase of land resources and can further consolidate the results of returning farmland to forest and grassland and natural ecological restoration. Gully protection needs to be comprehensively planned and implemented. A group of check dams can be used and combined with sediment barriers to form a multilayer interception system so as to prevent the gullies from being cut downward. For the gullies with bank expansion and serious headward erosion, vegetative measures should be taken on the bank slopes of the gullies to establish a complete protection system of the gullies.

The laws of nature should be respected, the self-repair capability of nature should be fully utilized, and the principle of harmonious coexistence between human beings and nature should be followed. As a result, efforts need to be made to first protect natural forests and grasslands and then carry out the ecological restoration to shrub woodland, sparse woodland, and moderate- and low-coverage grassland, promote large-scale vegetation restoration, and improve the ecological environment. Ecological restoration measures are mainly to block access and ban wood harvesting; demarcate the perimeter boundaries of the logging-prohibition zone; install prohibition signages in the clear areas of the restricted zone; take stewardship responsibility; adopt measures such as tending management, replanting, feeding livestock, biogas digester construction, grassland improvement, fence installation, and ecomigration; and restrict productive and operational activities in the prohibition zone such as wood harvesting, firewood gathering, grass cutting, and grazing so as to ensure that the forest, brush, and grass in the prohibition zone will be restored soon.

Sediment reduction in river channels and soil conservation are not contradictory. Soil conservation is aimed to protect the forest, grassland, and vegetation and preserve water sources' conservation function, cultivated land, and grain production, with the emphasis on preventing human activities from causing soil erosion.

8.6.2 *Sediment Management*

Before human activities caused soil erosion, sediments contained in the water of a river had been a natural phenomenon due to natural erosion and the evolution of the river. Turbid water is not necessarily a bad thing since turbid water indicates the normal transport of sediment, increases mineral and nutrients to the body of water, shapes the channel and estuarine delta of the river, and creates land adjacent to the river. Turbid water also manifests the river's function of the ecological environment. If the sediment is reduced abnormally, the riverbed will not only be scoured, and the estuary and coastal area will experience erosion in the long run, affecting the health of the ecosystem of the river and its estuary. Therefore, the sediment problem is a neutral issue, which is not good or bad, and the key is whether the increase or decrease of the sediment content in water is abnormal. For example, if sediments had not been transported to the estuarine delta over the past 3000 years, there would be no estuary area or Shanghai. Thus, sediment is not a simple problem of control or prevention, but it is an issue of how to distribute and manage rationally.

8.6.2.1 Shortage of Sediments in the Middle and Lower Reaches of the Yangtze River

The management of sediments in a river needs to consider the sediment yield of the river basin and the sediment demand of the river channel. According to Wang and others (2008), the Yangtze River basin not only produces large amounts of sediments, but the river channel also needs a certain amount of sediments to maintain the balance of the river's evolution. The normal sediment demand of the river mainly includes the amount of sediment needed for the riverbed in the middle and lower reaches of the Yangtze River to evolve toward a balanced longitudinal profile with the minimum energy consumption, the quantity of human sand quarrying, and the amount of sediment required for land expansion at the Yangtze Estuary. The lateral swing of the middle and lower reaches of the Yangtze River is restrained by the dikes along both banks, and the riverbed evolution is limited mainly to the adjustment of the longitudinal profile. Presently, the middle reaches are far from reaching the balanced longitudinal profile with the minimum energy consumption, and the long-term development trend should be mainly sedimentation. Therefore, the middle reaches would mainly undergo sedimentation under normal conditions. Since the impoundment of control reservoirs, such as the TGR, in the upper reaches, the evolution process has changed over a certain time period. In the last 20 years, the sediment in the upper reaches of the Yangtze River has been reduced gradually. Moreover, the impacts of various human activities, such as sediment interception in reservoirs, sand mining on river channels, etc., have intensified the decrease of sediment transport in the Yangtze River, while large amounts of sediments are required at the estuary, especially in Shanghai that needs large areas of new land to be developed at the estuary.

The sediments in the middle and lower reaches of the Yangtze River are mainly the waste load of non-bed quality during the flood season and are primarily bed-quality sediments during the non-flood season, and about 90% of the sediments transported to the estuary are the coarse-grained sediments suitable for land development. To achieve the goal of the planned land development at the estuary requires at least 300 million tonnes of sediments per year from the Yangtze River to maintain the balance, and the current annual amount of sediment shortage of the Yangtze River is about 100 million tonnes.

Due to the construction of cascade reservoirs and other hydraulic structures on the Rhine River in Germany, the sediment supply from the upper reaches to the lower reaches was blocked, and the riverbed was severely short of sediments. As a result, the riverbed was scoured by 2 m deep, and a number of hydraulic projects along the river lost their functions. In the meantime, the groundwater level also dropped. In order to control the scouring of the Rhine and maintain the stability of the river channel, Germany has artificially thrown 200,000 tonnes of selected bed-quality sand into the river each year to replenish sediment to the river section downstream of each hydropower station.

Because the sediment and water at the Yichang Station on the Yangtze River are from the upper reaches of the river, the amount of sediment transport is always proportional to the flowrate. However, at the Hankou and Datong Stations, the amount of sediments is only proportional to the flowrate in the non-flood season during January–June and October–December but has a poor correlation with the flowrate in the flood season during July–September. This is because the sediment in the middle and lower reaches of the Yangtze River in the non-flood season is mainly of the bed quality; the larger the flowrate, the more sediment will result from the riverbed scouring and the more amount of sediment transport. In the flood season, most of the sediments come from the upper reaches, and 50% of the water comes from the middle and lower reaches. As the sediments and water come from different reaches and most of the sediments are of non-bed quality, the correlation between the amount of sediment transport and the flowrate is very poor.

According to the estimates by Wang and others (2008), the current annual total amount of sand quarrying in the Yangtze River is at least 64 million tonnes. According to the 1951–2002 data, the annual amount of sediments transported to the Yangtze Estuary was about 433 million tonnes, of which 45 million tonnes were deposited in the Datong-Xuliujing section, 28 million tonnes in the North Branch north of Chongming Island, 4 million tonnes in the South Branch south of Chongming Island, 138 million tonnes at the mouth of the Yangtze Estuary that raised the elevation of the continental shelf, 180 million tonnes at the Qiantang Estuary, and the rest of the sediments into the deep sea. In the past 50 years, the land-making process has never stopped at the Yangtze Estuary where 800 km^2 of new land has been created.

Shanghai Municipality has planned to utilize the sediments from the Yangtze River to develop 1000 km^2 of new land in the Yangtze Estuary in the next 20+ years. As the tidal range in the Yangtze Estuary area is about 3.5 m, the area for developing new land must be filled and raised by 3.5 m. In theory, to achieve the Shanghai Municipal Government's goal of developing 1000 km^2 of new land, the required volume of sediments is about 2.59 billion m^3, but in practice as the boundary of the raised land is a slope rather than a vertical wall, additional amounts of sediments are required. It is estimated that 5.1 billion m^3 or about 7 billion tonnes of sediments are required for the planned land development with an average annual amount of 280 million tonnes.

Due to a significant reduction in the amount of sediments to the lower reaches and the estuary, it is not possible to meet the needs of sediments in the river. Presently, the shortage of sediments has brought new challenges to river researchers and sediment managers. Since the mid-1980s, the annual amount of sediment transport had decreased from 514 million tonnes to 382 million tonnes at the Yichang Station and from 427 million tonnes to 327 million tonnes at the Datong Station. During 2003-2004, the annual amount of sediment transport decreased to 100 million tonnes at the Yichang Station and to about 200 million tonnes at the Datong Station. On the other hand, a comparison of gradations between sediments transported to the estuary and those for land development indicates that about 10% of the transported sediments were fine particles that could not be used for land development but drifted to the sea. Only 90% of the transported suspended sediments were coarse material that could be used for land development or building materials. Thus, the annual amount of sediment demand for the balance of the Yangtze Estuary and land development is up to 300 million tonnes and the total shortage of sediments in the Yangtze River is about 100 million tonnes.

Due to the shortage of sediments, some underwater sandbars in the Yangtze Estuary area have changed from previous gradual expansion trend to gradual shrinking. For example, the area of Baimao sandbar started its development from scratch in the 1950s to more than 50 km^2 in 1990. However, since 1990, the area has begun to decrease to the current area of only 35 km^2. The total area of Qingcao and Ruifen sandbars increased to 120 km^2 in 1980, and since then, the two sandbars have gradually shrunk to their current total area of 60 km^2.

8.6.2.2 Causes of Sediment Reduction in the Yangtze River

Although the sediment yield of the Yangtze River basin is high, the total amount of sediment transport is not large, and the resulting ratio between the amount of sediment transport and yield is well below 1. According to the surveys conducted in the 1980s, the ratio was 0.338 at the Yichang Station and 0.1–0.5 for each tributary. One of the primary causes for this was that the eroded sediments generally deposited in the pediment areas at the feet of mountains, depression areas, and alluvial fans at the mouths of gullies. As a result, a certain amount of sediments was deposited in a short distance. The riverbeds of most tributaries in the basin consist mostly of rocks, pebbles, gravels, and coarse sands with limited replenishment along the course of the rivers, and the surface erosion material was mostly composed of coarse particles that were not readily available to be transported for a long distance. The second cause was that reservoirs, ponds, and agricultural facilities played a certain role in sediment interception.

The sediment in the mainstream of the Yangtze River originating from surface erosion is mainly suspended material with minimal bedload. For example, at the Cuntan Station, the average annual amount of suspended sediments was 439 million tonnes (1953–2000), but the average annual bedload of pebble, gravel, and sand was 510,000 tonnes that made up only about 1.2‰ of total sediment transport.

As the relationship between sediment yield and rainstorm is closer than that between runoff and rainstorm in the upper reaches, the distribution of sediment transport amount is more concentrated than that of runoff in the year. In the flood season during May–October, the amount of sediment transport in the mainstream and major tributaries in the upper reaches of the Yangtze River generally accounts for 95–99% of that of the entire year, while the runoff quantity was 77–87% of that of the entire year. The large-scale quarrying of gravels and pebbles since the 1980s has been one of the main causes for the decrease in the amount of pebble transport.

Under the "Key Prevention and Control Project for Soil Conservation in the Upper and Middle Reaches of the Yangtze River" (referred to as "Yangtze Control Project"), which began in 1989, the upper Yangtze was listed as the key prevention and control area for soil conservation, and it was decided to first carry out prevention and control measures in four key areas of severe soil erosion in Yunnan, Guizhou, Sichuan, Gansu, Shaanxi, and Hubei, including the lower reaches and the Bijie area of the Jinsha River, middle and lower reaches of the Jialing River, southern Gansu, southern Shaanxi, and the TGR area. By the end of 2003, 15.929 billion yuan had been spent to treat soil erosion areas of more than 80,000 km^2. Soil conservation had played an important role in soil erosion reduction. On the other hand, in the past 20 years, a large number of peasants have come to work in the urban areas within the basin, and their incomes have been much better than they would have earned if they had farmed on the hillslopes, which has reduced land disturbance. Therefore, it is the best soil conservation measure to change land use.

Although soil conservation has had an obvious effect on sediment reduction, the construction of cascade reservoirs has had more obvious effect on the reduction of sediment transport. The reasons are as follows: ① By 1999, the area of soil erosion in the Yangtze River basin had reduced from 622,200 km^2 to 530,700 km^2, only down by 14.7%. However, the treated area has still produced sediments, but only the sediment yield has been reduced. The reduction percentage of sediment yield in the upper and middle reaches of the Yangtze River had been even lower, generally not exceeding 10%, but in a considerable number of river sections, the amount of sediment transport had decreased by more than 30%. ② The Jinsha River was also one of the key areas in the "Yangtze Control Project," where the amount of sediment transport has not decreased but increased in the past 20 years. Although the increase might be attributed to natural factors, such as different types of soil erosion, an important factor was that no large reservoirs had been constructed in the mainstream and major tributaries of the Jinsha River to intercept sediment, and the Jinsha River and its major tributaries had basically maintained as natural rivers. ③ Whenever large reservoirs or large-scale cascade development occurred in a river, the amount of sediment transport was obviously reduced, and the time of sediment content reduction basically coincided with the construction time of the large reservoirs.

8.6.2.3 Main Scope of Sediment Management

The main scope of sediment management is to maintain, to the extent practically possible, the river channel's normal function of sediment transport and balance scouring or sedimentation suitable to the characteristics of specific river sections, alleviate overconcentrated sedimentation in reservoirs, and allow the river channel to function normally. The focuses should be ① to control soil erosion caused by human activities such as watershed development and utilization; ② to discharge sediments deposited in the reservoirs using floodwater to the extent practically possible during the flood season through scientific regulation, dissipate the kinetic energy of water flow from discharge of clear water, control the scouring of the riverbed, and transport sediment to the estuary area in a balanced manner; and ③ to strictly manage sand quarrying in river channels and scientifically establish the quarrying area and time limits, so as to achieve the control of the total amount of sand quarrying.

8.6.3 Protection of Agricultural Land

8.6.3.1 Existing Major Issues

Food production using traditional farming methods in China has maintained for thousands of years, which is a world miracle, and has fed the largest population of the world. However, due to the long-term extensive application of chemical fertilizers and pesticides on China's land and the use of highly multiple-cropping farming method, China's soil function has degraded, and soil contamination has been increasingly serious, which has not only affected the arability of the land but has also caused water pollution due to soil erosion. The current primary soil issues in China are as follows.

1. Rapid Loss and Degradation of Soil Resources

First, due to land occupation for construction, farmland abandonment for ecological restoration, the adjustment of the agricultural structure, and the destruction of natural hazards, China's cultivated area reduced from 1.3 million km^2 in 1996 to 1.224 million km^2 in 2004, a total reduction of 76,000 km^2 in 8 years, or 5.8%, and this trend continues. Second, soil quality has degraded. Soil degradation refers to the process of declining soil quality and productivity under unfavorable natural factors and the influence of human irrational soil use. Presently, the national soil erosion area has reached 3.56 million km^2, accounting for 37% of the national land area; the average annual soil loss is up to 4.5 billion tonnes; and the lost fertility of the soil is higher than the total fertilizers manufactured nationally. Third, soil salinization is prominent. Presently, the newly salinized land has reached nearly 370,000 km^2. After considering the naturally occurring salinized land, the total area of salinized arable land has already reached over 800,000 km^2. This phenomenon can also be attributed to the increased salt content in subsurface soil caused by flood irrigation through unscientific massive water diversion, which has resulted in a sharp decrease of crop production and even the abandonment of farming. Fourth, China's land of stony desertification is mainly distributed in Guizhou, Guangxi, Yunnan, Hubei, Hunan, Guangdong, Chongqing, and Sichuan. By the end of 2005, China's total land area of stony desertification had reached 129,600 km^2, and the area of stony desertification is still generally trending up. Fifth, the land area of soil acidification in China has accounted for over 40% of the total national land, and soil acidification is especially more serious in south China. Therefore, China's soil resources are faced with many prominent issues, such as soil erosion, declined fertility of arable land, land desertification, soil salinization, soil stony desertification, and soil acidification, which have constituted a serious threat to the ecological safety.

2. Accelerated Regional Soil Contamination

Surveys of soil quality conditions have been completed on an estimated 1.27 million km^2 of land, and the results indicate that China's watershed and regional geochemical anomaly or contamination of soil is unprecedented in extent. Heavy metal anomalies mainly of cadmium coupled with lead and mercury have occurred throughout the basins of the Yangtze, Pearl, Songhua, and Liao Rivers and the coastal economic zones. The high concentrations of fluorine and arsenic and the low concentrations of iodine have been identified to be extremely conspicuous in the Yellow River basin, and soil mercury anomalies are common throughout China's large- and medium-sized cities. A preliminary analysis of the data indicates that these watershed geochemical anomalies can be characterized by the high background level of natural geochemistry with additional anthropogenic contamination. The preliminary analysis also indicates that in the areas with heavy metal anomalies or contamination throughout the watershed, major geochemical hazards and potential contamination are prominent in some areas, and the deterioration process of soil geochemical conditions appears to have accelerated. Both direct soil contamination and the associated pollution of air, surface water, and groundwater will ultimately be harmful to animals and human beings.

3. Weak Public Consciousness of Soil Protection and Lack of Policies and Regulations

As for the protection of water bodies and the atmospheric environment, China's recognition of and attention to soil protection are lagging. The public awareness of soil protection in China is weak, and the consciousness and enthusiasm are not high either. The government decision-makers have insufficient knowledge of soil resources, soil quality, soil function, and soil's social value and don't have the awareness of promoting public consciousness or proactively protecting soil. The education program for the soil environment is very weak, and there are no associated professional agencies or publicity organizations.

China has not had any laws or regulations on soil pollution prevention, which makes it difficult for governmental departments, especially environmental protection departments, to play a substantial role in the prevention and control of soil contamination. Since the establishment of current soil environmental quality standards in the 1990s, China has not performed much research on the establishment of soil environmental quality benchmarks and standards and has not yet introduced revised national standards. The existing unsystematic, incomplete, and inadequate national standard system of soil environmental protection is far from meeting the needs of national soil utilization planning and environmental management.

8.6.4 China's Strategic Policies of Soil Conservation

Since the 1930s, the United States has begun to pay attention to soil quality protection and developed policies to support farmers to "retire" arable land from production (land had remained fallow from agricultural production for 10–30 years or even longer). Soil conservation has been combined with the protection of agriculture, forestry, animal husbandry, fishery, water resources, and wildlife habitat into a large category of natural resources and implemented as part of the overall protection of the environment and the conservation of natural resources. While the short-term cost seemed very high, soil protection has achieved remarkable success and created values for the long-term land use and the ecological environment. Therefore, the efforts have been beneficial overall. China has a large population with less land and the intensification level of farmland is low. As a result, China does not have the conditions to implement the protective measures similar to those of the United States, but China should take comprehensive measures suitable to China's existing conditions and protect land resources. Efforts should be made to the following.

4. Management Policies of Soil Protection

Clearly define the responsibilities of the parties for land management and utilization; formulate a soil conservation plan or action framework; strengthen legislation on administrative organizations for the soil environment; standardize the administrative behavior of governments; improve the administrative decision-making process; formulate and improve the laws and regulations of soil protection; and improve the public participating ability in environmental protection. Moreover, establish and improve the program for the use of pesticides, fertilizers, sludge, and agricultural waste; implement preventative and remedial programs for soil contamination and degradation; and apply the economic and ecological compensation mechanism and program for soil remediation. Furthermore, develop technical measures for the emergent treatment of soil contamination, and establish a coordinated management model for the environmental protection of soil, water, and air and the utilization of soil resources.

5. Standardized Measures for Soil Environmental Protection

Establish a system for the preparation of soil protection standards based on risk assessments; improve the standard system for soil protection; and strengthen the management of soil zoning and classification. Moreover, study the soil ecological toxicology of contaminants; determine soil environmental background levels and biogeochemical benchmarks; establish the assessment method for the soil ecological risk, health risk, and environmental risk; formulate management regulations for contaminated sites and technical specifications for risk assessments; revise and improve the national soil-quality standard system applicable to various land uses; and carry out the work for the establishment of soil quality standards for each province, autonomous region, and municipality.

6. Scientific and Technical Measures for Soil Conservation

Establish the mechanism of soil resource conservation for regional coordination, spatial optimization, and rational utilization, and continue to perform the dynamic census of the national soil resources' quantity and quality so as to timely comprehend the conditions of soil resources and the status quo of soil quality. Moreover, more efforts should be made to study the trend of change in quantity and quality of soil resources and the evaluation method of impacts; establish a national soil quality evaluation index system and monitoring network; achieve the scientific protection and informative management of soil resources; establish and develop a zoning management system for farmland soil quality suitable for agricultural production in China; establish a technical platform for the management of information sharing and technology transfer; and form an innovative mechanism for the effective promotion and application of soil fertility quality in rural areas.

Establish an integrated management model of soil conservation that combines hydraulic engineering, bioengineering, and agricultural technology together; establish a technical system to prevent soil desertification consisting of a socioeconomic-natural complex ecosystem model and ecological protection engineering; develop technical specifications for the ecological protection of forest soil; and implement ecological restoration projects. Moreover, develop a technical system for soil's stony desertification prevention and ecological conservation, construct a complete soil irrigation and drainage system to effectively control and prevent soil salinization, and proactively promote and demonstrate successful ecological management models.

Efforts should be made to study how to identify various pollution sources and new pollutants and investigate the biogeographic process, mechanism, and cycle of pollutants in the soil environment and the response of the ecological environment; improve the method for site investigations and risk assessments of soil contamination at abandoned mining sites, oil fields, industrial enterprises, the affected areas of large construction projects, farmland soils, waste storage and stacking sites, radioactive nuclides, etc., and devote more efforts to the research of monitoring equipment and new technologies for remediation; and step up the promotion and demonstration of soil remediation technologies.

Perform the in-depth study of the effects of global warming on soil process, biodiversity, and their ecological functions; devote more efforts to the research on the soil biogeochemical cycle based on carbon, nitrogen, phosphorus, and sulfur; establish a biogeochemical cycle model of soil effects caused by global warming and human activities; predict the potential soil responses under various climate change conditions; and develop soil carbon sequestration technologies.

7. Soil Conservation Measures in Key Regions

China is vast in area, complex in natural conditions, diversified in soil types, and prominent in spatial stratified heterogeneity and characteristic variability. As for the variability in the natural ecological environment and soil, strategic measures need to be developed for soil conservation based on site-specific conditions and regional variations. China can be divided into six key regions: eastern coastal economic developed region, central key grain-producing region, Inner Mongolia region of north China, northeastern old industrial base, western ecological fragile region and alpine region, and major measures should be developed for soil conservation in each key region.

References

Wang Z, Huang W, He Y (2008) Study on sediment demand in the Yangtze River. J Sed Res (1):26–34

Yangtze River Yearbook Editorial Board. 2011. Yangtze River yearbook 2011

Zhao Q, Luo Y, Teng Y (2009) Thoughts about macro-strategy of soil conservation in China. Acta Pedol Sin 46(6):1140–1145

Chapter 9
Future of the Yangtze River

Abstract Presently, as the Chinese government and local governments within the Yangtze River basin have realized the increasingly serious ecological and environmental problems of the Yangtze River, they will continue to adhere to the principle of "promoting protection but not engaging in large-scale development" when implementing the national strategy for the development of the Yangtze River economic zone. Additionally, they will consider restoring the ecological environment of the Yangtze River as a predominant priority. In the meantime, the governments have promoted long-term regulation of rivers and lakes so as to not follow the traditional way that involves transferring the zombie enterprises from the economically developed Yangtze Delta area to the more undeveloped regions in the upper and middle reaches with a healthier ecological environment in an attempt to make the Delta area greener. This method is not ideal, as doing so would be detrimental to the area in the upper and middle reaches. This chapter summarizes the natural evolution of the Yangtze River and expectations for the restoration of the Yangtze River's aquatic ecological environment and looks ahead into the establishment of a well-functioning integrated watershed management system and mechanisms to support it.

Keywords The Yangtze River · Changjiang River · Evolution of river system · Basin ecosystem · Water resources utilization · Floods and drought · Ecological and environmental protection · Basin management

9.1 Evolution of the Yangtze River

It is very difficult to predict the future. The future evolution of the Yangtze River will depend on natural evolution such as climate change, tectonic movement, biological succession, and the impact of human activities. The evolution on a long timescale (more than tens of thousands of years up to the length of a geological age) will mainly depend on natural changes. The evolution on a medium timescale (a millennium to 10 millennia) will depend on the impacts of both natural changes and human activities. The short-timescale evolution will depend largely on the impact of

J. Chen, *Evolution and Water Resources Utilization of the Yangtze River*,
https://doi.org/10.1007/978-981-13-7872-0_9

human activities. As the level of human productivity improves, the ability to regulate and control rivers, lakes, and other waters will be increasingly stronger and may even obstruct the trend of natural evolution on a short timescale.

Human beings can change the temporal and spatial distribution of the natural runoff by constructing reservoirs and water diversion projects to meet human needs, but it is very difficult to change the global climate and precipitation distribution. Although it can be intervened locally through artificial precipitation, it is not possible to carry out large-scale and longtime interventions in rainfall and climate. Additionally, tectonic movement cannot be interfered with. Although surface waters such as river runoff have a significant impact on the ecological environment, climate change and tectonic movement have a greater impact on the ecosystem of the earth. Therefore, the overall trend of the evolution of the Yangtze River will not change on a large scale or a long timescale, but if the ability for human activities to interfere with the natural ecosystem is very strong, or if human beings excessively interfere with the natural ecosystem, the progress and even the survival of human beings will inevitably be affected on a medium to short timescale. Futurists have predicted that it will only take thousands of years for human beings to disappear from the earth, and the civilizations created by human beings over thousands of years will also gradually vanish, but the earth will continue to evolve according to its own laws. It should be realized that if mankind does not exist, what is the importance of all human struggles today? Based on this logic, human activities will need to be more careful, preferably in accordance with the law of nature, so as to protect water resources and ecological environment and to avoid the possibility of self-destruction.

As long as the climate does not change greatly, the status of the Yangtze River as one of the world's great rivers will not change. As the mainstream and major tributaries in the upper reaches of the Yangtze River are situated in the mountains or canyons and the middle and lower reaches are located between increasingly stable dikes, coupled with the regulation of reservoirs, the channel position of the Yangtze River and the river-lake relationship will not change. With the construction of cascade reservoirs, it is worrying whether the morphology of the river and the diversity of the habitats for aquatic organisms will be well maintained. Other causes for worry are whether the water quality of lakes and reservoirs will be maintained to meet the management requirements of planned water function zones and whether the rare and endemic fish and bird species will continue to exist.

On a medium to short timescale, human activities will impact the evolution of the Yangtze River most and will affect the natural evolution of the entire basin in the long run. Climate change and biological succession, which interact with each other, will have the second most impact on the evolution of the Yangtze River. The third biggest impact on the evolution of the Yangtze River will be from large floods or large earthquakes and subsequent disasters. Since the upper reaches of the Yangtze River are in the frequent earthquake zone (tectonically active zone), if a strong earthquake occurs in the mountainous area, it will inevitably cause the collapse of the mountain mass or a landslide, causing barrier lakes to emerge, and even a local river channel to potentially change its course. A large flood will inevitably have a significant impact on the river channel and the river-lake relationship in the middle and lower reaches of the Yangtze River, which may cause not only major adjustments of the sediment distribution and sandbars/sandbanks in the river channel but also major adjustments of the river-lake relationship. The human development, utilization, and protection of water resources in the Yangtze River should fully consider the objective rule of natural evolution to achieve a multiplied effect.

9.2 Restoration of Aquatic Ecology and Environment of the Yangtze River

China has experienced rapid economic development for more than 30 years. According to the *National Water Resources Comprehensive Plan*, the development of water resources for another additional 20 years should be able to meet the water resources demand for sustainable socioeconomic development. Based on the experience of developed Western countries, the total amount of water use has not increased in most developed countries since the 1980s, but their economies have still been growing. Therefore, in 2030, China's total amount of water use should also be able to remain unchanged, while socioeconomic development will continue to expand. However, the aquatic environment and ecosystem continues to deteriorate, and the restoration of the aquatic environment and ecosystem will be a long-term endeavor of the state, various departments, and the entire society in the future.

The dikes along both sides of the mainstream and tributaries of the Yangtze River have turned the Yangtze system into controlled narrow rivers, and the densely distributed cascade hydropower stations and reservoirs in the rivers have transformed the Yangtze River into sections of seminatural and semiartificial reservoirs. In 100 years, vast majority of the reservoirs' dead storage capacity will be filled with sediments; various sized sluices on both sides of the river will separate lacustrine wetlands from the river; lakes, grasslands, and marshes along both banks of the Yangtze River will be separated from the river, resulting in unfavorable consequences to the ecological environment; fish species in the lakes will not be able to freely migrate into the Yangtze River for wintering and breeding; the juveniles of the FMCC and other fish species in the Yangtze River will not be able to enter lakes to feed and grow; and the FMCC left in lakes will have a difficult time breeding due to the lack of required ecological hydraulic conditions such as a certain flow velocity and flowrate. Consequently, the number of fish species in the river and lakes will gradually decrease. Due to the separation between the lakes and the river, the water exchange cycle of lakes will be lengthened, and the eutrophication and pollution of lakes will become increasingly more serious. Therefore, the restoration of river and lake aquatic ecosystems will be long-term and entail expensive systems engineering that may require up to hundreds of years. Figure 9.1 shows a straight excavated river for diverting water from the Yangtze River. Although it is straight and uniform, its ecological function is limited. In the future, the curvature of the river will need to be gradually restored (Fig. 9.2) to provide space for the floodwater and to carry out restoration for the riparian ecosystem. Figure 9.3 shows the green slope with stone

Fig. 9.1 Straight man-made river channel to divert water from the Yangtze River

Fig. 9.2 Naturally curvy river

revetment for the restoration of the floodland outside the dikes, which has not only ensured the stability of the riparian zone but has also restored the environmental function of the ecosystem in the riparian zone.

For the protection of fish species, the first step is to protect the habitats, followed by the provision of favorable breeding conditions and release of farm-raised larvae. When the conditions in the middle and lower reaches of the river permit, the construction of fish ladder facilities may be considered. Figure 9.4 shows a fish ladder facility constructed for a hydraulic project.

Fig. 9.3 Ecological slope protection

Fig. 9.4 Constructed fish ladder facility

In the past decades, the change of nutrient contents in the mainstream of the Yangtze River has shown an almost synchronous growth with the amount of fertilizers applied in the watershed. From the 1950s to the end of the twentieth century, nearly 120 red tides (harmful algal blooms) were recorded in the waters of the East China Sea, accounting for 36.3% of China's recorded offshore red tides. In addition, the Yangtze Estuary has a large oxygen-deficient water zone. For example, an inspection conducted in August 2006 identified an oxygen-deficient zone of more than 15,000 km^2, which had a significant impact on the estuarine ecosystem. The red tides and oxygen deficiency are caused by the eutrophication of the water body. Most of China's rivers and lakes have undergone continuous contamination and pollutant accumulation for decades. Therefore, the promotion of clean production and the collection and high-grade treatment of wastewater are the prerequisites to restore water quality and protect the aquatic environment. Then the aquatic ecosystem will need to be modified to secure the future of the aquatic environment.

Presently, water pollution of the lakes in the middle and lower reaches of the Yangtze River basin is prominent, especially the lakes surrounding urban areas that have mostly been seriously contaminated. The ecological and service functions of wetlands have declined or degraded, and the restoration of lacustrine wetlands' functions has become the focus of environmental protection work by watershed, regional, and urban governments. The Yangtze River-Tai Lake Water Diversion Project that has been in operation for years and the seasonal river-lake connection implemented by the fishery sectors, which involves "opening sluices to allow fish fries to migrate from the river to lakes," have had a positive effect on improving the water quality and restoring the aquatic ecology. The projects to be implemented, such as the Great East Lake Ecologic Water Network Project in Wuhan, Yangtze River-Cao Lake Water Diversion Project, and Lacustrine Wetland Restoration Projects in Various Urban Areas, will have a long-term impact on the restoration of wetland functions.

In the restoration process of wetlands' comprehensive function, attention should be paid to the following: ① Respect the objective evolution law of the wetland ecosystem, take full account of the relationship between the cycle of the wetland biological system and the hydrological process, and conditionally continue the effort to promote the river-lake connection under the premise to restore the structure and integrity of the wetland ecosystem. ② Protect not only the aquatic function of water zones but also the transition zone between land and water. Properly deal with the relationship between urban development and lake protection. A proper vegetative transition zone should be maintained along lakeshores and houses should not be built at watersides. ③ Enforce a longer fishing ban period, restrain overharvesting of wild fish species, and restore the wetland biodiversity. ④ Devote major efforts to the development of the ecotourism industry; carry out activities such as bird sightseeing, fishing, and exploration; and promote the establishment of an ecological compensation mechanism for important wetlands. ⑤ Implement the adaptive system and mechanism for comprehensive management based on the characteristics of the wetland ecosystem.

As a result of massive human use and consumption of valuable freshwater resources, the amount of water left to the natural environment has been greatly reduced, or the alteration of the hydrological process that was adapted to the aquatic ecosystem has caused serious damage to the ecosystem. For example, since the 1970s, the global population of the vertebrate species has declined to an average of 70%, of which two-thirds have been facing extinction, especially in the freshwater ecosystem where human activities have been intense. According to calculations of experts such as Falkenmark and Rockstar, in order to maintain the function and service value of the ecosystem, the amount of the water use by the ecological environment should account for 75%, and thus, the amount of direct human water use should only be 25%. It is very difficult for China to meet this requirement. However, if the three redlines of water resources management are strictly followed in the Yangtze River basin, it is possible that the requirement can be met.

The ecological restoration in the Yangtze River basin will be a long-term project, and it will take many years to see the effects of the restoration efforts. Therefore, there is a need for a long-term restoration strategy and plan. Because the watershed ecosystem is a whole, and there exist interactions between the upstream and downstream, right and left banks, and the mainstream and tributaries, there is a need to carry out an ecological rehabilitation plan starting from the primary water resources zone with the entire watershed in mind, gradually proceeding to secondary and tertiary water resources zones, and finally addressing smaller river reaches or water zones. As for the rehabilitation method, efforts should be made to best use the existing conditions and adhere to the principle of minimizing human interference and maximizing the natural self-repair capability, supplemented by necessary ecological rehabilitation projects. Because the ecosystem is very complex, the law of evolution is not fully understood. Reducing human interference and contamination and allowing nature to recuperate may be the most economical and reliable method. Of course, there is a need to devote efforts to scientific monitoring and constant adjustment of the idea and method of rehabilitation, so as to possibly achieve the effect of continuous improvement and avoid artificially creating an irrational man-made environment for the ecosystem.

9.3 Integrated Management of the Yangtze River Basin

If the past focus of the integrated management of the Yangtze River basin has been to ensure the safety of flood control and development of water resources to meet the needs of socioeconomic development, then the focus of the future work will be shifted to the scientific management of flood and drought hazards, as well as the protection of water resources, aquatic environment, and aquatic ecosystem, so as to guarantee sustainable socioeconomic development and improve the quality of human life in the watershed.

The integrated watershed management originated from laws and regulations. The 2002 *Water Law of the People's Republic of China* has prescribed the management system as the combined watershed management and regional management. However, the watershed management in the *Water Law* mainly involves the traditional water conservancy, such as flood control, drought relief, water supply, hydraulic engineering, etc. Other water-related laws include *Water Pollution Prevention and Control Law*, *Flood Control Law*, *Environmental Protection Law*, *Forest Law*, *Water and Soil Conservation Law*, *Fisheries Law*, etc. Regulations related to the watershed management include the *Regulations of Sand Mining in the Yangtze River Channel*, *Regulations on the Water Quantity Allocation in the Yellow River*, and *Regulations on Administration of the Tai Lake Basin*. In the past 10 years, the YRWRC and the YRWCC have carried out early research on the *Yangtze River Law* and the *Yellow River Law*, hoping to introduce the laws of the river basins. It is hard to say to what extent a watershed legislation can be implemented currently, but the major problems of existing relevant legislation include bias toward the sectoral law and overcoverage of administrative aspects. In fact, the integrated watershed management is the management of natural resources and therefore should emphasize the establishment of the coordination system and mechanism among all departments and stakeholders so as to allow stakeholders and the public to participate. Therefore, it is difficult for the water legislation to obtain acceptance from all parties, and it is also difficult to advance the legislature presently. Only by changing the starting point of legislation can an integrated watershed management be promoted. In addition, if it is difficult to legislate for each watershed, an *Integrated Watershed Management Law* may be initiated. While the idea of the legislation is to keep a few necessary administrative authorizations, the main contents should be to ensure the establishment of a coordination mechanism for water-related departments and stakeholders and the supervision and management of watershed protection.

Although the protection of the ecological environment in the Yangtze River basin still requires that many scientific and technological problems be researched and solved, if the scientific and rational management system and mechanism can be established early to carry out the necessary eco-environmental restoration and effective protection, the Yangtze River can still maintain a heathy ecosystem and provide sustainable water resources for our generation and future generations.

The manufacturer's authorised representative in the EU is Springer Nature Customer Service Centre GmbH, Europaplatz 3, 69115 Heidelberg, Germany. If you have any concerns regarding our products, please contact ProductSafety@springernature.com

Printed and bound by CPI Group (UK) Ltd, Croydon, CR0 4YY
15/07/2026
02167620-0003